ABSOLUTE VALUE

$$|a| = \begin{cases} a & \text{if } a \geq 0 \\ -a & \text{if } a < 0 \end{cases} \qquad \begin{aligned} |a + b| \leq |a| + |b| \\ |ab| = |a||b| \end{aligned}$$

$$|x| < c \quad \text{if and only if} \quad -c < x < c$$

QUADRATIC FORMULA

The solutions of $ax^2 + bx + c = 0$ are given by

$$x = \frac{-b \pm \sqrt{b^2 - 4ac}}{2a}$$

LAWS OF EXPONENTS

$$a^x a^y = a^{x+y}$$

$$(a^x)^y = a^{xy}$$

$$(ab)^x = a^x b^x$$

$$\left(\frac{a}{b}\right)^x = \frac{a^x}{b^x}$$

$$\frac{a^x}{a^y} = a^{x-y}$$

$$a^{-x} = \frac{1}{a^x}$$

$$a^0 = 1$$

$$a^1 = a$$

$$1^x = 1$$

LAWS OF LOGARITHMS

$$\log_a (xy) = \log_a x + \log_a y$$

$$\log_a \frac{1}{x} = -\log_a x$$

$$\log_a \frac{x}{y} = \log_a x - \log_a y$$

$$\log_a (x^c) = c \log_a x$$

$$\log_b x = \frac{\log_a x}{\log_a b}$$

$$\log_a 1 = 0$$

$$\log_a a = 1$$

$$a^{\log_a x} = x$$

$$\log_a (a^x) = x$$

FUNDAMENTALS OF
COLLEGE ALGEBRA
AND TRIGONOMETRY

FUNDAMENTALS OF COLLEGE ALGEBRA AND TRIGONOMETRY

Robert Ellis   Denny Gulick

University of Maryland at College Park

Harcourt Brace Jovanovich, Publishers

San Diego   New York   Chicago   Washington, D.C.   Atlanta

London   Sydney   Toronto

By the same authors:

Calculus With Analytic Geometry, **Second Edition**

College Algebra

College Algebra and Trigonometry, **Second Edition**

Photo Credits

Cover: Weaving by Eve Gulick

1 ©Blair Irwin.   **58, 87** ©Dan Helms/Duomo.   **125, 142** ©Barry L. Runk/Grant Heilman.
149 ©Tiers/Monkmeyer.   **179** ©Grant Heilman.   **212** Official U.S. Navy Photograph.   **213** The
Port Authority of New York and New Jersey.   **231, 261** UPI Photo.   **268** The Metropolitan Museum
of Art, New York.   **274, 295** ©Helen Faye.   **322** ©From SOUNDS OF MUSIC, by Charles Taylor,
BBC London.   **333** ©Harry W. Rinehart.   **334** HBJ Photo.   **358** From SOUNDS OF MUSIC, by
Charles Taylor, BBC London.   **367** California Dept. of Forestry, San Diego Ranger Unit.

Requests for permission to make copies of any part of this work should be
mailed to: Permissions, Harcourt Brace Jovanovich, Publishers, Orlando,
Florida 32887.

ISBN: 0-15-529350-8
Library of Congress Catalog Card Number: 83-82765
Printed in the United States of America

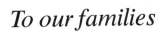

To our families

Preface

This book provides a thorough treatment of the central topics of algebra and trigonometry, and it can be used by anyone with the equivalent of two years of high school algebra. It is identical to the first eight chapters of *College Algebra and Trigonometry*, Second Edition. The book contains enough material for a one-semester or two-quarter course and is intended for those instructors who wish to present a basic course without any additional topics.

The initial chapter contains a review of basic algebraic concepts and can be covered quickly or slowly, depending on the preparation of the reader. Chapters 2 through 5 present the main topics of college algebra and related topics from analytic geometry. These include equations and inequalities, functions and equations and their graphs, and the exponential and logarithmic functions. In Chapters 6 through 8 we turn our attention to trigonometry and discuss trigonometric functions, trigonometric identities, and solutions of triangles.

The revisions to the first edition of *College Algebra and Trigonometry* were made in the light of comments from users of the book. The most noteworthy changes are the following:

1. Chapter 6, which is the introductory chapter on trigonometry, begins with a study of right triangles and trigonometric values of angles in right triangles. It then covers in detail the trigonometric functions and their graphs.

2. To help students prepare for calculus and other studies, we have added a section on polar coordinates (Section 8.4).

3. The chapters on trigonometry include a discussion of sound waves related to musical strings (see Sections 6.8 and 7.4).

4. Cumulative review exercises for Chapters 1 through 4 and Chapters 5 through 8 have been added.

Many of the expositions are easier to understand, and additional examples provide extra help. New problems related to applied mathematics—especially in the early chapters and Chapter 6—show how various concepts can be used in the real world.

One major goal has been to provide an exceptionally readable book on algebra and trigonometry. As a result, we have tried to present each new topic carefully, with plenty of explanation and illustrations, supporting examples, and cautionary remarks (which are set off from the text to enhance their impact). The examples thoroughly reinforce new concepts and provide computational practice. The symbol □ signals the completion of an example.

The exercise sets have been designed to afford ample practice on the topics appearing in the text. The exercises are graded, from routine to more sophisticated; the most difficult exercises are accompanied by an asterisk (*) for easy recognition. Generally the exercises are paired, odd and even, in order to give additional practice and because answers to the odd-numbered exercises are included in the back of the book. Where appropriate we have provided calculator exercises; these are clearly identified with the symbol ◧. In keeping with current trends, both text and exercises include numerous applications to such diverse areas as chemistry, geology, acoustics, motion, demography, forestry, economics, archeology, architecture, and games. Both the metric and the English systems of measurement are employed in applied exercises. Each chapter concludes with a list of key terms and formulas and a set of review exercises that give practice on the major topics of the chapter. For convenient reference, the major formulas and laws from algebra and trigonometry are also printed on the inside covers. An Answer Manual containing answers to the even-numbered exercises is also available to instructors.

We wish to thank our reviewers. Those who reviewed the first edition of *College Algebra and Trigonometry* are: Murray Cantor (Shell Development Company, Houston, Texas), Mary Farantonello (Radio Shack Computer Center, Houston, Texas), Daniel Mosenkis (City College of the City University of New York), Nancy Jim Poxon (California State University, Sacramento), Helen Salzberg (Rhode Island College), Erik Schreiner (Western Michigan University), and William Smith (University of North Carolina, Chapel Hill). Those who contributed suggestions for the Second Edition (which, of course, includes *Fundamentals of College Algebra and Trigonometry*) are: C. Patrick Collier (University of Wisconsin, Oshkosh), Helen Santiz (University of Michigan, Dearborn), and F. Eugene Tidmore (Baylor University).

We very much appreciate the help, encouragement, and vision of two fine editors at Harcourt Brace Jovanovich, Inc.: Marilyn Davis, who worked with us on the first edition, and Richard Wallis, who worked with us on this edition. In addition, we appreciate the assistance of the following staff members at Harcourt Brace Jovanovich for their endeavors on behalf of the Second Edition: Mary Kitzmiller, Marji James, Nancy Shehorn, Robin Risque, and Sharon Weldy.

ROBERT ELLIS
DENNY GULICK

Contents

8

**TRIGONOMETRY
AND ITS
APPLICATIONS, 367**

Tables

1

Basic Algebra

Chapter 1 is the first of four chapters that contain the core of our discussion of algebra. It is intended primarily as a review of the basic rules of algebra.

We begin the chapter with a description of the various kinds of real numbers and the basic properties of the arithmetic operations of addition, subtraction, multiplication, and division. Next we show how to associate the real numbers with the points on a line. Then we branch out and discuss exponents, radicals, polynomials, and rational expressions, all of which are derived in one way or another from the four basic arithmetic operations.

The results of this chapter form the foundation for the many subjects that will be developed in the remaining chapters. Moreover, the results and laws we discuss in this chapter can help in solving everyday problems such as the following one.

> An agency normally rents mid-size cars at $30 per day. It gives a 5% discount to students but raises the rental price 3% on holiday weekends. For such weekends, would the final cost be affected if the agent first took the discount and then added the 3% surcharge, rather than adding the 3% surcharge first and then taking the discount?

We will solve this problem in Example 3 of Section 1.1.

1.1
REAL
NUMBERS

Real numbers appear everywhere in our lives. Indeed, we encounter them anywhere from a grocery store to a doctor's office to a sports page. Let us begin by describing the various kinds of real numbers.

Kinds of Real Numbers

There are several basic collections of real numbers:

Natural numbers (or *counting numbers*): 1, 2, 3, 4, . . .

Even integers: integers that are multiples of 2, such as 0, 2, −2, 4, −4, 6, −6, . . .

Odd integers: integers that are not even, such as 1, −1, 3, −3, 5, −5, . . .

Rational numbers: numbers that can be written in the form a/b, where a and b are integers with $b \neq 0$; examples are 22/7, 11.6, $\dfrac{123}{49,265}$.

Irrational numbers: any number that is not rational, that is, that cannot be written as a/b with a and b integers.

Examples of irrational numbers are π (which is the ratio of the circumference of a circle to its diameter) and $\sqrt{2}$ (which is the ratio of the length of the diagonal of a square to the length of a side).

We remark that every integer a is a rational number, since a can be written as a/b with $b = 1$. Thus the collection of rational numbers includes the integers.

Every real number has a decimal expansion. For example, 5/4 has the decimal expansion 1.25. The decimal expansion of any rational number is either terminating, as is the decimal expansion 1.25 of 5/4, or repeating, as are the decimal expansions 0.6666 . . . of 2/3 and 3.545454 . . . of 39/11. The decimal expansion of an irrational number may or may not be possible to determine. For example, it is known that the decimal expansion of π begins

$$3.14159265358979 \ldots$$

but the expansion does not terminate, nor does it repeat, nor is there any known pattern to the digits in the expansion. However, the digits in the expansion of an irrational number *can* have an easily recognizable pattern (see Exercise 55).

Operations and Conventions

There are four fundamental operations of arithmetic. For two real numbers a and b, we denote the operations as follows:

Arithmetic Operations
Sum: $a + b$
Difference: $a - b$
Product: ab, or $a \cdot b$, or $(a)(b)$
Quotient: $\dfrac{a}{b}$, or a/b, or $a \div b$, with $b \neq 0$

The numbers a and b are *factors* of the product ab, and in the quotient a/b, a is the *numerator* and b the *denominator*. Frequently the quotient a/b of two real numbers is called a *fraction*.

Caution: You will notice that a/b is not defined for $b = 0$. Thus dividing by 0 is not permitted and the quotient $a/0$ is not defined. In particular, $0/0$ and $1/0$ are not defined.

It will be important to keep in mind two conventions that are universally applied in evaluating arithmetic expressions. First, in evaluating an expression such as $a + bc$, multiplication and division take precedence over addition and subtraction. Thus

$$3 + 4 \cdot 5 = 3 + 20 = 23$$

and

$$3 - 2 \div 6 = 3 - \frac{1}{3} = \frac{8}{3}$$

Second, we perform operations inside parentheses before those outside the parentheses. As a result,

$$4(-3 + 2 \cdot 5) = 4(7) = 28$$

More generally, computation progresses from the inside expressions outward. This implies that

$$5[-3 + (4 + 2 \cdot 6)] = 5(-3 + 16) = 5 \cdot 13 = 65$$

Properties of the Arithmetic Operations

We now list the most important properties of addition and multiplication.

Properties of Addition and Multiplication		
Addition	*Multiplication*	*Name*
$a + b = b + a$	$ab = ba$	Commutative properties
$(a + b) + c =$ $a + (b + c)$	$(ab)c = a(bc)$	Associative properties
$a + 0 = a =$ $0 + a$	$a \cdot 1 = a = 1 \cdot a$	Identity properties
For any number a there is a number $-a$ (called the *additive inverse* of a) such that $a + (-a) =$ $0 = (-a) + a$	For any number $a \neq 0$ there is a number $1/a$, or a^{-1} (called the *multiplicative inverse* of a), such that $a \cdot \frac{1}{a} = 1 = \frac{1}{a} \cdot a$	Inverse properties
$a(b + c) = ab + ac$ $(a + b)c = ac + bc$		Distributive properties

The associative property says that $(a + b) + c = a + (b + c)$, and as a result we can eliminate the parentheses and write $a + b + c$ for either $(a + b) + c$ or $a + (b + c)$. Thus by definition,

$$a + b + c = (a + b) + c = a + (b + c)$$

For example, the sum $6 + 4 + 1$ can be computed either as $(6 + 4) + 1$ or as $6 + (4 + 1)$, each of which equals 11.

In the same vein, we define abc by

$$abc = (ab)c = a(bc)$$

If we apply the distributive law several times, we find that

$$(a + b)(c + d) = (a + b)c + (a + b)d = ac + bc + ad + bd$$

Consequently

$$(a + b)(c + d) = ac + bc + ad + bd \tag{1}$$

Thus to multiply $a + b$ and $c + d$, we add together the four products that we can form by multiplying one of the numbers a and b from the first sum $(a + b)$ by one of the numbers c and d from the second sum $(c + d)$.

The use of formula (1) is illustrated in Example 1, in which we use the fact that the square a^2 of any number a is defined by

$$a^2 = a \cdot a$$

EXAMPLE 1.   Find the product $(3a + 2)(4a + 5)$.

Solution.   By (1),

$$\begin{aligned}
(3a + 2)(4a + 5) &= (3a)(4a) + (2)(4a) + (3a)(5) + (2)(5) \\
&= 12a^2 + 8a + 15a + 10 \\
&= 12a^2 + 23a + 10 \quad \square
\end{aligned}$$

A special case of (1) arises if $c = a$ and $d = b$:

$$(a + b)(a + b) = a \cdot a + b \cdot a + a \cdot b + b \cdot b$$

or more succinctly,

$$(a + b)^2 = a^2 + 2ab + b^2 \tag{2}$$

Similarly, other special cases of (1) are

$$(a - b)^2 = a^2 - 2ab + b^2 \tag{3}$$

and

$$(a + b)(a - b) = a^2 - b^2 \qquad\qquad (4)$$

EXAMPLE 2.   Using the fact that $(\sqrt{2})^2 = 2$, verify that $(3 + \sqrt{2})^2 = 11 + 6\sqrt{2}$.

Solution.   By (2),

$$(3 + \sqrt{2})^2 = 3^2 + 2(3)(\sqrt{2}) + (\sqrt{2})^2$$
$$= 9 + 6\sqrt{2} + 2 = 11 + 6\sqrt{2} \quad \square$$

We will use the following property of multiplication in solving equations.

Zero Property
If $ab = 0$, then either $a = 0$ or $b = 0$.

Before we move on to subtraction and division, let us see how the laws presented in this section can be used in solving everyday problems. Example 3 repeats the problem posed at the outset of this chapter.

EXAMPLE 3.   An agency normally rents mid-size cars at $30 per day. It gives a 5% discount to students but raises the rental price 3% on holiday weekends. For such weekends, would the final cost be affected if the agent first took the discount and then added the surcharge, rather than adding the surcharge first and then taking the discount?

Solution.   We observe that giving a 5% discount on $30 is equivalent to multiplying $30 by .95; attaching a 3% surcharge at this point amounts to multiplying (.95)(30) dollars by 1.03, yielding (1.03)[(.95)(30)] dollars. In contrast, if we first add the 3% surcharge on $30, we obtain (1.03)(30) dollars; discounting that amount by 5% yields (.95)[(1.03)(30)] dollars. But by the commutativity and associativity laws for multiplication,

$$(1.03)[(.95)(30)] = (.95)[(1.03)(30)]$$

so it does not matter which order the agent uses in computing the rental price to the student on a holiday weekend. $\square$

Properties of Subtraction and Division

The basic properties involving subtraction and division of real numbers are listed below and on the following page.

$a - b = a + (-b)$	$-(ab) = (-a)b = a(-b)$
$-(-a) = a$	$(-a)(-b) = ab$
$-a = (-1)a$	$a(b - c) = ab - ac$

$$\frac{a}{b} = \frac{c}{d} \text{ if and only if } ad = bc \qquad \frac{a}{b} + \frac{c}{b} = \frac{a+c}{b}$$

$$\frac{a}{b} = a\left(\frac{1}{b}\right) \qquad\qquad \frac{a}{b} \cdot \frac{c}{d} = \frac{ac}{bd}$$

$$\frac{ac}{bc} = \frac{a}{b}$$

$$-\left(\frac{a}{b}\right) = \frac{-a}{b} = \frac{a}{-b} \qquad \frac{1}{1/a} = a$$

$$\frac{a}{b} + \frac{c}{d} = \frac{ad+bc}{bd} \qquad \frac{a/b}{c/d} = \frac{a}{b} \cdot \frac{d}{c} = \frac{ad}{bc}$$

If a and b are integers with $b \neq 0$, then the fraction a/b is in ***lowest terms*** if a and b have no common integer factors besides 1 and -1. Thus $\frac{4}{7}$ is in lowest terms, whereas $\frac{6}{8}$ is not, because 2 is an integer factor of both 6 and 8. The law

$$\frac{ac}{bc} = \frac{a}{b}$$

assists us in reducing a fraction to lowest terms. When we wish to reduce a fraction to lowest terms we first determine an integer c that is a factor of both numerator and denominator, and then cancel c from both numerator and denominator.

EXAMPLE 4.   Reduce the following fractions to lowest terms.

a. $\dfrac{8}{20}$ b. $\dfrac{198}{54}$

Solution.
a. Since $8 = 4 \cdot 2$ and $20 = 4 \cdot 5$, it follows that

$$\frac{8}{20} = \frac{4 \cdot 2}{4 \cdot 5} = \frac{2}{5}$$

Because 2 and 5 have no common integer factors besides 1 and -1, $\frac{2}{5}$ is reduced to lowest terms.
b. Since $198 = 9 \cdot 22 = 9 \cdot 11 \cdot 2$ and $54 = 9 \cdot 3 \cdot 2$, it follows that

$$\frac{198}{54} = \frac{9 \cdot 11 \cdot 2}{9 \cdot 3 \cdot 2} = \frac{11}{3}$$

Because 11 and 3 have no common integer factors besides 1 and -1, $\frac{11}{3}$ is reduced to lowest terms.   □

If we add $\frac{3}{8}$ and $\frac{7}{12}$ by using the formula

$$\frac{a}{b} + \frac{c}{d} = \frac{ad + bc}{bd}$$

we obtain

$$\frac{3}{8} + \frac{7}{12} = \frac{3 \cdot 12 + 8 \cdot 7}{8 \cdot 12} = \frac{92}{96}$$

An equivalent way of obtaining the same result is to write $\frac{3}{8}$ and $\frac{7}{12}$ with the same denominator, called a **common denominator**:

$$\frac{3}{8} = \frac{3 \cdot 12}{8 \cdot 12} = \frac{36}{96} \quad \text{and} \quad \frac{7}{12} = \frac{7 \cdot 8}{12 \cdot 8} = \frac{56}{96}$$

Then

$$\frac{3}{8} + \frac{7}{12} = \frac{36}{96} + \frac{56}{96} = \frac{36 + 56}{96} = \frac{92}{96} \tag{5}$$

The common denominator 96 was obtained by taking the product of the denominators 8 and 12 of the fractions $\frac{3}{8}$ and $\frac{7}{12}$. However, 24 is also a common denominator of $\frac{3}{8}$ and $\frac{7}{12}$, because

$$\frac{3}{8} = \frac{3 \cdot 3}{8 \cdot 3} = \frac{9}{24} \quad \text{and} \quad \frac{7}{12} = \frac{7 \cdot 2}{12 \cdot 2} = \frac{14}{24}$$

Since the common denominator 24 is less than 96, the calculations required to add $\frac{3}{8}$ and $\frac{7}{12}$ are simpler if we use 24 instead of 96:

$$\frac{3}{8} + \frac{7}{12} = \frac{9}{24} + \frac{14}{24} = \frac{23}{24} \tag{6}$$

Of course $\frac{23}{24} = \frac{92}{96}$, so that the results of (5) and (6) are the same. In general it is simplest to use the smallest possible positive common denominator when adding two fractions. That common denominator is called the **least common denominator** (often abbreviated l.c.d.) of the two fractions. One way of obtaining the least common denominator of a/b and c/d is to determine the smallest multiple of b that is also a multiple of d.

EXAMPLE 5.   Find the least common denominator of $\frac{3}{16}$ and $\frac{5}{12}$; then write $\frac{3}{16} + \frac{5}{12}$ as one fraction with that denominator.

Solution.   Since the smallest multiple of 16 that is also a multiple of 12 is 48, the least common denominator is 48. Consequently

$$\frac{3}{16} + \frac{5}{12} = \frac{9}{48} + \frac{20}{48} = \frac{29}{48} \quad \square$$

EXAMPLE 6.   Write the sum $\dfrac{b}{a-b} + \dfrac{a}{a+b}$ as one fraction.

Solution.   Using the properties of fractions, we have

$$\frac{b}{a-b} + \frac{a}{a+b} = \frac{b(a+b) + (a-b)a}{(a-b)(a+b)}$$

$$= \frac{ba + b^2 + a^2 - ba}{a^2 - b^2}$$

$$= \frac{a^2 + b^2}{a^2 - b^2} \quad \square$$

Care in Using Rules of Algebra

When the rules of algebra are used correctly, they lead to valid conclusions. However, it is easy to be careless in using the rules or to manufacture rules that may look plausible but unfortunately are not valid. Below we list several incorrect formulas; beside each we state the corresponding correct formula:

Incorrect	*Correct*
$(a+b)^2 \overset{?}{=} a^2 + b^2$	$(a+b)^2 = a^2 + 2ab + b^2$
$a(b+c) \overset{?}{=} ab + c$	$a(b+c) = ab + ac$
$(ab)(ac) \overset{?}{=} abc$	$(ab)(ac) = a^2bc$
$a - (b-c) \overset{?}{=} a - b - c$	$a - (b-c) = a - b + c$
$\dfrac{-a}{-b} \overset{?}{=} -\dfrac{a}{b}$	$\dfrac{-a}{-b} = \dfrac{a}{b}$
$\dfrac{a}{b+c} \overset{?}{=} \dfrac{a}{b} + \dfrac{a}{c}$	$\dfrac{a}{b+c}$ remains $\dfrac{a}{b+c}$
$a/(b+c) \overset{?}{=} a/b + c$	$a/(b+c)$ remains $a/(b+c)$
$\sqrt{a+b} \overset{?}{=} \sqrt{a} + \sqrt{b}$	$\sqrt{a+b}$ remains $\sqrt{a+b}$
$\dfrac{a}{c} \cdot \dfrac{b}{c} \overset{?}{=} \dfrac{ab}{c}$	$\dfrac{a}{c} \cdot \dfrac{b}{c} = \dfrac{ab}{c^2}$

EXERCISES 1.1

In Exercises 1–6, evaluate the given expression.

1. $-5 + (5 \cdot 2 + 1)$ **2.** $3 \cdot 6 - 6$

3. $\frac{2}{3} - (\frac{1}{2} - \frac{5}{6})$

4. $-2[(4 - 2 \cdot 3) - (2 - 6) + 6]$

5. $[(2 - 3)5 + (-1 + 4)(-2)][4 - 2 \cdot 3]$

6. $2[(\frac{1}{2} - \frac{1}{3}) + \frac{1}{3} \cdot 4] - \frac{1}{2}$

In Exercises 7–16, find the given product or power.

7. $(a+1)(a+2)$ **12.** $(2 + \sqrt{2})(5 - 3\sqrt{2})$

8. $(2a - 1)(3a + 4)$ **13.** $(5 + \sqrt{2})^2$

9. $(-a + 1)(a + 1)$ **14.** $(\frac{1}{2} - \frac{2}{3}\sqrt{3})^2$

10. $(\frac{1}{2}a + 4)(a - \frac{1}{2})$ **15.** $(-3 + \frac{1}{3})^2$

11. $(1 - \sqrt{3})(1 + \sqrt{3})$ **16.** $(-4 - 2\sqrt{2})^2$

17. Write $(a + b + 2)(c + d - 3)$ without parentheses.

18. Write $(a + bc)(d + ef)$ without parentheses.

In Exercises 19–24, reduce the given fraction to lowest terms.

19. $\dfrac{15}{20}$

21. $\dfrac{63}{210}$

23. $-\dfrac{105}{147}$

20. $\dfrac{60}{42}$

22. $-\dfrac{286}{520}$

24. $\dfrac{480}{612}$

In Exercises 25–30, carry out the indicated operation. Then reduce your answer to lowest terms.

25. $\dfrac{3}{4} \cdot \dfrac{5}{6}$

28. $\dfrac{-\dfrac{1}{6}}{\dfrac{1}{12}}$

29. $\dfrac{\dfrac{2}{3}}{\dfrac{8}{9}}$

30. $\dfrac{-\dfrac{2}{15}}{-\dfrac{7}{75}}$

26. $\dfrac{21}{25} \cdot \dfrac{5}{9}$

27. $\left(-\dfrac{4}{9}\right) \cdot \left(-\dfrac{27}{10}\right)$

In Exercises 31–36, find the least common denominator of the two fractions and use it to perform the indicated operation on the two fractions.

31. $\dfrac{1}{6} + \dfrac{1}{8}$

33. $\dfrac{5}{9} + \dfrac{7}{15}$

35. $\dfrac{23}{30} - \dfrac{29}{36}$

32. $\dfrac{5}{12} - \dfrac{7}{9}$

34. $-\dfrac{11}{24} + \dfrac{13}{36}$

36. $\dfrac{2}{35} + \dfrac{3}{49}$

C In Exercises 37–40, use a calculator to approximate the given expression.

37. $\dfrac{3.487 - 2.3496}{48.63 + 3.012}$

39. $\dfrac{23}{247} - \dfrac{59}{1001}$

38. $\dfrac{(3.0107)(16.38)^2}{49.07}$

40. $\dfrac{\pi - \sqrt{2}}{\pi + \sqrt{2}}$

In Exercises 41–44, write the given expression as one fraction.

41. $\dfrac{a - 1}{a + 1} - \dfrac{a + 1}{a - 1}$

43. $\dfrac{2}{a} - \dfrac{3}{b} + \dfrac{4}{ab}$

42. $\dfrac{1}{a + b} + \dfrac{b}{a^2 - b^2}$

44. $\dfrac{1}{a + 2} + \dfrac{4}{a - 2} - \dfrac{2a}{a^2 - 4}$

In Exercises 45–52, correct the given incorrect formula.

45. $(a + 1)(b + 1) \overset{?}{=} ab + 1$

46. $\dfrac{1/a}{1/b} \overset{?}{=} \dfrac{1}{ab}$

47. $a - (b + c) \overset{?}{=} a - b + c$

48. $(-a)(-b) \overset{?}{=} -ab$

49. $(a + b)^3 \overset{?}{=} a^3 + b^3$

50. $(-a)^2 \overset{?}{=} -a^2$

51. $\dfrac{1}{a + b} \overset{?}{=} \dfrac{1}{a} + \dfrac{1}{b}$

52. $\dfrac{a}{b} + \dfrac{c}{d} \overset{?}{=} \dfrac{a + c}{b + d}$

53. Use (2) to calculate $(9.1)^2$.

54. Use (3) to calculate $(.99)^2$.

55. Explain why the number $0.10110111011110 \ldots$, whose decimal expansion has increasingly long strings of 1's, is irrational.

56. Use formula (1) to prove formulas (3) and (4).

57. Show that $(2a)^2 + (a^2 - 1)^2 = (a^2 + 1)^2$ for any real number a.

58. Show that

$$n = \left(\dfrac{n + 1}{2}\right)^2 - \left(\dfrac{n - 1}{2}\right)^2$$

for any real number n. Conclude that every odd integer is the difference of the squares of two integers.

59. a. Show that subtraction is neither commutative nor associative by finding numbers, a, b, and c such that $a - b \neq b - a$ and $a - (b - c) \neq (a - b) - c$.

b. Show that division is neither commutative nor associative by finding numbers a, b, and c such that $a/b \neq b/a$ and $a/(b/c) \neq (a/b)/c$.

60. Show that $a \cdot 0 = 0$ for any real number a. (*Hint:* First use the fact that $a \cdot 0 = a(0 + 0)$. Then use the distributive property. Finally, subtract $a \cdot 0$ from both sides of the resulting equation.)

61. Prove the Zero Property, that is, show that if a and b are real numbers with $ab = 0$, then either $a = 0$ or $b = 0$. (*Hint:* If $a \neq 0$, multiply both sides of the equation $ab = 0$ by a^{-1}.)

62. Show that if a is a real number with $a^2 = a$, then $a = 0$ or $a = 1$. (*Hint:* Rewrite the equation $a^2 = a$ first as $a^2 - a = 0$ and then as $a(a - 1) = 0$, and use Exercise 61.)

63. Let a be a real number. Show that $a = -a$ if and only if $a = 0$.

64. Show that if a and b are nonzero real numbers, then $ab(a^{-1} + b^{-1}) = a + b$.

65. Let a, b, and c be real numbers. Show that if $a + b = a + c$, then $b = c$.

66. Let a, b, and c be real numbers. Show that if $ab = ac$ and $a \neq 0$, then $b = c$.

67. Show that $(a + b + c)^2 = a^2 + b^2 + c^2 + 2ab + 2bc + 2ca$ for any real numbers a, b, and c.

**1.2
THE REAL
LINE**

It is possible to associate the real numbers with the points on a given line l so that every real number is associated with one and only one point on l, and so that, conversely, every point on l is associated with one and only one real number. To establish such an association we first select a particular line l, usually drawn horizontally as in Figure 1.1, and then choose an arbitrary point 0 on l to associate with the number 0. The point 0 is called the **origin**.

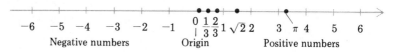

FIGURE 1.1

Next we select the points to associate with the positive integers $1, 2, 3, \ldots$, by marking off line segments of equal length to the right of 0 as in Figure 1.1, and points to associate with the negative integers $-1, -2, -3, \ldots$, by marking off similar line segments to the left of 0. To determine the points on l associated with the rational numbers, we subdivide appropriate portions of l into smaller line segments of equal length. For example, to determine the points to associate with $\frac{1}{3}$ and $\frac{2}{3}$ we subdivide the line segment determined by 0 and 1 into three segments of equal length (Figure 1.1). All points on l that are not associated with rational numbers are associated with irrational numbers. The points associated with the irrational numbers π and $\sqrt{2}$ are exhibited in Figure 1.1.

The **positive direction** on l, pointing from left to right, is indicated by an arrow on l (Figure 1.1). Those numbers corresponding to points to the right of 0 are positive numbers, and those numbers corresponding to points to the left of 0 are the negative numbers.

The number that is associated with an arbitrary point A and l is called the **coordinate** of A, and the association of the points on l with real numbers is frequently called a **coordinate system** for l. The line l with a coordinate system is often referred to as a **real number line**, or **real line**.

Inequalities

Let a and b be real numbers. If $a - b$ is positive, we say that a is **greater than** b and write $a > b$; or alternatively, we say that b is **less than** a and write $b < a$. Geometrically, $a > b$ means that the point corresponding to a on the real line in Figure 1.2 lies to the right of the point corresponding to b. Using this geometric interpretation and the fact that one of two distinct points on the real line must lie to the right of the other, we conclude that if a and b are real numbers, then exactly one of the following three possibilities is true:

FIGURE 1.2

$$a < b, \qquad a = b, \qquad \text{or} \qquad a > b$$

This result is called the ***trichotomy law***.* Taking $b = 0$ in the trichotomy law, we see that any real number a satisfies exactly one of the following:

$$a < 0, \qquad a = 0, \qquad \text{or} \qquad a > 0$$

The relations $a < b$ and $a > b$ are called ***inequalities***, and the symbols $<$ and $>$ are ***inequality signs***. Although we will return to the general laws governing inequalities in Sections 2.6 through 2.8, we will now present a few basic properties of inequalities. To simplify the statements of the properties, we say that two nonzero numbers a and b have the ***same sign*** if $a > 0$ and $b > 0$, or if $a < 0$ and $b < 0$. If $a > 0$ and $b < 0$, or if $a < 0$ and $b > 0$, then we say that a and b have ***opposite signs***. We are thus ready to give the properties:

$$ab > 0 \text{ if and only if } a \text{ and } b \text{ have the same sign.} \qquad (1)$$

$$ab < 0 \text{ if and only if } a \text{ and } b \text{ have opposite signs.} \qquad (2)$$

$$a > 0 \text{ if and only if } -a < 0, \text{ and } a < 0 \text{ if and only if } -a > 0. \qquad (3)$$

Caution: It is sometimes tempting to regard $-a$ as a negative number simply because the expression $-a$ contains a minus sign. However, as the second half of (3) indicates, if a is negative, then $-a$ is actually positive, For instance, if $a = -5$, then $-a = -(-5) = 5$, a positive number.

EXAMPLE 1.   Let $a \neq 0$. Show that a and $1/a$ have the same sign.

Solution.   Recall that

$$a \cdot \frac{1}{a} = 1 > 0$$

Since the product of a and $1/a$ is positive, a and $1/a$ must have the same sign by (1).   □

We write $a \geq b$ to mean that either $a > b$ or $a = b$, and express this by saying that a is ***greater than or equal to*** b. Alternatively, we can write $b \leq a$ and say that b is ***less than or equal to*** a. The symbols $\geq$ and $\leq$ are also called ***inequality signs***. If $a \geq 0$, then a is not negative, so we say that a is ***nonnegative***. For example, the number of miles a person travels during a given day is a nonnegative number.

If a, b, and c are three numbers, then the compound inequality $a < c < b$ means that $a < c$ and $c < b$, and we say that ***c is between a and b***. For example, $2 < 2.13 < \frac{5}{2}$. Thus 2.13 is between 2 and $\frac{5}{2}$. Other compound inequalities, such as $a < c \leq b$ and $a \leq c \leq b$, are defined analogously.

Caution: In any compound inequality, the inequality signs must all point in the same direction. We never write a compound inequality such as $3 > x \leq 5$, in which the inequality signs point in opposite directions.

* The word *trichotomy* comes from a Greek word meaning "threefold division."

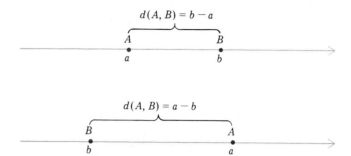

FIGURE 1.3

Absolute Value   One consequence of endowing a line *l* with a coordinate system is that we can measure the distance between two points on *l* by using their coordinates. To be precise, let *A* and *B* be points on *l* having coordinates *a* and *b*, respectively. We define the ***distance*** $d(A, B)$ between *A* and *B* to be either $a - b$ or $b - a$, whichever is nonnegative (Figure 1.3). Because we have identified the points *A* and *B* with the numbers *a* and *b*, we often write $d(a, b)$ instead of $d(A, B)$ and refer to $d(a, b)$ as the distance between *a* and *b*. Thus the distance $d(2, -5)$ between 2 and -5 is $2 - (-5) = 7$, and the distance $d(3, 9)$ between 3 and 9 is $9 - 3 = 6$ (Figure 1.4).

By our definition, the distance $d(0, a)$ between 0 and *a* is either *a* or $-a$, whichever is nonnegative. Of the two numbers *a* and $-a$, the one that is nonnegative is very important in mathematics and is called the ***absolute value*** $|a|$ of *a*:

$$|a| = \begin{cases} a & \text{if} \quad a \geq 0 \\ -a & \text{if} \quad a < 0 \end{cases}$$

In the absolute value notation, the distance between 0 and *a* is given by

$$d(0, a) = |a|$$

Since $b - a = -(a - b)$, it follows from the definition of absolute value that $|a - b|$ is the nonnegative number of the two numbers $a - b$ and $b - a$, so that the distance between *a* and *b* can be written in the succinct form

$$d(a, b) = |b - a|$$

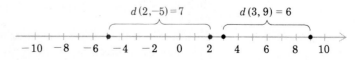

FIGURE 1.4

EXAMPLE 2. Find $|5|, |-5|, |0|, |\pi - 3|$, and $|\sqrt{2} - 7|$.

Solution. Since 5, 0, and $\pi - 3$ are nonnegative, whereas -5 and $\sqrt{2} - 7$ are negative, we have

$$|5| = 5 \qquad |-5| = -(-5) = 5 \qquad |0| = 0$$
$$|\pi - 3| = \pi - 3 \qquad |\sqrt{2} - 7| = -(\sqrt{2} - 7) = 7 - \sqrt{2} \quad \square$$

EXAMPLE 3. Find the distances between the following pairs of numbers on the real line.

a. $a = -9, b = 0$
c. $a = 7, b = -1$
b. $a = 2, b = 6$
d. $a = -2, b = -5$

Solution.
a. $d(a, b) = d(-9, 0) = |0 - (-9)| = |9| = 9$
b. $d(a, b) = d(2, 6) = |6 - 2| = |4| = 4$
c. $d(a, b) = d(7, -1) = |-1 - 7| = |-8| = 8$
d. $d(a, b) = d(-2, -5) = |-5 - (-2)| = |-3| = 3 \quad \square$

We complete the section with a list of several special properties of absolute values:

$|a| = 0$ if and only if $a = 0$	$|a + b| \le |a| + |b|$
$|a| \ge 0$	$|ab| = |a||b|$
$|a| = |-a|$	$\left|\dfrac{a}{b}\right| = \dfrac{|a|}{|b|}$
$|a - b| = |b - a|$	

From the property $|a + b| \le |a| + |b|$ it follows that either $|a + b| = |a| + |b|$ or $|a + b| < |a| + |b|$. See if you can find values of a and b for which $|a + b| = |a| + |b|$, and values for which $|a + b| < |a| + |b|$. Exercise 57 discusses exactly when equality holds.

EXERCISES 1.2

1. Draw a line, set up a coordinate system on the line, and locate the points corresponding to $-1, 2, -2, 3, -3, \frac{5}{2}, -\frac{5}{2}, \frac{7}{4}$, and $-\frac{7}{4}$.

In Exercises 2–12, write out the given statement using the symbols for inequalities.

2. x is less than 0.

3. x is greater than or equal to $\sqrt{2}$.

4. a is between 1 and 2.

5. $6 - r$ is between -1 and 1.

6. y is positive.

7. z is negative.

8. $x + 1$ is nonnegative.

9. $|x - 2|$ is less than 0.01.

10. $|x - 2|$ is greater than or equal to d.

11. c is less than or equal to $\frac{1}{10}$.

12. $4x$ is greater than or equal to 8.

In Exercises 13–20, write out the pairs of numbers in the form $a = b$, $a < b$, or $a > b$, whichever is correct.

13. $\sqrt{2}, 1$ **16.** $(-5)^2, 25$ **19.** $\sqrt{16}, 4$

14. $|-4|, 4$ **17.** $0, \pi - 3$ **20.** $\frac{5}{7}, 0.7$

15. $(-2)^2, 3$ **18.** $\frac{22}{7}, \pi$

In Exercises 21–28, find the distance between a and b.

21. $a = 0, b = -1$ **25.** $a = -2, b = -1$

22. $a = 5, b = 5$ **26.** $a = \frac{1}{2}, b = \frac{1}{4}$

23. $a = 6, b = 0$ **27.** $a = 9.6, b = 1.1$

24. $a = -1, b = 3$ **28.** $a = \pi, b = \sqrt{2}$

In Exercises 29–42, write the numbers without using absolute values.

29. $|4 - 9|$ **35.** $|3 - \pi| + 3$

30. $|5 + 1|$ **36.** $\frac{1}{2}|6 - 4|$

31. $3 - |2 - 4|$ **37.** $\frac{1}{3}|4 - 10|$

32. $|6 - 4| + |-2 - 5|$ **38.** $\dfrac{6}{|-4|}$

33. $-2 - |-2|$

34. $|4 - \sqrt{2}| - 5$ **39.** $|-7| + |-9|$

40. $\dfrac{-|-4|}{|12|}$ **41.** $|-5| + |5|$

42. $|-5| - |5|$

In Exercises 43–52, write the expression without absolute values.

43. $|x^2|$ **48.** $|a - 4|$ if $a < 4$

44. $|(-4 - x)^2|$ **49.** $|a - b|$ if $a \geq b$

45. $|x^2 + 1|$ **50.** $|a - b|$ if $a < b$

46. $|-2 - y^2|$ **51.** $|a - b| - |b - a|$

47. $|a - 4|$ if $a \geq 4$ **52.** $\dfrac{|a - b|}{|b - a|}$ if $a \neq b$

53. Show that $|a| = |-a|$.

54. Show that $|a^2| = |a|^2$.

55. Show that $-|a| \leq a \leq |a|$.

56. Show that $a^2 \geq 0$.

*57. Show that $|a + b| = |a| + |b|$ if and only if one of the following conditions is satisfied:
   a. $a = 0$ or $b = 0$
   b. a and b have the same sign

1.3 INTEGRAL EXPONENTS

Expressions of the form a^n, where a is a number and n an integer, appear in many calculations. For example, if $1 is deposited into a savings account that pays 7% interest compounded annually, then $(1.07)^4$ is the amount of money in the account at the end of four years. For another example, suppose that when a superball is dropped and bounces, it attains $\frac{6}{7}$ of the height from which it was dropped. After bouncing n times it will attain $(\frac{6}{7})^n$ of the height from which it was originally dropped.

In Section 1.1 we noted that

$$a^2 = a \cdot a$$

Similarly, the cube a^3 is defined by the formula

$$a^3 = a \cdot a \cdot a$$

In general, if n is any positive integer, the expression a^n stands for the product of n factors of a:

$$a^n = \overbrace{a \cdot a \cdot a \cdots a}^{n \text{ factors}}$$

with the understanding that $a^1 = a$. The expression a^n is read "a to the nth power," or "the nth power of a." Thus

$$2^5 = 2 \cdot 2 \cdot 2 \cdot 2 \cdot 2 = 32, \qquad \left(\frac{1}{3}\right)^2 = \frac{1}{3} \cdot \frac{1}{3} = \frac{1}{9}$$

and

$$(1.07)^4 = (1.07)(1.07)(1.07)(1.07) = 1.31079601$$

Next we define negative powers of numbers. Let n be a positive integer. We define a^{-n} by the formula

$$a^{-n} = \frac{1}{a^n}, \qquad \text{for} \quad a \neq 0 \tag{1}$$

EXAMPLE 1.   Compute

a. 2^{-3} b. $\left(\frac{1}{3}\right)^{-4}$

Solution.

a. By (1),

$$2^{-3} = \frac{1}{2^3} = \frac{1}{8}$$

b. By (1),

$$\left(\frac{1}{3}\right)^{-4} = \frac{1}{\left(\frac{1}{3}\right)^4} = \frac{1}{\frac{1}{81}} = 81 \quad \square$$

To complete the definition of integral powers of numbers, we define

$$a^0 = 1, \qquad \text{for} \quad a \neq 0 \tag{2}$$

For example, $(1.7)^0 = 1$ and $\left(-\frac{1}{6}\right)^0 = 1$

Caution: In (1) we did not define a^{-n} for $a = 0$, since the expression $\frac{1}{0}$ is meaningless. In (2) we did not define a^0 for $a = 0$ because no definition would be reasonable for all the various ways one might wish to assign a value to 0^0. Thus 0^n is not defined for any integer $n \leq 0$.

As a result of (1) and (2), a^n is defined for any integer n, with the restriction that $a \neq 0$ if $n \leq 0$. In the expression a^n, a is called the **base** and n the **exponent** or **power**. Thus a^n has the form

$$(\text{Base})^{\text{Exponent}}$$

Laws of Exponents

There are many ways of combining powers of numbers. First of all, let us consider the product $2^4 \cdot 2^3$, rearranged as follows:

$$2^4 \cdot 2^3 = \overbrace{(2 \cdot 2 \cdot 2 \cdot 2)}^{4\text{ factors}}\overbrace{(2 \cdot 2 \cdot 2)}^{3\text{ factors}} = \overbrace{(2 \cdot 2 \cdot 2 \cdot 2 \cdot 2 \cdot 2 \cdot 2)}^{7\text{ factors}}$$
$$= 2^7 = 2^{4+3}$$

Thus

$$2^4 \cdot 2^3 = 2^{4+3}$$

More generally, if a is any number and m and n are positive integers, then

$$a^m a^n = \overbrace{(a \cdot a \cdot a \cdots a)}^{m\text{ factors}}\overbrace{(a \cdot a \cdot a \cdots a)}^{n\text{ factors}} = \overbrace{a \cdot a \cdot a \cdots a}^{(m+n)\text{ factors}} = a^{m+n}$$

Therefore

$$a^m a^n = a^{m+n}$$

Actually, this formula is valid even if we remove the restriction that m and n be positive. This and several other formulas involving exponents are listed together below.

Laws of Integral Exponents

Let a and b be real numbers and m and n integers. Each of the following formulas is valid for all values of a and b for which both sides of the equation are defined.

i. $a^m a^n = a^{m+n}$

iv. $\left(\dfrac{a}{b}\right)^n = \dfrac{a^n}{b^n} = a^n b^{-n}$

ii. $(a^m)^n = a^{mn}$

v. $a^{-n} = \dfrac{1}{a^n}$

iii. $(ab)^n = a^n b^n$

vi. $\dfrac{a^m}{a^n} = a^{m-n}$

Recall that division by 0 is meaningless, as is raising 0 to a power that is 0 or negative. This implies, for example, that (iv) does not hold if $b = 0$, or if $a = 0$ and $n \le 0$.

EXAMPLE 2.   Simplify the following expressions.

a. $5^7 5^{-4}$ b. $[(-3)^3]^2$ c. $\left(\dfrac{2}{5}\right)^3\left(\dfrac{5}{4}\right)^3$ d. $\dfrac{4^{17}}{4^{13}}$

Solution.
a. By (i),

$$5^7 5^{-4} = 5^{7-4} = 5^3 = 125$$

b. By (ii),

$$[(-3)^3]^2 = (-3)^{3 \cdot 2} = (-3)^6 = 729$$

c. By (iii),

$$\left(\frac{2}{5}\right)^3 \left(\frac{5}{4}\right)^3 = \left(\frac{2}{5} \cdot \frac{5}{4}\right)^3 = \left(\frac{1}{2}\right)^3 = \frac{1}{8}$$

d. By (vi),

$$\frac{4^{17}}{4^{13}} = 4^{17-13} = 4^4 = 256 \quad \square$$

__EXAMPLE 3.__ Simplify the following expressions.

a. $\dfrac{(ab)^{-2}}{a^{-3}b^4}$

b. $\dfrac{x - y}{x^{-1} - y^{-1}}$

Solution.

a. Using the Laws of Integral Exponents, we have

$$\frac{(ab)^{-2}}{a^{-3}b^4} = \frac{a^{-2}b^{-2}}{a^{-3}b^4} = \frac{a^{-2}}{a^{-3}} \cdot \frac{b^{-2}}{b^4} = a^{-2-(-3)}b^{-2-4} = ab^{-6}$$

b. Using the Laws of Integral Exponents, we have

$$\frac{x - y}{x^{-1} - y^{-1}} = \frac{x - y}{\dfrac{1}{x} - \dfrac{1}{y}} = \frac{x - y}{\dfrac{y - x}{xy}} = (x - y)\left(\frac{xy}{y - x}\right)$$

$$= \frac{-(y - x)}{y - x}(xy) = -xy \quad \square$$

Caution: We have not included a law for evaluating $(a + b)^n$. It might be tempting to equate $(a + b)^n$ with $a^n + b^n$, but in general the two expressions are not equal. In fact, if $n \neq 1$, then $(a + b)^n \neq a^n + b^n$ if a, b, and $a + b$ are different from 0. Correct formulas for $(a + b)^2$ and $(a + b)^3$ appear in Section 1.6.

The exponential laws can be extended to include products of more than two numbers. For example,

$$a^m a^n a^p = a^{m+n+p}$$

$$(abc)^n = a^n b^n c^n$$

EXAMPLE 4.   Simplify the following expressions and write them with only positive exponents.

a. $(xy^2)^3(x^2z^3)^{-2}(xyz)^2$

b. $\dfrac{(2r^3s^{-2})^4}{(rs^{-2}t^{-3})^2}$

Solution.

a.
$$(xy^2)^3(x^2z^3)^{-2}(xyz)^2 = (x^3y^6)(x^{-4}z^{-6})(x^2y^2z^2)$$
$$= (x^3x^{-4}x^2)(y^6y^2)(z^{-6}z^2)$$
$$= x^{3-4+2}y^{6+2}z^{-6+2}$$
$$= x^1y^8z^{-4}$$
$$= \frac{xy^8}{z^4}$$

b.
$$\frac{(2r^3s^{-2})^4}{(rs^{-2}t^{-3})^2} = \frac{16r^{12}s^{-8}}{r^2s^{-4}t^{-6}}$$
$$= 16r^{12-2}s^{-8-(-4)}t^{-(-6)}$$
$$= 16r^{10}s^{-4}t^6 = \frac{16r^{10}t^6}{s^4} \quad \square$$

Let a be any real number. It follows from (1) in Section 1.2 that

$$a^2 \geq 0 \qquad\qquad\qquad (3)$$

In contrast,

$$a^3 \quad \text{has the same sign as} \quad a \qquad\qquad\qquad (4)$$

The statements in (3) and (4) are special cases of the following general results:

> a^n is nonnegative if n is an even integer.
>
> a^n has the same sign as a if n is an odd integer.

EXAMPLE 5.   Determine whether the given number is positive or negative.

a. $(-2.17)^3(-4.63)^{-2}$

b. $(-1)^5(-2)^{-3}(\tfrac{3}{4})^{-14}$

Solution.

a. Notice that

$(-2.17)^3$ is negative because 3 is odd and -2.17 is negative

$(-4.63)^{-2}$ is positive because -2 is even

Therefore the product $(-2.17)^3(-4.63)^{-2}$, being the product of a negative and a positive number, is negative.

b. Observe that

$$(-1)^5 \text{ is negative because 5 is odd and } -1 \text{ is negative}$$
$$(-2)^{-3} \text{ is negative because } -3 \text{ is odd and } -2 \text{ is negative}$$
$$(\tfrac{3}{4})^{-14} \text{ is positive because } -14 \text{ is even}$$

Consequently the number in (b) is the product of two negative numbers and a positive number and is therefore positive. ☐

Scientific Notation

Quantities in the physical world come in all sizes, from the microscopic to the astronomical. For instance, the mass of an electron is approximately 0.0000000000000000000000000000009 kilograms, and the mass of the sun is approximately 1,987,000,000,000,000,000,000,000,000,000 kilograms. Many, perhaps most, quantities that arise in the physical sciences are either very small or very large. To make writing such numbers more convenient, scientists have adopted the standard practice of writing any quantity, regardless of its size, as a product $b \times 10^n$, where $1 \le b < 10$ and n is an integer. This notation for the number is called the **scientific notation** for the number. In scientific notation the mass of the electron mentioned above is approximately 9×10^{-31} kilograms, and the mass of the sun is approximately 1.987×10^{30} kilograms. These numbers are obviously easier to write and remember when given in scientific notation.

Let a given positive number be equal to $b \times 10^n$ in scientific notation. On the one hand, if the number is greater than or equal to 1, then n is the number of places the decimal point must be moved to the *left* in order to make the decimal expansion of the resulting number lie between 1 and 9.9999 . . . . For example, $n = 3$ for the number 2341, since 2.341 lies between 1 and 9.9999 . . . . Thus

$$2341 = 2.341 \times 10^3$$

On the other hand, if the number is less than 1, then n is the negative of the number of places the decimal point must be moved to the right in order to make the decimal expansion of the resulting number lie between 1 and 9.9999 . . . . For example, $n = -4$ for the number 0.00073, since 7.3 lies between 1 and 9.9999 . . . . Thus

$$0.00073 = 7.3 \times 10^{-4}$$

EXAMPLE 6.   Write the following numbers in scientific notation.
    a. 14,753   b. 0.23   c. 0.00000912   d. 1,000,000

Solution.
a. $14{,}753 = 1.4753 \times 10^4$
b. $0.23 = 2.3 \times 10^{-1}$
c. $0.00000912 = 9.12 \times 10^{-6}$
d. $1{,}000{,}000 = 1 \times 10^6$ (or simply 10^6) ☐

Since most calculators display no more than ten digits, they normally use scientific notation in displaying very large or very small numbers. For example, in scientific notation, 2^{50} is approximately

$$1.125899907 \times 10^{15}$$

A calculator might display this as

$$1.125899907 \qquad 15$$

In contrast, 3^{-26} is approximately

$$3.934117957 \times 10^{-13}$$

and a calculator might display this as

$$3.934117957 \qquad -13$$

EXERCISES 1.3

In Exercises 1–8, compute the given number.

1. 2^0

2. 2^{-2}

3. $\left(\dfrac{1}{2}\right)^3$

4. $\left(\dfrac{1}{2}\right)^{-3}$

5. 4^{-1}

6. 10^3

7. 10^{-3}

8. $(0.03)^2$

In Exercises 9–22, simplify the given expression.

9. $4^2 \cdot 4^3$

10. $(-7)^4(-7)^8$

11. $2^4(-2)^5$

12. $\dfrac{2^3}{2^5}$

13. $\dfrac{(-7)^5}{7^6}$

14. $2^5 \div 2^3$

15. $2^5 \div 2^{-3}$

16. $3^{-6} \cdot 3^4$

17. $10^9 \cdot 10^{-11}$

18. $(3 \cdot 3^3)^2$

19. $4^4 \cdot 4^2 \cdot 4$

20. $\dfrac{\pi^2 \pi^5}{\pi^3}$

21. $[(-4)^5]^6$

22. $[(-3)^{-3}]^{-3}$

In Exercises 23–28, write the expression as a quotient a/b, where a and b are integers.

23. $2^3 \cdot 3^{-2}$

24. $\dfrac{4^2}{5^3}$

25. $\dfrac{3^4 2^{-3}}{3^2 2^{-1}}$

26. $\left(\dfrac{9}{5}\right)^3 \left(\dfrac{5}{9}\right)^4$

27. $(147)^5 \div (147)^6$

28. $(2^4 \cdot 5^3) \div 30$

In Exercises 29–44, simplify the expression.

29. $a^5 a^7$

30. $a^4 a^{-2}$

31. $y^{-2} y^{-6}$

32. $r^{-8} r^8$

33. $b^4 \dfrac{1}{b^2} b^3$

34. $\dfrac{b^3}{b^{-5}}$

35. $(-c^2)^4$

36. $(-c^2)^5$

37. $(xy^2)^3$

38. $(x^2 y^3 z)^4$

39. $(\tfrac{2}{3}x^4)^{-2}$

40. $(z^2/x^3)^5$

41. $rs^2(r^5 s^4)^3$

42. $(t^{-1})^{-1}$

43. $\dfrac{(st^{-1})^{-1}}{s^{-1} t^{-1}}$

44. $\dfrac{\dfrac{1}{s^{-1}} + \dfrac{1}{t^{-1}}}{s^{-1} t^{-1}}$

In Exercises 45–50, write the expression as a quotient involving only positive exponents.

45. $a^{-1} b^{-1}$

46. $a^{-1} + b^{-2}$

47. $(a^{-1} + b^{-1})^{-3}$

48. $(a + b)^{-1}(a^{-1} + b^{-1})$

49. $\dfrac{(a^{-1} + b^{-1})^{-1}}{(ab)^{-1}}$

50. $\dfrac{a^{-2}}{a^{-2} + b^{-2}}$

In Exercises 51–54, determine whether the given number is positive or negative.

51. $(3.1)^{-2}(2)^{-3}$

53. $(-1)^3(5.2)^3$

52. $(-3.2)^2 4^3$

54. $(-1.3)^{-2} 7^{-2}$

In Exercises 55–60, write the number in scientific notation.

55. 483.2

57. 1.009

59. 0.9999

56. 0.791

58. $891,134$

60. 0.0000134

⊂ In Exercises 61–72, use a calculator to approximate the given value and write the answer in scientific notation.

61. $(9876)(2751)$

66. $(0.00000007)(0.000009)$

62. $(3715)(0.0015)$

67. $(7.9 \times 10^3)(2.3 \times 10^5)$

63. $(0.4646)(0.3801)$

68. $(4.791 \times 10^7)(9.31 \times 10^{-8})$

64. $(55.55)(6.148)$

69. $9824 \div 112,344$

65. $(0.0012)(0.00025)$

70. $83.74 \div 0.012$

71. $(3.246 \times 10^{-6}) \div (4.158 \times 10^{-4})$

72. $(1.111 \times 10^8) \div (5.876 \times 10^{-9})$

In Exercises 73–79, write the statement as an equation.

73. The circumference C of a circle is 2π times the radius r.

74. The area A of a circle is π times the square of the radius r.

75. The area A of a square is the square of a side s.

76. The area A of a triangle is half the product of the base b and the height h.

77. The volume V of a sphere is $\frac{4}{3}\pi$ times the cube of the radius r.

78. The volume V of a cylinder is π times the product of the height h and the square of the radius r of the base.

79. The volume V of a cone is π times $\frac{1}{3}$ the product of the height h and the square of the radius r of the base.

80. a. Show that if a and b have the same sign, then $(ab)^n > 0$ for any integer n.

b. Show that if a and b have different signs, then $(ab)^n > 0$ if and only if the integer n is even.

81. Show that $(a^2 + b^2)(c^2 + d^2) = (ac + bd)^2 + (ad - bc)^2$.

82. Show that $3(2ab)^2 + (a^2 - 3b^2)^2 = (a^2 + 3b^2)^2$.

*83. a. For what values of a, m, and n does part (ii) of the Laws of Integral Exponents not hold?

b. For what values of a, b, and n does part (iii) of the Laws of Integral Exponents not hold?

84. The mass of the earth is approximately 5.98×10^{24} kilograms. Write this number in decimal form.

85. The mass of a proton is approximately 1.67×10^{-27} kilograms. Write this number in decimal form.

⊂ **86.** The average distance between the earth and the sun is approximately 149,000,000 kilometers. Assuming that 1 kilometer is equal to 0.621 miles, use a calculator to compute the average distance in miles between the earth and the sun. Write your answer in scientific notation.

⊂ **87.** The mass of a proton is approximately 1.673×10^{-24} grams, whereas the mass of an electron is approximately 9.109×10^{-28} grams. How many times more massive than an electron is a proton?

⊂ **88.** A *light year* is the distance light in a vacuum travels in one year (approximately $365\frac{1}{4}$ days).

a. Assuming that light in a vacuum travels 186,000 miles per second, use a calculator to compute the number of miles in one light year. Write your answer in scientific notation.

b. Sirius is approximately 5.106×10^{13} miles from earth. How many light years away from earth is Sirius?

⊂ **89.** Under ideal conditions, if a superball is dropped from a height of one meter above ground, the nth bounce will have a height of approximately $(\frac{6}{7})^n$ meter.

a. Compute the height of the 15th bounce.

b. Would the 15th bounce of the superball be greater than the 6th bounce of a ball whose nth bounce has a height of $(\frac{2}{3})^n$ meter? Explain your answer.

1.4 RADICALS

If the area of a square is 9 square units, then any side of the square has length 3 (Figure 1.5), and we write $\sqrt{9} = 3$. Similarly, for any $a \geq 0$ we define $\sqrt{a}$ to be the nonnegative number b whose square is a. Thus

$$\sqrt{a} = b \qquad \text{if and only if} \qquad b \geq 0 \quad \text{and} \quad b^2 = a$$

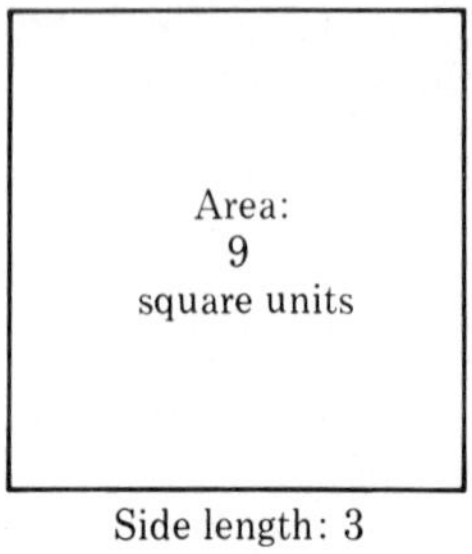

Area:
9
square units

Side length: 3

FIGURE 1.5

For example,

$$\sqrt{1.21} = 1.1 \qquad \text{because} \quad 1.1 \geq 0 \quad \text{and} \quad (1.1)^2 = 1.21$$

$$\sqrt{\frac{1}{4}} = \frac{1}{2} \qquad \text{because} \quad \frac{1}{2} \geq 0 \quad \text{and} \quad \left(\frac{1}{2}\right)^2 = \frac{1}{4}$$

We call $\sqrt{a}$ the **square root** of a, and in the expression $\sqrt{a}$ the a is the **radicand** and $\sqrt{}$ the **radical sign**. The expression $\sqrt{a}$ is also called a **radical**.

By definition, $\sqrt{a} = b$ only if $a = b^2 \geq 0$, so $\sqrt{a}$ is defined only for $a \geq 0$. That the number $\sqrt{a}$ exists for *every* $a \geq 0$ is proved in more advanced books. We observe also that

$$(-\sqrt{a})^2 = (\sqrt{a})^2 = a \qquad \text{for} \quad a \geq 0$$

so the squares of $-\sqrt{a}$ and of $\sqrt{a}$ are the same.

Two basic rules for combining radicals pertain to products and quotients:

$$\sqrt{ab} = \sqrt{a}\sqrt{b} \tag{1}$$

and

$$\sqrt{\frac{a}{b}} = \frac{\sqrt{a}}{\sqrt{b}} \tag{2}$$

Frequently we use (1) and (2) to simplify radicals and combinations of radicals.

EXAMPLE 1.   Simplify the following expressions.

   a. $\sqrt{128}$ b. $\sqrt{8}\sqrt{18}$ c. $\dfrac{\sqrt{54}}{\sqrt{24}}$

Solution.

a. $$\sqrt{128} = \sqrt{64 \cdot 2} \overset{(1)}{=} \sqrt{64}\sqrt{2} = 8\sqrt{2}$$

b. $$\sqrt{8}\sqrt{18} \overset{(1)}{=} \sqrt{8 \cdot 18} = \sqrt{144} = 12$$

c. $$\frac{\sqrt{54}}{\sqrt{24}} = \frac{\sqrt{9 \cdot 6}}{\sqrt{4 \cdot 6}} \overset{(1)}{=} \frac{\sqrt{9}\sqrt{6}}{\sqrt{4}\sqrt{6}} = \frac{\sqrt{9}}{\sqrt{4}} = \frac{3}{2} \quad \square$$

EXAMPLE 2.   Assume that all quantities appearing in the following radicals are positive. Simplify the expressions.

   a. $\sqrt{a^3 b^2}$ b. $\sqrt{6r^2 s}\sqrt{30r^4 s^3}$ c. $\dfrac{\sqrt{x^2 y^{-6} z^3}}{\sqrt{x^2 y^3 z}}$

Solution.

a. $\sqrt{a^3b^2} = \sqrt{a^2ab^2} = \sqrt{(ab)^2a} \overset{(1)}{=} \sqrt{(ab)^2}\sqrt{a} = ab\sqrt{a}$

b. $\sqrt{6r^2s}\sqrt{30r^4s^3} \overset{(1)}{=} \sqrt{(6r^2s)(6\cdot 5r^4s^3)} = \sqrt{6^2\cdot 5r^6s^4}$

$= \sqrt{(6r^3s^2)^2\cdot 5} = \sqrt{(6r^3s^2)^2}\sqrt{5} = 6r^3s^2\sqrt{5}$

c. $\dfrac{\sqrt{x^4y^{-6}z^3}}{\sqrt{x^2y^3z}} \overset{(2)}{=} \sqrt{\dfrac{x^4y^{-6}z^3}{x^2y^3z}} = \sqrt{\dfrac{x^2z^2}{y^9}} = \sqrt{\dfrac{x^2z^2}{y^8}\dfrac{1}{y}}$

$= \sqrt{\left(\dfrac{xz}{y^4}\right)^2\dfrac{1}{y}} \overset{(1)}{=} \sqrt{\left(\dfrac{xz}{y^4}\right)^2}\sqrt{\dfrac{1}{y}} = \dfrac{xz}{y^4}\sqrt{\dfrac{1}{y}}$

$= \dfrac{xz}{y^4}\dfrac{1}{\sqrt{y}} = \dfrac{xz}{y^4\sqrt{y}}$ □

Caution: Despite the product and quotient rules for radicals given in (1) and (2), a corresponding sum rule fails. In fact, $\sqrt{a+b} \neq \sqrt{a} + \sqrt{b}$ (unless $a = 0$ or $b = 0$; see Exercise 85). To support this claim, we notice that

$$\sqrt{16 + 9} = \sqrt{25} = 5 \quad \text{but} \quad \sqrt{16} + \sqrt{9} = 4 + 3 = 7$$

so that

$$\sqrt{16 + 9} \neq \sqrt{16} + \sqrt{9}$$

If a fraction such as $1/\sqrt{a}$ contains one or more radicals, it is common to alter it so that the denominator contains no radicals. The procedure, called *rationalizing the denominator*, involves multiplying the fraction by 1 written in a special way. For $1/\sqrt{a}$ we would multiply by 1 written as $\sqrt{a}/\sqrt{a}$, which yields

$$\frac{1}{\sqrt{a}} = \frac{1}{\sqrt{a}}\cdot\frac{\sqrt{a}}{\sqrt{a}} = \frac{\sqrt{a}}{\sqrt{a}\sqrt{a}} = \frac{\sqrt{a}}{a}$$

The denominator of the last fraction contains no radicals.

EXAMPLE 3.   Simplify the following fractions by rationalizing the denominator.

a. $\dfrac{3}{\sqrt{2}}$ b. $\dfrac{\sqrt{6}}{\sqrt{5}}$

Solution.

a. Using the strategy discussed above, we multiply the fraction by $\sqrt{2}/\sqrt{2}$, which gives us

$$\frac{3}{\sqrt{2}} = \frac{3}{\sqrt{2}}\cdot\frac{\sqrt{2}}{\sqrt{2}} = \frac{3\sqrt{2}}{2} = \frac{3}{2}\sqrt{2}$$

b. Here we multiply the fraction by $\sqrt{5}/\sqrt{5}$, which yields

$$\frac{\sqrt{6}}{\sqrt{5}} = \frac{\sqrt{6}}{\sqrt{5}} \cdot \frac{\sqrt{5}}{\sqrt{5}} = \frac{\sqrt{30}}{5} \quad \square$$

EXAMPLE 4.   Simplify the following.

a. $\dfrac{\sqrt{2}}{\sqrt{2}-1}$ b. $\dfrac{\sqrt{a}-\sqrt{b}}{\sqrt{a}+\sqrt{b}}$

Solution.

a. We multiply the fraction by $(\sqrt{2}+1)/(\sqrt{2}+1)$, obtaining

$$\frac{\sqrt{2}}{\sqrt{2}-1} = \frac{\sqrt{2}}{\sqrt{2}-1} \cdot \frac{\sqrt{2}+1}{\sqrt{2}+1}$$

$$= \frac{\sqrt{2}(\sqrt{2}+1)}{2-1} = 2 + \sqrt{2}$$

b. We multiply the fraction by $(\sqrt{a}-\sqrt{b})/(\sqrt{a}-\sqrt{b})$, which yields

$$\frac{\sqrt{a}-\sqrt{b}}{\sqrt{a}+\sqrt{b}} = \frac{\sqrt{a}-\sqrt{b}}{\sqrt{a}+\sqrt{b}} \cdot \frac{\sqrt{a}-\sqrt{b}}{\sqrt{a}-\sqrt{b}}$$

$$= \frac{a - 2\sqrt{a}\sqrt{b} + b}{a - b} = \frac{a - 2\sqrt{ab} + b}{a - b} \quad \square$$

Let us observe that the square root and the absolute value (defined in Section 1.2) are intimately related by the formula

$$\boxed{\sqrt{a^2} = |a| \quad \text{for any real number } a} \tag{3}$$

After all, if $a \geq 0$, then $\sqrt{a^2} = a = |a|$. But if $a < 0$, then since $-a > 0$ and $(-a)^2 = a^2$, we conclude that

$$\sqrt{a^2} = -a = |a|$$

More generally, one can prove that for any real number a and any positive integer m,

$$\boxed{\sqrt{a^{2m}} = |a|^m}$$

(See Exercise 84.) Thus

$$\sqrt{\pi^{12}} = \pi^6 \quad \text{and} \quad \sqrt{(-11)^{10}} = 11^5$$

Also

$$\sqrt{a^6} = |a|^3$$

nth Roots If the volume of a cube is 64 cubic units, than any side of the cube has length 4 (Figure 1.6), and we write $\sqrt[3]{64} = 4$. Similarly, for any real number a we define $\sqrt[3]{a}$ to be the number b such that $b^3 = a$, that is,

$$\sqrt[3]{a} = b \quad \text{if and only if} \quad b^3 = a$$

For example,

$$\sqrt[3]{-27} = -3 \quad \text{because} \quad (-3)^3 = -27$$

The number $\sqrt[3]{a}$ is called the **cube root** (or **third root**) of a. Notice that unlike the square root, the cube root is defined for *all* real numbers.

More generally, for any integer $n \geq 2$ we define the **nth root** $\sqrt[n]{a}$ of a as follows:

$$\sqrt[n]{a} = b \quad \text{if and only if} \quad b^n = a \quad \begin{cases} \text{for } a \geq 0 \text{ and } b \geq 0 \text{ if } n \text{ is even} \\ \text{for any real number } a \text{ if } n \text{ is odd} \end{cases}$$

If $n = 2$, then $\sqrt[n]{a}$ becomes $\sqrt[2]{a}$, which is normally written $\sqrt{a}$. The number n in $\sqrt[n]{a}$ is called the **index** of the root; the index of $\sqrt{a}$ is 2.

Caution: Notice carefully that when n is an even integer, $\sqrt[n]{a}$ is defined only for nonnegative values of a.

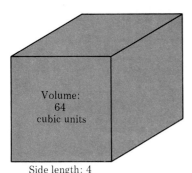

Volume: 64 cubic units

Side length: 4

FIGURE 1.6

EXAMPLE 5.   Simplify the following expressions.

  a. $\sqrt[4]{81}$ b. $\sqrt[5]{-32}$ c. $\sqrt[6]{64/729}$

Solution.

  a. $\sqrt[4]{81} = 3$, since $3^4 = 81$ and $3 \geq 0$.

  b. $\sqrt[5]{-32} = -2$, since $(-2)^5 = -32$.

  c. $\sqrt[6]{64/729} = 2/3$, since $(2/3)^6 = 64/729$ and $2/3 \geq 0$.   □

The same kinds of laws hold for general nth roots as for square roots.

Laws of nth Roots

Let a and b be real numbers and m and n positive integers. Each of the following formulas is valid for all values of a and b for which both sides of the equation are defined.

i. $(\sqrt[n]{a})^n = a$

ii. $\sqrt[n]{a^n} = \begin{cases} |a| & \text{if } n \text{ is even} \\ a & \text{if } n \text{ is odd} \end{cases}$

iii. $\sqrt[n]{ab} = \sqrt[n]{a}\,\sqrt[n]{b}$

iv. $\sqrt[n]{\dfrac{a}{b}} = \dfrac{\sqrt[n]{a}}{\sqrt[n]{b}}$

v. $\sqrt[m]{\sqrt[n]{a}} = \sqrt[mn]{a}$

Simplifying nth roots is similar to simplifying square root, but with nth roots we use the laws listed above and factor out nth powers of numbers or variables where possible.

EXAMPLE 6.   Simplify the following expressions.

a. $\sqrt[3]{-81}$ b. $\sqrt[4]{8}\,\sqrt[4]{162}$ c. $\sqrt{\sqrt[3]{729}}$ d. $\dfrac{\sqrt[4]{32x^8y^6}}{\sqrt[4]{x^2y^2}}$

Solution.

a. $\sqrt[3]{-81} = \sqrt[3]{(-27)3} = \sqrt[3]{(-3)^3 3} \overset{\text{(iii)}}{=} \sqrt[3]{(-3)^3}\,\sqrt[3]{3} \overset{\text{(ii)}}{=} -3\sqrt[3]{3}$

b. $\sqrt[4]{8}\,\sqrt[4]{162} \overset{\text{(iii)}}{=} \sqrt[4]{(2^3)(2 \cdot 3^4)} = \sqrt[4]{2^4 3^4} = \sqrt[4]{(2 \cdot 3)^4} \overset{\text{(ii)}}{=} 6$

c. $\sqrt{\sqrt[3]{729}} \overset{\text{(v)}}{=} \sqrt[6]{729} = \sqrt[6]{3^6} \overset{\text{(ii)}}{=} 3$

d. $\dfrac{\sqrt[4]{32x^8y^6}}{\sqrt[4]{x^2y^2}} \overset{\text{(iv)}}{=} \sqrt[4]{\dfrac{32x^8y^6}{x^2y^2}} = \sqrt[4]{32x^6y^4} = \sqrt[4]{(2^4x^4y^4)(2x^2)}$

$\overset{\text{(iii)}}{=} \sqrt[4]{2^4x^4y^4}\,\sqrt[4]{2x^2} \overset{\text{(ii)}}{=} 2|xy|\sqrt[4]{2x^2}$ $\square$

EXERCISES 1.4

In Exercises 1–8 simplify the expression.

1. $\sqrt{16}$

2. $\sqrt{81}$

3. $\sqrt{\dfrac{1}{25}}$

4. $\sqrt{\dfrac{4}{49}}$

5. $\sqrt{1.69}$

6. $\sqrt{2.25}$

7. $\sqrt{0.04}$

8. $\sqrt{0.0036}$

In Exercises 9–22, simplify the expression.

9. $\sqrt{48}$

10. $\sqrt{9 \cdot 16}$

11. $\sqrt{\dfrac{1}{4} \cdot \dfrac{1}{36}}$

12. $\sqrt{3 \times 10^{12}}$

13. $\sqrt{4 \times 10^{11}}$

14. $\sqrt{0.5}\,\sqrt{4.5}$

15. $\sqrt{6}\,\sqrt{12}$

16. $\sqrt{32}\,\sqrt{72}$

17. $\sqrt{5 \times 10^5}\,\sqrt{20 \times 10^7}$

18. $\dfrac{1}{\sqrt{18}}$

19. $\dfrac{4}{\sqrt{96}}$

20. $\dfrac{\sqrt{63}}{\sqrt{21}}$

21. $\dfrac{\sqrt{75}}{\sqrt{147}}$

22. $\dfrac{\sqrt{6 \times 10^5}}{\sqrt{2 \times 10^{11}}}$

In Exercises 23–40, assume that all letters denote positive numbers. Simplify the expressions.

23. $\sqrt{a^2b^3}$

24. $\sqrt{24x^6y^{-4}}$

25. $\sqrt{(-32)x^3(-y)^5z^2}$

26. $\sqrt{\dfrac{12}{p^4q^7}}$

27. $\sqrt{\dfrac{p^5}{27q^8}}$

28. $\sqrt{3rs^{-3}t^2}\,\sqrt{27r^3s^5t^6}$

29. $\sqrt{8r/s^2}\,\sqrt{16s^2/r^4}$

30. $\dfrac{\sqrt{c^2d^6}}{\sqrt{4c^3d^{-4}}}$

31. $\dfrac{\sqrt{c^{-10}d^{-12}}}{\sqrt{c^{14}d^{-3}}}$

32. $\sqrt{(x + y)^2}$

33. $\sqrt{x^2 + 2xy + y^2}$

34. $\sqrt{x^2 - 2xy + y^2}$

35. $\sqrt{(4a - 7b)^4}$

36. $\sqrt{(9a^2 - 11b^3)^4}$

37. $\sqrt{\sqrt{16a^4b^8}}$

38. $\sqrt{\sqrt{9a^8/b^{10}}}$

39. $\sqrt{ab}\left(\dfrac{1}{\sqrt{b}} + \dfrac{1}{\sqrt{a}}\right)$

40. $\left(\sqrt{x} - \dfrac{1}{\sqrt{x}}\right)^2$

In Exercises 41–52, simplify the expression.

41. $\sqrt[3]{27}$ **42.** $\sqrt[3]{-1/64}$ **43.** $\sqrt[3]{0.125}$

44. $\sqrt[3]{0.000125}$ **47.** $\sqrt[5]{243}$ **50.** $\sqrt[3]{32}$

45. $\sqrt[4]{256}$ **48.** $\sqrt[7]{-1/128}$ **51.** $\sqrt[4]{16 \cdot 8}$

46. $\sqrt[4]{0.0016}$ **49.** $\sqrt[3]{8 \cdot 2}$ **52.** $\sqrt[6]{64}$

In Exercises 53–62, assume that all letters denote positive numbers. Simplify the expressions.

53. $\sqrt[3]{a^3 b^6}$

54. $\sqrt[4]{16a^4 b^{12}}$

55. $\dfrac{1}{\sqrt[3]{-125x^3 y^6 z}}$

56. $\sqrt[3]{\dfrac{16}{x^2 y^5}}$

57. $\dfrac{\sqrt[4]{(x+y)^4}}{\sqrt[4]{81x^{12}y^8}}$

58. $\dfrac{\sqrt[5]{64r^8 s^9 t^{13}}}{\sqrt[5]{2r^3 s^{-6} t^{-7}}}$

59. $\sqrt[4]{t \sqrt[3]{t^9}}$

60. $\sqrt[3]{(3x-5)^3}$

61. $\sqrt[4]{(4x+2y)^8}$

62. $\sqrt[3]{a+b} \sqrt[3]{a^2 - ab + b^2}$

In Exercises 63–68, simplify the given expression. Do not assume that a, b, c, x, and y are necessarily positive.

63. $\sqrt[4]{a^{12} b^8}$

64. $\sqrt[4]{\dfrac{a^4}{b^{16} c^{20}}}$

65. $\sqrt[4]{\dfrac{a^9 b^7}{b^3 a}}$

66. $\sqrt[6]{(2x-3y)^6}$

67. $\sqrt[6]{\dfrac{x^5 y^{-2}}{x^{-7} y^4}}$

68. $\sqrt[6]{\dfrac{x^6 + y^6}{(x+y)^{18}}}$

In Exercises 69–78, simplify the given expression by rationalizing the denominator.

69. $\dfrac{1}{\sqrt{3}}$

70. $\dfrac{\sqrt{9}}{\sqrt{7}}$

71. $\dfrac{1}{1+\sqrt{2}}$

72. $\dfrac{1}{1-\sqrt{3}}$

73. $\dfrac{\sqrt{6}}{3+\sqrt{6}}$

74. $\dfrac{2}{\sqrt{3}-\sqrt{2}}$

75. $\dfrac{4+\sqrt{3}}{4-\sqrt{3}}$

76. $\dfrac{6-\sqrt{2}}{6+\sqrt{2}}$

77. $\dfrac{\sqrt{a}+\sqrt{b}}{\sqrt{a}-\sqrt{b}}$

78. $\dfrac{\sqrt{a}-2\sqrt{b}}{\sqrt{a}+2\sqrt{b}}$

In Exercises 79–81, write the statement as an equation.

79. The length s of a side of a cube is the cube root of the volume V.

80. The radius r of a sphere is equal to the square root of the quantity obtained by dividing the surface area S by 4π.

81. The radius r of a sphere is the cube root of the quantity obtained by dividing 3 times the volume V by 4π.

C In Exercises 82–83, use a calculator to see whether it gives the same value for each of the two numbers.

82. $\dfrac{\sqrt{(2.34)^4 (1.79)^3}}{\sqrt{(5.21)^3 (4.08)^2}}$ and $\sqrt{\dfrac{(2.34)^4 (1.79)^3}{(5.21)^3 (4.08)^2}}$

83. $\dfrac{\sqrt[3]{(3.29)^4 (-1.136)^5}}{\sqrt[3]{671.209}}$ and $\sqrt[3]{\dfrac{(3.29)^4 (-1.136)^5}{671.209}}$

84. Show that for any real number a and any positive integer m we have $\sqrt{a^{2m}} = |a|^m$.

85. Show that if $a > 0$ and $b > 0$, then

$$\sqrt{a+b} \neq \sqrt{a} + \sqrt{b}$$

(*Hint*: Square both $\sqrt{a+b}$ and $(\sqrt{a}+\sqrt{b})$.)

86. a. Show that $\sqrt{a^2 b^2} = |a||b|$ for any real numbers a and b.
   b. Use (1), (3), and part (a) to show that $|ab| = |a||b|$.

87. According to Einstein's Theory of Relativity, the mass of an object moving with velocity v is

$$\dfrac{m_0}{\sqrt{1 - v^2/c^2}}$$

where c is the velocity of light and where m_0 is the "rest mass" of the object, that is, the mass when the velocity is 0. Find the mass of an object moving with velocity $c/3$ if its rest mass is 5 grams.

88. In order for a satellite to go around the earth in a circular orbit of radius r, its velocity v in miles per hour must be given by

$$v = \sqrt{GM/r}$$

where G is the universal gravitational constant and M is the mass of the earth. Taking $GM = 1.237 \times 10^{12}$, find the velocity a satellite must have in order to go around the earth in a circular orbit with radius 4096 miles (so the satellite is approximately 100 miles above the surface of the earth).

1.5 FRACTIONAL EXPONENTS

In the preceding two sections we discussed integral powers and roots of real numbers. Now we will define rational powers of numbers. We begin by defining $a^{1/n}$ for any positive integer n. If the Laws of Integral Exponents are to remain valid for rational exponents, then we must have

$$(a^{1/n})^n = a^{(1/n)n} = a^1 = a = (\sqrt[n]{a})^n$$

which implies that $a^{1/n} = \sqrt[n]{a}$. Thus we make the following definition:

$$a^{1/n} = \sqrt[n]{a} \begin{cases} \text{for } a \geq 0 \text{ if } n \text{ is even} \\ \text{for any } a \text{ if } n \text{ is odd} \end{cases}$$

EXAMPLE 1.   Compute the following powers.

a. $4^{1/2}$ b. $27^{1/3}$ c. $1^{1/6}$ d. $\left(-\dfrac{1}{32}\right)^{1/5}$

Solution.

a. $4^{1/2} = \sqrt{4} = 2$
b. $27^{1/3} = \sqrt[3]{27} = 3$
c. $1^{1/6} = \sqrt[6]{1} = 1$

d. $\left(-\dfrac{1}{32}\right)^{1/5} = \sqrt[5]{-\dfrac{1}{32}} = -\dfrac{1}{2}$ $\square$

Recall that any rational number r can be written as m/n, where m and n are integers with $n > 0$ and m/n in lowest terms. Thus to define a^r it suffices to define $a^{m/n}$, where m and n are integers and $n > 0$. If the Laws of Integral Exponents are to remain valid for rational exponents, then we must have

$$(a^{1/n})^m = a^{(1/n)m} = a^{m/n} \quad \text{and} \quad (a^m)^{1/n} = a^{m(1/n)} = a^{m/n}$$

In the special case in which $m = 2$, the Laws of nth Roots imply that

$$(a^2)^{1/n} = \sqrt[n]{a^2} = \sqrt[n]{a} \cdot \sqrt[n]{a} = (\sqrt[n]{a})^2 = (a^{1/n})^2$$

More generally, one can show that if 2 is replaced by an arbitrary integer, the formula

$$(a^m)^{1/n} = (a^{1/n})^m$$

is valid. This leads us to the following definition:

$$a^{m/n} = (a^{1/n})^m = (a^m)^{1/n} \tag{1}$$

Thus the value of $a^{m/n}$ may be computed as either $(a^{1/n})^m$ or $(a^m)^{1/n}$. However, sometimes one of the expressions is easier to compute than the other, as we will see in the following example.

EXAMPLE 2.   Compute the following powers.

  a. $4^{3/2}$ b. $(-8)^{5/3}$ c. $(\sqrt{32})^{-2/5}$

Solution.

a. $4^{3/2} = (4^{1/2})^3 = 2^3 = 8$

b. $(-8)^{5/3} = [(-8)^{1/3}]^5 = (-2)^5 = -32$

c. $(\sqrt{32})^{-2/5} = [(\sqrt{32})^{-2}]^{1/5} = \left[\left(\dfrac{1}{\sqrt{32}}\right)^2\right]^{1/5} = \left(\dfrac{1}{32}\right)^{1/5} = \dfrac{1}{2}$ □

Observe that we computed $4^{3/2}$ as $(4^{1/2})^3$ in (a). We could easily have computed $4^{3/2}$ alternatively as $(4^3)^{1/2} = (64)^{1/2} = 8$. However, it would have been more difficult to compute $(-8)^{5/3}$ in (b) as $[(-8)^5]^{1/3}$, because 8^5 is such a large number.

Caution: If n is even and m is odd, the number $a^{m/n}$ is defined only for $a \geq 0$ (so that the middle expression of (1) is meaningful). Similarly, if $m \leq 0$, then a must be different from 0 in order for $a^{m/n}$ to be defined.

After studying logarithms in Chapter 5, we will be able to define a^r when r is *any real number*, irrational as well as rational. But in the meantime we will assume that r is rational.

All the previous laws of exponents remain valid for rational exponents. They are given next.

Laws of Rational Exponents

Let a and b be real numbers and r and s rational. Each of the following formulas is valid for all values of a and b for which all expressions in the formula are defined.

  i. $a^r a^s = a^{r+s}$ iii. $(ab)^r = a^r b^r$ v. $a^{-r} = \dfrac{1}{a^r}$

  ii. $(a^r)^s = a^{rs} = (a^s)^r$ iv. $\left(\dfrac{a}{b}\right)^r = \dfrac{a^r}{b^r}$ vi. $\dfrac{a^r}{a^s} = a^{r-s}$

EXAMPLE 3.   Simplify the following expressions.

  a. $x^{2/5}y^{-2/3}(x^6y^3)^{4/3}$ b. $\left(\dfrac{x^{2/3}}{y^{1/5}}\right)^{15/7}$

Solution.

a. $x^{2/5}y^{-2/3}(x^6y^3)^{4/3} \overset{\text{(iii)}}{=} x^{2/5}y^{-2/3}(x^6)^{4/3}(y^3)^{4/3}$

  $\overset{\text{(ii)}}{=} x^{2/5}y^{-2/3}x^8y^4 = x^{2/5+8}y^{-2/3+4} \overset{\text{(i)}}{=} x^{42/5}y^{10/3}$

b. $\left(\dfrac{x^{2/3}}{y^{1/5}}\right)^{15/7} \overset{\text{(iv)}}{=} \dfrac{(x^{2/3})^{15/7}}{(y^{1/5})^{15/7}} \overset{\text{(ii)}}{=} \dfrac{x^{10/7}}{y^{3/7}}$ □

As was the case with integral exponents, the Laws of Rational Exponents can be extended. For example,

$$(abc)^r = a^r b^r c^r$$

and

$$\left(\frac{ab}{c}\right)^r = \frac{a^r b^r}{c^r}$$

EXAMPLE 4.   Simplify $\left(\dfrac{9x^4 y^7}{z^2}\right)^{2/5} \left(\dfrac{3x^{1/2}}{yz}\right)^{4/5}$.

Solution.   Using the Laws of Rational Exponents and their extensions, we find that

$$\left(\frac{9x^4 y^7}{z^2}\right)^{2/5} \left(\frac{3x^{1/2}}{yz}\right)^{4/5} = \left(\frac{9^{2/5} x^{8/5} y^{14/5}}{z^{4/5}}\right) \left(\frac{3^{4/5} x^{2/5}}{y^{4/5} z^{4/5}}\right)$$

$$= \frac{9^{2/5} 3^{4/5} x^{8/5} x^{2/5} y^{14/5}}{y^{4/5} z^{4/5} z^{4/5}}$$

$$= \frac{(3^{4/5} 3^{4/5}) x^{10/5} y^{10/5}}{z^{8/5}}$$

$$= \frac{3^{8/5} x^2 y^2}{z^{8/5}} \quad \square$$

Sometimes radicals and exponents appear in the same expression. In order to simplify such an expression it is usually convenient to convert all radicals to exponents by using the relation

$$\sqrt[n]{a} = a^{1/n}$$

EXAMPLE 5.   Simplify $\left(\dfrac{\sqrt{x}\,\sqrt[3]{y^2}}{z}\right)^{12} \left(\dfrac{\sqrt[4]{z}}{xy}\right)^2$.

Solution.

$$\left(\frac{\sqrt{x}\,\sqrt[3]{y^2}}{z}\right)^{12} \left(\frac{\sqrt[4]{z}}{xy}\right)^2 = \left(\frac{x^{1/2}(y^2)^{1/3}}{z}\right)^{12} \left(\frac{z^{1/4}}{xy}\right)^2$$

$$= \frac{x^6 (y^2)^4}{z^{12}} \cdot \frac{z^{1/2}}{x^2 y^2} = \frac{x^6 y^8}{z^{12}} \cdot \frac{z^{1/2}}{x^2 y^2} = \frac{x^4 y^6}{z^{23/2}} \quad \square$$

EXERCISES 1.5

In Exercises 1–18, simplify the given expression.

1. $8^{5/3}$	**3.** $125^{-2/3}$	**5.** $(-64)^{-4/3}$	**7.** $64^{-5/6}$	**10.** $(10^{-5})^{2/5}$	**13.** $50^{3/2}$
2. $16^{3/2}$	**4.** $64^{3/2}$	**6.** $64^{5/6}$	**8.** $(-1000)^{5/3}$	**11.** $(16)^{0.25}$	**14.** $81^{4/3}$
			9. $\left(\frac{9}{25}\right)^{-3/2}$	**12.** $\left(\frac{1}{4}\right)^{-1.5}$	**15.** $(2.25)^{-3/2}$

16. $(\frac{9}{8})^{3/2}$ **17.** $2^{-1.5}$ **18.** $5^{2.5}$

In Exercises 19–38, assume that all letters denote positive numbers. Simplify the given expression.

19. $x^{1/3}x^{2/5}x^{4/15}$

20. $x^{-4/3}x^{7/5}x^{-13/9}$

21. $(36a^3b^4)^{3/2}$

22. $(x^3y^6z^9)^{2/3}$

23. $(x^3y^6z^8)^{2/3}$

24. $\left(\frac{8a^4}{27b^2}\right)^{2/3}$

25. $\left(\frac{16}{z^3}\right)^{-3/4}$

26. $\left(\frac{x^4y^7}{z^5}\right)^{-3/7}$

27. $\left(\frac{x^3y^7z^{-3}}{x^2y^{-5}z^{-10}}\right)^{2/5}$

28. $\left(\frac{x^3y^7z^{-3}}{x^2y^{-5}z^{-10}}\right)^0$

29. $(32a\sqrt{a^3})^{2/5}$

30. $[(a^{-2}b^3)^3]^{-2/3}$

31. $(p^2q)^{2/3}(pq^2)^{-2/3}$

32. $(\sqrt{pq})^{2/3}$

33. $(\sqrt[3]{p^2+q^2})^{3/2}$

34. $\left(\frac{\sqrt[3]{x}\,y^2}{\sqrt[4]{z}}\right)^6$

35. $\left(\frac{z^{1/3}\sqrt{x-y}}{2(x-y)}\right)^6$

36. $\sqrt[3]{\frac{b}{27a^3}}$

37. $\sqrt{b^3}\sqrt[3]{b^2}$

38. $(a^2+b^2)^{2/3}-\frac{a^2}{\sqrt[3]{a^2+b^2}}$

In Exercises 39–44, use a calculator to approximate the given expression.

39. $4^{2/5}$ **42.** $(\sqrt{2})^{1/5}$

40. $10^{-1/4}$ **43.** $(1.27)^{-4/9}$

44. $\frac{(2.45)^{1.7}}{(3.97)^{2.13}}$

41. $\pi^{1/3}$

45. In computing orbits of earth satellites the number

$$\left(\frac{1.237\times10^{12}}{4\pi^2}\right)^{1/3}$$

appears. Use a calculator to approximate its value.

46. Kepler's Third Law of Motion says in effect that if a satellite is in orbit around the earth and has a period of T hours, then the average distance in miles between the satellite and the center of the earth is $cT^{2/3}$, where c is approximately 3152.6. In order for the satellite to orbit exactly once every 24 hours, determine its average distance from the center of the earth.

47. When a diatomic gas (such as oxygen) undergoes an adiabatic change (one in which no heat is lost from the gas), the pressure p and volume V of the gas must change in such a way that $pV^{1.4}$ remains constant. If the volume doubles, how must the pressure change?

1.6 POLYNOMIALS

We have already used letters such as a, b, c, x, y, and z to represent real numbers. In the remaining sections of this chapter we will focus on combinations of expressions involving such letters.

The basic building block of the expressions we will consider is the **monomial** ax^k, where a is a **constant** (that is, a specific, preassigned number), x is a **variable** (that is, any real number, neither specified nor preassigned), and k is a nonnegative integer. The constant a is called the **coefficient** of the monomial. Examples of monomials are

$$16x^2,\qquad -\sqrt{2}x^6,\quad \text{and}\quad -\frac{3}{5}x^{58}$$

Since $x^1=x$ and $x^0=1$ for $x\neq0$, we normally write

$$ax \quad\text{for}\quad ax^1 \quad\text{and}\quad a \quad\text{for}\quad ax^0$$

When we add the two monomials ax^k and bx^m, we obtain the **binomial**

$$ax^k+bx^m$$

It follows that

$$x-2,\qquad 2x^2+x,\quad\text{and}\quad x^8-3x^3$$

are binomials.

More generally, the sum of a finite number of monomials is a *polynomial* (or *polynomial expression*). The general form of a polynomial is

$$a_n x^n + a_{n-1} x^{n-1} + \cdots + a_1 x + a_0$$

where n is a nonnegative integer, x is a variable, and the coefficients a_n, $a_{n-1}, \ldots, a_1, a_0$ are constants. Examples of polynomials are

$$5, \qquad x - \frac{4}{3}, \qquad x^2 + 6x + 9, \qquad x^3 - 27, \quad \text{and} \quad -\sqrt{7}x^6 - x^4$$

Notice that whereas the number n in x^n is an exponent, the number n in a_n simply indicates that a_n is the coefficient of x^n.

The monomials that make up a polynomial are called *terms* of the polynomial. Thus the terms of $x^2 - 6x + 1$ are x^2, $-6x$, and 1. If $a_n \neq 0$, then the coefficient a_n of $a_n x^n$ is the *leading coefficient* of the polynomial and n is the *degree* of the polynomial. Thus the degree of $x^3 - 27$ is 3 and the leading coefficient is 1; likewise, the degree of $-\sqrt{7}x^6 - x^4$ is 6 and the leading coefficient is $-\sqrt{7}$. Finally, the number a_0 is called the *constant term* of the polynomial. The constant term may be 0, as in the polynomial $2x^3 - x^2 + x$.

Polynomials are classified according to their degrees. The polynomial 0 has no degree attached to it. For the other polynomials we have:

Polynomial	*Degree*	*Form*	*Example*
constant	0	$a_0 \; (a_0 \neq 0)$	$-\dfrac{4}{7}$
linear	1	$a_1 x + a_0 \; (a_1 \neq 0)$	$2x - 8$
quadratic	2	$a_2 x^2 + a_1 x + a_0 \; (a_2 \neq 0)$	$4x^2 - 6x + \sqrt{3}$
cubic	3	$a_3 x^3 + a_2 x^2 + a_1 x + a_0 \; (a_3 \neq 0)$	$x^3 - \dfrac{5}{2}x^2 + \pi$
nth-degree	n	$a_n x^n + a_{n-1} x^{n-1} + \cdots + a_1 x + a_0 \; (a_n \neq 0)$	$x^n + 1$

Caution: In a polynomial, the exponents of the powers of x must be integers; rational powers of x are not allowed. Thus $4x^{3/2} + 3x - 7$ is not a polynomial.

We say that two polynomials are *equal* if they have the same degree and if the corresponding powers of x have the same coefficients. In particular, the two polynomials

$$x^2 - 5x + 4 \quad \text{and} \quad ax^2 + bx + c$$

are equal if and only if $a = 1$, $b = -5$, and $c = 4$.

In a given polynomial, if we replace the variable x by a particular real number, the resulting expression is meaningful and represents a real number. Thus if we replace x by 2 in the polynomial $x^2 + 6x + 9$, then the polynomial becomes $2^2 + 6(2) + 9$, which simplifies to $4 + 12 + 9$, or 25. The process of replacing x by a particular real number is called **substitution**. The number obtained when a real number c is substituted into a polynomial is called the **value** of the polynomial for $x = c$.

EXAMPLE 1.   Find the values of $x^2 + 6x + 9$ for $x = 0$, $x = -\frac{4}{3}$, and $x = \sqrt{3}$.

Solution.   We find in turn that

for $x = 0$, $x^2 + 6x + 9 = 0^2 + 6(0) + 9 = 0 + 0 + 9 = 9$

for $x = -\dfrac{4}{3}$, $x^2 + 6x + 9 = \left(-\dfrac{4}{3}\right)^2 + 6\left(-\dfrac{4}{3}\right) + 9$

$$= \frac{16}{9} - 8 + 9 = \frac{25}{9}$$

for $x = \sqrt{3}$, $x^2 + 6x + 9 = (\sqrt{3})^2 + 6(\sqrt{3}) + 9$

$$= 3 + 6\sqrt{3} + 9 = 12 + 6\sqrt{3} \quad \square$$

So far we have used the letter x for the variable of a polynomial. However, it is possible to use other letters as well. Thus $y^2 + 6y + 9$ is a polynomial, as is $z^2 + 6z + 9$. Occasionally we refer to a polynomial whose variable is y as a **polynomial in y**. In this vein, $y^2 + 6y + 9$ is a polynomial in y. Next we observe that if we substitute a number such as $-\frac{4}{3}$ for y in $y^2 + 6y + 9$, we obtain

$$\left(-\frac{4}{3}\right)^2 + 6\left(-\frac{4}{3}\right) + 9$$

which has the value $\frac{25}{9}$. This is the same value we found in Example 1. The reason it is the same is that $x^2 + 6x + 9$ and $y^2 + 6y + 9$ have the same degree and the same coefficients for corresponding powers. The fact that different letters have been used for the variable in the two polynomials is immaterial to the value of the polynomial when a real number is substituted.

Addition and Subtraction of Polynomials

We can combine numbers by adding, subtracting, multiplying, and dividing them. We can do the same with polynomials because they represent numbers. All the rules (including the commutative, associative, and distributive laws) that apply to real numbers may be applied to polynomials as well. In particular, the sum of any two polynomials in x can be obtained by adding the coefficients of like powers of x, which frequently can be done most easily by lining up terms with like powers of x vertically.

EXAMPLE 2. Find the sum $(x^3 - 2x^2 + 4x - 1) + (-2x^5 - 3x^3 + \sqrt{5}x^2 + 3)$.

Solution. By lining up terms with like powers of x vertically and using the laws mentioned above, we find that

$$
\begin{array}{r}
x^3 \qquad - \ 2x^2 + 4x - 1 \\
-2x^5 - 3x^3 \qquad + \sqrt{5}\,x^2 \qquad + 3 \\
\hline
-2x^5 - 2x^3 + (-2 + \sqrt{5})x^2 + 4x + 2 \quad \square
\end{array}
$$

Adding polynomials can also be accomplished in a horizontal format. In doing so we combine all terms with like powers of x.

$$
\begin{aligned}
(x^3 - 2x^2 &+ 4x - 1) + (-2x^5 - 3x^3 + \sqrt{5}x^2 + 3) \\
&= -2x^5 + (x^3 - 3x^3) + (-2x^2 + \sqrt{5}x^2) + 4x + (-1 + 3) \\
&= -2x^5 - 2x^3 + (-2 + \sqrt{5})x^2 + 4x + 2
\end{aligned}
$$

Which format we use, vertical or horizontal, depends on personal preference. It is probably easier to keep coefficients of like powers of the variable straight with the vertical format, but it takes more space on the page of a book. Therefore, except in this section, we will generally adopt the horizontal format.

We subtract polynomials by subtracting terms with like powers.

EXAMPLE 3. Find the difference $(x^4 - 6x^3 - 2x + 1) - (x^3 - 2x^2 - 3x - 5)$.

Solution. Subtracting terms with like powers, we find that

$$
\begin{array}{r}
x^4 - 6x^3 \qquad - 2x + 1 \\
x^3 - 2x^2 - 3x - 5 \\
\hline
x^4 - 7x^3 + 2x^2 + \ x + 6 \quad \square
\end{array}
$$

Multiplication of Polynomials   To multiply two polynomials we use the law of exponents

$$
x^m x^n = x^{m+n}
$$

in conjunction with the distributive law.

EXAMPLE 4. Find the product $(3x - 2)(4x + 5)$.

Solution. We multiply in a vertical fashion, as we would multiply a pair of two-digit numbers:

$$
\begin{array}{r}
3x - 2 \\
4x + 5 \\
\hline
15x - 10 \\
12x^2 - \ 8x \\
\hline
12x^2 + \ 7x - 10 \quad \square
\end{array}
$$

The product multiplied out in Example 4 is a special case of the more general formula

$$(x + a)(x + b) = x^2 + (a + b)x + ab$$

EXAMPLE 5.   Find the product $(2x^2 - 3x - 4)(x^3 - 6x^2 + 1)$.

Solution.   Using the same format, we have

$$
\begin{array}{r}
2x^2 - 3x - 4 \\
x^3 - 6x^2 + 1 \\
\hline
2x^2 - 3x - 4 \\
-12x^4 + 18x^3 + 24x^2 \\
2x^5 - 3x^4 - 4x^3 \\
\hline
2x^5 - 15x^4 + 14x^3 + 26x^2 - 3x - 4 \quad \square
\end{array}
$$

Observe that the degree of the product in Example 5 is 5, which is the sum $2 + 3$ of the degrees 2 and 3 of the factors $2x^2 - 3x - 4$ and $x^3 - 6x^2 + 1$, respectively. In general, the degree of the product of two polynomials is the sum of the degrees of the two polynomials.

Polynomials in Two Variables

A polynomial in the two variables x and y is a sum of monomials (or terms) of the form $cx^k y^m$, where c is a constant, x and y are variables, and k and m are nonnegative integers. Examples are

$$2x + y^2, \qquad x^2 + 2xy + y^2, \quad \text{and} \quad x^4 + 4x^3y^2 + 9xy^3 + y^6$$

One can define a polynomial in the three variables x, y, and z, or in more than three variables, in the same way. A polynomial in more than one variable is usually referred to as a ***polynomial in several variables***.

Adding and multiplying polynomials of several variables proceeds as with ordinary polynomials. To compute a sum we add the terms with like powers of all the variables.

EXAMPLE 6.   Find the sum $(2x^2 - 6x^2y + 3xy^2 - 6x + y^2) + (-x^2 + 3x^2y + xy^2 + y^3 - y^2)$.

Solution.   We add vertically, lining up terms with like powers of x and like powers of y:

$$
\begin{array}{l}
2x^2 - 6x^2y + 3xy^2 - 6x + y^2 \\
-x^2 + 3x^2y + xy^2 - y^2 + y^3 \\
\hline
x^2 - 3x^2y + 4xy^2 - 6x + y^3 \quad \square
\end{array}
$$

To compute a product we again use the distributive law and then add terms with like powers of all the variables.

EXAMPLE 7.   Find the product $(x^2 - xy + y)(3x^2 - 4y)$.

Solution.   We find that

$$
\begin{array}{r}
x^2 - xy + y \\
3x^2 - 4y \\
\hline
-4x^2y + 4xy^2 - 4y^2 \\
3x^4 - 3x^3y + 3x^2y \\
\hline
3x^4 - 3x^3y - x^2y + 4xy^2 - 4y^2
\end{array}
$$ □

Notice that in Examples 6 and 7 we have systematically written the answers with the x's before the y's in each term. It is very helpful to make such a choice about order and then stick to it.

Of the infinite collection of products of polynomials, several recur frequently. We list a few of these here. The first three essentially appeared as (2), (3), and (4), respectively, in Section 1.1.

$$(x + y)^2 = x^2 + 2xy + y^2 \tag{1}$$

$$(x - y)^2 = x^2 - 2xy + y^2 \tag{2}$$

$$(x + y)(x - y) = x^2 - y^2 \tag{3}$$

$$(x + y)^3 = x^3 + 3x^2y + 3xy^2 + y^3 \tag{4}$$

$$(x - y)^3 = x^3 - 3x^2y + 3xy^2 - y^3 \tag{5}$$

In each of these formulas, x or y may be replaced by another letter, by a specific number, or even by a more complicated expression.

EXAMPLE 8.   Find the following products.
    a. $(x - 3y)^2$ b. $(x^2 + 2y^5)^3$

Solution.
a. By (2), with $3y$ substituted for y, we have

$$(x - 3y)^2 = x^2 - 2x(3y) + (3y)^2 = x^2 - 6xy + 9y^2$$

b. By (4), with x^2 substituted for x and $2y^5$ substituted for y,

$$
\begin{aligned}
(x^2 + 2y^5)^3 &= (x^2)^3 + 3(x^2)^2(2y^5) + 3(x^2)(2y^5)^2 + (2y^5)^3 \\
&= x^6 + 6x^4y^5 + 12x^2y^{10} + 8y^{15} \quad □
\end{aligned}
$$

EXERCISES 1.6

In Exercises 1–4, find the value of the given polynomial for $x = 0$, $x = 2$, and $x = -1$.

1. $2x^2 - 5x + 3$

2. $x^3 - 3x^2 + 3x - 1$

3. $x^4 - 5x^2 + \sqrt{5}$

4. $x(x^4 - 5x^2 + \sqrt{5})$

In Exercises 5–26, perform the indicated operations and then simplify.

5. $(3x^2 - 2x - 1) + (4x^2 + 2x - 5)$

6. $(5x^4 + 3x^2 - 7) + (6x^3 + 5x^2 - 4x + 9)$

7. $(2x^3 + \frac{1}{2}x^2 + 4x) - (3x^3 - \frac{1}{2}x^2 + 2x - 4)$

8. $(\sqrt{3}x^3 - \sqrt{2}x^2) - (2x^3 - \sqrt{2}x^2 - x - 1)$

9. $(2x - 3)(4x - 5)$

10. $(-2x + 1)(7x - 3)$

11. $(\frac{1}{2}x + 3)(\frac{1}{3}x + 4)$

12. $(-4x - 2)(-3x + 6)$

13. $(3x - 2)(-x + 5) + (2x - 1)(5x - 7)$

14. $(x + 3)^2 - (x - 3)^2$

15. $(2x - 1)^2 - 4(x - 2)^2$

16. $3y(y^2 - 1)$

17. $\sqrt{2}y(y^2 - \sqrt{2})$

18. $y^2(4y^3 - y^2) + (y^2 - 1)^2$

19. $(r + 3)^2 + (r - 4)^2$

20. $(2 - s)(s - 4)^2$

21. $(2 - 3x)^2(3x - 1)^2$

22. $(x^2 + 2x + 4)(x - 2)$

23. $(2y^4 - 4y^2 + 8)(y^2 + 2)$

24. $(y^3 + 3y - 1)(2y^3 - y^2 - 2)$

25. $(x^{15} + 3x^{10} - 2x^5)(4x^{10} - 5x^5)$

26. $(-2x^{34} - 7x^{17} + 1)(x^{34} - 4x^{17} - 1)$

In Exercises 27–48, perform the indicated operations and then simplify.

27. $(x^2 + 2xy + y^2) + (3x^2 - xy + y^2)$

28. $(-x^2y^2 + xy + y^2) + (2x^2y^2 - x^2y + x + 2y^2)$

29. $(x^2 + 2xy) - (xy + y^2)$

30. $-(-7x^3y + 2xy + 3y^2)$

31. $-2(x^2y^2 - 3xy + 6y^2)$

32. $(2x^2 - y^3)^2$

33. $(\frac{1}{2}x + \frac{1}{4}y)^2$

34. $(2x - 3y)^3$

35. $(4x - y^2)^3$

36. $(p + 2q^2)^3$

37. $(x + h)^2 - x^2$

38. $(x + h)^3 - x^3$

39. $(x - y)(x^2 + xy + y^2)$

40. $(x + y)(x^2 - xy + y^2)$

41. $(r^2 - s^2)(r^3s + rs^3)$

42. $(x + y + z)(x + y - z)$

43. $(x + 2y - 3z)^2$

44. $(-2x - 5y + z)(4x - 3y - z)$

45. $(u^{1/2} + v^{1/2})^2$

46. $(u^{1/3} + v^{1/3})^3$

47. $\left(\frac{1}{r} - \frac{1}{s}\right)^2$ 48. $\left(\frac{1}{r} + \frac{1}{s}\right)^2$

49. Show that
$$x(x + y)(x + 2y)(x + 3y) = (x^2 + 3xy + y^2)^2 - y^4$$

50. Show that $x^3 - y^3 = (x - y)^3 + 3xy(x - y)$.

51. Show that if x is replaced by $y - b/3a$, then the cubic polynomial $ax^3 + bx^2 + cx + d$ is transformed into a cubic polynomial of the form $ay^3 + ey + f$ (so the coefficient of y^2 has become 0).

52. Two baseballs are thrown straight downward at the same instant, one from a window 50 feet above ground, and the other from a window 100 feet above ground. The first ball is thrown with a speed of 30 feet per second, so while it descends, its height t seconds later is given by $-16t^2 - 30t + 50$. The second ball is thrown with a speed of 45 feet per second, so while it descends, its height t seconds later is given by $-16t^2 - 45t + 100$. Find an expression for the distance between the two baseballs until the first hits the ground.

53. A pizzeria figures that if it sets the price of a pizza at x dollars, its daily revenue from the sale of pizzas will be $100x - 10x^2$ dollars, and its daily cost of producing the pizzas sold would be $-2x + 150$ dollars. Find an expression for the daily profit from selling pizzas at x dollars. (*Hint:* Profit equals revenue minus cost.)

1.7 FACTORING POLYNOMIALS

In the preceding section we multiplied polynomials. Now we consider the opposite procedure: writing a polynomial as a product of other polynomials, called *factors*. For example, since

$$4x^2 + 12x + 8 = 4(x + 2)(x + 1)$$

it follows that 4, $x + 2$, and $x + 1$ are factors of $4x^2 + 12x + 8$. Similarly,

$$x^4 - 6x^3 + 9x^2 = x^2(x - 3)^2$$

so that x^2 and $(x - 3)^2$ are factors of $x^4 - 6x^3 + 9x^2$. Of course, since $x^2 = x \cdot x$ and $(x - 3)^2 = (x - 3)(x - 3)$, we have

$$x^4 - 6x^3 + 9x^2 = x \cdot x(x - 3)(x - 3)$$

so that x and $x - 3$ are also factors of $x^4 - 6x^3 + 9x^2$. The process of rewriting a polynomial as the product of factors is called **factoring**, and factoring is important in the analysis of properties of polynomials and quotients of polynomials. Our interest lies in finding factors of degree 1 or higher, which are called **nontrivial factors**.

Factors of Certain Quadratic Polynomials

To determine which polynomials have nontrivial factors is a difficult problem, and actually finding them can range from easy to impossible (and is likely to be hard if the degree of the polynomial is large). Even when the degree of the polynomial is 2, finding factors can be involved; the general discussion in this case must wait until Section 2.3. Since at this point we wish to become familiar with factors of polynomials, we will now look at polynomials that have the following special features:

(a) The polynomial is of degree 2.
(b) The coefficients of the polynomial are integers.
(c) Two (possibly identical) nontrivial factors exist and have integer coefficients.

Let us set out to factor $x^2 + 9x + 8$ so that

$$x^2 + 9x + 8 = (x + a)(x + b) \tag{1}$$

Since $(x + a)(x + b) = x^2 + (a + b)x + ab$, we can write (1) alternatively as

$$x^2 + 9x + 8 = x^2 + (a + b)x + ab$$

which implies that $a + b = 9$ and $ab = 8$. Since $ab = 8$, the possible choices for a and b as a pair are 1 and 8, -1 and -8, 2 and 4, or -2 and -4. But $a + b = 9$, so our choices is restricted to 1 and 8. Thus $a = 1$ and $b = 8$ (or $a = 8$ and $b = 1$). Either possibility yields the factors $x + 1$ and $x + 8$, so that

$$x^2 + 9x + 8 = (x + 1)(x + 8)$$

Notice that at first many choices for a and b seemed possible, but by studying the situation carefully we were able to reduce the number of possibilities dramatically.

In general, if we wish to factor a polynomial of the form $x^2 + dx + e$ in such a way that the factors have integer coefficients, then we must find integers a and b such that

$$x^2 + dx + e = (x + a)(x + b) \tag{2}$$

Since

$$(x + a)(x + b) = x^2 + (a + b)x + ab \tag{3}$$

we can write (2) alternatively as

$$x^2 + dx + e = x^2 + (a + b)x + ab$$

which implies that

$$a + b = d \quad \text{and} \quad ab = e \tag{4}$$

EXAMPLE 1.   Factor the following polynomials.
  a. $x^2 - 4x - 5$ b. $x^2 - 7x + 12$

Solution.
a. If a and b are integers such that

$$x^2 - 4x - 5 = (x + a)(x + b)$$

then (4) becomes

$$a + b = -4 \quad \text{and} \quad ab = -5$$

Since $ab = -5$, the possibilities for a and b as a pair are 1 and -5, and -1 and 5. Of these, only 1 and -5 satisfy $a + b = -4$. Therefore $a = 1$ and $b = -5$ (or $a = -5$ and $b = 1$). We conclude that

$$x^2 - 4x - 5 = (x + 1)(x - 5)$$

b. If a and b are integers such that

$$x^2 - 7x + 12 = (x + a)(x + b)$$

then the equations in (4) become

$$a + b = -7 \quad \text{and} \quad ab = 12$$

Since $ab = 12$, the possibilities for a and b as a pair are 1 and 12, -1 and -12, 2 and 6, -2 and -6, 3 and 4, and -3 and -4. Of these, only -3 and -4 satisfy $a + b = -7$. Thus $a = -3$ and $b = -4$ (or $a' = -4$ and $b = -3$). We conclude that

$$x^2 - 7x + 12 = (x - 3)(x - 4) \quad \square$$

Two special forms of $x^2 + dx + e$ are

$$x^2 + 2ax + a^2 \quad \text{(where } d = 2a \text{ and } e = a^2\text{)}$$

and $\qquad\qquad x^2 - a^2 \quad \text{(where } d = 0 \text{ and } e = -a^2\text{)}$

Using (1) and (3) of Section 1.6, with a instead of y, we can find factors for these special polynomials:

$$x^2 + 2ax + a^2 = (x + a)^2 \tag{5}$$

$$x^2 - a^2 = (x + a)(x - a) \tag{6}$$

EXAMPLE 2. Factor the following polynomials.
    a. $x^2 + 6x + 9$ b. $x^2 - 16$ c. $49 - y^2$

Solution.
a. By (5), with $a = 3$, we have

$$x^2 + 6x + 9 = x^2 + 2(3x) + 3^2 = (x + 3)^2$$

b. By (6), with $a = 4$, we have

$$x^2 - 16 = x^2 - 4^2 = (x + 4)(x - 4)$$

c. By (6), with $x = 7$ and y substituted for a, we find that

$$49 - y^2 = 7^2 - y^2 = (7 + y)(7 - y) \quad \square$$

In spite of the results of Examples 1 and 2, there are polynomials of degree 2 that do not have any nontrivial factors with real coefficients. To prove this we consider the polynomial $x^2 + c^2$, with $c \neq 0$. If $x - a$ were a factor $x^2 + c^2$, then we would have

$$x^2 + c^2 = (x - a)(\text{polynomial})$$

But the right side is 0 for $x = a$, whereas the left side is positive for every value of x (because $c \neq 0$). Consequently $x^2 + c^2$ has no non-trivial factors. Therefore none of the polynomials

$$x^2 + 1 \,(= x^2 + 1^2), \qquad x^2 + 3 \,(= x^2 + (\sqrt{3})^2),$$
$$\text{and} \quad x^4 + 2 \,(= (x^2)^2 + (\sqrt{2})^2)$$

has a nontrivial factor. More generally, the monomial $x - a$ is a factor of a given polynomial if and only if the value of the polynomial for $x = a$ is 0.

EXAMPLE 3.   Show that $x + \frac{1}{2}$ is a factor of $3x^2 - \frac{13}{2}x - 4$.

Solution.   First we observe that $x + \frac{1}{2} = x - (-\frac{1}{2})$. Then we find the value of $3x^2 - \frac{13}{2}x - 4$ for $x = -\frac{1}{2}$:

$$3\left(-\frac{1}{2}\right)^2 - \frac{13}{2}\left(-\frac{1}{2}\right) - 4 = \frac{3}{4} + \frac{13}{4} - 4 = 0$$

By the comment preceding the example, $x + \frac{1}{2}$ is a factor of $3x^2 - \frac{13}{2}x - 4$. $\square$

A two-variable version of (3) is

$$x^2 + (a + b)xy + aby^2 = (x + ay)(x + by)$$

as you can verify by multiplying out the right side. Thus if we wish to factor $x^2 + dxy + ey^2$ so as to have

$$x^2 + dxy + ey^2 = (x + ay)(x + by)$$

we must once again have

$$a + b = d \quad \text{and} \quad ab = e$$

which appeared in (4).

EXAMPLE 4.   Factor $x^2 - 2xy - 8y^2$.

Solution.   If a and b are integers such that

$$x^2 - 2xy - 8y^2 = (x + ay)(x + by)$$

then $ab = -8$. It follows that the choices for a and b as a pair are 1 and -8, -1 and 8, 2 and -4, and -2 and 4. Since $a + b = -2$, we conclude that $a = 2$ and $b = -4$ (or $a = -4$ and $b = 2$). Either possibility yields the factors $x + 2y$ and $x - 4y$, so that

$$x^2 - 2xy - 8y^2 = (x + 2y)(x - 4y) \quad \square$$

If the leading coefficient of the polynomial is not 1, then similar techniques can be used.

EXAMPLE 5.   Factor $3x^2 + 2x - 1$.

Solution.   If the factors are to have integer coefficients, then the choices for the coefficients of x as a pair must be 3 and 1 or -3 and -1, and the constant terms must be 1 or -1. By trial and error we find that

$$3x^2 + 2x - 1 = (3x - 1)(x + 1) \quad \square$$

Factoring General Polynomials

Although it is impossible to give instructions for factoring polynomials in general, we can give guidelines in certain cases. In one type of polynomial there is a nontrivial factor of every term. Such a factor is called *common factor*.

EXAMPLE 6.   Factor the following polynomials.

a. $x^3 - 7x^2 + 12x$ b. $x^4 + 3x^2$

Solution.

a. The factor common to each term is x, so we factor it out using the distributive law:

$$x^3 - 7x^2 + 12x = x(x^2 - 7x + 12)$$

By part (b) of Example 1, we know that

$$x^2 - 7x + 12 = (x - 3)(x - 4)$$

Consequently

$$x^3 - 7x^2 + 12x = x(x - 3)(x - 4)$$

b. Here x^2 is a common factor, and we factor it out, obtaining

$$x^4 + 3x^2 = x^2(x^2 + 3)   \quad \square$$

In another type of polynomial there is a common factor not of every term but of each of several groups of terms together comprising the polynomial. Such common factors can then be factored out by grouping them together. The method is known as *factoring by grouping*.

EXAMPLE 7.   Factor the following polynomials by grouping.

a. $(x^4 + 2)x^2 - 3x(x^4 + 2)$ b. $x^3 - 3x^2 + 2x - 6$

c. $2x^3 + 5x^2 - 6x - 15$ d. $ax^2 + bxy - axy - by^2$

Solution.

a. The polynomial $x^4 + 2$ is a common factor, and we factor it out:

$$(x^4 + 2)x^2 - 3x(x^4 + 2) = (x^4 + 2)(x^2 - 3x)$$

Next we notice that x is a common factor of $x^2 - 3x$, so it can also be factored out, giving us

$$(x^4 + 2)x^2 - 3x(x^4 + 2) = x(x^4 + 2)(x - 3)$$

b. We observe that $x - 3$ is a factor of $x^3 - 3x^2$ and of $2x - 6$, and consequently

$$x^3 - 3x^2 + 2x - 6 = (x^3 - 3x^2) + (2x - 6)$$
$$= x^2(x - 3) + 2(x - 3)$$
$$= (x^2 + 2)(x - 3)$$

c. Since

$$2x^3 + 5x^2 = x^2(2x + 5) \quad \text{and} \quad -6x - 15 = -3(2x + 5)$$

it follows that $2x + 5$ is a common factor and that

$$\begin{aligned} 2x^3 + 5x^2 - 6x - 15 &= (2x^3 + 5x^2) + (-6x - 15) \\ &= x^2(2x + 5) - 3(2x + 5) \\ &= (x^2 - 3)(2x + 5) \end{aligned}$$

d. We find that

$$ax^2 + bxy = x(ax + by) \quad \text{and} \quad -axy - by^2 = -y(ax + by)$$

Consequently

$$\begin{aligned} ax^2 + bxy - axy - by^2 &= x(ax + by) - y(ax + by) \\ &= (x - y)(ax + by) \quad \square \end{aligned}$$

Observe that in the solution of (c) we could further factor $x^2 - 3$ and obtain

$$x^2 - 3 = (x - \sqrt{3})(x + \sqrt{3})$$

but since we are interested only in factors with integer coefficients at this time, we left the answer as it was.

Finally, we factor the polynomials $x^n - a^n$ and $x^n + a^n$, which appear frequently in mathematics:

$$\begin{aligned} x^n - a^n = (x - a)(x^{n-1} + ax^{n-2} + a^2x^{n-3} + \cdots \\ + a^{n-2}x + a^{n-1}) \quad \text{for } n \text{ positive} \end{aligned} \tag{7}$$

$$\begin{aligned} x^n + x^n = (x + a)(x^{n-1} - ax^{n-2} + a^2x^{n-3} - \cdots \\ - a^{n-2}x + a^{n-1}) \quad \text{for } n \text{ positive and odd} \end{aligned} \tag{8}$$

For $n = 3$, these formulas become

$$x^3 - a^3 = (x - a)(x^2 + ax + a^2) \tag{9}$$

$$x^3 + a^3 = (x + a)(x^2 - ax + a^2) \tag{10}$$

EXAMPLE 8.   Factor
   a. $x^3 - 27$ b. $x^5 + 32$ c. $x^6 - 2^6$

Solution.
a. Using (9) with $a = 3$, we obtain

$$x^3 - 27 = x^3 - 3^3 = (x - 3)(x^2 + 3x + 9)$$

b. Using (8) with $n = 5$ and $a = 2$, we have

$$x^5 + 32 = x^5 + 2^5 = (x + 2)(x^4 - 2x^3 + 4x^2 - 8x + 16)$$

c. There are several ways to factor $x^6 - 2^6$:
  i. Use (7) with $n = 6$ and $a = 2$. This yields

$$x^6 - 2^6 = (x - 2)(x^5 + 2x^4 + 4x^3 + 8x^2 + 16x + 32)$$

  ii. Use (9) with $a = 4$ and x replaced by x^2, and then (6) with $a = 2$. This yields

$$x^6 - 2^6 = (x^2)^3 - (2^2)^3 = (x^2)^3 - 4^3 = (x^2 - 4)(x^4 + 4x^2 + 16)$$
$$= (x - 2)(x + 2)(x^4 + 4x^2 + 16)$$

  iii. First we use (6) with $a = 8$ and x replaced by x^3:

$$x^6 - 2^6 = (x^3)^2 - (2^3)^2 = (x^3)^2 - 8^2 = (x^3 - 8)(x^3 + 8)$$

   Then we apply (9) to $x^3 - 8$ and (10) to $x^3 + 8$ and conclude that

$$x^6 - 2^6 = (x^3 - 8)(x^3 + 8)$$
$$= (x - 2)(x^2 + 2x + 4)(x + 2)(x^2 - 2x + 4) \quad \square$$

As part (c) of Example 8 illustrates, there are sometimes several ways of setting out to factor a polynomial. Some yield more complete factorizations than others.

Caution: Notice that (8) is valid only if n is an odd integer. If n is even, then $x + a$ is *not* a factor of $x^n + a^n$, although there may be nontrivial factors of $x^n + a^n$ (see Exercise 77).

EXERCISES 1.7

In Exercises 1–22, factor the given polynomial.

1. $x^2 + 8x + 12$
2. $x^2 - 2x - 3$
3. $x^2 - 7x + 6$
4. $t^2 + t - 12$
5. $2t^2 - 6t - 8$
6. $x^2 - 5x + 6$
7. $y^2 + 13y + 36$
8. $4y^2 - 1$
9. $x^2 - 19x + 60$
10. $x^2 + 10x + 9$
11. $21 - 10b + b^2$
12. $x^2 - 4x + 4$
13. $x^2 - 4$
14. $100 - 4x^2$
15. $a^2 - 14a + 49$
16. $a^2 - 5a + \frac{25}{4}$
17. $x^2 + 11x + \frac{121}{4}$
18. $49z^2 - 36$
19. $16 - 9z^2$
20. $x^2 + 20x - 800$
21. $x^2 - 2\sqrt{2}x + 2$
22. $x^2 + 2\sqrt{6}x + 6$

In Exercises 23–30, factor the given polynomial.

23. $2x^2 + 7x + 3$
24. $5x^2 + 4x - 1$
25. $-x^2 + 5x - 6$
26. $7x^2 + 8x - 12$
27. $7x^2 + 17x - 12$
28. $18t^2 - 15t - 18$
29. $6t^2 + 16t - 6$
30. $40x^2 + 14x - 45$

In Exercises 31–40, factor the given polynomial.

31. $x^2 + 2xy + y^2$ **36.** $12x^2 + xy - y^2$

32. $5x^2 + 10xy - 40y^2$ **37.** $x^4 - 2x^2y^2 + y^4$

33. $x^2 - 4y^2$ **38.** $x^4 + 5x^2y^2 + 6y^2$

34. $a^2 - 3ab - 4b^2$ **39.** $5x^2 - 14xy - 3y^2$

35. $144a^2 - b^2$ **40.** $2x^2 - 5xy + 3y^2$

In Exercises 41–56, factor the given polynomial by finding a common factor or by grouping.

41. $x^2 + 2x$

42. $x^3 - 5x^2 + 4x$

43. $x^7 - x^6$

44. $15x^4 + 3x^3$

45. $(3x + 5)x^2 + (3x + 5)x$

46. $x^2 - 1 + 3(x - 1)$

47. $(x^2 - 5)^2 - 8(x^2 - 5) + 16$

48. $x^2 - 16 - 2(x + 4)$

49. $x^3 - 2x^2 - x + 2$

50. $6x^3 + 12x^2 + 24x + 48$

51. $x^3 + x^2 - x - 1$

52. $x^5 + 2x^4 - x^3 - 2x^2$

53. $(3x + 2y)^2 + (3x + 2y) - 12$

54. $4x^2 + 4xy + y^2 - 16$

55. $x^2 - 10x + 25 - 4y^2$

56. $x^2 - 12x + 36 - 16y^2$

In Exercises 57–70, factor the given polynomial.

57. $x^3 + 1$ **60.** $8x^3 - 27$ **63.** $32x^5 - 1$

58. $x^3 - 1$ **61.** $8x^3 - 1$ **64.** $t^5 + 1$

59. $x^3 - 8$ **62.** $y^4 - 16$

65. $64x^6 - 729y^6$ (*Hint:* Use (6), (9), and (10).)

*66. $x^6 + y^6$ **68.** $(x + y)^3 - z^3$

67. $x^{10} - 1$ **69.** $(x - 2y)^4 - 1$

70. $(x + y + z)^2 - (x + y - z)^2$

In Exercises 71–74, determine whether the given polynomial has the factor $x - a$ for the specified value of a.

71. $5x^5 + 4x^2 + 1;$ $a = -1$

72. $x^3 - 3x^2 - 6x + 9;$ $a = 3$

73. $x^6 - 7x^3 - 3x - 4;$ $a = 2$

74. $-x^4 - 3x^3 - 4x^2 - 5x - 2;$ $a = -2$

75. Using (6) and treating 899 as 900−1, factor 899.

76. Show that $x^4 + 1 = (x^2 - \sqrt{2}x + 1)(x^2 + \sqrt{2}x + 1)$.

77. Show that

$$x^4 + a^4 = (x^2 - \sqrt{2}ax + a^2)(x^2 + \sqrt{2}ax + a^2).$$

*78. Show that $x^2 + x + 1$ cannot be factored as $(x + c)(x + d)$, where c and d are real numbers, even if c and d are not required to be integers. (*Hint:* Show that if $x^2 + x + 1 = (x + c)(x + d)$, then $c + d = 1$ and $cd = 1$. Then show from these equations that $0 < c < 1$ and $0 < d < 1$. But then $cd < 1$, which contradicts the equation $cd = 1$.)

1.8
RATIONAL
EXPRESSIONS

In Section 1.6 we added, subtracted, and multiplied polynomials, and in each case the result was a polynomial. When we divide one polynomial by another, the result in general is not a polynomial. Quotients of polynomials are *rational expressions*. Some examples are

$$\frac{1}{x + 1}, \quad \frac{x^2 - 5x + 4}{x - 3}, \quad \text{and} \quad \frac{xy}{3x^2 + 4y^2}$$

     A rational expression bears the same relationship to a polynomial as a rational number does to an integer. As with rational numbers, the polynomial appearing on top of a rational expression is its *numerator*, and the polynomial on the bottom is its *denominator*.

We have already discussed substituting real numbers into polynomials. We can also substitute real numbers into rational expressions, but we must use caution, since division by 0 is undefined. For example, it is not possible to substitute the number -1 into

$$\frac{1}{x+1}$$

since the result would be $\frac{1}{0}$, which is meaningless. Similarly, since $x^3 + 3x^2 + 2x = x(x+1)(x+2)$, it is not possible to substitute 0, -1, or -2 into

$$\frac{-2x+1}{x^3+3x^2+2x}$$

In general we can substitute into a rational expression any number for which the denominator is not 0. For example, we can substitute 2 for x in the preceding rational expression and obtain

$$\frac{-2(2)+1}{2^3+3(2)^2+2(2)} = -\frac{3}{24} = -\frac{1}{8}$$

The set of all numbers that can be substituted for the variable in a rational expression is called the ***domain*** of the rational expression. Thus the domain of

$$\frac{1}{x+1}$$

consists of all real numbers except -1, and the domain of

$$\frac{-2x+1}{x^3+3x^2+2x}$$

consists of all real numbers except 0, -1, and -2. In contrast, the domain of

$$\frac{1}{x^2+1}$$

consists of all real numbers, since $x^2 + 1 \neq 0$ for all x.

A polynomial can be represented as a rational expression whose denominator is the constant polynomial 1. For example, the polynomial $2x^4 - 3x - 7$ can be represented as the rational expression

$$\frac{2x^4-3x-7}{1}$$

The domain of any polynomial consists of all real numbers.

In the same way that we normally reduce a fraction to lowest terms, we can also reduce a rational expression to lowest terms by canceling factors common to numerator and denominator. In reducing a rational expression we first identify factors common to numerator and denominator and then use the law

$$\frac{ac}{bc} = \frac{a}{b} \tag{1}$$

to cancel the common factors.

EXAMPLE 1.   Reduce the following rational expressions to lowest terms.

a. $\dfrac{x^2 - 4}{x - 2}$ b. $\dfrac{x^2 - x - 6}{x^3 + 3x^2 + 2x}$ c. $\dfrac{x^3 + 3x^2 + 3x + 1}{(x + 1)(x^2 - 1)}$

Solution.
a. By (6) in Section 1.7,

$$x^2 - 4 = (x + 2)(x - 2)$$

Therefore we use (1) to cancel, obtaining

$$\frac{x^2 - 4}{x - 2} = \frac{(x + 2)(x - 2)}{x - 2} = x + 2$$

b. We notice that

$$x^2 - x - 6 = (x + 2)(x - 3)$$

and

$$x^3 + 3x^2 + 2x = x(x^2 + 3x + 2) = x(x + 2)(x + 1)$$

Thus by (1),

$$\frac{x^2 - x - 6}{x^3 + 3x^2 + 2x} = \frac{(x + 2)(x - 3)}{x(x + 2)(x + 1)} = \frac{x - 3}{x(x + 1)}$$

c. We observe that

$$x^3 + 3x^2 + 3x + 1 = (x + 1)^3$$

and

$$(x + 1)(x^2 - 1) = (x + 1)(x + 1)(x - 1) = (x + 1)^2(x - 1)$$

Consequently by (1),

$$\frac{x^3 + 3x^2 + 3x + 1}{(x + 1)(x^2 - 1)} = \frac{(x + 1)^3}{(x + 1)^2(x - 1)} = \frac{x + 1}{x - 1} \quad \square$$

Combinations of Rational Expressions

All the rules for adding, subtracting, multiplying, and dividing rational numbers hold for rational expressions. Multiplying and dividing rational expressions are usually less complicated than adding and subtracting them, because we do not need to employ common denominators. For multiplication of rational expressions we apply the formula

$$\frac{a}{b} \cdot \frac{c}{d} = \frac{ac}{bd} \tag{2}$$

and then cancel all common factors in order to reduce the expression to lowest terms.

EXAMPLE 2.   Find the product $\dfrac{x^3 - x^2}{x - 2} \cdot \dfrac{x + 3}{x^3 - 2x^2 + x}$ and reduce it to lowest terms.

Solution.   Notice that

$$x^3 - x^2 = x^2(x - 1)$$

and $x^3 - 2x^2 + x = x(x^2 - 2x + 1) = x(x - 1)^2$

Using (2) and then canceling common factors yields

$$\frac{x^3 - x^2}{x - 2} \cdot \frac{x + 3}{x^3 - 2x^2 + x} = \frac{(x^3 - x^2)(x + 3)}{(x - 2)(x^3 - 2x^2 + x)}$$

$$= \frac{x^2(x - 1)(x + 3)}{(x - 2)x(x - 1)^2}$$

$$= \frac{x(x + 3)}{(x - 2)(x - 1)} \quad \square$$

When we divide rational expressions we use the rule

$$\frac{a}{b} \div \frac{c}{d} = \frac{ad}{bc} \tag{3}$$

and as before we cancel all common factors to reduce the expression to lowest terms.

EXAMPLE 3.   Find the quotient $\dfrac{x + 1}{x - 1} \div \dfrac{x^2 + 2x + 1}{x^3 - 1}$ and reduce it to lowest terms.

Solution. First we notice that

$$x^2 + 2x + 1 = (x + 1)^2 \quad \text{and} \quad x^3 - 1 = (x - 1)(x^2 + x + 1)$$

Using (3), we find that

$$\frac{x + 1}{x - 1} \div \frac{x^2 + 2x + 1}{x^3 - 1} = \frac{(x + 1)(x^3 - 1)}{(x - 1)(x^2 + 2x + 1)}$$

$$= \frac{(x + 1)(x - 1)(x^2 + x + 1)}{(x - 1)(x + 1)^2}$$

Finally, we cancel common factors in the last expression, obtaining

$$\frac{x + 1}{x - 1} \div \frac{x^2 + 2x + 1}{x^3 - 1} = \frac{x^2 + x + 1}{x + 1} \quad \square$$

When we add or subtract rational expressions whose denominators have no nontrivial common factors, we first use the rule

$$\frac{a}{b} + \frac{c}{d} = \frac{ad + bc}{bd} \tag{4}$$

or

$$\frac{a}{b} - \frac{c}{d} = \frac{ad - bc}{bd} \tag{5}$$

and then simplify if possible.

<u>**EXAMPLE 4.**</u>   Find the sum $\dfrac{x + 1}{x - 2} + \dfrac{x + 2}{x + 3}$, and simplify.

Solution.   Using (4), we have

$$\frac{x + 1}{x - 2} + \frac{x + 2}{x + 3} = \frac{(x + 1)(x + 3) + (x - 2)(x + 2)}{(x - 2)(x + 3)}$$

$$= \frac{(x^2 + 4x + 3) + (x^2 - 4)}{(x - 2)(x + 3)}$$

$$= \frac{2x^2 + 4x - 1}{(x - 2)(x + 3)} \quad \square$$

When adding rational expressions with common factors in the denominators, we may simplify the algebraic computations by using a *least common denominator* of the denominators in the original rational expressions—that is, a polynomial of lowest degree that is a multiple of the denominators of both rational expressions. (This is reminiscent of the procedure applied when we added fractions by utilizing the least common denominator of two numbers in Section 1.1.)

EXAMPLE 5.   Express $\dfrac{2}{5x} + \dfrac{3}{x^2}$ as a rational expression in lowest terms.

Solution.   A least common denominator is $5x^2$, so we rewrite each fraction so its denominator will be $5x^2$, and then we combine:

$$\frac{2}{5x} + \frac{3}{x^2} = \frac{2x}{5x^2} + \frac{15}{5x^2} = \frac{2x + 15}{5x^2} \quad \square$$

EXAMPLE 6.   Express $\dfrac{1}{x^2 + 5x + 4} - \dfrac{1}{x^2 + 8x + 16}$ as a rational expression in lowest terms.

Solution.   Since

$$x^2 + 5x + 4 = (x + 4)(x + 1)$$

and

$$x^2 + 8x + 16 = (x + 4)^2$$

a least common denominator is $(x + 4)^2(x + 1)$. Consequently

$$\frac{1}{x^2 + 5x + 4} - \frac{1}{x^2 + 8x + 16} = \frac{1}{(x + 4)(x + 1)} - \frac{1}{(x + 4)^2}$$

$$= \frac{x + 4}{(x + 4)^2(x + 1)} - \frac{x + 1}{(x + 4)^2(x + 1)}$$

$$= \frac{(x + 4) - (x + 1)}{(x + 4)^2(x + 1)}$$

$$= \frac{3}{(x + 4)^2(x + 1)} \quad \square$$

For the sum of three or more rational expressions the technique is similar.

EXAMPLE 7.   Express $\dfrac{2x - 1}{x^2 + 4x - 5} + \dfrac{3}{x^2} - \dfrac{x - 3}{x^2 + 5x}$ as a rational expression in lowest terms.

Solution.   First notice that the denominators are

$$x^2 + 4x - 5 = (x + 5)(x - 1), \qquad x^2, \qquad x^2 + 5x = x(x + 5)$$

Therefore a least common denominator is $x^2(x + 5)(x - 1)$, so we proceed as follows:

$$\frac{2x - 1}{x^2 + 4x - 5} + \frac{3}{x^2} - \frac{x - 3}{x^2 + 5x}$$

$$= \frac{2x - 1}{(x + 5)(x - 1)} + \frac{3}{x^2} - \frac{x - 3}{x(x + 5)}$$

$$= \frac{(2x - 1)x^2}{x^2(x + 5)(x - 1)} + \frac{3(x + 5)(x - 1)}{x^2(x + 5)(x - 1)} - \frac{x(x - 3)(x - 1)}{x^2(x + 5)(x - 1)}$$

$$= \frac{(2x - 1)x^2 + 3(x + 5)(x - 1) - x(x - 3)(x - 1)}{x^2(x + 5)(x - 1)}$$

$$= \frac{2x^3 - x^2 + 3x^2 + 12x - 15 - x^3 + 4x^2 - 3x}{x^2(x + 5)(x - 1)}$$

$$= \frac{x^3 + 6x^2 + 9x - 15}{x^2(x + 5)(x - 1)} \quad \square$$

Similar techniques apply if there are more than one variable.

EXAMPLE 8.   Express $\dfrac{1}{x^2y} + \dfrac{1}{xy^2}$ as a rational expression in lowest terms.

Solution.   The common denominator we use is x^2y^2. As a result,

$$\frac{1}{x^2y} + \frac{1}{xy^2} = \frac{y}{x^2y^2} + \frac{x}{x^2y^2} = \frac{y + x}{x^2y^2} \quad \square$$

Combinations of expressions that are not necessarily polynomials but are nevertheless obtained by adding, subtracting, multiplying, dividing, or taking roots of polynomials are called **algebraic expressions**. The same general rules apply when combining algebraic expressions. In the following example, we will reduce the numerator and the denominator separately and then combine, noting that

$$\frac{a/b}{c/d} = \frac{a}{b} \div \frac{c}{d} = \frac{a}{b} \cdot \frac{d}{c} \tag{6}$$

EXAMPLE 9.   Simplify $\dfrac{1 - \dfrac{1}{x + 1}}{1 + \dfrac{1}{x - 1}}$

Solution.   First we rearrange the numerator:

$$1 - \frac{1}{x + 1} = \frac{x + 1}{x + 1} - \frac{1}{x + 1} = \frac{(x + 1) - 1}{x + 1} = \frac{x}{x + 1}$$

The denominator is altered similarly:

$$1 + \frac{1}{x-1} = \frac{x-1}{x-1} + \frac{1}{x-1} = \frac{(x-1)+1}{x-1} = \frac{x}{x-1}$$

Therefore by (6),

$$\frac{1 - \dfrac{1}{x+1}}{1 + \dfrac{1}{x-1}} = \frac{\dfrac{x}{x+1}}{\dfrac{x}{x-1}} = \frac{x}{x+1} \cdot \frac{x-1}{x} = \frac{x-1}{x+1} \quad \square$$

If the denominator of an algebraic expression contains radicals, then one can sometimes rationalize the denominator as we did in Section 1.4 by multiplying by 1 written in a special way. For example, if the denominator had the form $\sqrt{x} - \sqrt{a}$, we would multiply by

$$\frac{\sqrt{x} + \sqrt{a}}{\sqrt{x} + \sqrt{a}}$$

<u>**EXAMPLE 10.**</u>   Rationalize the denominator of $\dfrac{1}{\sqrt{x} - \sqrt{2}}$.

Solution.   We multiply by $\dfrac{\sqrt{x} + \sqrt{2}}{\sqrt{x} + \sqrt{2}}$ and obtain

$$\frac{1}{\sqrt{x} - \sqrt{2}} = \frac{1}{\sqrt{x} - \sqrt{2}} \cdot \frac{\sqrt{x} + \sqrt{2}}{\sqrt{x} + \sqrt{2}} = \frac{\sqrt{x} + \sqrt{2}}{(\sqrt{x} - \sqrt{2})(\sqrt{x} + \sqrt{2})}$$

$$= \frac{\sqrt{x} + \sqrt{2}}{x - 2} \quad \square$$

EXERCISES 1.8

In Exercises 1–4, determine the domain of the rational expression, and find the value of the rational expression at the given number.

1. $\dfrac{-4}{x^2 + 5x + 6}; 0$

2. $\dfrac{x^2 - 3x}{x^3 + 2x^2 - 15x}; -1$

3. $\dfrac{x^2 + 1}{x^3 - 1}; \dfrac{1}{2}$

4. $\dfrac{x^2 + 7x + 12}{x^2 + 3}; \sqrt{3}$

In Exercises 5–14, reduce the rational expression to lowest terms.

5. $\dfrac{x + 1}{x^2 + 5x + 4}$

6. $\dfrac{x^2 + 3x - 28}{x^2 - 3x + 4}$

7. $\dfrac{y^2 - 5y - 24}{y^2 - 9}$

8. $\dfrac{9y^2 - 4}{3y^2 - 3y + \frac{2}{3}}$

9. $\dfrac{2b^2 + b - 3}{8b^2 + 2b - 15}$

10. $\dfrac{s^4 - 8s^2 + 16}{s^2 - s - 2}$

11. $\dfrac{s^4 - 13s^2 + 36}{s^2 - 7s + 12}$

12. $\dfrac{s^4 - 5s^2 - 36}{s^2 - 7s + 12}$

13. $\dfrac{x^2 - y^2}{x^3 - y^3}$

14. $\dfrac{x^2 + y^2}{x^4 - y^4}$

In Exercises 15–26, write as one rational expression in simplified form.

15. $\dfrac{x^2 - 1}{x^2 + x - 2} \cdot (x^2 - 4)$

16. $\dfrac{x^2 - 16}{(x^2 + 3x - 28)^2} \cdot (x^2 - 49)$

17. $\dfrac{y^2}{y + 3} \cdot \dfrac{y^2 + y - 6}{y}$

18. $\dfrac{y + 3}{y - 2} \cdot \dfrac{y^2 - 4}{y + 3}$

19. $\dfrac{z^2 - 9}{z^2 - 4} \cdot \dfrac{z^2 + 6z + 9}{z^2 - 4z + 3}$

20. $\dfrac{x - 2}{x - 4} \div \dfrac{x}{x - 4}$

21. $\dfrac{x + 3}{x - 1} \div \dfrac{x - 1}{x + 3}$

22. $\dfrac{a^2 + 5a - 50}{a^2 - 1} \div \dfrac{a^2 - 7a + 10}{a^2 + 6a + 5}$

23. $\dfrac{b^2 + 3b + 2}{b^2 - 2} \div \dfrac{b^2 - 7b - 18}{b^3 - b^2 - 2b + 2}$

24. $\dfrac{x^2 + y^2}{x^3 + y^3} \div \dfrac{x^2 + y^2}{x + y}$

25. $\dfrac{x^2 - 2xy + y^2}{x^2 + xy + y^2} \div \dfrac{x^2 - y^2}{x^3 - y^3}$

26. $\dfrac{x^2 + 2xy + y^2}{x^2 + xy + y^2} \div \dfrac{x^2 + y^2}{x^3 - y^3}$

In Exercises 27–52, write as one rational expression in simplified form.

27. $\dfrac{1}{x} + \dfrac{2}{x - 1}$

28. $\dfrac{x}{x - 3} + \dfrac{4}{x + 2}$

29. $\dfrac{5}{y^2 - 9} + \dfrac{3}{y + 3}$

30. $\dfrac{1}{y^2 - 7y + 6} + \dfrac{1}{y^2 - 2y - 24}$

31. $\dfrac{y}{y^2 + 5y - 24} + \dfrac{1}{3 - y}$

32. $\dfrac{z + 1}{z - 1} + \dfrac{z - 1}{z + 1}$

33. $\dfrac{2z - 3}{6z + 1} + \dfrac{3z - 1}{z + 4}$

34. $\dfrac{t^2 + 1}{t^2 - 1} + \dfrac{1}{1 - t}$

35. $\dfrac{4}{t - 1} + \dfrac{t + 7}{t^2 - 4t + 3}$

36. $\dfrac{3u}{u - 2} - \dfrac{4u}{u + 5}$

37. $\dfrac{2u + 1}{3u - 1} - \dfrac{3 - u}{2u - 3}$

38. $\dfrac{v - 1}{v + 1} - \dfrac{1 - 7v}{v^2 - 2v - 3}$

39. $\dfrac{v + 5}{v - 4} - \dfrac{12v + 6}{v^2 - 2v - 8}$

40. $\dfrac{1}{v + 2} - \dfrac{1}{v^2 + 3v} - \dfrac{1}{v^2 + 5v + 6}$

41. $\dfrac{a}{b} + \dfrac{b}{a}$

42. $\dfrac{a}{a - b} + \dfrac{b}{b - a}$

43. $1 - \dfrac{a^2}{a^2 + b^2}$

44. $1 + \dfrac{a^2}{a^2 + b^2}$

45. $\dfrac{1}{b} - \dfrac{2}{ab^2} + \dfrac{4}{ab^3}$

46. $\dfrac{1}{p - q} + \dfrac{1}{p + q} - \dfrac{1}{p^2 - q^2}$

47. $\dfrac{1}{p} + \dfrac{1}{q} + \dfrac{1}{r}$

48. $\dfrac{q}{p} + \dfrac{r}{q} + \dfrac{p}{r}$

49. $\dfrac{1}{h}\left(\dfrac{1}{x + h} - \dfrac{1}{x}\right)$

50. $\dfrac{1}{h}\left[\dfrac{1}{(x + h)^2} - \dfrac{1}{x^2}\right]$

51. $\left(\dfrac{1}{y} - \dfrac{1}{x}\right) \div \left(\dfrac{1}{y} + \dfrac{1}{x}\right)$

52. $\left(\dfrac{x}{y} + \dfrac{y}{x}\right) \div \dfrac{x^2 + y^2}{xy}$

In Exercises 53–62, simplify the expression.

53. $\dfrac{\dfrac{1}{x} + x}{\dfrac{2}{x} + 1}$

54. $\dfrac{\dfrac{1}{x} - 1}{x^2 + 2}$

55. $\dfrac{x^2 + \dfrac{4}{x}}{x + \dfrac{4}{x^2}}$

56. $\dfrac{\dfrac{x+4}{x} - 1}{3 - \dfrac{4-x}{x}}$

57. $\dfrac{\dfrac{z-2}{z+5} - \dfrac{z-1}{z+1}}{\dfrac{z-3}{z+1} - \dfrac{z}{z+4}}$

58. $\dfrac{\dfrac{x-y}{x}}{\dfrac{x^2 - y^2}{xy}}$

59. $\dfrac{\dfrac{x}{y} - \dfrac{y}{x}}{\dfrac{1}{x} + \dfrac{1}{y}}$

60. $\dfrac{\dfrac{x^2}{y} - \dfrac{y^2}{x}}{\dfrac{1}{x} + \dfrac{1}{y}}$

61. $\dfrac{1}{\sqrt{x}} - \dfrac{1}{\sqrt{y}}$

62. $\dfrac{x}{\sqrt{y}} + \dfrac{y}{\sqrt{x}}$

In Exercises 63–69, rationalize the denominator.

63. $\dfrac{1}{\sqrt{x} - \sqrt{3}}$

64. $\dfrac{1}{\sqrt{x} - \sqrt{5}}$

65. $\dfrac{1}{\sqrt{x} - 9}$

66. $\dfrac{1}{\sqrt{x} + 6}$

67. $\dfrac{x - 3}{\sqrt{x} + \sqrt{3}}$

68. $\dfrac{1}{\sqrt{x} - \sqrt{y}}$

69. $\dfrac{\sqrt{x} - \sqrt{y}}{\sqrt{x} + \sqrt{y}}$

70. If p denotes the object distance and q the image distance of a simple lens, then the focal length is

$$\dfrac{1}{\dfrac{1}{p} + \dfrac{1}{q}}$$

Simplify this expression.

71. If three resistors having resistances R_1, R_2, and R_3, are connected in parallel, the resistance of the combination is

$$\dfrac{1}{\dfrac{1}{R_1} + \dfrac{1}{R_2} + \dfrac{1}{R_3}}$$

Simplify this expression.

KEY TERMS

real number	base	degree
natural number	exponent	constant term
integer	scientific notation	polynomial in several variables
rational number	square root	rational expression
irrational number	radical	numerator
factor	cube root	denominator
real line	nth root	domain
origin	polynomial	rational expression in lowest
positive direction	monomial	terms
coordinate	coefficient	least common denominator
distance	binomial	rationalizing the denominator
absolute value	leading coefficient	algebraic expression

KEY FORMULAS

$(a + b)(c + d) = ac + bc + ad + bd$
$(a + b)^2 = a^2 + 2ab + b^2$
$(a - b)^2 = a^2 - 2ab + b^2$

$$\frac{a}{b} + \frac{c}{d} = \frac{ad + bc}{bd}$$

$$\frac{a}{b} \cdot \frac{c}{d} = \frac{ac}{bd}$$

$$\frac{a}{b} \div \frac{c}{d} = \frac{ad}{bc}$$

$x^2 - a^2 = (x + a)(x - a)$
$x^3 - a^3 = (x - a)(x^2 + ax + a^2)$
$x^3 + a^3 = (x + a)(x^2 - ax + a^2)$

$x^n - a^n = (x - a)(x^{n-1} + ax^{n-2} + a^2x^{n-3} + \cdots + a^{n-2}x + a^{n-1})$
$x^n + a^n = (x + a)(x^{n-1} - ax^{n-2} + a^2x^{n-3} - \cdots - a^{n-2}x + a^{n-1})$ (n odd)

$a^r a^s = a^{r+s}$
$(a^r)^s = a^{rs} = (a^s)^r$
$(ab)^r = a^r b^r$

$$a^{-r} = \frac{1}{a^r}$$

$$|a| = \begin{cases} a \text{ if } a \geq 0 \\ -a \text{ if } a < 0 \end{cases}$$

REVIEW EXERCISES

In Exercises 1–12, calculate the value of the given expression.

1. $\dfrac{|7 - 9|}{|6 - 3|} - \dfrac{|-1|}{|4 - 8|}$

2. $\dfrac{3}{10} - \dfrac{4}{35} \cdot \dfrac{7}{2}$

3. $9^{-3/2}$

4. $\left(-\dfrac{1}{8}\right)^{5/3}$

5. $\dfrac{2^{-3}}{2^{-7}}$

6. $(\sqrt[3]{-3})^6$

7. $\sqrt{0.09}$

8. $9^{2.5}$

9. $3^4 \cdot 3^{-2}$

10. $\dfrac{(\sqrt{2})^3(\sqrt{2})^{-4}}{(\sqrt{2})^5}$

11. $\sqrt[3]{\dfrac{-8}{27}}$

12. $\sqrt[5]{\dfrac{1}{32}}$

In Exercises 13–14, write out the given statement using the symbols for inequalities.

13. $a - \sqrt{5}$ is nonnegative.

14. 0.2 is greater than $|3x + 1|$.

In Exercises 15–16, find the distance between a and b.

15. $a = 2.7, b = -1.6$

16. $a = -\frac{4}{5}, b = -\frac{2}{3}$

In Exercises 17–20, write the given number in scientific notation.

17. 159,000

18. 0.00314

19. $(16)^{3/2} \times 10^{-5}$

20. $\dfrac{231 \times 10^9}{11 \times 10^{-3}}$

In Exercises 21–36, simplify the given expression.

21. $\left(\dfrac{1}{2}a^{-2}\right)^3 a^4$

22. $(\sqrt{3} + \sqrt{2})^2$

23. $\sqrt{2} + \sqrt{50}$

24. $\dfrac{1}{2}(\sqrt{a + b} + \sqrt{a - b})^2$

25. $\dfrac{(15a^3)^2}{(15a^2)^3}$

26. $\dfrac{(a^3 b^{-3} c^{-1})^6}{(b^{-5} c^7)^{-2}}$

27. $\dfrac{(x - 5)^3}{|x - 5|}$

28. $\dfrac{\sqrt{x}}{\sqrt[5]{x}}$

29. $(\sqrt[3]{2x})^9$

30. $\sqrt{3x}\ \sqrt[4]{9x^2}$

31. $\dfrac{\sqrt[3]{32a^5b^{-5}}}{\sqrt[3]{4a^2b^{-4}}}$

32. $(x^5y^{-2}z^6)^{3/2}$

33. $\dfrac{(a+b)^2 - (a-b)^2}{(a+b)^2 + (a-b)^2}$

34. $[(a-b)^{-1} - (a+b)^{-1}]^{-1}$

35. $\dfrac{a - \dfrac{a^2}{a+b}}{b - \dfrac{b^2}{a+b}}$

36. $\dfrac{\dfrac{x}{x+y} + \dfrac{y}{x-y}}{x^2 + y^2}$

In Exercise 37–44, perform the indicated operation and simplify.

37. $(2a - 4)(5a - 1)$

38. $(\tfrac{1}{2}a - \tfrac{2}{3})^2$

39. $(2x - 9)(5x + 2) - 3(-2x + 7)(4x - 2)$

40. $(3x + 4)^2 - (3x - 5)^2$

41. $(x^{1/3} - x^{-1/3})^3$

42. $(y/x - 2x/y)^2$

43. $(2x^2 + 5y)^3$

44. $(x^2 + y^2 - 2z^2)^2$

In Exercises 45–46, simplify the given expression by rationalizing the denominator.

45. $\dfrac{1}{\sqrt{5} - 2}$

46. $\dfrac{\sqrt{2a} - \sqrt{8b}}{\sqrt{2a} + \sqrt{8b}}$

In Exercises 47–48, assume that $a > 0$, and determine whether the expression is positive or negative.

47. $a - \sqrt{a^2 + 4}$

48. $a - \sqrt{a^2 - 1}$

In Exercises 49–64, factor the given polynomial.

49. $x^2 - 6x - 27$

50. $y^2 + 8y - 105$

51. $t^2 - 20t + 100$

52. $4u^2 + 4u + 1$

53. $12x^2 - 11x + 2$

54. $x^3 - 2x^2 + 2x - 4$

55. $(y - 1)^5 - 4(y - 1)^3$

56. $x^3 - x^2 - 90x$

57. $z^4 + z^2 - 2$

58. $z^6 - z^4$

59. $x^3 + x^2 + x + 1$

60. $x^3y + xy^3$

61. $x^2 + 2xy - 35y^2$

62. $9x^4 - 6x^2y^2 + y^4$

63. $16x^4 - 1$

64. $x^8 - 256$

In Exercises 65–66, find the domain of the given expression.

65. $\dfrac{-7x^2 + 2}{x^2 + 9x + 20}$

66. $\dfrac{2x - 3}{x(4x - 3)^3}$

In Exercises 67–70, find the value of the given expression at the given number.

67. $4x^3 - 2x^2 + 3x - 19;\ 1$

68. $\dfrac{3x - 5}{2x + 1};\ 2$

69. $\dfrac{3x^2 + 1}{-4x^3 - 11};\ -3$

70. $\dfrac{x^2}{x^6 - 2};\ \sqrt{2}$

In Exercises 71–86, write as one rational expression in simplified form.

71. $\dfrac{x^2 - 6x + 5}{x^2 - 3x - 10}$

72. $\dfrac{x^2 + 2xy + y^2}{x^3 - x^2y - y^2x + y^3}$

73. $\dfrac{4x^2 - 1}{x^2 + x - 12} \cdot \dfrac{x^2 - 5x + 6}{6x^2 + x - 1}$

74. $\left(\dfrac{1}{x^2} - \dfrac{1}{x^3}\right)\left(\dfrac{1}{x} + \dfrac{1}{x^2}\right)$

75. $\dfrac{x^2 - 4}{x^2 - x - 6} \div \dfrac{x^2 + 2x - 8}{x^2 + x - 12}$

76. $\dfrac{3}{2x - 1} + \dfrac{2}{3x - 2}$

77. $\dfrac{2x - 3}{x - 2} + \dfrac{x}{x + 2}$

78. $\dfrac{x}{(x - 1)^2} - \dfrac{1}{x^2 - 1}$

79. $\dfrac{3x - 4}{x^2 + x} - \dfrac{5x - 2}{x^2 - x}$

80. $\dfrac{5}{x^2 + 3x - 10} + \dfrac{3}{x^2 - 3x + 2}$

81. $1 - \dfrac{x}{x - y}$

82. $\dfrac{x}{x - y} - \dfrac{y}{x + y}$

83. $\dfrac{\dfrac{2}{x} + \dfrac{3}{y}}{\dfrac{1}{x} - \dfrac{2}{y}}$

85. $\dfrac{\dfrac{1}{x} - \dfrac{1}{x^2}}{\dfrac{1}{x} + \dfrac{1}{x^2}}$

84. $\dfrac{\dfrac{1}{x} - \dfrac{1}{y}}{(x - y)^2}$

86. $\dfrac{\sqrt{x}}{\sqrt{x} - \sqrt{y}} + \dfrac{\sqrt{y}}{\sqrt{x} + \sqrt{y}}$

In Exercises 87–89, write the statement as an equation.

87. The surface area S of a rectangular box with height h and square base of side s is equal to the sum of twice the square of s and four times the product of s and h.

88. The surface area S of a cylindrical can (including top and bottom) of radius r and height h is the sum of 2π times the radius squared and 2π times the product of the radius and height.

89. The area A of an equilateral triangle is the product of $\sqrt{3}/4$ and the square of the length s of a side.

90. The age of Rob is x, and Rachel is 3 years less than twice as old as Rob. What is the sum of their ages?

91. Jill's salary is $30,000. Susan's salary can be determined by dividing Jill's salary by 8 and then adding to the result the product of 1982 and 13. Who has the larger salary?

92. What are the possible values of $\dfrac{a - b}{|a - b|}$?

93. Prove that

$$\left(\frac{x - y}{2}\right)^2 = \left(\frac{x + y}{2}\right)^2 - xy$$

94. A positive integer is called a **perfect number** if it is the sum of its positive divisors less than itself. Show that the following are perfect numbers.
    a. 6   b. 28   c. 496

95. Let a and b be real numbers with $b \neq -1$, and let

$$x = \frac{a^2}{1 + b^3}, \qquad y = bx, \quad \text{and} \quad z = ax$$

Show that $x^3 + y^3 = z^2$.

96. Let u and v be positive numbers with $u > v$, and let

$$a = 2uv, \qquad b = u^2 - v^2, \quad \text{and} \quad c = u^2 + v^2$$

The numbers $a, b,$ and c are known as a **Pythagorean triple**.
    a. Show that $c^2 = a^2 + b^2$. (This is the reason for the name of the triple; after all, it follows that there is a right triangle whose sides have lengths a, b and c.)
    b. Find the Pythagorean triples arising from the choices
       i. $u = 2, \quad v = 1$ iii. $u = 3, \quad v = 2$
       ii. $u = 3, \quad v = 1$

2
Equations and Inequalities

Consider the following two problems:

1. Suppose a swimmer dives from a platform 64 feet above a deep pool. If we disregard air resistance, how long would it take the swimmer to reach the pool?
2. A human fever is generally regarded to be an oral temperature exceeding 98.6 degrees Fahrenheit. What temperatures in degrees Celsius correspond to a fever?

The answers to these problems are not obvious. Although you might be able to solve them by trial and error, there is no guarantee that you would guess the right answers. In Chapter 2 we use equations and inequalities to solve such problems. (See Example 1 in Section 2.4 and Example 7 in Section 2.6.)

Equations arise by setting two given algebraic expressions equal to one another; inequalities arise by making one given algebraic expression less than another. As you see, equations and inequalities are close relatives of one another. They are also extremely important in mathematics. In fact, it could almost be said that higher mathematics revolves about the study of equations and inequalities. Moreover, there are many applications of equations and inequalities outside mathematics. These applications range from the solutions of simple, everyday problems to the formulation of the deepest principles of modern science. In this chapter we will illustrate some of the simpler applications of equations and inequalities.

2.1
LINEAR
EQUATIONS

Before we begin our discussion of linear equations, we need some terminology concerning equations in general. First of all, an *equation* is a statement that two expressions are equal. In Chapter 1 we encountered equations such as

$$x^2 + 6x + 9 = (x + 3)^2 \quad \text{and} \quad \frac{x^2 - 4}{x - 2} = x + 2 \tag{1}$$

and in the present chapter we will see equations such as

$$2x + 7 = 5x - 3 \quad \text{and} \quad 3x^2 - 4x + 1 = 0 \tag{2}$$

However, the role the variable x plays in (1) is different from its role in (2). Indeed, the first equation in (1) is valid for all values of x, and the second equation in (1) is valid for all values of x except 2 (for which the expression on the left side is undefined). In contrast, the equations in (2) are valid only for very special values of x. In the course of this chapter we will show that the equation $2x + 7 = 5x - 3$ is valid only for $x = \frac{10}{3}$ and that the equation $3x^2 - 4x + 1 = 0$ is valid only for $x = 1$ or $x = \frac{1}{3}$.

We call an equation an *identity* if it is valid for all values of the variable that make each constituent expression meaningful. Thus the equations in (1) are identities. In contrast, an equation is called *conditional* if it is valid only for special values of the variable. This is true of both equations in (2).

Two conditional equations are said to be *equivalent* if they are valid for precisely the same values of the variable. For example, the following pairs of equations are equivalent:

$x^2 - 6x = 0 \quad \text{and} \quad t^2 - 6t = 0$ (Only the variable letters are different.)

$\dfrac{9}{5}x + 32 = 68 \quad \text{and} \quad \dfrac{9}{5}x = 36$ (The same constant is added to both sides of one equation to obtain the other equation.)

$3x = 17 \quad \text{and} \quad x = \dfrac{17}{3}$ (Both sides of one equation are multiplied by the same constant to obtain the other equation.)

The equations of main interest in this chapter will be conditional equations. Because conditional equations are valid only for special values of the variable, our goal will be to determine those special values of the variable, called *solutions* (or sometimes *roots*) of the equation. In this terminology, 1 and $\frac{1}{3}$ are solutions of $3x^2 - 4x + 1 = 0$, because if we substitute 1 or $\frac{1}{3}$ for x in the equation, the resulting equation is valid. The process of finding all the solutions of a given equation is called *solving the equation*. Generally speaking, when we set out to solve a given equation whose solutions are not obvious, we will try to obtain an equivalent equation that is more easily solved.

Linear Equations

The simplest of all conditional equations is a *linear equation*, that is, an equation that is equivalent to either

$$ax = b \quad \text{or} \quad ax + b = 0$$

where a and b are fixed real numbers and $a \neq 0$. The following are examples of linear equations:

$$x = 3, \quad 5x = 2, \quad \pi x - \frac{7}{2} = 0, \quad \text{and} \quad 2x + 7 = 5x - 3$$

Presently we will show that every linear equation has a solution. Moreover, the solution is unique, that is, there is only one value of x for which the equation is valid.

It may happen that a given equation has the form $x = c$, in which case the unique solution is c. For example, the solution of $x = 3$ is 3. However, for other linear equations of the form $ax = b$ we usually find the solution by the following procedure, consisting of a chain of equivalent equations:

$$
\begin{aligned}
&\text{a.} \quad ax = b \quad &&\text{(given equation)} \\[1em]
&\text{b.} \quad \frac{ax}{a} = \frac{b}{a} \quad &&\text{(division by } a \neq 0) \\[1em]
&\text{c.} \quad x = \frac{b}{a} \quad &&\text{(equation (b) simplified)}
\end{aligned}
$$

Because (a) and (c) are equivalent, we see that the unique solution of the equation $ax = b$ is b/a.

In actually solving equations of the form $ax = b$, we frequently combine steps (b) and (c), passing straight from $ax = b$ to $x = b/a$. Example 1 illustrates this procedure.

EXAMPLE 1.   Solve the equation $-2x = 6$ for x.

Solution.   We divide both sides of the equation by -2 and simplify:

$$-2x = 6$$

$$x = \frac{6}{-2} = -3$$

Check: $-2(-3) = 6$

Thus -3 is the solution.   □

Caution: It is always possible to make mistakes. For that reason it is a good idea to check the proposed solution by substituting it into the original equation, as we have done in Example 1 and will continue to do in the future.

Sometimes the linear equation to be solved is not given in the form $ax = b$ but can be put into that form by algebraic manipulation and then solved for x. The general procedure is:

> i. Put all terms containing x on one side of the equation.
> ii. Put all other terms on the other side of the equation.
> iii. Simplify the resulting equation to solve for x.

EXAMPLE 2.   Solve the equation $\frac{9}{5}x + 32 = 68$ for x.

Solution.   We first subtract 32 from both sides:

$$\frac{9}{5}x + 32 = 68$$

$$\left(\frac{9}{5}x + 32\right) - 32 = 68 - 32$$

$$\frac{9}{5}x = 36$$

Now we have an equation in the form $ax = b$, so we divide by $\frac{9}{5}$ (or, equivalently, multiply by $\frac{5}{9}$) and simplify:

$$\frac{9}{5}x = 36$$

$$\frac{5}{9}\left(\frac{9}{5}x\right) = \frac{5}{9}(36)$$

$$x = 20$$

Check: $\frac{9}{5}(20) + 32 = 36 + 32 = 68$

Thus 20 is the solution.   □

The equation solved in Example 2 could arise in the conversion from degrees Fahrenheit to degrees Celsius. If F represents degrees Fahrenheit and C represents degrees Celsius, then F and C are related by the equation

$$\frac{9}{5}C + 32 = F \tag{3}$$

Suppose we wish to determine the number of degrees Celsius corresponding to 68 degrees Fahrenheit. In that case we can substitute 68 for F in (3) to obtain the linear equation

$$\frac{9}{5}C + 32 = 68$$

which is equivalent to the equation solved in Example 2. Its solution is 20, so 20 degrees Celsius corresponds to 68 degrees Fahrenheit.

EXAMPLE 3.   Solve the equation $2x + 7 = 5x - 3$ for x.

Solution.   In order to have all terms containing x on the left side and all other terms on the right, we first subtract $5x$ from both sides:

$$2x + 7 = 5x - 3$$

$$(2x + 7) - 5x = (5x - 3) - 5x$$

$$-3x + 7 = -3$$

Then we subtract 7 from both sides, and finally solve the resulting equations as in previous examples:

$$(-3x + 7) - 7 = -3 - 7$$

$$-3x = -10$$

$$x = \frac{-10}{-3} = \frac{10}{3}$$

$$Check: \quad 2\left(\frac{10}{3}\right) + 7 = \frac{20}{3} + 7 = \frac{41}{3} \quad \text{and}$$

$$5\left(\frac{10}{3}\right) - 3 = \frac{50}{3} - 3 = \frac{41}{3}$$

Therefore the solution is $\frac{10}{3}$.   □

The next two examples involve equations that do not look at all like linear equations. Yet by appropriate manipulations they turn out to be equivalent to linear equations, which can be solved by the process outlined in (i)–(iii).

EXAMPLE 4.   Solve $(x - 3)(x - 4) = (x + 1)(x - 2)$ for x.

Solution.   We multiply out both sides and then simplify:

$$(x - 3)(x - 4) = (x + 1)(x - 2)$$

$$x^2 - 7x + 12 = x^2 - x - 2$$

$$-7x + 12 = -x - 2$$

$$-6x = -14$$

$$x = \frac{-14}{-6} = \frac{7}{3}$$

Check: $\left(\dfrac{7}{3} - 3\right)\left(\dfrac{7}{3} - 4\right) = \left(-\dfrac{2}{3}\right)\left(-\dfrac{5}{3}\right) = \dfrac{10}{9}$ and

$$\left(\dfrac{7}{3} + 1\right)\left(\dfrac{7}{3} - 2\right) = \left(\dfrac{10}{3}\right)\left(\dfrac{1}{3}\right) = \dfrac{10}{9}$$

Thus $\frac{7}{3}$ is the solution. ☐

EXAMPLE 5. Solve $\dfrac{x + 2}{x - 1} = 2 - \dfrac{x - 1}{x + 3}$ for x.

Solution. First we multiply both sides by $(x - 1)(x + 3)$ in order to remove the terms $x - 1$ and $x + 3$ from the denominators. Then we manipulate the resulting equation in order to obtain the solution:

$$\frac{x + 2}{x - 1} = 2 - \frac{x - 1}{x + 3}$$

$$\left(\frac{x + 2}{x - 1}\right)(x - 1)(x + 3) = 2(x - 1)(x + 3) - \left(\frac{x - 1}{x + 3}\right)(x - 1)(x + 3)$$

$$(x + 2)(x + 3) = 2(x - 1)(x + 3) - (x - 1)^2$$

$$x^2 + 5x + 6 = 2(x^2 + 2x - 3) - (x^2 - 2x + 1)$$

$$x^2 + 5x + 6 = 2x^2 + 4x - 6 - x^2 + 2x - 1$$

$$x^2 + 5x + 6 = x^2 + 6x - 7$$

$$-x = -13$$

$$x = 13$$

Check: $\dfrac{13 + 2}{13 - 1} = \dfrac{15}{12} = \dfrac{5}{4}$ and

$$2 - \frac{13 - 1}{13 + 3} = 2 - \frac{12}{16} = 2 - \frac{3}{4} = \frac{5}{4}$$

Thus 13 is the solution. ☐

Equations Containing Absolute Values

Consider the equation $|x - 3| = 2$. If the absolute value were not present, the equation would be $x - 3 = 2$, which we know how to solve. Nevertheless, the equation $|x - 3| = 2$ is not linear, so we need a new method to solve it. Since $|x - 3| = x - 3$ or $|x - 3| = -(x - 3)$, depending on whether $x - 3 \geq 0$ or $x - 3 < 0$, it follows that in order to solve the equation $|x - 3| = 2$ we need to consider two cases: $|x - 3| = x - 3$ and $|x - 3| = -(x - 3)$. Each of these cases yields a linear equation, which we can solve by the method already discussed.

EXAMPLE 6.   Solve the equation $|x - 3| = 2$ for x.

Solution.   As we suggested above, we consider the following two cases:

a. $$|x - 3| = x - 3$$

which yields

$$x - 3 = 2$$
$$x = 5$$

b. $$|x - 3| = -(x - 3)$$

which yields

$$-(x - 3) = 2$$
$$-x + 3 = 2$$
$$-x = -1$$
$$x = 1$$

Check: $|5 - 3| = |2| = 2$
$|1 - 3| = |-2| = 2$

Thus 5 and 1 are the solutions of the given equation.   ☐

EXAMPLE 7.   Solve $|3x - 4| = |2x + 5|$ for x.

Solution.   Since $|3x - 4| = 3x - 4$ or $|3x - 4| = -(3x - 4)$, and similarly, since $|2x + 5| = 2x + 5$ or $|2x + 5| = -(2x + 5)$, there appear to be four cases: $3x - 4 = 2x + 5$, $3x - 4 = -(2x + 5)$, $-(3x - 4) = 2x + 5$, and $-(3x - 4) = -(2x + 5)$. But notice that the equations $3x - 4 = 2x + 5$ and $-(3x - 4) = -(2x + 5)$ are equivalent, and the equations $-(3x - 4) = 2x + 5$ and $3x - 4 = -(2x + 5)$ are themselves equivalent. Therefore we only have to consider two cases:

a. $$3x - 4 = 2x + 5$$

which yields

$$x = 9$$

b. $$3x - 4 = -(2x + 5)$$

which yields

$$3x - 4 = -2x - 5$$
$$5x = -1$$
$$x = -\frac{1}{5}$$

Check: $|3(9) - 4| = 23$ and $|2(9) + 5| = |23| = 23$

$$\left|3\left(-\frac{1}{5}\right) - 4\right| = \left|-\frac{23}{5}\right| = \frac{23}{5} \quad \text{and}$$

$$\left|2\left(-\frac{1}{5}\right) + 5\right| = \left|\frac{23}{5}\right| = \frac{23}{5}$$

Thus 9 and $-\frac{1}{5}$ are the two solutions of the given equation. □

Equations with Two Variables

Consider the equation

$$2x + 5 = 3y - 7 \tag{4}$$

This equation contains two variables, x and y, and thus is quite different from the other equations we have studied in this section. But suppose we assign a numerical value, say 6, to y. Then (4) becomes

$$2x + 5 = 3(6) - 7$$

which reduces to

$$2x + 5 = 11$$

itself a linear equation in x that can be solved by the methods of this section. In fact, no matter what numerical value we assign to y, we obtain a linear equation in x that can be solved. But rather than solve every such equation, we can treat y in (4) as a constant. We use this idea as we solve the equation in (4) in our final example.

EXAMPLE 8.   Solve the equation

$$2x + 5 = 3y - 7$$

for x.

Solution.   We treat y as a constant and solve the equation for x:

$$2x + 5 = 3y - 7$$
$$(2x + 5) - 5 = (3y - 7) - 5$$
$$2x = 3y - 12$$
$$x = \frac{1}{2}(3y - 12) = \frac{3}{2}y - 6 \quad \square$$

EXERCISES 2.1

In Exercises 1–12, solve the equation.

1. $2x = 0$

2. $-4x = 0$

3. $-x = \frac{1}{3}$

4. $x - 2 = 5$

5. $4y + 5 = 9$

6. $\sqrt{2}y + \dfrac{1}{\sqrt{2}} = 3\sqrt{2}$

7. $z + 3 = 2z - 4$

8. $4z - 2 = 3 - 7z$

9. $2(1 + z) = 3z + 5$

10. $\frac{1}{2}(2 - t) = \frac{3}{2} + 6t$

11. $1.3t + 5.2 = 2.6 - 7.8t$

12. $\frac{1}{3}(1 - 2t) + \frac{1}{6}(3t + 6) = 0$

In Exercises 13–30, solve the equation.

13. $x^2 - 5x + 6 = x^2 + 3x + 2$

14. $x^2 - 6 = -(3x - x^2)$

15. $(x - 1)(x + 2) = (x + 3)(-4 + x)$

16. $(y + 2)^2 = y^2 - 4$

17. $(y - 3)^2 = y^2 + 6y + 8$

18. $(y + 2)^3 = (y + 1)^3 + 3y^2 - 1$

19. $(y - 1)^3 + 3(y + 2)^2 = y^3 + 2y + 4$

20. $2 - \dfrac{3}{x} = 4 + \dfrac{2}{3x}$

21. $\dfrac{1}{x} + 3 = \dfrac{2}{x} + 4$

22. $\dfrac{1}{1 - t} = \dfrac{1}{1 + t}$

23. $\dfrac{3}{3 + t} = \dfrac{5}{4t - 1}$

24. $\sqrt{u} + \dfrac{4}{\sqrt{u}} = 2\sqrt{u} - \dfrac{3}{2\sqrt{u}}$

25. $u^{1/3} - u^{-2/3} = 4u^{1/3} + 6u^{-2/3}$

26. $\dfrac{2w + 1}{2w - 5} = \dfrac{w}{w - 6}$

27. $\dfrac{w - 2}{w + 3} = \dfrac{w + 1}{w - 1}$

28. $1 - \dfrac{z}{2z + 1} = \dfrac{\frac{1}{2}z - 1}{z + 3}$

29. $\dfrac{z - 2}{z - 1} + 3 = \dfrac{4z + 1}{z + 2}$

30. $\dfrac{1}{1 - t} + \dfrac{2}{1 - t^2} = \dfrac{3}{1 + t}$

In Exercises 31–40, solve the given equation.

31. $|x - 4| = 3$

32. $|x + 7| = 2$

33. $|x + 1| = \frac{5}{2}$

34. $|-2x + 5| = 7$

35. $|\frac{2}{3}x + 4| = 2$

36. $|2x - 1| = |-x + 5|$

37. $|\frac{1}{2}x + 3| = |\frac{2}{3}x - 3|$

38. $|x| = |x + 1|$

39. $|2x - 3| = x$

40. $|3x - 1| = x$

In Exercises 41–50, solve for x.

41. $x - y = 0$

42. $x + y = 1$

43. $2x = 6y - 1$

44. $3x - 2 = 4y$

45. $4x + 2y = 5$

46. $-\frac{1}{2}x + 1 = 3y + 5$

47. $\frac{1}{4}x - \frac{1}{2}y - 3 = 0$

48. $\frac{4}{7}y = \frac{2}{3}x - \frac{5}{2}$

49. $0.3x - 0.2y = 1.8x + 3.3y - 0.1$

50. $1.7x - 2.4y + 1.2 = -8.3x - 3.7y - 2.3$

In Exercises 51–54, solve for y.

51. $x = 2y + 7$

52. $6x + 3y = -1$

53. $\frac{1}{3}x - \frac{1}{2}y - 2 = 0$

54. $1.4x - 2.3y + 1.8 = 5.3x - 4.9y - 0.3$

55. Show that the equation $|2x + 3| = x$ has no solution.

56. Determine those values of a for which $|2x + a| = x$ has a solution.

57. Find a value of a such that -4 is a solution of the equation $2x + 3 - 4a = x + 7$.

58. Find a value of a such that 3 is a solution of the equation

$$\frac{1}{2 - x} + \frac{a}{x + 2} = \frac{1}{x^2 - 4}$$

59. Find a value of a such that $2x + 4 = a$ and $3 - x = 1$ are equivalent.

60. What relationship must a have to b in order for $\frac{1}{2}$ to be a solution of $ax + b = 0$?

61. If x denotes the length of an object in inches and y the length in centimeters, then x and y satisfy the equation

$$y = 2.54x$$

   a. What is the length in centimeters of a foot-long hot dog?
   b. How long in inches is a baby that is 63.5 centimeters long?
   c. Use a calculator to approximate the number of inches in one meter (100 centimeters).

62. If the temperature is 86 degrees Fahrenheit, what is the temperature in degrees Celsius? (*Hint:* Use (3).)

63. If the temperature is 100 degrees Celsius, what is the temperature in degrees Fahrenheit? (*Hint:* Use (3).)

64. Use (3) to determine a number T for which

$$T \text{ degrees Fahrenheit} = T \text{ degrees Celsius}$$

65. By solving equation (3) for C in terms of F, express the temperature in degrees Celsius in terms of the temperature in degrees Fahrenheit.

66. Light of sufficiently high frequency can dislodge electrons from their associated atoms. This effect is known as the **photoelectric effect.** The kinetic energy E of an electron so dislodged is given by

$$E = hv - \omega$$

where h is Planck's constant, v is the frequency of the light, and ω is the binding energy of the electron. Solve the equation for the frequency v.

67. The object distance p, image distance q, and focal length f of a simple lens satisfy the equation

$$\frac{1}{p} + \frac{1}{q} = \frac{1}{f}$$

Solve the equation for p.

2.2 APPLICATIONS OF LINEAR EQUATIONS

One of the reasons algebra is so important is that very often it can be used to help solve problems in such disciplines as physics, engineering, economics, and geometry. Problems like these are frequently formulated verbally. For that reason they are sometimes called *word problems*, but we prefer to call them *applied problems*. Although you may think that some of the problems we will present are rather contrived, such problems are part of mathematical folklore, and serve to illustrate the fact that mathematics can indeed be applied in everyday life.

To use the mathematics we have developed in solving an applied problem, we will translate the problem into mathematical language and then solve the resulting mathematical problem, which in this section will always be a linear equation. Because of the extreme breadth of applied problems, there is no hope of prescribing a single detailed procedure that works for all of them. After discussing and solving various applied problems, we will give a synopsis of our general method of attacking such problems. It should serve as a guideline to help you in solving other applied problems.

An Averaging Problem

EXAMPLE 1. A student has scores of 67, 90, 76, and 82 on the first four tests in a history course. What grade must the student achieve on the fifth test in order for the average of the scores on all five tests to be 80?

Solution. The average of five tests is the sum of the scores on the five tests divided by 5. If we let

$$x = \text{the score on the fifth test}$$

then the average on the five tests is

$$\frac{67 + 90 + 76 + 82 + x}{5}$$

Since

$$\frac{67 + 90 + 76 + 82 + x}{5} = \frac{315 + x}{5} = 63 + \frac{1}{5}x$$

it follows that the average will be 80 if

$$63 + \frac{1}{5}x = 80$$

Solving for x, we obtain

$$\frac{1}{5}x = 17$$

$$x = 85$$

$$Check: \quad \frac{67 + 90 + 76 + 82 + 85}{5} = \frac{400}{5} = 80$$

Consequently the student must achieve an 85 on the fifth test in order for the average of all five tests to be 80. □

A Simple Interest Problem

Suppose that one invests a sum of money P (called the ***principal***) in a savings account that draws simple interest at an annual rate r. This means that after n years the interest I earned by the account is given by the formula

$$I = Prn \tag{1}$$

Four letters appear in (1). If three of them are assigned specific numerical values, then we can solve for the fourth. Normally the interest rate is expressed as a percentage. Thus if the interest rate is 7%, then $r = 0.07$. As an example of the use of (1), suppose that $2000 is invested in a savings account which draws simple interest at an annual rate of 7%. Then after 1 year the interest I earned is given by

$$I = (2000)(0.07)(1) = 140$$

and after 5 years the interest earned is given by

$$I = (2000)(0.07)(5) = 700$$

EXAMPLE 2. Suppose that $16,000 is deposited in a bank in the following way: $10,000 into a U.S. Treasury note earning 9% simple interest annually, and $6000 into a passbook account earning 5% simple interest annually. How long will it take to earn $12,000 in interest?

Solution. Let

n = the number of years required to earn $12,000 in interest

After n years the amount of interest in dollars earned from the Treasury note will be $(10,000)(0.09)(n)$, and the amount earned from the passbook account will be $(6000)(0.5)(n)$. Since the total amount of interest to be earned is $12,000, we have the linear equation

$$(10,000)(0.09)(n) + (6000)(0.05)(n) = 12,000$$

This simplifies to

$$900n + 300n = 12,000$$
$$1200n = 12,000$$
$$n = 10$$

Check: After 10 years the Treasury note will have earned $(10,000)(0.09)(10)$ dollars in interest, and the passbook account will have earned $(6000)(0.05)(10)$ dollars in interest. Therefore their combined earned interest will be given by

$$(10,000)(0.09)(10) + (6000)(0.05)(10) = 9000 + 3000 = 12,000$$

It will take 10 years to earn $12,000 interest from the two deposits. □

A Geometric Problem

In our next example we discuss a geometric problem to which algebraic techniques apply.

EXAMPLE 3. A rectangular plot of land has a perimeter of 270 feet, and its length is twice its width (Figure 2.1). Find the dimensions of the plot.

Solution. Let

x = the length in feet of the plot
y = the width in feet of the plot

Since the perimeter $2x + 2y$ is 270 feet, we have the equation

$$2x + 2y = 270 \tag{2}$$

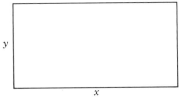

Length = 2 × Width
Perimeter = 270 feet

FIGURE 2.1

What we are told about the length and width implies that

$$x = 2y \tag{3}$$

When we substitute $2y$ for x in (2), we obtain the linear equation

$$2(2y) + 2y = 270$$

which reduces to

$$6y = 270$$
$$y = 45$$

Now it follows from (3) that $x = 2(45) = 90$.

$$\text{Check:}\quad \text{perimeter} = 2x + 2y = 2(90) + 2(45)$$
$$= 180 + 90 = 270$$
$$\text{length} = x = 90 = 2y = 2 \cdot \text{width}$$

Therefore the plot is 90 feet long and 45 feet wide.   ☐

Rate Problems

If an object (such as a train or a spaceship) travels at a constant speed or rate r, then the distance d traveled by the object during time t is given by

$$d = rt \tag{4}$$

For example, if a car travels at a rate of 55 miles per hour, then the distance it travels in 2 hours is (55)(2), that is, 110 miles. In order for (4) to apply, the units in which r is measured must be compatible with the units in which d and t are measured. In the problems of this section, distance will be measured in miles, time in hours, and the rate in miles per hour. Notice that if two of the three letters are given specific values, then we can solve for the third.

EXAMPLE 4.   A passenger train averaging 75 miles per hour begins the 200-mile trip from New York to Boston at 2:00 P.M. A freight train traveling at 45 miles per hour sets out from Boston at 4:00 P.M. the same day and travels toward New York on an adjacent set of tracks. At what time will they meet?

Solution.   Since we must determine when the trains meet and since they meet sometime after 4 P.M., we let

t = the elapsed time in hours after 4 P.M. until the trains meet

If we let d_1 be the distance in miles the passenger train will have traveled from New York when the trains meet and d_2 the distance in miles the freight train will have traveled from Boston, then $d_1 + d_2$ must be 200, the distance in

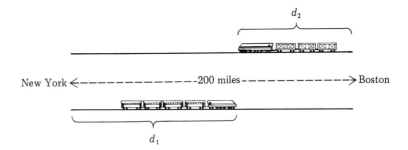

FIGURE 2.2

miles between New York and Boston (Figure 2.2). Thus we have

$$d_1 + d_2 = 200 \qquad (5)$$

Now since the passenger train (which left New York at 2 P.M.) will have traveled $t + 2$ hours at 75 miles per hour and the freight train (which left Boston at 4 P.M.) will have traveled t hours at 45 miles per hour, (4) tells us that

$$d_1 = 75(t + 2) \quad \text{and} \quad d_2 = 45t \qquad (6)$$

Substituting the expressions given in (6) for d_1 and d_2 into (5), we obtain the linear equation

$$75(t + 2) + 45t = 200$$

which is solved as follows:

$$75t + 150 + 45t = 200$$

$$120t = 50$$

$$t = \frac{50}{120} = \frac{5}{12} \quad \square$$

Check: $\frac{5}{12}$ hour after 4 P.M. the passenger train is

$$\left(2 + \frac{5}{12}\right)75 = 150 + \frac{125}{4} = 181\frac{1}{4}$$

miles from New York, and the freight train is

$$\frac{5}{12}(45) = \frac{75}{4} = 18\frac{3}{4}$$

miles from Boston. Since New York and Boston are 200 miles apart, when the freight train is $18\frac{3}{4}$ miles from Boston, it is

$$200 - 18\frac{3}{4} = 181\frac{1}{4}$$

miles from New York and thus meets the passenger train at that time.

Thus the trains indeed meet $\frac{5}{12}$ hour after 4 P.M., that is, at 4:25 P.M. □

Next, suppose that a certain job (such as mowing a lawn) is performed at a constant rate. If r is the rate, then r is the fraction of the job that is performed in one hour (or in some other convenient unit of time). It follows that the amount W of work (that is, the fraction of the job) done in t hours is given by

$$W = rt \tag{7}$$

(Notice the similarity between equations (4) and (7).) Again the units in which W, r, and t are measured must be compatible, as they will be in our problems. If we solve equation (7) for r, we find that

$$r = \frac{W}{t} \tag{8}$$

The work W required to perform the entire job is 1, so that by (8),

$$r = \frac{1}{\text{the time required to perform the job}} \tag{9}$$

EXAMPLE 5.   If it takes an adult 2 hours to mow a lawn and a child 3 hours, how long would it take them to mow the lawn at the same time with two lawn mowers?

Solution.   Let r_a and r_c be the rates at which the adult and the child can mow the grass, respectively. Then (9) implies that

$$r_a = \frac{1}{2} \quad \text{and} \quad r_c = \frac{1}{3} \tag{10}$$

This means that in 1 hour the adult can finish one half of the lawn and the child one third of the lawn. Let

$$t = \text{the time in hours needed for the adult and the child}$$
$$\text{to mow the whole lawn simultaneously}$$

If W_a denotes the amount of work done by the adult in mowing his portion of the lawn and W_c the corresponding amount of work done by the child, then

$$1 = W_a + W_c \tag{11}$$

By (7) and (10) we have

$$W_a = r_a t = \frac{1}{2}t \quad \text{and} \quad W_c = r_c t = \frac{1}{3}t$$

Substituting for W_a and W_c in (11) yields the linear equation

$$1 = \frac{1}{2}t + \frac{1}{3}t$$

which simplifies to

$$1 = \frac{5}{6}t$$

Thus

$$t = \frac{1}{\frac{5}{6}} = \frac{6}{5}$$

> *Check:* The work done by the adult and the child in $\frac{6}{5}$ hours is given by
>
> $$\frac{1}{2}\left(\frac{6}{5}\right) + \frac{1}{3}\left(\frac{6}{5}\right) = \frac{3}{5} + \frac{2}{5} = 1$$

Therefore it would take $\frac{6}{5}$ hours (that is, 72 minutes) for the adult and the child to mow the lawn simultaneously.   □

A Mixture Problem

EXAMPLE 6.   A radiator contains 6 quarts of fluid consisting of 40% antifreeze and 60% water (by volume). How much of the mixture should be drained off and replaced by pure antifreeze in order to obtain a mixture containing 60% antifreeze?

Solution.   Let

$$x = \text{the amount of fluid to be drained off}$$

Then the amount of antifreeze remaining is 40% of $6 - x$ quarts, that is, $(0.4)(6 - x)$ quarts. With x quarts of antifreeze added, the total antifreeze in the radiator will be $x + (0.4)(6 - x)$ quarts, and the goal is for this to be 60% of the total of 6 quarts, that is, $(0.6)(6)$ quarts. Thus we obtain the linear equation

$$x + (0.4)(6 - x) = (0.6)(6)$$

which is solved as follows:

$$x + 2.4 - 0.4x = 3.6$$
$$0.6x = 1.2$$
$$x = \frac{1.2}{0.6} = 2$$

Check: If 2 quarts are removed and replaced by antifreeze, then the amount of antifreeze afterwards will be

$$2 + (0.4)(6 - 2) = 2 + 2.4 - 0.8 = 3.6$$

quarts. Since there are 6 quarts in the radiator and $(3.6)/6 = 0.6$, the percentage of antifreeze in the radiator at the end will be 60%.

Therefore 2 quarts must be drained and replaced by antifreeze to make 60% of the mixture antifreeze.   □

Solving Applied Problems

The following guidelines should help you solve other applied problems, including the exercises that follow.

> 1. After reading the problem carefully, choose a variable for the quantity to be determined. If necessary, choose auxiliary variables for other quantities appearing in the problem (as in Examples 3, 4, and 5).
> 2. From the information given in the problem, write down any equations that must be satisfied by the variables. This step may require ingenuity. A picture is sometimes helpful (as in Examples 3 and 4).
> 3. Eliminate all the auxiliary variables from the equations. The goal is to obtain an equation that contains only the variable for the quantity to be determined (see Examples 3, 4, and 5).
> 4. Solve the equation obtained in step 3.
> 5. Check the solution to be sure that it really is a solution of the given applied problem.

Before going on to the exercises, you might find it helpful to reread the examples in this section and their solutions, in order to see how we followed the guidelines listed above. And when you attempt to solve applied problems, don't give up too quickly. Translating applied problems and the information in them into mathematical equations takes time, and the more you try and practice, the more successful you will become at solving them.

EXERCISES 2.2

1. A student's grades from three examinations are 82, 64, and 91. In order to obtain an average of 80, what must the student's grade on the fourth test be?

2. In order to raise the average of the three tests in the preceding exercise by 3 points, what must the score on the fourth test be?

3. A student has a 72 average on four examinations. If the score on the final examination is to count double, what score will yield an average of 80?

4. Jack's mother is three times as old as Jack. In 14 years she will be twice as old as Jack is then. How old are they both now?

5. Karen's father is four times as old as Karen. In 6 years his age will be 10 years more than double Karen's age at that time. How old are they both now?

6. The sum of the ages of two children is 14. Two years ago the age of one child exceeded twice the age of the other by one year. How old are they now?

7. Find two numbers whose sum is 29 and whose difference is 5.

8. Find three consecutive odd integers whose sum is 147.

9. Receipts from 6500 tickets to a basketball game totaled $14,800. If student tickets cost $2 each and nonstudent tickets cost $3 each, how many of each were sold?

10. A collection of 64 coins containing only nickels and dimes is worth $4.50. How many of each are there?

11. A newspaper boy collects $112 in dollar bills, quarters, dimes, and nickels. If he has 5 times as many dollar bills as quarters, $\frac{2}{5}$ as many quarters as dimes, and 10 more dimes than nickels, determine the number of each denomination.

12. In a local election, 75% of the eligible voters voted for more primary schools, 15% of those eligible voted against more primary schools, and the remaining 92 eligible voters did not vote on the issue. How many eligible voters were there?

13. A professor has a list of problems to assign to the students in a class. If 3 problems are assigned to each student, there will be 33 problems left over, but 15 more problems would be needed in order to be able to assign 5 problems to each person in the class. How many students are there in the class?

14. A football coach rewards a quarterback with 50¢ for each correctly solved mathematics problem and fines the quarterback 30¢ for each incorrectly solved problem. After working 100 problems, the quarterback has a net gain of $22.00. How many problems were solved correctly?

15. Suppose the cost of a digital clock, including a 4% sales tax, is $26.52. Find the cost of the clock without the tax.

16. According to the provisions of a will, a sum of money is to be divided among four people. One person is to receive one half of the sum, a second person is to receive one third of the sum, and a third is to receive one twelfth of the sum. The fourth person is to receive the remainder, which amounts to $1500. How large is the sum of money?

17. In a basketball game the high scorer had 30 points and the rest of the team scored $\frac{11}{16}$ of the team's total points. How many points did the team score?

18. A mutual fund pays an 11% dividend per year. If the dividend is $467.50 after the first year, how much was originally invested?

19. Suppose the amount to be invested in two savings accounts with simple interest is $3000, and assume that the accounts pay 6% and 8%, respectively. If the interest accumulated in one year is to be $196, how much must be deposited in each account?

20. Suppose that $6000 is to be invested in two savings accounts, one at 7% simple interest and the other at 8% simple interest. How much should be put into each account to have both accounts yield the same amount of interest?

21. The interest rate on an investment of $4200 is 1% greater than on a second investment of $2400. If the total annual interest from the two investments is $471, what are the respective interest rates?

22. A bricklayer gets a 6% increase in salary, and this amounts to $1116.
   a. What was the original salary?
   b. What is the new salary?

23. The perimeter of a rectangle is 100 inches, and the width is two thirds of the length. Find the dimensions of the rectangle.

24. The perimeter of a rectangular driveway is 106 feet, and the difference between the length and width is 23 feet. Find the length and width of the driveway.

25. The perimeter of a rectangle is 160 meters. If a new rectangle is formed by doubling the length of one pair of sides and decreasing the length of the remaining pair of sides by 30 meters, the perimeter remains the same. What were the dimensions of the original rectangle?

26. The perimeter of a rectangular rose garden is 56 feet. Determine its dimensions if three such gardens placed side by side would form a square.

27. The perimeter of an isosceles triangle is 33, and one side is three fourths as long as the other two sides. Determine the length of the shortest side.

28. The perimeter of an isosceles trapezoid is 42 inches. The shortest two sides, which are the same length, are 7 inches shorter than the next longer side and 15 inches shorter than the longest side. Determine the length of the shortest sides.

29. The ratio of the earth's area of land to area of sea is approximately $\frac{7}{18}$, and the total area of the earth is approximately 197,000,000 square miles. Determine the approximate number of square miles of land on the earth.

30. Two bicyclists are 4 miles apart and travel toward each other. One bicycles at the rate of 8 miles per hour, and the other bicycles at 12 miles per hour. After how long will they meet?

31. A child hikes along a trail through a forest at 2.5 miles per hour. If a parent sets out after the child half an hour later at a pace of 4 miles per hour, after how long will the parent overtake the child?

32. A jet traveled from Philadelphia to San Francisco at an average velocity of 500 miles per hour, and because of the west-to-east tailwind it made the return trip at an average velocity of 600 miles per hour. The return trip took 48 minutes less than the trip west. What is the distance between Philadelphia and San Francisco?

33. A farmer sets out to walk to town 13 miles away at a rate of 4 miles per hour. After a while the farmer is picked up and driven the rest of the way at an average speed of 40 miles per hour. If the total trip takes 1 hour, how far does the farmer walk?

34. Two runners in the 26-mile Boston marathon travel at rates of 6 and 8 miles per hour, respectively. Suppose that they begin at the same instant.
    a. How far back will the slower runner be when the faster one completes the course?
    b. How long will it take for the two runners to be 4.5 miles apart?

35. One picker can harvest a strawberry patch in 2 hours, and a second picker can harvest the patch in 2.5 hours. How long would it take for both pickers to harvest the patch together?

36. One gasoline truck can fill a storage tank in 20 minutes. Another can fill the same tank in 30 minutes. How long would it take to fill the tank using both trucks simultaneously?

37. Suppose it takes 4 hours for Pam to shovel the snow from the driveway and 4.5 hours for Sam. How long would it take them to do it together, with one shovel each?

38. Suppose Bram joins Pam and Sam, and Bram alone can shovel the driveway in 6 hours. How long would it take the three to shovel the driveway together, with three shovels?

39. Suppose it would take Mary and Jane 3 hours and 45 minutes to paint a porch together. If Mary could paint it alone in 6 hours, how long would it take Jane to paint it alone?

40. Suppose that Mary, Jane, and Barbara can paint a porch in 3 hours together. If it would take 12 hours for Mary to paint it alone and 9 hours for Jane alone, how long would it take Barbara to paint the porch alone?

41. Suppose in Example 6 that 80% antifreeze is added rather than pure antifreeze. How much of the radiator mixture should be drained off and replaced by the new 80% antifreeze liquid to obtain a mixture containing 60% antifreeze?

42. A sample of 20 pounds of sea water has a 10% salt content. How much fresh water must be added to produce a mixture with a 6% salt content?

43. One alloy is 40% silver, and another alloy is 30% silver. How much of each should be used to produce 50 pounds of an alloy that is 36% silver?

44. How much pure alcohol must be added to 4 gallons of a solution that is 60% alcohol to achieve a solution that is 84% alcohol?

45. How much water must be evaporated from 8 gallons of a saline solution with a 20% salt content in order for it to have a 25% salt content?

46. Unleaded gasoline for cars has octane ratings (percent by volume of iso-octane in one gallon of gasoline). Normally, the octane ratings have been approximately 86 for regular gasoline and 92 for super, or ethyl, gasoline. Suppose a gas tank contains 6 gallons of regular with an octane rating of 86. To achieve a mixture with an octane rating of 90, how many gallons of super with an octane rating of 92 must be added?

47. A gambler began with a certain sum of money. In the first investment $\frac{1}{3}$ of the money was lost. In the second investment $\frac{1}{4}$ of the remaining money was lost. In the third investment $\frac{1}{5}$ of the money remaining after the second investment was lost. If a total of $36,000 was lost, how large was the original sum of money?

48. A silversmith starts with a collection of necklaces and a small amount of savings, and he sells his wares at three crafts shows, where he also purchases other items of silver. At the first show, the silversmith doubles his money and then spends $20. At the second, he triples his new sum of money and then spends $40. At the third show, he quadruples his new

sum of money and then spends $150. If the silver-smith now has $650, how much did he originally have?

49. Little is known about the personal life of the math-ematician Diophantus of Alexandria, who is thought to have lived in the third century A.D. However, in an epitaph to Diophantus, it is stated that he lived one sixth of his life in childhood, one twelfth in his youth, and one seventh more as a bachelor. Five years after his marriage a son was born who died four years before Diophantus, at half of Diophantus's final age. Assuming that all this is true, determine the age of Diophantus when he died.

50. The *harmonic mean* of two numbers a and b with nonzero sum is

$$\frac{2ab}{a+b}$$

*a. Suppose a car travels from A to B with speed v_1 and returns with speed v_2. Show that the round trip could be made in the same length of time if the car traveled at a constant speed equal to the harmonic mean of v_1 and v_2.

b. If v_1 in part (a) is 40 miles per hour and v_2 is 60 miles per hour, find the constant speed that yields the same time for the round trip.

2.3 QUADRATIC EQUATIONS AND THE QUADRATIC FORMULA

A *quadratic equation* is an equation of the form

$$ax^2 + bx + c = 0, \quad \text{with } a \neq 0 \tag{1}$$

or any equation that can be put into the form of (1) by shifting all nonzero terms to the left side of the equation. Thus

$$x^2 = 9, \quad x^2 - 4x - 3 = 0, \quad \text{and} \quad 3t^2 = 2t - 1$$

are quadratic equations.

Quadratic equations occur in physical applications. For instance, suppose that a ball is thrown upward at 64 feet per second from a dormitory roof 48 feet above the ground. If the air resistance is negligible, the laws of motion imply that until it hits the ground, the ball's height above the ground x seconds after it is thrown is $-16x^2 + 64x + 48$ feet. Now when the ball hits the ground, its height is 0, so x satisfies

$$-16x^2 + 64x + 48 = 0$$

or equivalently,

$$x^2 - 4x - 3 = 0 \tag{2}$$

But this is a quadratic equation. What value(s) of x satisfy the equation in (2)? You might see if you can answer the question before reading on.

Our goal is to give a method of determining all real solutions of an arbitrary quadratic equation. In contrast to a linear equation, which always has one and only one real solution, a quadratic equation can have two, one, or no real solutions, depending on the constants occurring in the equation.

If we can find real numbers p and q such that $x - p$ and $x - q$ are factors of $ax^2 + bx + c$, that is,

$$ax^2 + bx + c = a(x - p)(x - q) \tag{3}$$

then p and q are evidently solutions of (1), because

$$ap^2 + bp + c = a(\overbrace{p - p}^{0})(p - q) = 0$$

and

$$aq^2 + bq + c = a(q - p)(\overbrace{q - q}^{0}) = 0$$

Morever, p and q are the only solutions of the given equations, because if

$$ax^2 + bx + c = a(x - p)(x - q) = 0$$

then since $a \neq 0$ by hypothesis, the Zero Property (see Section 1.1) implies that either $x - p = 0$ or $x - q = 0$, that is, $x = p$ or $x = q$.

EXAMPLE 1.   Solve the following quadratic equations.
  a. $x^2 + 2x - 3 = 0$ b. $2x^2 - x - 1 = 0$

Solution.
  a. Since

$$x^2 + 2x - 3 = (x - 1)(x + 3)$$

the given equation is equivalent to

$$(x - 1)(x + 3) = 0$$

Therefore the proposed solutions are 1 and -3.

$$Check: \quad 1^2 + 2(1) - 3 = 1 + 2 - 3 = 0$$
$$(-3)^2 + 2(-3) - 3 = 9 - 6 - 3 = 0$$

Consequently 1 and -3 are the solutions of the given equation.
  b. Since

$$2x^2 - x - 1 = (2x + 1)(x - 1) = 2\left(x + \frac{1}{2}\right)(x - 1)$$

the given equation is equivalent to

$$2\left(x + \frac{1}{2}\right)(x - 1) = 0$$

Therefore the proposed solutions are $-\frac{1}{2}$ and 1.

$$Check: \quad 2\left(-\frac{1}{2}\right)^2 - \left(-\frac{1}{2}\right) - 1 = \frac{1}{2} + \frac{1}{2} - 1 = 0$$

$$2(1)^2 - 1 - 1 = 2 - 1 - 1 = 0$$

Consequently $-\frac{1}{2}$ and 1 are the solutions of the given equation.   □

Solving a quadratic equation by factorization, as we did in Example 1, depends on our ability to factor a quadratic polynomial. Now we will derive a formula for the solutions of a quadratic equation that does not depend on factoring but nevertheless is easily applied.

The Quadratic Formula

In order to find the solutions of the equation

$$x^2 - 4x - 3 = 0 \tag{4}$$

in (2), we begin in what may at first seem a roundabout way by adding 3 to both sides, to eliminate the constant term on the left side of the equation. This yields

$$x^2 - 4x = 3$$

Then we add a suitable constant to both sides so that the left side is the square $(x + r)^2$ of a linear polynomial. In this case the constant to add is 4, because $x^2 - 4x + 4 = (x - 2)^2$. Adding 4 to both sides, we obtain

$$x^2 - 4x + 4 = 3 + 4$$

which is equivalent to

$$(x - 2)^2 = 7 \tag{5}$$

To solve for x is now simple. By taking square roots of both sides of (5), we find that

$$x - 2 = \sqrt{7} \quad \text{or} \quad x - 2 = -\sqrt{7}$$

which means that

$$x = 2 + \sqrt{7} \quad \text{or} \quad x = 2 - \sqrt{7}$$

Thus $2 + \sqrt{7}$ and $2 - \sqrt{7}$ are the solutions of the equation in (4).

If x represents time and $x^2 - 4x - 3$ represents the motion of the ball mentioned at the outset of this section, then the time at which the ball hits the ground is a nonnegative number that satisfies the equation $x^2 - 4x - 3 = 0$. Since $2 - \sqrt{7} < 0$ and thus cannot represent any time after the ball was thrown, it follows that the ball hits the ground $2 + \sqrt{7}$ seconds (approximately 4.6 seconds) after it was thrown.

The process leading from the given equation in (4) to the equation in (5) is called *completing the square*, because the terms involving x^2 and x become a part of the complete square of the form $(x + r)^2$.

EXAMPLE 2.   Find all real solutions of the quadratic equation

$$x^2 - 7x + 6 = 0$$

by first completing the square.

Solution.   We follow the procedure detailed above:

$$x^2 - 7x = -6$$

$$x^2 - 7x + \frac{49}{4} = -6 + \frac{49}{4} = -\frac{24}{4} + \frac{49}{4} = \frac{25}{4}$$

$$\left(x - \frac{7}{2}\right)^2 = \frac{25}{4} = \left(\frac{5}{2}\right)^2$$

Taking square roots, we find that

$$x - \frac{7}{2} = \frac{5}{2} \quad \text{or} \quad x - \frac{7}{2} = -\frac{5}{2}$$

so that

$$x = \frac{7}{2} + \frac{5}{2} = \frac{12}{2} = 6 \quad \text{or} \quad x = \frac{7}{2} - \frac{5}{2} = \frac{2}{2} = 1$$

Therefore the proposed solutions are 6 and 1.

> *Check:* $6^2 - 7(6) + 6 = 36 - 42 + 6 = 0$
>
> $1^2 - 7(1) + 6 = 1 - 7 + 6 = 0$

Consequently 6 and 1 are the solutions of the given equation.   □

Notice that the answer obtained in Example 2 by completing the square is the same as would be obtained by factoring using methods of Section 1.7. However, by completing the square we need not use trial and error to arrive at the solution.

To obtain all solutions of a quadratic equation

$$ax^2 + bx + c = 0, \qquad \text{with} \quad a \neq 0 \tag{6}$$

we factor out a from the terms involving x^2 and x, and then complete the square as we did in Example 2:

$$ax^2 + bx + c = 0$$

$$ax^2 + bx = -c$$

$$a\left(x^2 + \frac{b}{a}x\right) = -c$$

$$a\left(x^2 + \frac{b}{a}x + \frac{b^2}{4a^2} - \frac{b^2}{4a^2}\right) = -c$$

$$a\left(x^2 + \frac{b}{a}x + \frac{b^2}{4a^2}\right) - \frac{b^2}{4a} = -c$$

$$a\left(x + \frac{b}{2a}\right)^2 = \frac{b^2}{4a} - c = \frac{b^2 - 4ac}{4a}$$

$$\left(x + \frac{b}{2a}\right)^2 = \frac{b^2 - 4ac}{4a^2} \tag{7}$$

If $b^2 - 4ac \geq 0$, we can find the real solutions of (7) by taking square roots of both sides of (7), which yields

$$x + \frac{b}{2a} = \frac{\sqrt{b^2 - 4ac}}{2a} \quad \text{or} \quad x + \frac{b}{2a} = -\frac{\sqrt{b^2 - 4ac}}{2a}$$

or equivalently,

$$x = -\frac{b}{2a} + \frac{\sqrt{b^2 - 4ac}}{2a} \quad \text{or} \quad x = -\frac{b}{2a} - \frac{\sqrt{b^2 - 4ac}}{2a}$$

Thus if $b^2 - 4ac \geq 0$, the real solutions of (7) and hence of (6) are

$$x = \frac{-b + \sqrt{b^2 - 4ac}}{2a} \quad \text{and} \quad x = \frac{-b - \sqrt{b^2 - 4ac}}{2a} \tag{8}$$

The formulas in (8) are often combined to give one formula:

The Quadratic Formula

$$x = \frac{-b \pm \sqrt{b^2 - 4ac}}{2a}$$

Because the square root has been defined only for nonnegative numbers, the quadratic formula yields real solutions of the quadratic equation $ax^2 + bx + c = 0$ only if $b^2 - 4ac$, called the **discriminant** of the quadratic equation, is nonnegative.

The number of real solutions of $ax^2 + bx + c = 0$ given by the quadratic formula depends on the value of the discriminant:

> i. If $b^2 - 4ac > 0$, the quadratic formula provides two real solutions.
> ii. If $b^2 - 4ac = 0$, the quadratic formula provides one real solution.
> iii. If $b^2 - 4ac < 0$, the quadratic formula provides no real solutions.

In practice, we determine the real solutions of $ax^2 + bx + c = 0$ by applying the quadratic formula in the following way. We first substitute the values of a, b, and c into the formula

$$x = \frac{-b \pm \sqrt{b^2 - 4ac}}{2a} \tag{9}$$

regardless of the value of $b^2 - 4ac$, and then we simplify $b^2 - 4ac$, noting whether it is positive, zero, or negative. If $b^2 - 4ac \geq 0$, we further simplify the right side of (9) when possible to find the value(s) of x; if $b^2 - 4ac < 0$, we conclude that no real solutions exist.

The power of the quadratic formula is that it enables us by numerical computation to tell how many real solutions a given quadratic equation has and to compute the values of any real solutions that exist. The following examples illustrate this point.

EXAMPLE 3.   Find all real solutions of the following equations.
 a. $3x^2 - 4x + 1 = 0$ b. $6x^2 + \sqrt{2}x - 2 = 0$

Solution.
a. By the quadratic formula, any real solutions are given by

$$x = \frac{-(-4) \pm \sqrt{(-4)^2 - 4(3)(1)}}{2(3)} = \frac{4 \pm \sqrt{16 - 12}}{6}$$

$$= \frac{4 \pm \sqrt{4}}{6} = \frac{4 \pm 2}{6}$$

Since $\quad \dfrac{4 + 2}{6} = 1 \quad$ and $\quad \dfrac{4 - 2}{6} = \dfrac{1}{3}$

there are two proposed solutions of the given equation: 1 and $\frac{1}{3}$.

Check: $3(1)^2 - 4(1) + 1 = 3 - 4 + 1 = 0$

$$3\left(\frac{1}{3}\right)^2 - 4\left(\frac{1}{3}\right) + 1 = \frac{1}{3} - \frac{4}{3} + 1 = 0$$

Consequently 1 and $\frac{1}{3}$ are the solutions of the given equation.
b. By the quadratic formula, any real solutions are given by

$$x = \frac{-\sqrt{2} \pm \sqrt{(\sqrt{2})^2 - 4(6)(-2)}}{2(6)} = \frac{-\sqrt{2} \pm \sqrt{2 + 48}}{12}$$

$$= \frac{-\sqrt{2} \pm \sqrt{50}}{12} = \frac{-\sqrt{2} \pm \sqrt{25 \cdot 2}}{12}$$

$$= \frac{-\sqrt{2} \pm \sqrt{25}\sqrt{2}}{12} = \frac{-\sqrt{2} \pm 5\sqrt{2}}{12}$$

Since

$$\frac{-\sqrt{2} + 5\sqrt{2}}{12} = \frac{4\sqrt{2}}{12} = \frac{1}{3}\sqrt{2}$$

and

$$\frac{-\sqrt{2} - 5\sqrt{2}}{12} = \frac{-6\sqrt{2}}{12} = -\frac{1}{2}\sqrt{2}$$

the two proposed solutions of the given equation are $\frac{1}{3}\sqrt{2}$ and $-\frac{1}{2}\sqrt{2}$.

Check: $6\left(\frac{1}{3}\sqrt{2}\right)^2 + \sqrt{2}\left(\frac{1}{3}\sqrt{2}\right) - 2 = 6\left(\frac{2}{9}\right) + \frac{2}{3} - 2 = 0$

$$6\left(-\frac{1}{2}\sqrt{2}\right)^2 + \sqrt{2}\left(-\frac{1}{2}\sqrt{2}\right) - 2$$

$$= 6\left(\frac{1}{2}\right) - 1 - 2 = 0$$

Consequently $\frac{1}{3}\sqrt{2}$ and $-\frac{1}{2}\sqrt{2}$ are the solutions of the given equation. □

EXAMPLE 4.   Find all real solutions of the following equations.
  a. $16x^2 + 24x + 9 = 0$ b. $3t^2 = 2t - 1$

Solution.
a. By the quadratic formula, any real solutions are given by

$$x = \frac{-24 \pm \sqrt{(24)^2 - 4(16)(9)}}{2(16)}$$

$$= \frac{-24 \pm \sqrt{576 - 576}}{32} = \frac{-24 \pm 0}{32} = -\frac{3}{4}$$

Thus there is one proposed solution of the given equation: $-\frac{3}{4}$.

$$Check: \quad 16\left(-\frac{3}{4}\right)^2 + 24\left(-\frac{3}{4}\right) + 9$$

$$= 16\left(\frac{9}{16}\right) - 18 + 9 = 0$$

Consequently $-\frac{3}{4}$ is the solution of the equation.

b. If we subtract $2t - 1$ from both sides of the given equation, we obtain the equivalent equation

$$3t^2 - 2t + 1 = 0$$

whose solutions, if any, are precisely the solutions of the given equation. But by the quadratic formula, any real solutions are given by

$$t = \frac{-(-2) \pm \sqrt{(-2)^2 - 4(3)(1)}}{2(3)}$$

$$= \frac{2 \pm \sqrt{4 - 12}}{6} = \frac{2 \pm \sqrt{-8}}{6}$$

Since $-8 < 0$, there are no real solutions. $\square$

One consequence of the quadratic formula is the fact that any quadratic equation

$$ax^2 + bx + c = 0$$

for which a and c have opposite signs has two real solutions. The reason is that in this case, $-4ac > 0$, so that the discriminant $b^2 - 4ac > 0$. In contrast, if a and c have the same sign, then the equation can have two, one, or no real solutions. These possibilities occur in Examples 3 and 4.

Occasionally we encounter an equation that is not quadratic but can be transformed into an equivalent quadratic equation by means of an algebraic manipulation.

EXAMPLE 5.   Find all real solutions of $1 + \dfrac{1}{x^2} = \dfrac{3}{x}$.

Solution.   We multiply both sides of the equation by x^2 and then solve the resulting quadratic equation:

$$x^2 + 1 = 3x$$

$$x^2 - 3x + 1 = 0$$

By the quadratic formula,

$$x = \frac{-(-3) \pm \sqrt{(-3)^2 - 4(1)(1)}}{2(1)} = \frac{3 \pm \sqrt{9 - 4}}{2} = \frac{3 \pm \sqrt{5}}{2}$$

Therefore the proposed solutions are $(3 + \sqrt{5})/2$ and $(3 - \sqrt{5})/2$. We could check here that these are actually the solutions of the given equation, but the calculations are tedious. $\square$

In conclusion, we emphasize the close connection between nontrivial factors of the quadratic polynomial $ax^2 + bx + c$ and solutions of the quadratic equation $ax^2 + bx + c = 0$. Indeed, if

$$ax^2 + bx + c = a(x - p)(x - q)$$

then p and q are the solutions of the equation $ax^2 + bx + c = 0$. If $b^2 - 4ac \geq 0$, then conversely, the solutions p and q of the equation $ax^2 + bx + c = 0$ that are given by

$$p = \frac{-b + \sqrt{b^2 - 4ac}}{2a} \quad \text{and} \quad q = \frac{-b - \sqrt{b^2 - 4ac}}{2a}$$

provide the factorization

$$ax^2 + bx + c = a(x - p)(x - q)$$

of the quadratic polynomial $ax^2 + bx + c$.

EXERCISES 2.3

In Exercises 1–24, solve the given equation by rearranging (if necessary) and then factoring.

1. $x^2 = 16$

2. $x^2 - 121 = 0$

3. $4x^2 - 36 = 0$

4. $3x^2 - 25 = 0$

5. $x^2 + 4x + 4 = 0$

6. $x^2 - x - 2 = 0$

7. $x^2 - 50x + 625 = 0$

8. $y^2 - 2y = 3$

9. $9y^2 - 12y + 4 = 0$

10. $y^2 + 3y - 4 = 0$

11. $y^2 - 2\sqrt{2}y = -2$

12. $2y^2 + 18 = -12y$

13. $4x^2 + 2x + \frac{1}{4} = 0$

14. $9x^2 + \frac{1}{9} = 2x$

15. $5x + 4 = -x^2$

16. $6x - 9x^2 = 1$

17. $x^2 + 3x + 5 = 15$

18. $t^2 + t = 0$

19. $2t^2 + 42 = 20t$

20. $3x + 18 = x^2$

21. $5x = \frac{3}{x}$

22. $16 + \frac{25}{x^2} = \frac{40}{x}$

23. $2 + \frac{1}{x^2} = \frac{3}{x}$

24. $25x + \frac{4}{x} = 20$

In Exercises 25–30, solve the given equation by completing the square.

25. $x^2 - 2x - 6 = 0$

26. $y^2 = 4y + 12$

27. $y^2 - 10y = 2$

28. $3t^2 - 6t = 12$

29. $2x^2 + 4x - 3 = 0$

30. $4x - x^2 = -3$

In Exercises 31–54, find all real solutions (if any) of the given equation by using the quadratic formula.

31. $x^2 - 3x + 1 = 0$

32. $x^2 + 3x + 1 = 0$

33. $x^2 + x + 1 = 0$

34. $x^2 + x - 4 = 0$

35. $x^2 - 2x = 4$

36. $x^2 = 4\sqrt{3}x - 12$

37. $3 + 2x = 2x^2$

38. $2x^2 + 3x - 5 = 0$

39. $1 - 6w + 3w^2 = 0$

40. $6w^2 + 5w = 1$

41. $6x^2 + 5x = -1$

42. $3x^2 - 4\sqrt{3}x + 4 = 0$

43. $3x^2 - 4\sqrt{3}x - 4 = 0$

44. $5y^2 + 7y + 5 = 0$

45. $4y^2 + 4\sqrt{2}y = 2$

46. $4y^2 + 4\sqrt{2}y + 2 = 0$

47. $t^2 - t + 3 = 0$

48. $t + \dfrac{1}{t} = 6$

51. $\dfrac{x-1}{2x-3} = \dfrac{x+2}{x-2}$

49. $\dfrac{5}{t^2} + \dfrac{3}{t} - 1 = 0$

52. $\dfrac{x}{x^2-1} = 4$

50. $\dfrac{3}{t^2} - \dfrac{5}{t} = 2$

53. $(x-1)(x-2) = 3(x+1) + 4$

54. $(x+3)(x+7) = 3(x-2) + 5$

55. Find the values of b for which there is exactly one solution of the equation $x^2 + bx + 9 = 0$.

56. Find all values of b such that $x^2 + bx + 5 = 0$ has exactly one real solution.

57. Find the value of b such that the real solutions of $x^2 + bx - 8 = 0$ are negatives of each other.

58. Show that no matter what the value of b is, there are two distinct real solutions of $x^2 + bx - 9 = 0$.

59. Find the values of b for which the real solutions of $x^2 + bx - 8 = 0$ differ by 6.

60. Show that if x_1 and x_2 are real solutions of the equation $x^2 + bx + c = 0$, then $x_1 + x_2 = -b$ and $x_1 x_2 = c$.

▣ In Exercises 61–64, use a calculator and the quadratic formula to approximate the solutions of the given equation.

61. $3x^2 - 17x - 23 = 0$

62. $4x^2 + 100x + 3 = 0$

63. $3.14x^2 - 1.3x - 4.59 = 0$

64. $x^2 - 3\sqrt{2}x + \sqrt{5} = 0$

65. The area A of a circle of radius r is given by $A = \pi r^2$. Solve the equation for r.

66. The surface area S of a sphere of radius r is given by $S = 4\pi r^2$. Solve the equation for r.

67. The volume V of a cylinder of radius r and height h is given by $V = \pi r^2 h$. Solve the equation for r.

68. The area A of an equilateral triangle of side s is given by $A = \sqrt{3}s^2/4$. Solve the equation for s.

69. The ancient Greeks thought that the rectangles with the most pleasing shapes are those for which the ratio of the width to the length is the same as the ratio of the length to the sum of the length and the width, that is,

$$\frac{w}{l} = \frac{l}{l+w} = \frac{1}{1 + \dfrac{w}{l}}$$

where l is the length and w is the width of the rectangle. If we let $x = w/l$, the above equation becomes

$$x = \frac{1}{1+x} \qquad (10)$$

The ratio x is called the **golden ratio** (or **golden section**) because of the interpretation the ancient Greeks associated with it. Determine the golden ratio by solving (10).

**2.4
APPLICATIONS
OF QUADRATIC
EQUATIONS**

In Section 2.2 we solved applied problems that led to linear equations when translated into mathematical language. However, not all applied problems can be solved by means of linear equations. In this section we will consider problems that lead to quadratic equations.

   The general method of solving applied problems used in Section 2.2 still applies. However, in order to facilitate working the examples, we will not write out our checks of the proposed solutions of the equations. Nevertheless, we will need to see which of any proposed solutions qualify as solutions of the given applied problems.

Vertical Motion

Suppose that an object is thrown or propelled or dropped, and from that moment on moves vertically under the sole influence of gravity. For all practical purposes, this would be the case, for example, if a person jumps from a diving tower or throws a rock straight up.

Let us suppose that time is measured in seconds and that at some convenient moment (such as when the object begins moving under the sole influence of gravity) the time is 0. Suppose also that the height is measured in feet, with h_0 the **initial height** above ground level at the instant $t = 0$, and finally, assume that the velocity is measured in feet per second, with v_0 denoting the velocity at time $t = 0$. Then v_0 is the **initial velocity** of the object. We take v_0 to be positive if the object is moving upward at time $t = 0$ and negative if the object is moving downward at that time. For example, if a ball is hurled upward at 40 feet per second, then $v_0 = 40$, whereas if the ball is thrown downward at 40 feet per second, then $v_0 = -40$. If the ball is dropped (from rest), then $v_0 = 0$.

The laws of motion imply that the height h of such an object above ground level at time $t \geq 0$ is given by

$$h = -16t^2 + v_0 t + h_0 \qquad (1)$$

It follows that if the constants h_0 and v_0 are known and if h is given, then (1) is a quadratic equation in the variable t. Of course, (1) is valid only while the object is under the sole influence of gravity (for instance, before it hits the ground). By using (1) we can solve many problems concerning vertical motion.

EXAMPLE 1.   Suppose a swimmer dives from a platform 64 feet above a deep pool. How long will the swimmer fall before touching the water?

Solution.   We take $t = 0$ at the instant the swimmer begins to descend, the initial velocity v_0 to be 0, and the initial height h_0 to be 64. Since the height h will be 0 at the instant the swimmer first touches the water, we must find the time at which $h = 0$. To accomplish this we use (1) with $h = 0$, $v_0 = 0$, and $h_0 = 64$:

$$0 = -16t^2 + (0)t + 64$$

which reduces to

$$16t^2 = 64$$

or

$$t^2 = 4$$

Therefore $t = 2$ or $t = -2$. Consequently 2 and -2 are proposed solutions of the problem. Since $t = 0$ when the swimmer starts the descent, the proposed solution -2, which corresponds to a time *before* the dive, does not apply. Thus the swimmer touches the water after 2 seconds.   □

Olympic champion diver Greg Louganis executing a high dive.

Caution: In Example 1 we derived the mathematical equation $t^2 = 4$ from the given physical conditions. After we found the solutions 2 and -2 of the equation, we rejected the solution -2. It was not rejected because it was not a solution of the equation, but because it did not satisfy the physical condition that $t \geq 0$. Therefore in solving an applied problem it is critical that we scrutinize solutions arising from the mathematics in order to see which if any are actually solutions of the given applied problem, that is, which solutions actually satisfy the conditions relating to the given applied problem.

EXAMPLE 2.   A rocket loaded with fireworks is to be shot vertically upward from the ground with an initial velocity of 160 feet per second. The fireworks are to be detonated at a height of 384 feet while they are still on the rise. How long after takeoff should detonation occur?

Solution.   Let $t = 0$ at takeoff. Our problem is to find the time at which the height h of the rocket is 384 feet and the rocket is rising. We are assuming that the rocket is shot upward from ground level, which means that $h_0 = 0$. By assumption the initial velocity is given by $v_0 = 160$, so that by (1) we need to solve the equation

$$384 = -16t^2 + 160t + 0$$

or

$$16t^2 - 160t + 384 = 0 \qquad (2)$$

To solve (2) we have the following equivalent equations:

$$16(t^2 - 10t + 24) = 0$$
$$t^2 - 10t + 24 = 0$$
$$(t - 4)(t - 6) = 0$$

Therefore $t = 4$ or $t = 6$. Thus the rocket is at a height of 384 feet at 4 seconds and again at 6 seconds. Since the rocket is ascending at the former time and descending at the latter time, detonation should occur 4 seconds after the rocket is launched.   □

Geometric Applications

One of the most important results known from antiquity is the Pythagorean Theorem, named after the outstanding Greek mathematician Pythagoras, who lived around 550 B.C. Recall that a right triangle is a triangle two of whose sides are perpendicular to each other. Assume that the lengths of these two sides, called the **legs** of the triangle, are a and b (Figure 2.3) and the length of the third side, called the **hypotenuse** of the triangle, is c. Then the lengths of the three sides are related by the following formula:

Pythagorean Theorem
$a^2 + b^2 = c^2$

The Pythagorean Theorem is basic to arguments involving right triangles.

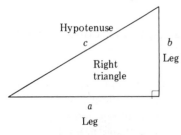

The Pythagorean Theorem:
$$a^2 + b^2 = c^2$$

FIGURE 2.3

EXAMPLE 3.   Suppose that the lengths of the legs of a right triangle differ by one inch, and the hypotenuse is one inch longer than the longer leg. Determine the lengths of all the sides of the triangle.

Solution.   Let

$$x = \text{the length in inches of the shorter leg}$$

Then the other leg is $x + 1$ inches long, and the hypotenuse is $x + 2$ inches long. By the Pythagorean Theorem,

$$x^2 + (x + 1)^2 = (x + 2)^2$$

We solve this equation by using the following equivalent equations:

$$x^2 + (x^2 + 2x + 1) = x^2 + 4x + 4$$

$$2x^2 + 2x + 1 = x^2 + 4x + 4$$

$$x^2 - 2x - 3 = 0$$

$$(x - 3)(x + 1) = 0$$

Therefore $x = 3$ or $x = -1$. Since all lengths must be nonnegative, it follows that the shorter leg is 3 inches long. Consequently the other leg is 4 inches long, and the hypotenuse is 5 inches long.   □

Miscellaneous Applications

EXAMPLE 4.   A rectangular painting has an area of 200 square inches. A frame 1 inch wide is added to each side of the painting, making the total area of painting and frame 261 square inches (Figure 2.4). What were the dimensions of the painting before the addition of the frame?

Solution.   Let

$$x = \text{the width of the painting before framing}$$

$$y = \text{the height of the painting before framing}$$

Then

$$xy = 200$$

so that

$$y = \frac{200}{x} \tag{3}$$

After the frame has been added, the width becomes $x + 2$ and the height becomes $y + 2$. Since the area of the painting plus the frame is by assumption 261 square inches, this means that

$$(x + 2)(y + 2) = 261 \tag{4}$$

Equation (4) has two variables, x and y. Since they are related by (3), we substitute for y in (4) to obtain

$$(x + 2)\left(\frac{200}{x} + 2\right) = 261$$

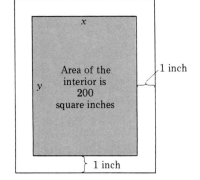

Area of the interior is 200 square inches

1 inch

1 inch

x

y

FIGURE 2.4

which is an equation in the single variable x, and hence an equation we know how to treat. Multiplying out and then multiplying both sides by x, we obtain the following equivalent equations:

$$200 + \frac{400}{x} + 2x + 4 = 261$$

$$2x - 57 + \frac{400}{x} = 0$$

$$2x^2 - 57x + 400 = 0$$

By the quadratic formula,

$$x = \frac{-(-57) \pm \sqrt{(-57)^2 - 4(2)(400)}}{2(2)} = \frac{57 \pm \sqrt{3249 - 3200}}{4}$$

$$= \frac{57 \pm \sqrt{49}}{4} = \frac{57 \pm 7}{4}$$

Since

$$\frac{57 + 7}{4} = \frac{64}{4} = 16 \quad \text{and} \quad \frac{57 - 7}{4} = \frac{50}{4} = \frac{25}{2}$$

it follows that $x = 16$ or $x = \frac{25}{2}$. Now by (3),

$$\text{if } x = 16, \quad \text{then } y = \frac{200}{16} = \frac{25}{2}$$

$$\text{if } x = \frac{25}{2}, \quad \text{then } y = \frac{200}{25/2} = 16$$

Consequently the dimensions of the painting before framing were 16 inches by $\frac{25}{2}$ inches.   □

EXAMPLE 5.   Pat drives the 432 miles between Boston and Washington, D.C., in one hour less than Dean and at an average speed of 6 miles per hour more than Dean. How fast does each drive?

Solution.   Let

$$v = \text{Dean's speed}$$

so that Pat's speed is $v + 6$. Since $d = rt$ (see (4) of Section 2.2), the time required for Dean to drive 432 miles is $432/v$, and the time for Pat is $432/(v + 6)$. Since Pat drives the distance in one hour less than Dean, this means that

$$\frac{432}{v + 6} = \frac{432}{v} - 1$$

Multiplying each side by $v(v + 6)$, we find that

$$432v = 432(v + 6) - v(v + 6)$$
$$432v = 432v + 2592 - v^2 - 6v$$
$$v^2 + 6v - 2592 = 0$$

By the quadratic formula,

$$v = \frac{-6 \pm \sqrt{6^2 - 4(1)(-2592)}}{2(1)} = \frac{-6 \pm \sqrt{36 + 10{,}368}}{2}$$
$$= \frac{-6 \pm \sqrt{10{,}404}}{2} = \frac{-6 \pm \sqrt{4 \cdot 9 \cdot 289}}{2} = \frac{-6 \pm 102}{2}$$

Since

$$\frac{-6 + 102}{2} = \frac{96}{2} = 48 \quad \text{and} \quad \frac{-6 - 102}{2} = -\frac{108}{2} = -54$$

it follows that $v = 48$ or $v = -54$. Because speed must be nonnegative, we conclude that Dean travels at a rate of 48 miles per hour. Pat therefore travels at the rate of $48 + 6 = 54$ miles per hour. $\square$

EXERCISES 2.4

1. A ball is dropped from a balcony 48 feet above the ground. How long will it take for the ball to hit the ground?

2. An acrobat drops from a platform 180 feet above a deep pool. How long will it take for the acrobat to reach the pool?

3. A skyscraper window cleaner loses a pail 256 feet above the ground. How many seconds later does it pass by a window at the 112-foot level?

4. A rock is thrown down from a bridge 96 feet above the water. If the initial velocity is -16 feet per second, how long does it take the rock to hit the water?

5. A paintbrush is thrown straight up with a velocity of 48 feet per second toward a painter 32 feet higher up. How long a wait does the painter have before catching it?

6. A ball is thrown downward from a bridge 192 feet above a river. After 1 second the ball has traveled 80 feet. After how many more seconds will it hit the water?

7. The legs of a right triangle have lengths 5 and 6 inches. What is the length of the hypotenuse?

8. The length of the hypotenuse of a certain right triangle is 1 inch longer than one leg and is 8 inches longer than the other leg. Find the lengths of all three sides of the triangle.

9. The hypotenuse of a right triangle is 10 meters long, and the length of one leg exceeds the length of the other by one meter. How long is each leg?

10. A baseball diamond is a square 90 feet on a side. What is the distance from home plate to second base?

11. A plot of land is in the shape of a right triangle whose hypotenuse is 1300 feet long and one of whose legs is 1200 feet long. How many feet of fencing are required to enclose the plot?

12. A ladder 13 feet long is placed so that its base lies 5 feet from a wall and its top is 12 feet above the ground. If the ladder slips so that the base is 7 feet from the wall, how far does the top of the ladder fall?

13. A car travels 8 miles per hour faster than a truck and travels 160 miles in one hour less than the truck. How fast is the car travelling?

14. Two jets leave St. Louis at the same time, one traveling north at 520 miles per hour and the other traveling west at 450 miles per hour. How far apart are they after 2 hours?

15. Two airplanes begin 1000 miles apart and fly along lines that intersect at right angles. One plane flies an average of 100 miles per hour faster than the other. If the planes meet after 2 hours, how fast do the airplanes fly?

16. A jet leaves Chicago at noon and travels south at 600 miles per hour. One hour later a second jet leaves Chicago and travels east at 520 miles per hour. How far apart are the jets at 2:30 P.M.?

17. A pilot wishes to make a round trip between Los Angeles and San Francisco in 5 hours. The distance between the airports is 420 miles, and it is anticipated that there will be a head wind of 30 miles per hour as the plane flies north and a tail wind of 40 miles per hour when the plane returns. At what constant air speed must the plane be flown to achieve the goal?

18. A baker must make a delivery by truck to a store 12.5 miles from the bakery. By increasing the normal speed of the truck by 5 miles per hour, the baker could reduce the delivery time by 5 minutes. How fast does the baker normally drive?

19. The average speed of a commuter on a 20-mile trip into downtown Chicago is 16 miles per hour slower at rush hour than at midday, and the trip takes 20 minutes longer. What are the two rates?

20. Find two consecutive odd integers whose product is 195.

21. Find two consecutive integers whose product is 1056.

22. One positive number is 3 more than twice a second positive number and the sum of their squares is 194. What are the numbers?

23. Find the two points on the y axis that are a distance of 6 units from the point $(4, 0)$ on the x axis.

24. A rectangular garden has an area of 1750 square feet and is 15 feet longer than it is wide. Find its dimensions.

25. The perimeter of a rectangle is 32 inches, and the area is 63 square inches. What are the dimensions of the rectangle?

26. A rectangular corral adjoins a barn. The corral has an area of 2352 square feet and is enclosed by a fence 140 feet long. If the barn forms one side of the corral, find the possible dimensions of the corral.

27. A rectangular lawn is 80 feet long and 60 feet wide. How wide a strip must be mowed around the lawn for half of the lawn to be cut?

28. A rectangular cloth measures 20 inches by 24 inches. We wish to embroider a strip of equal width on each side of the cloth in such a way that the cloth with embroidery will be rectangular with an area of 672 square inches. How wide a strip must we embroider?

29. A wire 44 inches long is cut into two pieces, each of which is bent into the form of a square. If the sum of the areas of the squares is 65 square inches, how long were the pieces of wire?

30. A square sheet of metal has sides of length 8 inches. A square piece is cut from each of the corners, and the edges are folded up to form a pan (Figure 2.5). If the area of the base of the pan is equal to the sum of the areas of the sides of the pan, what is the length of the sides of the squares that were cut out?

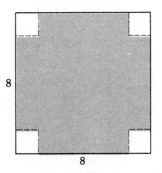

8

8

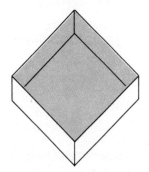

FIGURE 2.5

31. The volume of a cylindrical tin can is 48π cubic inches. If the can is 3 inches tall, find the radius of the base. (*Hint:* The volume is given by $V = \pi r^2 h$, where V, r, and h denote the volume, radius of the base, and height, respectively.)

32. A circular pool covers an area of 400π square feet. A path of constant width is to surround the pool. If the enlarged area of pool and path is 676π square feet, how wide must the path be?

33. On a certain rainy day the manager of a store estimates that if the price of umbrellas is set at p dollars, then $10(10 - p)$ umbrellas will be sold. Suppose the store acquires umbrellas at a cost of $3 each.

    a. What are the prices at which the store can sell umbrellas and neither make nor lose money on the sale (that is, at which the profit is 0)? How many umbrellas will be sold at those prices?

    b. What is the minimum price at which the store can sell umbrellas and make a profit of exactly $100? How many umbrellas will be sold at that price?

34. One can calculate the distance from ground level to water level in a well by dropping a stone from the ground and measuring the time t_0 elapsed until the splash is heard at ground level. Now $t_0 = t_1 + t_2$, where t_1 is the time it takes for the stone to hit the water and t_2 is the time it takes the echo to return to ground level. If the distance is s, then $s = 16t_1^2$, so that

$$t_1 = \sqrt{\frac{s}{16}} = \frac{1}{4}\sqrt{s}$$

With the assumption that sound travels at 1100 feet per second, the distance is also given by $s = 1100t_2$, which means that

$$t_2 = \frac{s}{1100}$$

Consequently

$$t = t_1 + t_2 = \frac{1}{4}\sqrt{s} + \frac{s}{1100}$$

    a. Calculate the distance if the time between drop and echo is $6\frac{9}{11}$ seconds.

    c b. Use a calculator to approximate the distance if the time between drop and echo is 5 seconds.

35. When an 18-foot tall stalk of bamboo is broken, the top portion of the stalk bends over and touches the ground 6 feet from the base of the stalk (Figure 2.6). How far from the ground was the bamboo broken? (This problem appeared in a Chinese algebra book in 1261.)

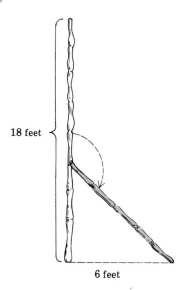

18 feet

6 feet

FIGURE 2.6

36. If a cannon is fired at an angle of $30°$ with respect to the ground, then until the ball hits the ground, the height y of the cannon ball after t seconds is given by $y = \frac{1}{2}v_0 t - 16t^2$, where v_0 is the initial speed of the ball. Determine how long the cannon ball is in the air before hitting the ground.

*37. A geometric proof of the Pythagorean Theorem is suggested by Figure 2.7. Let the hypotenuse have length c and the legs lengths a and b with $a \le b$.

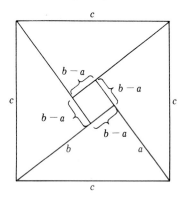

c

$b - a$

$b - a$

c

c

$b - a$

$b - a$

b

a

c

FIGURE 2.7

Assuming that a square of side c can be disected into four copies of the right triangle and a square of side $b - a$, and using the fact that the area of the triangle is $\frac{1}{2}ab$, prove that $c^2 = a^2 + b^2$.

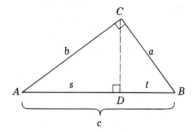

FIGURE 2.8

*38. A proof of the Pythagorean Theorem that uses similar triangles is suggested by Figure 2.8. Let the given right triangle have sides of length a, b, and c, with c the length of the hypotenuse. Using the fact that triangle ABC is similar to each of triangles ADC and BCD, prove that $c^2 = a^2 + b^2$. (*Hint:* Find expressions for a^2 and b^2 using similar triangles, and then add them, using the fact that $s + t = c$.)

2.5 OTHER TYPES OF EQUATIONS

In the first four sections of this chapter we discussed linear and quadratic equations and found formulas for their solutions. Rather than moving on to third-degree equations, whose solutions are generally much more difficult to obtain, we will devote this section to several general types of equations that can be modified and then solved by methods we have already described.

Quadratic-Type Equations

Each of the equations

$$(3 + x)^2 - 5(3 + x) + 4 = 0,$$

$$x^4 - 7x^2 + 10 = 0,$$

and

$$x - 4\sqrt{x} + 4 = 0$$

has the feature that if we substitute u for a suitable expression in x, the resulting equation in u is a quadratic equation. As a result, such an equation is called an **equation of quadratic type**. It is solved by first solving the associated quadratic equation in u and then solving for x in terms of u.

EXAMPLE 1.   Find all solutions of $(3 + x)^2 - 5(3 + x) + 4 = 0$.

Solution.   If we substitute u for $3 + x$, the equation becomes

$$u^2 - 5u + 4 = 0$$

which can be factored to yield

$$(u - 4)(u - 1) = 0$$

Therefore $u = 4$ or $u = 1$. Since $u = 3 + x$, we find that $x = u - 3$. Consequently if $u = 4$ then $x = 4 - 3 = 1$, and if $u = 1$ then $x = 1 - 3 = -2$. It follows that 1 and -2 are the proposed solutions of the given equation.

Check: $(3 + 1)^2 - 5(3 + 1) + 4 = 16 - 20 + 4 = 0$

$(3 + (-2))^2 - 5(3 + (-2)) + 4 = 1 - 5 + 4 = 0$

Thus 1 and -2 are the solutions. $\square$

EXAMPLE 2.   Find all solutions of $x^4 - 7x^2 + 10 = 0$.

Solution.   This time we substitute u for x^2, which in particular implies that $u \geq 0$. The given equation becomes

$$u^2 - 7u + 10 = 0, \quad \text{with} \quad u \geq 0$$

which yields $(u - 5)(u - 2) = 0, \quad \text{with} \quad u \geq 0$

The solutions of the associated equation in u are 5 and 2, both of which are positive. Since $u = x^2$, it follows that $x = \sqrt{u}$ or $x = -\sqrt{u}$, so that the proposed solutions of the given equation are $\sqrt{5}$, $-\sqrt{5}$, $\sqrt{2}$, and $-\sqrt{2}$.

$$\begin{aligned}
\textit{Check:} \quad (\sqrt{5})^4 - 7(\sqrt{5})^2 + 10 &= 25 - 35 + 10 = 0 \\
(-\sqrt{5})^4 - 7(-\sqrt{5})^2 + 10 &= 25 - 35 + 10 = 0 \\
(\sqrt{2})^4 - 7(\sqrt{2})^2 + 10 &= 4 - 14 + 10 = 0 \\
(-\sqrt{2})^4 - 7(-\sqrt{2})^2 + 10 &= 4 - 14 + 10 = 0
\end{aligned}$$

Thus $\sqrt{5}$, $-\sqrt{5}$, $\sqrt{2}$, and $-\sqrt{2}$ are the solutions of the given equation. □

If the equation given in Example 2 had been

$$x^4 + 7x^2 + 10 = 0 \tag{1}$$

then we would have substituted u for x^2 as before, and the associated equation in u would have been

$$u^2 + 7u + 10 = 0, \quad \text{with} \quad u \geq 0$$

so that

$$(u + 5)(u + 2) = 0, \quad \text{with} \quad u \geq 0 \tag{2}$$

But both of the solutions -5 and -2 of the equation in (2) are negative, so there are no nonnegative solutions that satisfy (2). Consequently there are no real solutions of the equation in (1).

EXAMPLE 3.   Find all solutions of $x - 4\sqrt{x} + 4 = 0$.

Solution.   If we substitute u for $\sqrt{x}$ and note that $u \geq 0$, then we find that the given equation is transformed into

$$u^2 - 4u + 4 = 0, \quad \text{with} \quad u \geq 0$$

which yields $(u - 2)^2 = 0, \quad \text{with} \quad u \geq 0$

Therefore $u = 2$. Since $u = \sqrt{x}$, we have $x = u^2$, so that if $u = 2$, then $x = 2^2 = 4$. Consequently the proposed solution of the given equation is 4.

Check: $4 - 4\sqrt{4} + 4 = 4 - 8 + 4 = 0$

Thus 4 is the only solution of the given equation.   □

Equations Involving Radicals

As illustrated in Example 3, an equation involving radicals occasionally arises. Often the solutions are most easily ascertained if we first eliminate the radicals by substitution or by squaring both sides of the equation and then solve the new equation. But we must be careful to check each solution of the new equation to determine which, if any, yield solutions of the original equation.

EXAMPLE 4.   Find all solutions of $\sqrt{3x + 4} = 5$.

Solution.   Let us square both sides of the equation and then solve for x:

$$\sqrt{3x + 4} = 5$$
$$3x + 4 = 25$$
$$3x = 21$$
$$x = \frac{21}{3} = 7$$

Check: $\sqrt{3(7) + 4} = \sqrt{25} = 5$

Therefore 7 is the only solution of the given equation.   □

EXAMPLE 5.   Find all solutions of the equation $\sqrt{4 - 3x} = x$.

Solution.   Following the same procedure of squaring both sides of the equation, we obtain

$$\sqrt{4 - 3x} = x$$
$$4 - 3x = x^2$$
$$x^2 + 3x - 4 = 0$$
$$(x - 1)(x + 4) = 0$$

Therefore the proposed solutions are 1 and -4.

Check: $\sqrt{4 - 3(1)} = 1$

$\sqrt{4 - 3(-4)} = \sqrt{16} = 4 \neq -4$, so -4 is not a solution.

Consequently the only real solution of the given equation is 1.   □

In working the preceding example, we altered the original equation and found two solutions of a new equation, but one of these turned out not to be a real solution of the original equation! What went wrong? The answer is that we squared both sides of the original equation, and as frequently happens, squaring both sides of the equation did *not* lead to an equivalent equation. For a simple example in which squaring leads to a nonequivalent equation, observe that

$$x = 2$$

is *not* equivalent to the equation obtained when both sides are squared:

$$x^2 = 4$$

Indeed, $x = 2$ has one solution, 2, whereas $x^2 = 4$ has two solutions, 2 and -2.

All solutions of a given equation are retained when we square both sides, but other candidates, which actually are not solutions, may be introduced by squaring. A solution of an altered equation that does not satisfy the original equation is called an ***extraneous solution***. Thus -4 is an extraneous solution of the equation $\sqrt{4 - 3x} = x$ of Example 5. As we have seen, extraneous solutions can be introduced when we square both sides of an equation. The same is true if we raise both sides of an equation to any *even* power (see Exercise 63).

Caution: Because of the possibility of introducing extraneous solutions when we raise both sides of a given equation to an even power, it is doubly important to check proposed solutions in these cases.

In the next example we will cube each side of an equation. Yet no extra candidates for solutions will be introduced. More generally, when both sides of an equation are raised to an *odd* power, no extraneous real solutions are introduced.

EXAMPLE 6. Find all solutions of $\sqrt[3]{x^2 - 8} = 2$.

Solution. We cube both sides of the equation and then simplify:

$$\sqrt[3]{x^2 - 8} = 2$$
$$x^2 - 8 = 8$$
$$x^2 = 16$$

Therefore the proposed solutions are 4 and -4.

Check: $\sqrt[3]{4^2 - 8} = \sqrt[3]{8} = 2$
$$\sqrt[3]{(-4)^2 - 8} = \sqrt[3]{8} = 2$$

Consequently 4 and -4 are the solutions of the given equation. □

In the following example we will square both sides of an equation twice, because squaring once does not clear the equation of radicals.

EXAMPLE 7.   Find the solutions of $\sqrt{x+2} - \sqrt{3x-5} = 1$.

Solution.   First we alter the equation so that one radical appears on each side of the equation. Then we square both sides and simplify:

$$\sqrt{x+2} - \sqrt{3x-5} = 1$$
$$\sqrt{x+2} = 1 + \sqrt{3x-5}$$
$$x+2 = 1 + 2\sqrt{3x-5} + (3x-5)$$
$$-2x + 6 = 2\sqrt{3x-5}$$
$$-x + 3 = \sqrt{3x-5}$$

Squaring again, we obtain

$$x^2 - 6x + 9 = 3x - 5$$
$$x^2 - 9x + 14 = 0$$
$$(x-2)(x-7) = 0$$

Thus the proposed solutions are 2 and 7.

$$Check: \quad \sqrt{2+2} - \sqrt{3(2)-5} = \sqrt{4} - \sqrt{1} = 2 - 1 = 1$$
$$\sqrt{7+2} - \sqrt{3(7)-5} = \sqrt{9} - \sqrt{16}$$
$$= 3 - 4 = -1 \neq 1$$

Therefore 2 is a genuine solution of the given equation, whereas 7 is an extraneous solution. We conclude that the given equation has one real solution, 2.   □

Equations Involving Integral Exponents

Some equations involving integral exponents can be solved for a variable by taking roots. The next two examples illustrate this feature.

EXAMPLE 8.   Solve the following equations for x.
  a. $(x^2 - 1)^3 = 27$ b. $81x^4 = 16$

Solution.
  a. We take cube roots and then solve for x:

$$x^2 - 1 = \sqrt[3]{27} = 3$$
$$x^2 = 3 + 1 = 4$$
$$x = 2 \text{ or } -2$$

Check: $(2^2 - 1)^3 = 3^3 = 27$
$[(-2)^2 - 1]^3 = 3^3 = 27$

Consequently the solutions are 2 and -2.

b. Here we divide both sides by 81 and then take fourth roots:

$$x^4 = \frac{16}{81}$$

$$|x| = \sqrt[4]{\frac{16}{81}} = \sqrt[4]{\frac{2^4}{3^4}} = \frac{2}{3}$$

Therefore the proposed solutions are $\frac{2}{3}$ and $-\frac{2}{3}$.

Check: $81\left(\frac{2}{3}\right)^4 = 81\left(\frac{16}{81}\right) = 16$
$81\left(-\frac{2}{3}\right)^4 = 81\left(\frac{16}{81}\right) = 16$

Consequently the solutions are $\frac{2}{3}$ and $-\frac{2}{3}$. $\square$

EXAMPLE 9.   Let T be the time in hours required for a planet to orbit its sun once, and let a be the average distance in miles between the planet and its sun. Kepler's Third Law states that $T^2 = ca^3$, where c is a nonzero constant. Solve the equation for a.

Solution.   From the equation $T^2 = ca^3$ we have

$$a^3 = \frac{T^2}{c}$$

Therefore by taking cube roots of both sides, we obtain

$$a = \left(\frac{T^2}{c}\right)^{1/3} = \frac{T^{2/3}}{c^{1/3}} \square$$

Other Equations   In this final part of the section we will analyze equations that do not fall into any of the earlier categories, but whose solutions can be found by means of factorization and the Zero Property (which appeared in Section 1.1).

EXAMPLE 10.   Find all solutions of $x^3 - 6x^2 + 8x = 0$.

Solution.   We can factor out an x from the equation, which yields

$$x(x^2 - 6x + 8) = 0$$

and then factor $x^2 - 6x + 8$ to obtain

$$x(x - 2)(x - 4) = 0$$

Thus by the Zero Property the proposed solutions are 0, 2, and 4.

Check: $0^3 - 6(0)^2 + 8(0) = 0$

$2^3 - 6(2)^2 + 8(2) = 8 - 24 + 16 = 0$

$4^3 - 6(4)^2 + 8(4) = 64 - 96 + 32 = 0$

Therefore 0, 2, and 4 are the solutions of the given equation. ☐

EXAMPLE 11.   Find all solutions of $(x + 2)(2x^2 - 5x - 7) = 0$.

Solution.   By the Zero Property the solutions of this equation consist of all solutions of $x + 2 = 0$ and all those of $2x^2 - 5x - 7 = 0$. Now for $x + 2 = 0$ we have the solution -2. For $2x^2 - 5x - 7 = 0$ we use the quadratic formula (or factor directly, if you wish):

$$x = \frac{-(-5) \pm \sqrt{(-5)^2 - 4(2)(-7)}}{2(2)} = \frac{5 \pm \sqrt{25 + 56}}{4}$$

$$= \frac{5 \pm \sqrt{81}}{4} = \frac{5 \pm 9}{4}$$

Thus $x = \frac{14}{4} = \frac{7}{2}$ or $x = -\frac{4}{4} = -1$, so the proposed solutions of the given equation are $-2, \frac{7}{2}$, and -1.

Check: $(-2 + 2)[2(-2)^2 - 5(-2) - 7] = (0)(8 + 10 - 7) = 0$

$$\left(\frac{7}{2} + 2\right)\left[2\left(\frac{7}{2}\right)^2 - 5\left(\frac{7}{2}\right) - 7\right] = \left(\frac{7}{2} + 2\right)\left(\frac{49}{2} - \frac{35}{2} - 7\right)$$

$$= \left(\frac{7}{2} + 2\right)(0) = 0$$

$$(-1 + 2)[2(-1)^2 - 5(-1) - 7] = (-1 + 2)(2 + 5 - 7)$$

$$= 1(0) = 0$$

Therefore $-2, \frac{7}{2}$, and -1 are the solutions of the given equation. ☐

EXAMPLE 12.   Find all solutions of $x^{3/2} = -2x^{1/2}$.

Solution.   We alter the equation so that it becomes

$$x^{3/2} + 2x^{1/2} = 0$$

which is equivalent to

$$x^{1/2}(x + 2) = 0$$

Thus the proposed solutions are 0 and -2.

Check: $0^{3/2} = 0$ and $(-2)(0)^{1/2} = 0$

Since $(-2)^{3/2}$ is undefined, -2 is an extraneous solution.

Consequently 0 is the only real solution of the given equation. ☐

EXERCISES 2.5

In Exercises 1–22, find all real solutions (if any) of the given equation.

1. $(x + 2)^2 + 11(x + 2) + 18 = 0$

2. $(x^2 - 3)^2 - 5(x^2 - 3) - 14 = 0$

3. $(x - 2)^2 + x - 32 = 0$

4. $(x^2 + x)^2 - 5(x^2 + x) - 6 = 0$

5. $x^4 - 6x^2 + 8 = 0$

6. $x^4 - 8x^2 + 15 = 0$

7. $x^4 - 8x^2 - 9 = 0$

8. $x^4 - 4x^2 - 12 = 0$

9. $x^4 + 5x^2 + 6 = 0$

10. $x^6 + 7x^3 - 8 = 0$

11. $x^6 - 27x^3 - 28 = 0$

12. $x^8 - 4x^4 - 12 = 0$

13. $x - \sqrt{x} - 12 = 0$

14. $x + 7\sqrt{x} + 10 = 0$

15. $x - 5\sqrt{x} - 6 = 0$

16. $x + 6\sqrt{x} - 3 = 0$

17. $x^{1/2} + 3x^{1/4} - 18 = 0$

18. $x^{1/2} - 2x^{1/4} + 1 = 0$

19. $x^{1/2} + 3x^{1/4} - 10 = 0$

20. $x^{2/3} - 2x^{1/3} + 1 = 0$

21. $x^{2/3} - 3x^{1/3} + 2 = 0$

22. $x^{4/3} - 5x^{2/3} + 6 = 0$

In Exercises 23–38, find all real solutions of the given equation.

23. $\sqrt{2x + 4} = 6$

24. $\sqrt{1 - x} = 4$

25. $\sqrt{4 + x^2} = 3$

26. $\sqrt{x^2 + 2x} = 2\sqrt{2}$

27. $\sqrt{x^2 - 5x} = \sqrt{14}$

28. $\sqrt{2x + 3} = x$

29. $\sqrt{6x - 1} = 3x$

30. $\sqrt[3]{x^2 + 2} = 3$

31. $\sqrt[3]{3x^2 - 1} = 2$

32. $\sqrt{3 - x} = \sqrt{5 - x^2}$

33. $\sqrt{4x - 5} = \sqrt{x^2 - 2x}$

34. $\sqrt{2x - 1} = 3 + \sqrt{x - 5}$

35. $\sqrt{5 - x} + 1 = \sqrt{7 + 2x}$

36. $\sqrt{x + 1} + \sqrt{x - 1} = \sqrt{2x + 1}$

37. $\sqrt{3 - 2\sqrt{x}} = \sqrt{x}$

38. $\sqrt{10 + 3\sqrt{x}} = \sqrt{x} + 2$

In Exercises 39–52, find all real solutions of the given equation.

39. $8x^3 = 27$

40. $6x^3 = -16$

41. $27x^3 - 10 = 0$

42. $16x^4 - 0.0081 = 0$

43. $x^n = 2$

44. $n^n = 3^n x^n$

45. $x^{2n-1} + 1 = 0$

46. $(x^2 + 1)^2 = \dfrac{9}{4}$

47. $(x^2 - 1)^2 = 4$

48. $(x^2 - 9)^2 = 4$

49. $(x^3 - 27)^3 = -64$

50. $(16x^2 - 9)^4 = 0$

51. $x^9 + x^4 = 0$

52. $\left(1 + \dfrac{1}{x}\right)^{100} = 2$

In Exercises 53–58, find all real solutions of the given equation.

53. $x^5 + x^3 - 2x = 0$

54. $x^3 - 16x^2 + 48x = 0$

55. $(x^2 - 4)(x^2 - 6x + 8) = 0$

56. $(x^2 - 6)(x^2 + x - 2) = 0$

57. $(x^2 - 9)(x^2 + 16) = 0$

58. $(x^3 + 8)(4x^4 - 2x^2 + \frac{1}{4}) = 0$

In Exercises 59–62, determine all real solutions of the given equation by finding common factors.

59. $x^3 - x^2 + x - 1 = 0$

60. $x^3 - x^2 - x + 1 = 0$

61. $x^4 - 3x^3 - 4x^2 + 12x = 0$

62. $x^4 + 5x^3 - 9x^2 - 45x = 0$

63. a. Determine the extraneous solution of the equation

$$\sqrt[4]{2x^2 - 1} = x$$

that we introduce if we solve it by raising both sides to the fourth power.

b. Let m be a positive integer. Determine the extraneous solution of the equation

$$\sqrt[4m]{2x^{2m} - 1} = x$$

that we introduce if we solve it by raising both sides to the $(4m)$th power.

64. The volume V of a sphere of radius r is given by $V = \frac{4}{3}\pi r^3$. Solve the equation for r.

65. The following equation occurs in the study of electricity:

$$E^2 = \frac{Q^2}{(1 + a^2)^3}$$

Solve the equation for a.

66. The equation

$$y = \frac{100\,kx^n}{1 + kx^n}$$

appears in the study of the saturation of hemoglobin with oxygen. Solve the equation for x.

67. The equation

$$V = \frac{\pi p r^4}{8\eta l}$$

expresses the rate of volume flow V (volume per unit time) of a fluid through a cylindrical tube in terms of the radius r, the length l, the pressure difference p at the ends of the tube, and the viscosity η of the fluid. Solve the equation for r.

68. About 150 B.C., the ancient Greek astronomer Hipparchus developed a scale for measuring the brightness of stars. The scale gave what is called the **apparent magnitude** of the stars visible to the naked eye, from first magnitude to sixth magnitude. The brightest 20 stars were assigned the first magnitude, the next fainter group the second magnitude, and so on until the very faintest stars visible to the eye were assigned the sixth magnitude. It turns out that, on the average, a first magnitude star is approximately 100 times as bright as a sixth-magnitude star, and moreover, there is a number c such that the average brightness of the stars of any given magnitude is c times the average brightness of stars of the next fainter magnitude.

a. Determine the number c.

b. Let d be the ratio of the average brightness of a second-magnitude star to a fifth-magnitude star. Solve for d in terms of c.

c. Use a calculator to approximate the numbers c and d.

**2.6
INEQUALITIES**

So far in this chapter we have studied equations and their solutions. Although equations are fundamental to mathematics, so are inequalities. In the remainder of this chapter we will study inequalities and their solutions. We will first introduce notation to facilitate the description of the solutions.

Intervals The four basic inequalities involving the real numbers a and b are

$$a < b, \quad a \le b, \quad a > b, \quad \text{and} \quad a \ge b$$

From Section 1.2 we know that

$$a < x < b \quad \text{means} \quad a < x \quad \text{and} \quad x < b \quad \text{simultaneously}$$

and

$$a > x \ge b \quad \text{means} \quad a > x \quad \text{and} \quad x \ge b \quad \text{simultaneously}$$

Other compound inequalities, which are defined similarly, include combinations of $<$ and $\le$ and combinations of $>$ and $\ge$.

Caution: We *never* use either < or ≤ together with > or ≥ in the same compound inequality. Thus we never write an expression like $2 < 5 \geq 3$. In all compound inequalities the inequality signs must open in the same direction (< and ≤, for example).

With the basic inequalities and the basic compound inequalities we can describe nine categories of special sets of real numbers called ***intervals***. They are listed below. In our list we use the symbols ∞ (read "infinity") and −∞ (read "minus infinity" or "negative infinity"). These two symbols *do not* represent real numbers but merely help us represent certain kinds of intervals.

Type of Interval	*Notation*	*Description*
Open interval	(a, b)	all x such that $a < x < b$
	(a, ∞)	all x such that $a < x$
	$(-\infty, a)$	all x such that $x < a$
	$(-\infty, \infty)$	all real numbers
Closed interval	$[a, b]$	all x such that $a \leq x \leq b$
	$[a, \infty)$	all x such that $a \leq x$
	$(-\infty, a]$	all x such that $x \leq a$
Half-open interval	$(a, b]$	all x such that $a < x \leq b$
	$[a, b)$	all x such that $a \leq x < b$

In the list, the symbols [and] indicate that the corresponding endpoint is included in the set, whereas the symbols (and) indicate that the corresponding endpoint (if there is one) is not included in the set. The various kinds of intervals are described graphically on the real line in Figure 2.9.

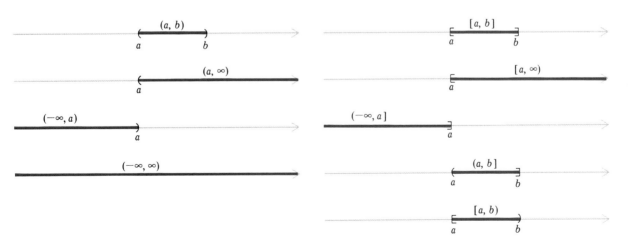

FIGURE 2.9

The four intervals (a, b), $[a, b]$, $(a, b]$, and $[a, b)$ are **bounded intervals**, and the remaining ones, each of which involves ∞ or $-\infty$, are **unbounded intervals**.

EXAMPLE 1.   Determine whether each of the following intervals is open, closed, or half-open, and whether it is bounded or unbounded. Then locate the interval on the real line.

a. $(0, 3]$ b. $(-\infty, -2)$ c. $[-4, 1]$

d. $[-1, \infty)$ e. $\left(-\frac{3}{2}, -\frac{1}{3}\right)$

Solution.

a. $(0, 3]$ is half-open and bounded.

b. $(-\infty, -2)$ is open and unbounded.

c. $[-4, 1]$ is closed and bounded.

d. $[-1, \infty)$ is closed and unbounded.

e. $\left(-\frac{3}{2}, -\frac{1}{3}\right)$ is open and bounded.

The five intervals are shown in Figure 2.10.   □

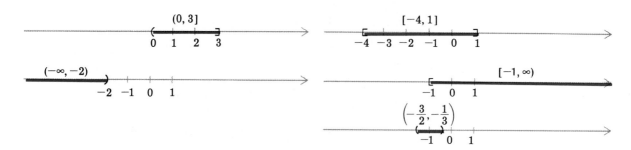

FIGURE 2.10

Basic Laws of Inequalities   In Section 1.2 we presented the law of trichotomy, which says that for any two given real numbers a and b, either $a < b$ or $a = b$ or $a > b$. We are ready now to present four more laws that will form the basis of mathematical computation involving inequalities. Throughout we will assume that a, b, c, and d are real numbers.

Laws of Inequalities
If $a < b$ and $b < c$, then $a < c$.
If $a < b$, then $a + c < b + c$.
If $a < b$ and $c > 0$, then $ac < bc$.
If $a < b$ and $c < 0$, then $ac > bc$.

Interchanging $<$ and $>$, we obtain the following versions.

If $a > b$ and $b > c$, then $a > c$.
If $a > b$, then $a + c > b + c$.
If $a > b$ and $c > 0$, then $ac > bc$.
If $a > b$ and $c < 0$, then $ac < bc$.

The laws listed above remain valid if $<$ is replaced by $\leq$ and if $>$ is replaced by $\geq$.

Caution: Note carefully that when both sides of an inequality are multiplied by a *negative* number, the sense of the inequality must be reversed (from $<$ to $>$, from $\leq$ to $\geq$, from $>$ to $<$, or from $\geq$ to $\leq$).

Solutions of Linear Inequalities

Consider the inequality

$$5x + 2 < -3$$

with x a variable as usual. If $x = -2$, then x satisfies $5x + 2 < -3$, since

$$5(-2) + 2 = -10 + 2 = -8 < -3$$

so we say that -2 is a solution of the inequality. By contrast, if $x = 1$, then x does not satisfy $5x + 2 < -3$, because

$$5(1) + 2 = 7 > -3$$

so we say that 1 is not a solution of $5x + 2 < -3$. More generally, the *solutions* of a given inequality are the values of the variable that make the inequality valid. We will show below that the solutions of $5x + 2 < -3$ consist of all x such that $x < -1$, that is, they comprise the interval $(-\infty, -1)$. For simplicity, in this chapter we will only consider inequalities whose solutions consist of either an interval or a finite collection of intervals.

When we determine the solutions of a given inequality, we say that we *solve the inequality*. This often involves finding a series of inequalities that are equivalent to the original inequality, by which we mean that they all have the same set of solutions as the original one has.

In this section we will solve *linear inequalities*, that is, inequalities that are equivalent to an inequality having one of the forms

$$ax < b, \qquad ax \leq b, \qquad ax > b, \quad \text{or} \quad ax \geq b$$

We solve a linear inequality in a way analogous to the way we solved linear equations:

> i. Put all terms containing x on one side of the inequality.
> ii. Put all other terms on the other side of the inequality.
> iii. Simplify the resulting inequality to solve for x.

EXAMPLE 2. Solve the inequality $x - 6 > 0$.

Solution. Since the only term containing x is on the left side, we need only add 6 to each side and then simplify:

$$x - 6 > 0$$
$$(x - 6) + 6 > 0 + 6$$
$$x > 6$$

Since $x - 6 > 0$ is equivalent to $x > 6$, it follows that x is a solution of the given equation if $x > 6$. Therefore the solutions form the interval $(6, \infty)$. □

EXAMPLE 3.   Solve the inequality $5x + 2 < -3$.

Solution.   Again, the only term containing x is on the left side, so we need only subtract 2 from each side and then simplify:

$$5x + 2 < -3$$
$$(5x + 2) - 2 < -3 - 2$$
$$5x < -5$$
$$\frac{1}{5}(5x) < \frac{1}{5}(-5)$$
$$x < -1$$

Since $5x + 2 < -3$ is equivalent to $x < -1$, it follows that x is a solution of the given equation if $x < -1$. In other words, the solutions form the interval $(-\infty, -1)$. □

EXAMPLE 4.   Solve the inequality $3x - 2 \geq 8 + 5x$.

Solution.   Using (i)–(iii), we obtain the following equivalent inequalities:

$$3x - 2 \geq 8 + 5x$$
$$(3x - 2) - 5x \geq (8 + 5x) - 5x$$
$$-2x - 2 \geq 8$$
$$(-2x - 2) + 2 \geq 8 + 2$$
$$-2x \geq 10$$
$$\left(-\frac{1}{2}\right)(-2x) \leq \left(-\frac{1}{2}\right)(10)$$
$$x \leq -5$$

In other words, the solutions of the original inequality consist of all $x \leq -5$, which means that they form the interval $(-\infty, -5]$. □

Observe that we reversed the sense of the inequality when we multiplied by the negative number $-\frac{1}{2}$ in the solution of Example 4.

Solutions of Composite Inequalities

Recall that the composite inequality

$$-2 < \frac{5 - x}{3} \leq 6x - 1 \tag{1}$$

is a shorthand way of writing the pair of inequalities

$$-2 < \frac{5-x}{3} \quad \text{and} \quad \frac{5-x}{3} \le 6x - 1 \tag{2}$$

Therefore x is a solution of the composite inequality in (1) if and only if it is a solution of both inequalities in (2) simultaneously.

EXAMPLE 5.   Solve the inequality $-2 < \dfrac{5-x}{3} \le 6x - 1$.

Solution.   We solve the given composite inequality by working separately on the two inequalities in (2):

$$-2 < \frac{5-x}{3} \qquad\qquad \frac{5-x}{3} \le 6x - 1$$

$$(-2)(3) < \frac{5-x}{3}(3) \qquad\qquad \frac{5-x}{3}(3) \le (6x-1)(3)$$

$$-6 < 5 - x \qquad\qquad 5 - x \le 18x - 3$$

$$-6 + x < (5 - x) + x \qquad (5 - x) + x \le (18x - 3) + x$$

$$x - 6 < 5 \qquad\qquad 5 \le 19x - 3$$

$$(x - 6) + 6 < 5 + 6 \qquad\qquad 5 + 3 \le (19x - 3) + 3$$

$$x < 11 \qquad\qquad 8 \le 19x$$

$$\left(\frac{1}{19}\right)(8) \le \left(\frac{1}{19}\right)(19x)$$

$$\frac{8}{19} \le x$$

Consequently the solutions consist of all values of x that satisfy $x < 11$ and $\frac{8}{19} \le x$ simultaneously, that is, $\frac{8}{19} \le x < 11$. Thus the solutions form the interval $\left[\frac{8}{19}, 11\right)$. $\square$

Sometimes we can solve a composite inequality by performing the same operations on all members of the inequality. Of course, the basic laws of inequalities must be carefully observed. The procedure is illustrated in the next example.

EXAMPLE 6.   Solve the composite inequality $-3 \le \dfrac{2x+3}{-4} < 7$.

Solution.   We have the following equivalent inequalities:

$$-3 \le \frac{2x+3}{-4} < 7$$

$$(-3)(-4) \ge \left(\frac{2x+3}{-4}\right)(-4) > (7)(-4)$$

$$12 \geq 2x + 3 > -28$$

$$12 - 3 \geq (2x + 3) - 3 > -28 - 3$$

$$9 \geq 2x > -31$$

$$\frac{9}{2} \geq \frac{2x}{2} > -\frac{31}{2}$$

$$\frac{9}{2} \geq x > -\frac{31}{2}$$

Thus the solutions of the given inequality form the interval $\left(-\frac{31}{2}, \frac{9}{2}\right]$. □

Our final example is an applied problem whose solution involves inequalities.

EXAMPLE 7.   By common agreement a fever is any oral temperature greater than 98.6 degrees Fahrenheit. What temperatures in degrees Celsius correspond to a fever?

Solution.   Recall from (3) in Section 2.1 that if F and C represent degrees Fahrenheit and Celsius respectively, then F and C are related by the formula

$$F = \frac{9}{5}C + 32$$

Then a fever corresponds to any Fahrenheit temperature F such that $F > 98.6$. The corresponding Celsius temperature must satisfy

$$\frac{9}{5}C + 32 > 98.6 \tag{3}$$

We need to determine the values of C for which (3) is valid, and this we do with the following equivalent inequalities:

$$\frac{9}{5}C + 32 > 98.6$$

$$\left(\frac{9}{5}C + 32\right) - 32 > 98.6 - 32$$

$$\frac{9}{5}C > 66.6$$

$$\frac{5}{9}\left(\frac{9}{5}C\right) > \frac{5}{9}(66.6)$$

$$C > 37$$

Therefore a fever is any temperature greater than 37 degrees Celsius. □

In the next section we will continue solving inequalities.

EXERCISES 2.6

In Exercises 1–14, identify the intervals as open, closed, or half-open, and as bounded or unbounded.

1. $(-1, 2)$

2. $(-7, -6]$

3. $(-\infty, 4]$

4. $[6, 6.1)$

5. $(6, 6.01]$

6. $(-1, \infty)$

7. $[7, 7]$

8. $(-\pi, \infty)$

9. $(1.9, 2.1)$

10. $[0, \infty)$

11. all x such that $-2 \leq x < 4$

12. all x such that $0 \leq x \leq 0.01$

13. all x such that $x > \dfrac{\sqrt{3}}{4}$

14. all x such that $-4 \leq x < \infty$

In Exercises 15–22, write the inequality in interval form.

15. $-4 < x \leq 3$

16. $5 \leq x \leq 7$

17. $-1.1 < x < -0.9$

18. $-1.01 < x \leq 0.99$

19. $x > -8$

20. $3 \leq x < \infty$

21. $-1 \leq x < 1$

22. $x \leq -2$

In Exercises 23–36, solve for x and then express the solutions as an interval.

23. $2x \leq 6$

24. $-4x < 8$

25. $-12x \geq -3$

26. $0 > 4x - 15$

27. $-7x - 2 \geq 0$

28. $2x + 7 \leq 5x - 3$

29. $4 - 3x > -1 - x$

30. $\frac{1}{2} + 2x \leq \frac{4}{3} - 5x$

31. $12 - 2x < 4(x - 6)$

32. $\frac{1}{3}(2x - 3) > 3(x + \frac{1}{3})$

33. $\dfrac{1 - x}{2} \geq \dfrac{2 + x}{-3}$

34. $\dfrac{5 - 2x}{7} \leq \dfrac{3x + 4}{2}$

35. $x^2 \geq (x + 3)^2$

36. $(x - 1)^2 < (x + 2)^2$

In Exercises 37–46, solve for x and then express the solutions as an interval.

37. $3 > x + 5 > 0$

38. $-2 \leq x - 1 \leq 4$

39. $0 \leq 5(x + 3) < 10$

40. $0 < \frac{1}{2}(2x + 4) < \frac{1}{3}$

41. $6 \geq \dfrac{3 - 3x}{12} \geq 4$

42. $\dfrac{1}{5} > \dfrac{2 - x}{-15} > \dfrac{1}{10}$

43. $-0.01 < x - 2 < 0.01$

44. $-10^{-4} < x - 2 < 10^{-4}$

45. $0 < x - a < d$

46. $-d < x - a < d$

47. Let $a, b, c,$ and d be real numbers with b and d positive.

a. Prove that

$$\frac{a}{b} < \frac{c}{d} \quad \text{if and only if} \quad ad < bc$$

(*Hint:* Multiply both sides of the first inequality by bd.)

b. Use (a) to show that

$$\frac{22}{59} < \frac{3}{8}$$

48. Which temperatures in degrees Celsius correspond to the temperatures larger than 32 and smaller than 212 degrees Fahrenheit?

49. A snack bar managar estimates that if the price of hot dogs is set at x cents, where $20 \leq x \leq 200$, then $1000 - 5x$ hot dogs will be sold daily. If the manager must sell at least 400 hot dogs each day at a cost of at least 20 cents, what are the possible prices for hot dogs?

50. A farmer is willing to sell s bushels of corn if the price of a bushel is $2 + (s/100,000)$ dollars. If the government sets a ceiling of $4.50 on the price of a bushel of corn, what are the possible numbers of bushels the farmer would sell?

51. A store manager figures that $10(10 - p)$ umbrellas can be sold at p dollars per umbrella. If at least 45 umbrellas are to be sold, what are the possible prices that can be charged for each one?

52. A certain car holds 21 gallons of gasoline and gets 22 miles per gallon. If the car runs out of gasoline after having traveled at least 330 miles during a day, what are the possible amounts of gasoline that were in the tank at the start of the day?

53. Assuming that he charges x dollars per gallon, an ice cream factory manager estimates that he can sell $50,000 - 10,000x$ gallons of marshmallow ice cream per month. If he wishes to sell at least 30,000 gallons of it per month, what are the possible prices per gallon that be can charge?

54. A student pays $25 per week for a room. Another room is available for $22 per week. If it would cost $15 to move to the second room, for what lengths of stay would it pay for the student to make the transfer to the new room?

**2.7
MORE ON
INEQUALITIES**

This section begins where the preceding section ended. The method of solving the inequalities appearing in this section will rely heavily on the following rules:

$ac > 0$ if $a > 0$ and $c > 0$,	or if $a < 0$ and $c < 0$ (1)
$ac < 0$ if $a < 0$ and $c > 0$,	or if $a > 0$ and $c < 0$ (2)

These rules, which appeared in a slightly different form in Section 1.2, are consequences of the inequality laws of Section 2.6.

We begin by solving **quadratic inequalities**, that is, inequalities that are equivalent to one of the following:

$$ax^2 + bx + c > 0, \qquad ax^2 + bx + c \geq 0,$$
$$ax^2 + bx + c < 0, \quad \text{or} \quad ax^2 + bx + c \leq 0 \tag{3}$$

EXAMPLE 1.   Solve the inequality $x^2 + x - 12 < 0$.

Solution.   Since $x^2 + x - 12 = (x + 4)(x - 3)$, the given inequality is equivalent to

$$(x + 4)(x - 3) < 0$$

To solve this inequality we first find the intervals on which the components $x + 4$ and $x - 3$ are positive and those on which they are negative. Then we use this information along with (1) and (2) to determine the intervals on which the product $(x + 4)(x - 3)$ is positive and those on which it is negative. It is convenient to display the results in a diagram such as the one shown in Figure 2.11.

From the diagram we see that $(x + 4)(x - 3) < 0$ for x in $(-4, 3)$. The endpoints of $(-4, 3)$ are not included because $(x + 4)(x - 3) = 0$ for $x = -4$

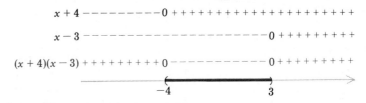

FIGURE 2.11

and for $x = 3$. Thus the solutions of the given inequality form the interval $(-4, 3)$.   □

If a quadratic inequality is not given in one of the forms listed in (3), we put it into such a form by transposing all terms to the left side of the inequality.

EXAMPLE 2.   Solve the inequality $x^2 - 7 \geq 6x$.

Solution.   We subtract $6x$ from both sides and then factor the left side:

$$x^2 - 7 \geq 6x$$

$$(x^2 - 7) - 6x \geq 6x - 6x$$

$$x^2 - 6x - 7 \geq 0$$

$$(x + 1)(x - 7) \geq 0$$

Now we set up a diagram (Figure 2.12) to determine the intervals on which $(x + 1)(x - 7) \geq 0$. From Figure 2.12 we see that $(x + 1)(x - 7) \geq 0$ for x in $(-\infty, -1]$ or $[7, \infty)$. The endpoints -1 and 7 are included because $(x + 1)(x - 7) = 0$ for $x = -1$ and $x = 7$. Thus the solutions of the given inequality form the intervals $(-\infty, -1]$ and $[7, \infty)$.   □

The rules in (1) and (2) concern products. Since any quotient a/c can be written as the product $a(1/c)$, and since $1/c$ has the same sign as c, we have the following rules for quotients:

$$\frac{a}{c} > 0 \quad \text{if} \quad a > 0 \quad \text{and} \quad c > 0, \quad \text{or if} \quad a < 0 \quad \text{and} \quad c < 0 \qquad (4)$$

$$\frac{a}{c} < 0 \quad \text{if} \quad a < 0 \quad \text{and} \quad c > 0, \quad \text{or if} \quad a > 0 \quad \text{and} \quad c < 0 \qquad (5)$$

EXAMPLE 3.   Solve the inequality $\dfrac{2x - 1}{5x + 3} > 0$.

Solution.   Notice that $2x - 1 = 0$ for $x = \frac{1}{2}$ and that $5x + 3 = 0$ for $x = -\frac{3}{5}$. Using this information along with (4), we prepare the diagram

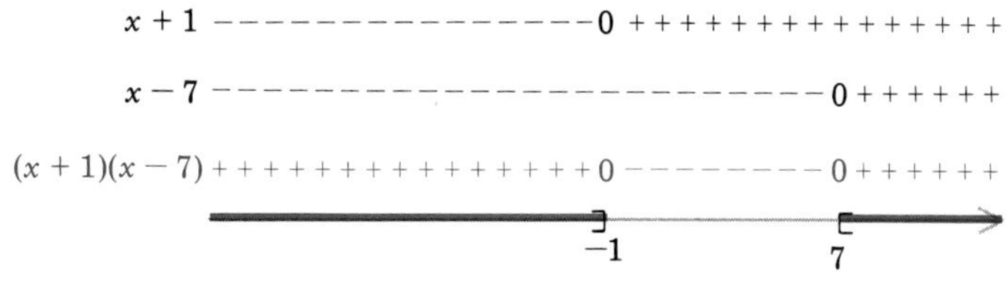

FIGURE 2.12

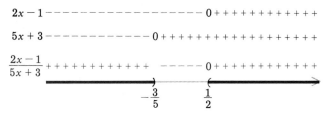

FIGURE 2.13

shown in Figure 2.13. The endpoint $-\frac{3}{5}$ is not a solution, since $\dfrac{2x-1}{5x+3}$ is not defined for $x = -\frac{3}{5}$. The endpoint $\frac{1}{2}$ is not a solution because

$$\frac{2\left(\dfrac{1}{2}\right) - 1}{5\left(\dfrac{1}{2}\right) + 3} = \frac{1 - 1}{\dfrac{5}{2} + 3} = 0$$

From these observations and the diagram we conclude that the solutions of the given inequality form the intervals $(-\infty, -\frac{3}{5})$ and $(\frac{1}{2}, \infty)$. □

EXAMPLE 4.   Solve the inequality $\dfrac{2x-1}{5x+3} \geq 0$.

   Solution.   From Example 3 we know that $(2x - 1)/(5x + 3) > 0$ if and only if x is in $(-\infty, -\frac{3}{5})$ or in $(\frac{1}{2}, \infty)$. Since

$$\frac{2x-1}{5x+3} = 0$$

if and only if $2x - 1 = 0$, that is, if and only if $x = \frac{1}{2}$, we conclude that the set of solutions of the given inequality consists of $\frac{1}{2}$ along with the intervals $(-\infty, -\frac{3}{5})$ and $(\frac{1}{2}, \infty)$. Thus the solutions of the given inequality form the intervals $(-\infty, -\frac{3}{5})$ and $[\frac{1}{2}, \infty)$. □

EXAMPLE 5.   Solve the inequality $\dfrac{6}{x-2} \leq 2$.

   Solution.   We subtract 2 from both sides so that the right side is 0, and then simplify:

$$\frac{6}{x-2} \leq 2$$

$$\frac{6}{x-2} - 2 \leq 0$$

$$\frac{6 - 2(x - 2)}{x - 2} \le 0$$

$$\frac{10 - 2x}{x - 2} \le 0$$

$$\frac{2(5 - x)}{x - 2} \le 0$$

Now we prepare the diagram shown in Figure 2.14. From the diagram we see that

$$\frac{2(5 - x)}{x - 2} \le 0$$

for x in $(-\infty, 2)$ and $[5, \infty)$. Thus the solutions of the given inequality form the intervals $(-\infty, 2)$ and $[5, \infty)$. □

Another way of solving inequalities such as the one in Example 5 involves eliminating the denominator. Since $x - 2$ can be either positive or negative, we multiply by $(x - 2)^2$, which is always nonnegative, and then rearrange:

$$\frac{6}{x - 2} \le 2$$

$$\frac{6}{x - 2}(x - 2)^2 \le 2(x - 2)^2$$

$$6(x - 2) \le 2(x^2 - 4x + 4)$$

$$6x - 12 \le 2x^2 - 8x + 8$$

$$0 \le 2x^2 - 14x + 20$$

$$0 \le x^2 - 7x + 10$$

$$0 \le (x - 5)(x - 2)$$

From a diagram analogous to Figure 2.14 we obtain the same solution as before (noting that 2 is not a solution of the original inequality).

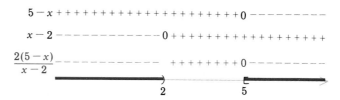

FIGURE 2.14

So far all of our examples have contained two factors. However, the same general method works for any number of factors. In the next example, there are three factors.

<u>EXAMPLE 6.</u>   Solve the inequality $(x + 3)(x + 1)(x - 2) < 0$.

   Solution.   We prepare the diagram shown in Figure 2.15. From it we see that the solutions of the given inequality form the intervals $(-\infty, -3)$ and $(-1, 2)$.   □

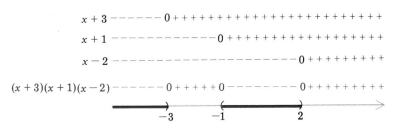

FIGURE 2.15

Applications of Inequalities   In Sections 2.2 and 2.4 we solved applied problems by means of equations. However, some applied problems can be solved by means of inequalities, as the next two examples illustrate.

<u>EXAMPLE 7.</u>   A rock is thrown vertically upward from a height of 6 feet above ground with an initial velocity of 96 feet per second. During what time interval will the rock be more than 134 feet above ground?

   Solution.   By (1) in Section 2.4, with $v_0 = 96$ and $h_0 = 6$, the height h of the rock at time t is given by

$$h = -16t^2 + 96t + 6$$

We wish to find the values of t for which $h > 134$, which means that we must solve the inequality

$$-16t^2 + 96t + 6 > 134$$

Proceeding as in earlier solutions of inequalities we find that

$$-16t^2 + 96t - 128 > 0$$
$$-16(t^2 - 6t + 8) > 0$$
$$t^2 - 6t + 8 < 0$$
$$(t - 2)(t - 4) < 0$$

Using this inequality and the facts that $t - 2 = 0$ if $t = 2$ and $t - 4 = 0$ if $t = 4$, we prepare the diagram shown in Figure 2.16. From the diagram we

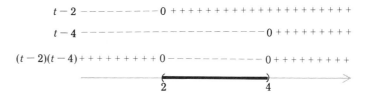

FIGURE 2.16

see that the solutions of $(t - 2)(t - 4) < 0$ form the interval $(2, 4)$. Thus the rock will be more than 134 feet high during the time interval between 2 seconds and 4 seconds. $\square$

EXAMPLE 8.   A store manager figures that x chess sets can be sold each month if the price per set is $10 - 0.1x$ dollars. Suppose that chess sets can be purchased from the wholesaler at \$3 per set. If the manager wishes to make a profit of at least \$120 per month from the sale of chess sets, how many sets must be sold per month?

Solution.   If x chess sets are sold per month, then by hypothesis the resulting monthly revenue is $x(10 - 0.1x)$ and the monthly cost to the store is $3x$. Therefore the monthly profit, which is the difference between the monthly revenue and the monthly cost, is given by

$$x(10 - 0.1x) - 3x$$

Since the manager wishes the monthly profit to be at least \$120, this means that x must satisfy

$$x(10 - 0.1x) - 3x \geq 120$$

which is equivalent to the following inequalities:

$$10x - 0.1x^2 - 3x \geq 120$$
$$-0.1x^2 + 7x - 120 \geq 0$$
$$x^2 - 70x + 1200 \leq 0$$
$$(x - 30)(x - 40) \leq 0$$

Using this inequality and the facts that $x - 30 = 0$ for $x = 30$ and $x - 40 = 0$ for $x = 40$, we prepare the diagram shown in Figure 2.17. From

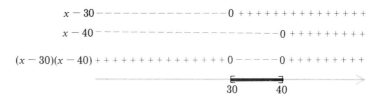

FIGURE 2.17

the diagram we see that the solutions of the inequality $(x - 30)(x - 40) \leq 0$ form the interval $[30, 40]$. We conclude that between 30 and 40 chess sets must be sold monthly in order to have a monthly profit of at least \$120.   ☐

If we wished to determine the possible prices that would create a profit of at least \$120, we would use the information gained in the preceding solution. The number x of chess sets satisfies $30 \leq x \leq 40$, and the price is $10 - 0.1x$. Thus we would need to determine the possible values of $10 - 0.1x$ for $30 \leq x \leq 40$. Now if $30 \leq x \leq 40$, then $3 \leq 0.1x \leq 4$, so that $-4 \leq -0.1x \leq -3$, and consequently $6 \leq 10 - 0.1x \leq 7$. Therefore the manager must price each chess set at between \$6 and \$7 in order to produce a profit of at least \$120.

EXERCISES 2.7

In Exercises 1–44, solve the given inequality and express the solutions in terms of intervals.

1. $(x - 1)(x - 2) \geq 0$

2. $(x + 5)(x - 1) < 0$

3. $(x + 1)(x + 3) \leq 0$

4. $(x - \sqrt{2})(x + \sqrt{3}) \leq 0$

5. $(x - 3)^2 > 0$

6. $(x + 2)^2 \geq 0$

7. $x^2 < 4$

8. $x^2 \geq 9$

9. $(x - 2)^2 \leq 1$

10. $(x^2 - 3)^2 \leq 9$

11. $(x^2 - 5)^2 < 16$

12. $(x^2 - 3)^2 < 16$

13. $x^2 - 5x + 6 < 0$

14. $x^2 + 3x - 10 \geq 0$

15. $x^2 - 2x - 15 > 0$

16. $x^2 + 5x \leq 14$

17. $x^2 \leq -9x$

18. $x^2 > \sqrt{2}\,x$

19. $-2x^2 + 7x > -4$

20. $2x^2 + x < 1$

21. $\dfrac{x - 4}{x + 5} > 0$

22. $\dfrac{2 - x}{3 + x} < 0$

23. $\dfrac{2x - 3}{3x + 6} \leq 0$

24. $\dfrac{2x - 1}{x} \geq 5$

25. $\dfrac{x}{2x - 1} \geq 5$

26. $\dfrac{-x + 5}{2x + 1} < -2$

27. $\dfrac{-5x + 3}{15x} > 1$

28. $\dfrac{1}{x - 1} > \dfrac{1}{x + 1}$

29. $\dfrac{x}{x - 1} > \dfrac{x}{x + 1}$

30. $\dfrac{2x}{x - 3} \leq \dfrac{2x}{x - 6}$

31. $x^3 \geq 0$

32. $x^3 < 8$

33. $(2 - x)^3 < 0$

34. $(x - 1)(x - 2)(x - 3) > 0$

35. $x(x + 3)(x + 5) \leq 0$

36. $(x - 7)(x^2 + 4) > 0$

37. $(x^2 - 1)(x^2 - 9) \leq 0$

38. $-5(x - 1)(x + \tfrac{3}{2})(x + 2)(x - 3) > 0$

39. $(x^2 - x)(x^2 - x + 2) < 0$

40. $\dfrac{(2 - x)(3 + x)^2}{x - \frac{1}{2}} \geq 0$

41. $\dfrac{(3x + 4)(\frac{1}{6} - x)}{7x + 2} < 0$

42. $\dfrac{x + 1}{x^2 - x} < 0$

43. $\dfrac{x + 1}{x^2 - x} \geq -1$

44. $\dfrac{2x + 3}{6x^2 + 1} \leq 3$

In Exercises 45–52, solve the given inequality and express the solutions in terms of intervals. You may need to use the quadratic formula to factor the left side.

45. $x^2 + x < 1$

46. $x^2 - 3x + 1 \geq 0$

47. $2x^2 + 4x + 1 < 0$

48. $3x^2 - 2x - 1 \leq 0$

49. $\dfrac{1}{2}x^2 + \sqrt{2}\,x - 7 > 0$

50. $\dfrac{2 - x}{1 + x^2} > -1$

*51. $\dfrac{3x - 1}{2x^2 - 1} < 1$

*52. $\dfrac{3 + 4x}{2 - x^2} \leq 1$

53. Find the values of a such that 2 is a solution of the inequality

$$\frac{x - a}{x + a} \leq 3$$

54. Find the values of a such that -3 is a solution of the inequality

$$\frac{-3x - 2a}{x - 2a} < -4$$

55. Show that if $a < b$, then $a < (a + b)/2 < b$. (The number $(a + b)/2$ is called the **arithmetic mean** of a and b.)

56. Let $0 < a < b$.
a. Show that $(\sqrt{b} - \sqrt{a})^2 > 0$.

b. Using part (a), prove that $a < \sqrt{ab} < (a + b)/2$. (The number $\sqrt{ab}$ is called the **geometric mean** of a and b.)

57. a. Show that $x^2 > x$ for $x > 1$.
b. Show that $x^2 < x$ for $0 < x < 1$.

58. a. If $x^2 \geq 36$, is it necessarily true that $x \geq 6$? Explain.
b. If $x^3 \geq 64$, is it necessarily true that $x \geq 4$? Explain.

59. For what values of x is $1/x < x$?

60. A ball is thrown vertically upward from a height of 100 feet with an initial velocity of 80 feet per second. During what time interval is the height of the ball at least 4 feet?

61. A ball is dropped from a window 100 feet above ground. Between what times will the ball be between 84 and 36 feet above ground?

62. If the ball in Exercise 61 were thrown upward at 48 feet per second, when would the ball be between 36 and 100 feet above ground?

63. In Example 8, if the manager would be satisfied to make a profit of at least $100, what would be the range in the number of chess sets that could be sold?

64. Suppose a bicycle shop can sell x bicycle replacement seats per month if the price is set at $12-0.2x$ dollars per seat. If the purchase price from the wholesaler is $4 per seat and the shop manager desires to make a profit of at least $75 per month, what are the possible numbers of seats that can be sold monthly, and at what prices?

65. Assume, as in Exercise 64, that the price per seat is $12-0.2x$ and the purchase price from the wholesaler is $4 per seat. Show that the manager cannot make a profit of more than $80.

66. A baker estimates that he can bake up to 1800 loaves of rye bread during a week and that he can sell x loaves weekly if he charges $100 - (1/20)x$ cents per loaf. If it costs 25 cents per loaf to produce rye bread, what are the possible numbers of loaves that will net a profit for the baker?

67. A rectangular sheet of metal is 16 inches long and 8 inches wide. A pan is to be made from the sheet by cutting out four square pieces, one from each corner of the sheet (Figure 2.18). If the area of the base is to be at least 48 square inches, what are the possible heights of the sides created by folding up the edges?

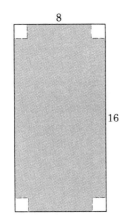
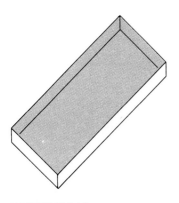

FIGURE 2.18

**2.8
INEQUALITIES
INVOLVING
ABSOLUTE
VALUES**

In addition to the kinds of inequalities we have already encountered, inequalities involving absolute values appear from time to time in advanced mathematics, primarily because of the relation of absolute value to distance between real numbers. In this section we will discuss and solve inequalities involving absolute values.

Recall from Section 1.2 that

$$|x| = \begin{cases} x & \text{if } x \geq 0 \\ -x & \text{if } x < 0 \end{cases} = \text{the distance between } x \text{ and } 0 \qquad (1)$$

It follows from (1) that if c is any positive number, then

$$|x| < c \text{ means that} \begin{cases} x < c & \text{if } x \geq 0 \\ -x < c & \text{if } x < 0 \quad \text{(so } x > -c \text{ if } x < 0\text{)} \end{cases} \qquad (2)$$

Combining the two parts of the right side of (2), we conclude that

$$|x| < c \quad \text{if and only if} \quad -c < x < c \qquad (3)$$

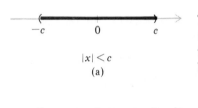

$|x| < c$
(a)

$|x| \leq c$
(b)

FIGURE 2.19

The inequality $|x| < c$ simply says that the distance between x and 0 is less than the positive number c (Figure 2.19a). For example, from (3) it follows that $|x| < 3$ means that $-3 < x < 3$, or equivalently, that the distance between x and 0 is less than 3.

By similar reasoning we find that for any nonnegative number c,

$$|x| \leq c \quad \text{if and only if} \quad -c \leq x \leq c \qquad (4)$$

The inequality $|x| \leq c$ means that the distance between x and 0 is less than or equal to c (Figure 2.19b).

EXAMPLE 1.   Solve the following inequalities.
   a. $|x| < \frac{1}{3}$ b. $|x| \leq 2$

Solution.
   a. By (3), the inequality $|x| < \frac{1}{3}$ is equivalent to $-\frac{1}{3} < x < \frac{1}{3}$, so the solutions form the open interval $(-\frac{1}{3}, \frac{1}{3})$.
   b. By (4), the inequality $|x| \leq 2$ is equivalent to $-2 \leq x \leq 2$, so the solutions form the closed interval $[-2, 2]$.   □

Now we turn to the inequalities $|x| > c$ and $|x| \geq c$. Since $|x| > c$ means that $|x| \leq c$ is false, (4) implies that for any nonnegative number c,

$$|x| > c \quad \text{if and only if} \quad x > c \text{ or } x < -c \qquad (5)$$

Similarly, (3) implies that for any nonnegative number c,

$$|x| \geq c \quad \text{if and only if} \quad x \geq c \text{ or } x \leq -c \qquad (6)$$

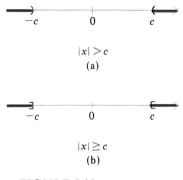

$|x| > c$
(a)

$|x| \geq c$
(b)

FIGURE 2.20

The inequality $|x| > c$ means that the distance between x and 0 is greater than c (Figure 2.20a), and the inequality $|x| \geq c$ means that the distance between x and 0 is greater than or equal to c (Figure 2.20b).

EXAMPLE 2. Solve the following inequalities.

a. $|x| > 4$ b. $|x| \geq \frac{1}{2}$

Solution.

a. By (5), the inequality $|x| > 4$ is equivalent to the statement that $x > 4$ or $x < -4$, so the solutions form the two open intervals $(-\infty, -4)$ and $(4, \infty)$.

b. By (6), the inequality $|x| \geq \frac{1}{2}$ is equivalent to the statement that $x \geq \frac{1}{2}$ or $x \leq -\frac{1}{2}$, so the solutions form the two closed intervals $(-\infty, -\frac{1}{2}]$ and $[\frac{1}{2}, \infty)$. □

Recall that $|x - a|$ is the distance between the numbers x and a. Thus geometrically the inequality $|x - a| < c$ means that the distance between x and a is less than c. There are analogous geometric interpretations of the inequalities

$$|x - a| \leq c, \qquad |x - a| > c, \quad \text{and} \quad |x - a| \geq c$$

EXAMPLE 3. Solve the following inequalities.

a. $|x - 7| < 2$ b. $|x - 7| \leq 2$

Solution by the algebraic method.

a. By (3), the inequality $|x - 7| < 2$ is equivalent to

$$-2 < x - 7 < 2$$

Adding 7 to all three expressions, we obtain

$$5 < x < 9$$

Thus the solutions form the open interval $(5, 9)$.

b. By (4), the inequality $|x - 7| \leq 2$ is equivalent to

$$-2 \leq x - 7 \leq 2$$

Adding 7 to each expression yields

$$5 \leq x \leq 9$$

Thus the solutions form the closed interval $[5, 9]$. □

Solution by the geometric method.

a. Notice that x satisfies the inequality $|x - 7| < 2$ if and only if the distance between x and 7 is less than 2 units. On the real line we locate 7 and mark off 2 units to either side (Figure 2.21a). The numbers 5 and 9

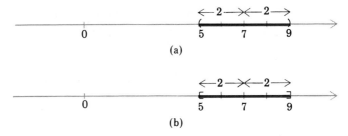

FIGURE 2.21

are not solutions, since they both lie exactly 2 units from 7, but all numbers between 5 and 9 are solutions. Thus the solutions form the open interval $(5, 9)$.

b. Notice that x satisfies the inequality $|x - 7| \leq 2$ if and only if the distance between x and 7 is less than or equal to 2. Thus 5 and 9 are solutions of the inequality, as are all numbers in between. Consequently the solutions comprise the closed interval $[5, 9]$ (Figure 2.21b). □

Other inequalities involving absolute values can be solved by similar methods.

EXAMPLE 4.   Solve the inequality $|3x - 5| < \frac{1}{2}$.

Solution.   By (3), the inequality $|3x - 5| < \frac{1}{2}$ is equivalent to

$$-\frac{1}{2} < 3x - 5 < \frac{1}{2}$$

We can either treat this compound inequality as the pair of inequalities

$$-\frac{1}{2} < 3x - 5 \quad \text{and} \quad 3x - 5 < \frac{1}{2}$$

and solve the pair for x separately, or we can perform our alterations on the compound inequality itself. We will do the latter:

$$-\frac{1}{2} < 3x - 5 < \frac{1}{2}$$

$$-\frac{1}{2} + 5 < (3x - 5) + 5 < \frac{1}{2} + 5$$

$$\frac{9}{2} < 3x < \frac{11}{2}$$

$$\frac{9}{2}\left(\frac{1}{3}\right) < 3x\left(\frac{1}{3}\right) < \frac{11}{2}\left(\frac{1}{3}\right)$$

$$\frac{3}{2} < x < \frac{11}{6}$$

Therefore the solutions comprise the open interval $\left(\frac{3}{2}, \frac{11}{6}\right)$. □

EXERCISES 2.8

In Exercises 1–6, solve the given inequality and locate the solutions on the real line.

1. $|x| < 4$

2. $|x| < 1.5$

3. $|x| \leq \frac{1}{5}$

4. $|x| \leq 3.2$

5. $|x| \geq \frac{9}{2}$

6. $|-x| > 3$

In Exercises 7–38, solve the given inequality.

7. $|x| > 0.01$

8. $|x| \geq 100$

9. $|x - 5| < 3$

10. $|x - 10| < 2$

11. $|x + 3| \leq 3$

12. $|x + 1| \leq 4$

13. $|7 - x| > 1$

14. $|\frac{1}{2} - x| > \frac{1}{4}$

15. $|x - 2| \geq \frac{1}{3}$

16. $|4 + x| \geq 6$

17. $|\frac{5}{2} - x| \geq \frac{3}{4}$

18. $|\frac{5}{2} + x| \geq \frac{3}{4}$

19. $|2x - 1| < 3$

20. $|3x - 2| < \frac{1}{2}$

21. $|3x - 2| > 0$

22. $|3x - 2| \geq 0$

23. $|3x - 2| > -1$

24. $|\frac{1}{2}x + 5| \leq 1$

25. $|\frac{2}{3}x - \frac{1}{6}| \leq \frac{1}{4}$

26. $|5x + 3| > 1$

27. $\left|\dfrac{3x - 4}{2}\right| > 7$

28. $\left|\dfrac{5 + 4x}{3}\right| \leq 2$

29. $|4x - 7| \geq 3$

30. $|-2x + 5| \geq 3$

31. $|-\frac{1}{3}x - 2| < 4$

32. $|x^2 - 5| < 4$

33. $|x^2 - 9| < 27$

34. $|x^3 - 13| < 14$

35. $1 \leq |x - 5| < 2$

36. $0 < |x + 3| \leq 4$

37. $1 \leq |6x + 4| \leq 3$

38. $\frac{1}{2} < |\frac{3}{2} - 5x| < \frac{3}{4}$

39. Find all values of b for which the equation $x^2 + bx + 5 = 0$ has
   a. no real solution   b. two real solutions

40. Find all values of b for which the equation $x^2 - 2bx + 6 = 0$ has
   a. no real solution   b. two real solutions

KEY TERMS

equation	interval	solution (root) of an equation
identity	open interval	extraneous solution
conditional equation	closed interval	completing the square
equivalent equations	half-open interval	solution of an inequality
linear equation	bounded interval	linear inequality
quadratic equation	unbounded interval	quadratic inequality

KEY FORMULAS

$$x = \frac{-b \pm \sqrt{b^2 - 4ac}}{2a}$$ quadratic formula $a^2 + b^2 = c^2$ Pythagorean Theorem

---

KEY LAWS

If $a < b$ and $c > 0$, then $ac < bc$.

If $a < b$ and $c < 0$, then $ac > bc$.

$ac > 0$ if $a > 0$ and $c > 0$, or if $a < 0$ and $c < 0$.

$ac < 0$ if $a < 0$ and $c > 0$, or if $a > 0$ and $c < 0$.

REVIEW EXERCISES

In Exercises 1–34, find all real solutions (if any) of the given equation.

1. $\frac{1}{2}x - 7 = 6$

2. $-3t + 2 = -4$

3. $\frac{1}{3}(x - 3) + 4(-x + 2) = -4$

4. $\frac{3}{x + 3} = \frac{-9}{x - 1}$

5. $\frac{3x + 8}{8x - 3} = -2$

6. $\frac{4x - 5}{3 - 7x} = 2$

7. $(3y + 5)(-2y + 1) = 0$

8. $2\sqrt{y} = \frac{1}{\sqrt{y}}$

9. $|3x - 5| = 2$

10. $|\frac{1}{2}x + \frac{1}{3}| = \frac{2}{5}$

11. $5x^2 - 15 = 0$

12. $s^2 - s - 380 = 0$

13. $3y^2 + y + 7 = 0$

14. $4s^2 = 13s + 3$

15. $2s^2 - s - 2 = 0$

16. $4x^2 - 20x + 25 = 0$

17. $\frac{5}{x - 2} = x + 2$

18. $3x - \frac{2}{x} = 6$

19. $2x(x + 4) = (x - 1)(x + 3)$

20. $2x(x + 1) = (x - 1)(x + 3)$

21. $x^4 - x^2 = 2$

22. $x^{5/2} + 2x^{3/2} + \sqrt{x} = 0$

23. $(2x - 1)^2 - 2x + 1 = 0$

24. $\sqrt{2x - 1} = 1 - \sqrt{4x - 1}$

25. $\sqrt{2 - 3x} = 1 + \sqrt{1 - 2x}$

26. $t^3 - 2t^2 = t$

27. $t^3 + 6t^2 = 9t$

28. $(3x - 5)(2x^2 + 13x - 7) = 0$

29. $2x^4 - x^3 - 2x^2 + x = 0$

30. $\frac{4x^2 - 3}{-17x + 7} = 0$

31. $(6 - x^2)^2 - 4 = 0$

32. $(\frac{1}{8} - x^3)^3 = -729$

33. $(x^2 - 16)^4 = 256$

34. $2^{n-1} = x^n$

In the Exercises 35–52, solve the given inequality and express the solutions in terms of intervals.

35. $4x - 7 \le 3$

36. $-2x + 3 > 6$

37. $-2x + 3 > -6$

38. $-7 \le -3x + 2 < 0$

39. $y^2 \le (y + 1)^2$

40. $(2y + 3)(y - \sqrt{3}) < 0$

41. $x^2 - 9x + 20 > 0$

42. $x^2 - 4x + 2 \le 0$

43. $3t^2 + 7t < -2$

44. $t(2t - 1) \ge (t + 1)^2 + 3$

45. $\frac{2x + 1}{-5x + 2} \ge 0$

46. $\frac{5x - 2}{2x + 3} < -1$

47. $\frac{1 + 8x}{x - 3} \ge 2$

48. $(2y + 1)(y + 2)(y - 5) > 0$

49. $|x| \ge \sqrt{2}$

50. $|x - 3| < 0.5$

51. $|-2x + 5| \le 13$

52. $4 < |-3x - 8| \le 6$

53. Find the values of a such that 0 is a solution of the equation

$$\frac{1}{2x - 1} + \frac{3a}{x + 1} = 4a$$

54. Find the values of a such that 1 is a solution of the equation $a^2x - 2ax = 3$.

55. For what values of a is -1 a solution of the following inequality?

$$\frac{2x - a}{3x + a} < -2$$

56. If the sales tax is 5% and amounts to $3.18 on a pair of sunglasses, what is the total cost of the glasses (item cost plus tax)?

57. The hottest and coldest outdoor temperatures ever recorded on the surface of the earth were 136.4 and −126.9 degrees Fahrenheit, respectively. What are the corresponding temperatures in degrees Celsius? Round your answer off to the nearest tenth of a degree. (*Hint:* Use (3) of Section 2.1.)

58. Two boats traveling at the same speed depart from the same port at the same time. One travels north for a while and then heads straight for a second port. The other travels 10 miles south and then 20 miles west and arrives at the second port at exactly the same time as the first boat. How far north did the first boat travel before heading for the second port?

59. A Coast Guard boat is 5 miles north of Bermuda, and a boat suspected of carrying smugglers is 25 miles west of Bermuda and moving directly toward Bermuda. If both boats travel at the same speed, how far will the Coast Guard boat have to travel in order to intercept the other boat? (*Hint:* See Figure 2.22.)

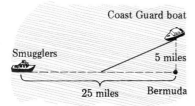

FIGURE 2.22

60. It takes two hours longer for an airplane flying at a constant air speed to fly 750 miles against a 50 mph headwind than it does to make the return trip at the same air speed with a 50 mph tailwind. Determine the air speed of the airplane.

61. If 3 people all mowing at the same rate can mow 4 acres in 8 hours, how long will it take 4 people all mowing at the same rate to mow 3 acres?

62. Bill and John can do a job in 20 minutes. If Bill worked at twice his normal rate, they could do the job in 15 minutes. How long would it take John to do the job alone?

63. A ball is thrown vertically upward from a height of 12 feet with an initial velocity of 48 feet per second. During what time interval is the ball at least 44 feet high?

64. Two square pictures have a combined area of 1000 square inches. The length of a side on one picture is 100 inches less than two thirds the length of a side of the other picture. Use a calculator to find the area of the larger picture.

65. A group of students plans to buy a refrigerator costing $180 and to divide the cost equally. If they can convince two more students to join them (with all paying equal amounts), the cost to each of the original students will decrease by $15. How many students are in the original group?

66. A plane has a cruising speed of 420 miles per hours, and the wind velocity averages 60 miles per hour. Assuming that the flight is with the wind in one direction and against it in the other, how far can the plane travel in a round trip of 7 hours?

67. A printer charges for producing any desired number of copies of a book in the following way. The printing charge depends only on the number of copies printed, and the charge for typesetting is fixed, independent of the number of books to be printed. If the printer is willing to make 20,000 copies of a certain book for $26,000 and to make 30,000 copies of the same book for $34,000, find the cost of typesetting the book.

* 68. A circular pond has a radius of 8 feet, and there is a reed in the middle of the pond that protrudes 4 feet above the water. When the top of the reed is pulled to the side (without bending), the tip just reaches the edge of the pond. How tall is the reed?

69. A walkway cuts diagonally across a 40-foot square yard (see Figure 2.23). If the area of the walkway is 304 square feet, find the width of the walkway.

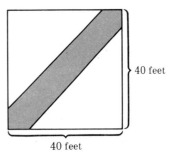

FIGURE 2.23

*70. A person delivers telephone books to three apartment buildings. The first building requires one more than half of all the telephone books. The second building requires one more than half of all the remaining telephone books. The third building requires one more than half of all the telephone books still remaining. If there are five telephone books left over, how many were there to begin with?

71. A tank contained 4 gallons of a mixture of antifreeze and water. After 1 gallon of the mixture was drained off and replaced by 1 gallon of pure antifreeze, the new mixture consisted of 60% antifreeze and 40% water. What percentage of the original mixture was antifreeze?

72. The volume V of a cone of radius r and height h is given by $V = \frac{1}{3}\pi r^2 h$. Solve the equation for r.

73. If p denotes the object distance, q the image distance, and f the focal length of a simple lens, then
$$f = \frac{pq}{p + q}.$$
   a. If $p = 5$ and $f = 3$, determine q.
   b. Suppose p, q, and f are given, with $q = 4$. Assume that if p were tripled, then f would be doubled. Find the value of p.

74. Let $0 < a < b$. Then the harmonic mean h of a and b is given by the equation
$$\frac{1}{h} = \frac{1}{2}\left(\frac{1}{a} + \frac{1}{b}\right)$$
   Solve the equation for b.

75. Recall from Section 2.2 that if P dollars are deposited into an account earning simple interest at an annual rate r, then the amount of interest after n years in Prn, and therefore the amount A_n in the account after n years is given by
$$A_n = P + Prn$$
   Solve the equation for r.

76. If, in Exercise 75, the interest were compounded annually, then the formula for A_n would be
$$A_n = P(1 + r)^n$$
   Solve the equation for r.

77. The heat Q that results when x moles of sulfuric acid are mixed with y moles of water is given by
$$Q = \frac{17{,}860xy}{1.798x + y}$$
   Solve the equation for x.

78. Under certain conditions the percentage efficiency E of an internal combustion engine is given by
$$E = 100\left(1 - \frac{v}{V}\right)^{0.4}$$
   where V and v are, respectively, the maximum and minimum volumes of air in each cylinder.
   a. Solve the equation for v.
   b. Solve the equation for V.

3

Functions and Their Graphs

It is common to associate various pairs of quantities with one another. For example, one can associate the volume of a spherical balloon with the corresponding radius. Thus by knowing the radius, one can compute the volume of the balloon. Similarly, the weather service at an airport such as O'Hare International Airport charts the temperature during the day. To determine what the temperature was at the airport yesterday noon, we would only need to locate yesterday noon on the chart and then read off the temperature at that instant. The dependence of one quantity on another, such as volume on radius or temperature on the time of day, is described mathematically by a function, which is the main topic of this chapter.

One of the most important ways of describing a function involves a mathematical picture called a graph. So before we define and study functions, we will discuss notions related to graphs.

3.1 CARTESIAN COORDINATES FOR THE PLANE

In Section 1.2 we associated the points on a line with real numbers. Now we will associate the points in a plane with pairs of real numbers. This association will aid us in describing functions.

We begin by drawing two perpendicular lines in the plane, a horizontal one called the *x axis* and a vertical one called the *y axis* (Figure 3.1). These are the *coordinate axes*, and they cross at the *origin*, denoted by 0. We now set up a coordinate system for the x axis so that points on it to the right of the origin correspond to positive numbers and points on it to the left of the origin

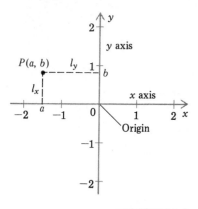

FIGURE 3.1

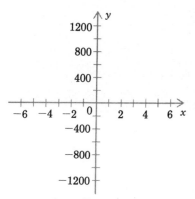

A coordinate system
with different scales on the axes

FIGURE 3.2

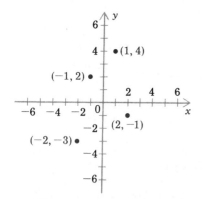

FIGURE 3.3

correspond to negative numbers. Similarly we set up a coordinate system for the y axis so that points on it above the origin correspond to positive numbers and points on it below the origin correspond to negative numbers. The positive directions on the coordinate axes are indicated by arrows (Figure 3.1).

To associate points in the plane with pairs of real numbers, we consider an arbitrary point P in the plane. Through P we draw the line l_x perpendicular to the x axis and the line l_y perpendicular to the y axis. Then l_x crosses the x axis at a point that corresponds to a real number a, called the **x coordinate** (or **abscissa**) of P. Similarly, l_y crosses the y axis at a point corresponding to a real number b, called the **y coordinate** (or **ordinate**) of P (Figure 3.1). Thus P determines the **ordered pair** of numbers a and b, written (a, b) and called an ordered pair because the x coordinate, a, precedes the y coordinate, b, in the pair.

Conversely, if (a, b) is a given ordered pair of real numbers, then the vertical line l_x that crosses the x axis at a and the horizontal line l_y that crosses the y axis at b meet at a single point P, and the x and y coordinates of P are a and b, respectively. Thus we have a correspondence between points in the plane and ordered pairs of numbers. If (a, b) corresponds to P, then we refer to a and b as the **coordinates** of P; we will occasionally write $P(a, b)$ for the point P whose coordinates are (a, b) and refer to (a, b) as the coordinates of P.

Such an association of the points in a plane with ordered pairs of real numbers is called a **Cartesian** (or **rectangular**) **coordinate system**, and a plane endowed with a Cartesian coordinate system is called a **Cartesian plane**. The name "Cartesian" honors René Descartes (1596–1650), the French mathematician who became famous for applying algebraic techniques to the discipline of geometry. Until Section 8.4 the only planes we will consider will have Cartesian coordinate systems, and we will simply call them planes. Moreover, we will frequently refer to points and ordered pairs of real numbers interchangeably. For example, we might refer to the ordered pair $(3, 2)$ as a point.

The spacing between successive integers on an axis of the plane is called the **scale** of the axis. We can adjust the scales of the two coordinate axes to suit our needs; as Figure 3.2 illustrates, the scales of the two axes may be very different from one another.

When we draw a point in the plane corresponding to a given ordered pair of numbers, we say that we **plot**, or **sketch**, the point. In our first few examples, we will plot single points and also sets of points in the plane,

<u>**EXAMPLE 1.**</u>   Determine the x and y coordinates of the following points, and then plot the points.

  a. $(1, 4)$ b. $(-1, 2)$ c. $(2, -1)$ d. $(-2, -3)$

Solution.

  a. The x coordinate is 1; the y coordinate is 4.
  b. The x coordinate is -1; the y coordinate is 2.
  c. The x coordinate is 2; the y coordinate is -1.
  d. The x coordinate is -2; the y coordinate is -3.

The points are plotted in Figure 3.3.   ☐

Turning to sets of points in the plane, we first observe that there are infinitely many points (x, y) such that $x = 0$ (that is, whose x coordinate is 0), and they constitute the y axis. Likewise, the points (x, y) such that $y = 0$ (that is, whose y coordinate is 0) constitute the x axis.

EXAMPLE 2.   Sketch the set of points (x, y) in the plane satisfying
   a. $x = 2$ b. $y = -1$

Solution.
   a. If $x = 2$, then the x coordinate of (x, y) is 2, so the point lies on the vertical line 2 units to the right of the y axis (Figure 3.4a).
   b. If $y = -1$, then the y coordinate of (x, y) is -1, so the point lies on the horizontal line 1 unit below the x axis (Figure 3.4b).   □

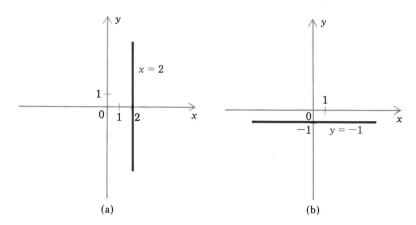

(a)   (b)

FIGURE 3.4

In general, if a is any real number, then the set of all points (x, y) satisfying the equation $x = a$ is a vertical line, and the set of all points (x, y) satisfying the equation $y = a$ is a horizontal line.

EXAMPLE 3.   Sketch the set of points (x, y) in the plane satisfying $(x - 2)(y + 1) = 0$.

Solution.   Observe that $(x - 2)(y + 1) = 0$ if and only if either $x - 2 = 0$ or $y + 1 = 0$, that is, if and only if $x = 2$ or $y = -1$. But the points (x, y) that satisfy $x = 2$ and those that satisfy $y = -1$ are sketched in Figure 3.4a and b. Thus (x, y) satisfies the given equation if and only if it lies on either of the two lines shown in Figure 3.5.   □

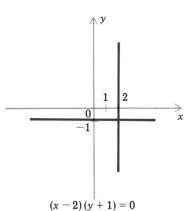

$(x - 2)(y + 1) = 0$

FIGURE 3.5

We can describe more extensive regions of the plane by means of inequalities involving coordinates of the points.

EXAMPLE 4.   Sketch the set of points (x, y) in the plane satisfying
   a. $x > 2$ b. $y > -1$

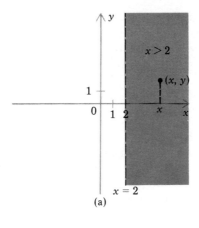

(a)

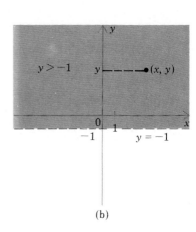

(b)

FIGURE 3.6

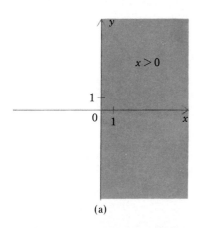

(a)

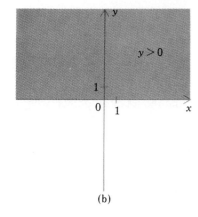

(b)

FIGURE 3.7

Solution.

a. If (x, y) satisfies $x > 2$, then the x coordinate of (x, y) is greater than 2, so that (x, y) lies to the right of the vertical line $x = 2$. Thus the points (x, y) satisfying $x > 2$ are the points to the right of, but not on, the line $x = 2$ (Figure 3.6a).

b. If $y > -1$, then the y coordinate of (x, y) is greater than -1, so that (x, y) lies above the horizontal line $y = -1$. Thus the points (x, y) satisfying $y > -1$ are the points above, but not on, the line $y = -1$ (Figure 3.6b).   □

Had we desired the points (x, y) satisfying $x \geq 2$ in part (a) of Example 4, we would have found them to be the points either to the right of or on the line $x = 2$.

The points (x, y) satisfying $x > 0$ are the points to the right of the y axis (Figure 3.7a), and the points satisfying $y > 0$ are the points above the x axis (Figure 3.7b). Thus the points in the plane satisfying *both* $x > 0$ and $y > 0$ are the points in the upper righthand fourth of. the plane, determined by the coordinate axes and called the *first quadrant*, or *quadrant I*. The other three quadrants are also identified by the positivity or negativity of the coordinates; all four quadrants are identified in Figure 3.8.

EXAMPLE 5.   Sketch the set of points (x, y) in the plane satisfying $(x - 2)(y + 1) > 0$.

Solution.   Notice that $(x - 2)(y + 1) > 0$ if and only if either $x - 2 > 0$ and $y + 1 > 0$, or $x - 2 < 0$ and $y + 1 < 0$. This is equivalent to $x > 2$ and $y > -1$, or $x < 2$ and $y < -1$. From Figure 3.6a and b we deduce that the points (x, y) satisfying $x > 2$ and $y > -1$ are those in region I of Figure 3.9. In a similar way we find that the points (x, y) satisfying $x < 2$ and $y < -1$ are

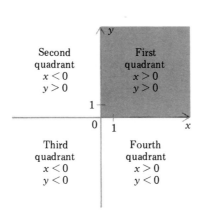

FIGURE 3.8

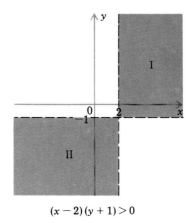

$(x - 2)(y + 1) > 0$

FIGURE 3.9

those in region II of Figure 3.9. Combining the information obtained so far, we find that the points satisfying the given inequality are those shaded in Figure 3.9.   ☐

The Distance Between Two Points

If two points are on the same horizontal or vertical line in the plane, then the distance between them is defined to be the distance between their x coordinates or their y coordinates, respectively. Thus the distance between $P(x_1, y_1)$ and $Q(x_2, y_1)$ is $|x_2 - x_1|$, and the distance between $P(x_1, y_1)$ and $R(x_1, y_2)$ is $|y_2 - y_1|$ (Figure 3.10).

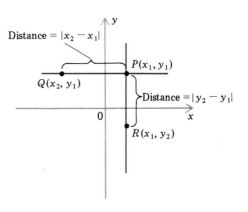

FIGURE 3.10

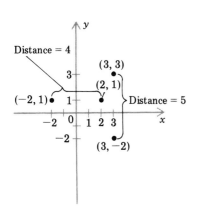

FIGURE 3.11

EXAMPLE 6.   Find the distances between the following pairs of points, illustrated in Figure 3.11.

    a. $(-2, 1)$ and $(2, 1)$ b. $(3, 3)$ and $(3, -2)$

Solution.

    a. The distance between $(-2, 1)$ and $(2, 1)$ is $|2 - (-2)| = |4| = 4$.

    b. The distance between $(3, 3)$ and $(3, -2)$ is $|-2 - 3| = |-5| = 5$.   ☐

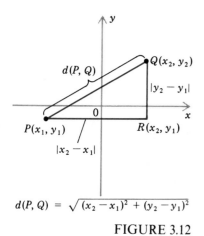

$$d(P, Q) = \sqrt{(x_2 - x_1)^2 + (y_2 - y_1)^2}$$

FIGURE 3.12

We define the **distance** $d(P, Q)$ between any two points $P(x_1, y_1)$ and $Q(x_2, y_2)$ in the plane by the following formula:

Distance Formula

$$d(P, Q) = \sqrt{(x_2 - x_1)^2 + (y_2 - y_1)^2}$$

This definition is based on the Pythagorean Theorem, which relates the lengths of the sides of a right triangle (see Figure 3.12).

EXAMPLE 7.   Find the distance between $(2, 6)$ and $(4, -1)$.

Solution.   By the distance formula, with $P = (2, 6)$ and $Q = (4, -1)$, we have

$$d(P, Q) = \sqrt{(4 - 2)^2 + (-1 - 6)^2}$$
$$= \sqrt{2^2 + (-7)^2}$$
$$= \sqrt{53} \quad \square$$

The Midpoint of a Line Segment

Let $P(x_1, y_1)$ and $Q(x_2, y_2)$ be two distinct points in the plane, so that P and Q determine a line segment (Figure 3.13). Then the coordinates (x, y) of the midpoint M of that line segment can be determined from $x_1, y_1, x_2,$ and y_2. We will not derive the formula for the coordinates but will simply state it:

Midpoint Formula

$$M(x, y) = \left(\frac{1}{2}(x_1 + x_2), \frac{1}{2}(y_1 + y_2)\right)$$

Thus the x coordinate of the midpoint is the average of the x coordinates of the two points P and Q, and the y coordinate of the midpoint is the average of the y coordinates of the two points.

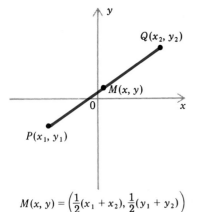

$$M(x, y) = \left(\frac{1}{2}(x_1 + x_2), \frac{1}{2}(y_1 + y_2)\right)$$

FIGURE 3.13

EXAMPLE 8.   Find the coordinates of the midpoint of the line segment joining $(3, 4)$ and $(5, -2)$ (Figure 3.14).

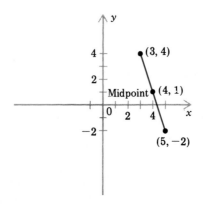

FIGURE 3.14

Solution. By the midpoint formula the coordinates are given by

$$\left(\frac{1}{2}(3+5), \frac{1}{2}[4+(-2)]\right)$$

which reduces to (4, 1). $\square$

EXERCISES 3.1

1. Set up a coordinate system, and plot the points $(-2, 0)$, $(1, 2)$, $(-3, 1)$, $(-1.5, -2.5)$, $(0, -\sqrt{2})$, $(1, -2)$.

2. Set up a coordinate system, and plot the points (100, 200), $(-150, 100)$, $(-200, 200)$, $(300, -150)$, (400, 0), $(0, -300)$.

In Exercises 3–22, set up a coordinate system and sketch the set of points (x, y) in the plane satisfying the given equation or inequality.

3. $y = 5$

4. $x = -2$

5. $x > 1$

6. $y < -1$

7. $xy = 0$

8. $(x - 3)(y + 2) = 0$

9. $\left(\frac{1}{2}x + 1\right)(2 - y) = 0$

10. $x^2 = 1$

11. $y^2 = 4$

12. $x^2 + x = 0$

13. $y^3 - y = 0$

14. $y^2 - 8y + 16 = 0$

15. $x^2 - 3x - 4 = 0$

16. $x^2 + 5x - 14 = 0$

17. $y^2 = 4y + 12$

18. $x^2 \geq 1$

19. $x^2 < 1$

20. $xy > 0$

21. $xy < 0$

22. $x^2 + y^2 = 0$

In Exercises 23–34, find the distance between the two given points.

23. $(-1, 2)$ and $(-1, 4)$

24. (10, 12) and (5, 12)

25. (5, 12) and (0, 0)

26. (3, 3) and (5, 1)

27. $(4, -3)$ and $(-2, -5)$

28. $(-1, -3)$ and $(-7, 4)$

29. $(4, -3)$ and $(-1, 9)$

30. $\left(\frac{1}{4}, \frac{1}{4}\right)$ and $\left(-\frac{1}{4}, -\frac{1}{4}\right)$

31. $(3, \sqrt{2})$ and $(-\sqrt{2}, 4)$

32. $(0.1, -0.1)$ and (0.4, 0.3)

33. (0, 0) and $(3a, 4a)$, where $a \geq 0$

34. (0, 0) and $(3a, 4a)$, where $a < 0$

In Exercises 35–42, find the coordinates of the midpoint of the line segment joining the two given points.

35. (4, 2) and (8, 10)

36. $(-1, 3)$ and $(5, -1)$

37. (0, 4) and (0, 10)

38. $(-2, 4)$ and (3, 10)

39. (3.2, 1.4) and $(-1.8, 4.3)$

40. $(-2.2, -0.8)$ and (3.6, 2.5)

41. $(-a, a)$ and (a, a)

42. (1, 1) and $\left(\frac{2}{3}a, \frac{1}{4}a\right)$

43. Determine the points on the x axis that are 2 units from the point (4, 1).

44. Determine the points on the y axis that are 3 units from the point (2, 0).

45. Let $P(a, b)$ and $Q(c, d)$ be distinct points. Find a formula for the coordinates of the point two thirds of the distance from P to Q.

46. A *rhombus* is a polygon whose four sides are equal in length. Show that the points (0, 0), (3, 0), $(2, \sqrt{5})$, and $(5, \sqrt{5})$ are the vertices of a rhombus.

47. A triangle is *isosceles* if two of its sides have equal length. Show that the points $(0, 4)$, $(3, -1)$, and $(3 + 3\sqrt{2}, 3)$ are the vertices of an isosceles triangle.

Exercises 48–50 illustrate how algebra can be used to prove geometric facts.

48. Show that the line segments joining the midpoints of the sides of an equilateral triangle form an equilateral triangle. (*Hint:* Let the three vertices of the given triangle be (0, 0), $(a, 0)$, and $(a/2, \sqrt{3}\,a/2)$.)

49. Show that the diagonals of a square intersect at their midpoints. (*Hint:* Set up a coordinate system as in Figure 3.15. Then compute the coordinates of the midpoint M_1 of the diagonal joining P_1 and P_3, and the midpoint M_2 of the diagonal joining P_2 and P_4. Finally, show that $M_1 = M_2$.)

50. The following instructions were found at the intersection of Walnut and Broad Streets in Philadelphia: "To find the treasure chest, go west 2 kilometers, then south 1 kilometer, then east 3 kilometers, then south 2 kilometers, and, finally, go half the distance in the direction of the starting point at the intersection of Walnut and Broad." What is the (straight-line) distance between the intersection and the treasure chest?

51. The infield in a baseball field is square, 90 feet on a side. What is the distance from home plate to second base?

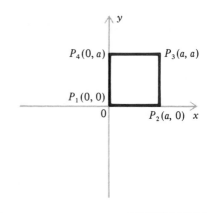

FIGURE 3.15

3.2
FUNCTIONS

Frequently temperature in degrees Fahrenheit is associated with the corresponding temperature in degrees Celsius, and the area of a circle is associated with the radius of the circle. In mathematics such associations are called functions.

DEFINITION 3.1

A ***function*** consists of a domain and a rule. The ***domain*** is a collection of real numbers. The ***rule*** assigns to each number in the domain one and only one number.

The total collection of numbers that a function assigns to the numbers in its domain is called the ***range*** of the function.

Functions are normally denoted by the letters f or g and occasionally by other letters such as h, F, G, or H (in the same part of the alphabet). The value assigned by a function f to a member x of its domain is written $f(x)$ and is read "f of x" or "the value of f at x." If f assigns $\sqrt{2}$ to the number -1, we would write $f(-1) = \sqrt{2}$, whereas if f assigns $\frac{2}{3}$ to 0.4, then we would write $f(0.4) = \frac{2}{3}$. A function f is like a machine that applies the rule of f to each number x in the domain of f and thereby produces the number $f(x)$ in the range of f (Figure 3.16).

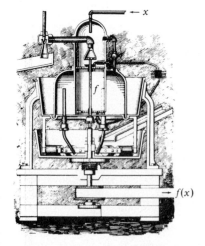

FIGURE 3.16

Caution: A function must make an assignment to *each* number in its domain. For example, if the domain of a function f is $(-\infty, \infty)$, then f must make an assignment to each real number. Moreover, a function assigns *only one* number to any given number in its domain. Thus a function cannot assign both -2 and 3 to a single number in its domain. Nor can a function assign $\pm \sqrt{x}$ to each nonnegative number x, since each positive number x would then be assigned two numbers, $\sqrt{x}$ and $-\sqrt{x}$. Simply remember: A function assigns *one and only one* number to each number in its domain.

Examples of Functions

Our first example of a function f is obtained by assigning the number 4 to each real number. Thus we write

$$f(x) = 4 \quad \text{for all real } x \tag{1}$$

For the particular values $-\pi$, 0, and $\sqrt{3}$ of x we have

$$f(-\pi) = 4, \qquad f(0) = 4, \quad \text{and} \quad f(\sqrt{3}) = 4$$

The range of f consists of the single number 4, which is assigned to each number.

More generally, if a function f assigns the same number c to each real number x, then we write

$$f(x) = c \quad \text{for all real } x$$

and call f a **constant function**. Its range consists of the single number c. The function described in (1) is thus a constant function, with $c = 4$. Notice that although Definition 3.1 stipulates that no single member of the domain of a function can be assigned more than one number in the range, it does permit more than one number in the domain of a function to be assigned the same number in the range, as a constant function amply illustrates.

Now let f be the function that assigns $x^2 + 3x - 1$ to each real number x. Then we write

$$f(x) = x^2 + 3x - 1 \quad \text{for all } x \tag{2}$$

<u>EXAMPLE 1.</u>   For the function given in (2), determine the value of
a. $f(-4)$ b. $f(x + h)$

Solution.
a. We let $x = -4$ in (2):

$$f(-4) = (-4)^2 + 3(-4) - 1 = 16 - 12 - 1 = 3$$

b. We substitute $x + h$ for x in (2):

$$f(x + h) = (x + h)^2 + 3(x + h) - 1$$
$$= x^2 + 2xh + h^2 + 3x + 3h - 1 \quad \square$$

Next let f be the function that assigns $\sqrt{x}$ to each nonnegative real number x. Then f is given by

$$f(x) = \sqrt{x} \quad \text{for all } x \geq 0 \tag{3}$$

This is the **square root function**. Its domain consists of all nonnegative numbers, as does its range.

EXAMPLE 2.   For the function f given in (3), determine the value of
a. $f(2)$ b. $f(\frac{1}{9})$ c. $f(x^2)$

Solution.
a. Letting $x = 2$ in (3), we have $f(2) = \sqrt{2}$.
b. Letting $x = \frac{1}{9}$ in (3), we have

$$f\left(\frac{1}{9}\right) = \sqrt{\frac{1}{9}} = \frac{1}{3}$$

c. Substituting x^2 for x in (3), we have

$$f(x^2) = \sqrt{x^2} = |x| \quad \square$$

We remark that although $\sqrt{x}$ is defined for nonnegative values of x, by contrast $\sqrt{x^2}$ is defined for all numbers x. Thus from part (c) of Example 2 we find that

$$\sqrt{(-3)^2} = |-3| = 3$$

For another example of a function, let g be the function that assigns $(x - 1)/(x + 3)$ to each real number x except -3. Then

$$g(x) = \frac{x - 1}{x + 3} \quad \text{for} \quad x \neq -3 \tag{4}$$

EXAMPLE 3.   For the function g given in (4), determine the value of

a. $g(0)$ b. $g\left(\frac{1}{x}\right)$

Solution.
a. Here we let $x = 0$ in (4) and obtain

$$g(0) = \frac{0 - 1}{0 + 3} = -\frac{1}{3}$$

b. We substitute $1/x$ for x in (4):

$$g\left(\frac{1}{x}\right) = \frac{\dfrac{1}{x} - 1}{\dfrac{1}{x} + 3} = \frac{1 - x}{1 + 3x} \quad \square$$

Notice that -3 is not in the domain of the function g defined by (4), so $g(-3)$ is undefined. If we attempted to compute $g(-3)$ by means of the formula in (4), we would obtain

$$g(-3) \overset{?}{=} \frac{-3 - 1}{-3 + 3} \overset{?}{=} \frac{-4}{0}$$

but $-4/0$ is undefined.

We say that two functions are **equal** if they have the same domains and if to any given number in their common domains their rules associate the same number. Thus the functions g and h given by

$$g(x) = \frac{x - 1}{x + 3} \quad \text{for} \quad x \neq -3 \quad \text{and} \quad h(t) = \frac{t - 1}{t + 3} \quad \text{for} \quad t \neq -3$$

are equal, since their domains both consist of all real numbers different from -3 and their rules associate the same number to any given number in their common domain. As these examples illustrate, the specific letter that is used for the variable in describing a function is irrelevant.

Two functions are *not equal* if either their domains or their rules are distinct. Consequently if

$$g(x) = \frac{x - 1}{x + 3} \quad \text{for} \quad x \neq -3$$

and (5)

$$G(x) = \frac{x - 1}{x + 3} \quad \text{for} \quad x \geq 0$$

then g and G are *not* equal, because the domain of g contains all negative numbers except -3, whereas the domain of G contains no negative numbers.

The rule of a function can be given in two or more parts. For example, consider the function f whose rule is given by

$$f(x) = \begin{cases} x^5 - 3 & \text{for } x < 2 \\ \sqrt{x} & \text{for } x \geq 4 \end{cases}$$ (6)

The domain of f consists of the intervals $(-\infty, 2)$ and $[4, \infty)$. The rule by which a number is assigned to a number x in the domain of f depends on whether x is in $(-\infty, 2)$ or in $[4, \infty)$.

EXAMPLE 4. For the function f given by (6), find the value of
 a. $f(-1)$ b. $f(16)$

Solution.
 a. Since $-1 < 2$, we let $x = -1$ in the top expression on the right of (6):

$$f(-1) = (-1)^5 - 3 = -4$$

 b. Since $16 > 4$, we let $x = 16$ in the bottom expression on the right of (6):

$$f(16) = \sqrt{16} = 4 \quad \square$$

An Alternative Way of Describing a Function

Suppose f is the function that associates the area of a circle with its radius x and is defined by

$$f(x) = \pi x^2 \quad \text{for} \quad x \geq 0$$

Then a second way of describing f is obtained by writing y for $f(x)$:

$$y = \pi x^2 \quad \text{for} \quad x \geq 0 \tag{7}$$

In (7) each nonnegative value of x determines a particular value of y, and in this way y "depends" on x, so y is called a ***dependent variable***. In contrast, x is an ***independent variable***.

Functions that describe physical relationships are frequently presented in variable notation; moreover, the letters used for variables usually relate to the physical aspects of the quantities. Since (7) represents the area corresponding to any given radius, the function is frequently given in variable notation by

$$A = \pi r^2 \quad \text{for} \quad r \geq 0 \tag{8}$$

where A stands for area and r for radius. In a like manner, we could describe the function that assigns the temperature C in degrees Celsius to any given temperature F in degrees Fahrenheit by

$$C = \frac{5}{9}(F - 32) \quad \text{for} \quad F \geq -459.67 \tag{9}$$

Notice that the domain of this function consists of all numbers greater than or equal to -459.67, which represents "absolute zero," theoretically the lowest Fahrenheit temperature possible. The variables in (8) and (9) were chosen to remind us of the physical quantities they represent.

Most functions we will encounter express $f(x)$ (or y) in terms of x. When giving such a formula, we normally specify the numbers in the domain directly after the formula, as in (1)–(9). However, when the domain is to consist of all numbers for which the formula is meaningful, we normally omit mention of the domain. Thus we may write

$$f(x) = x^2 + 3x - 1 \quad \text{and} \quad g(x) = \frac{x - 1}{x + 3} \tag{10}$$

respectively, for the functions presented in (2) and (4), since each of their domains consists of all numbers for which the expression on the right side of the equation makes sense. Nevertheless we cannot omit mention of the domain of G in

$$G(x) = \frac{x - 1}{x + 3} \quad \text{for} \quad x \geq 0$$

(which appeared in (5)) because $(x - 1)/(x + 3)$ makes sense for negative as well as nonnegative numbers. If the domain is understood and the rule of the function is simple, we sometimes designate the function by an expression only. Thus we could refer to the functions in (10) as $x^2 - 3x + 1$ and $(x - 1)/(x + 3)$, respectively.

When we do not specify the domain outright, we still need to be aware of the domain. Sometimes it is important to determine precisely those real numbers that are members of the domain.

EXAMPLE 5.   Let

$$f(x) = \frac{x + 2}{x^2 - 9}$$

Determine the domain of f.

   Solution.   A number x is in the domain of f if and only if it does not make the denominator 0, that is, if and only if $x^2 - 9 \neq 0$. Since $x^2 - 9 = 0$ means that $x^2 = 9$, and thus $x = 3$ or $x = -3$, it follows that the domain consists of all real numbers except 3 and -3.   □

EXAMPLE 6.   Find the domain of the function given by

$$f(x) = \frac{1}{\sqrt{x^2 - 16}}$$

   Solution.   Since the square root is defined only for nonnegative numbers, $\sqrt{x^2 - 16}$ is defined only for x such that $x^2 - 16 \geq 0$. But

$$\frac{1}{\sqrt{x^2 - 16}}$$

is not defined when the denominator is 0, that is, when $x^2 - 16 = 0$. Consequently the domain of f consists of all x such that $x^2 - 16 > 0$, or equivalently, $x^2 > 16$. This set consists of all numbers in the intervals $(-\infty, -4)$ and $(4, \infty)$.   □

Example 6 illustrates two facts that help in determining domains of functions defined by formulas:
   1. Division by zero is undefined.
   2. Square roots of negative numbers are undefined.

An Alternative Definition of Function

An alternative definition of function that is in common use employs the notion of ordered pair introduced in Section 3.1. To set the stage for the alternative definition, let f be a function. For any number x in the domain of f, we combine the numbers x and $f(x)$ to form the ordered pair $(x, f(x))$. Notice that because f is a function, there is precisely one pair whose first entry is x, namely, $(x, f(x))$. The collection of all such pairs $(x, f(x))$ for x in the domain of f identifies the function f.

Conversely, any set of ordered pairs (x, y) of numbers with the property that the first entries in any two distinct pairs are different defines a unique function f: The domain of f is the collection of first entries in the pairs, and for each such pair (x, y), $f(x)$ is the number y.

These comments lead to the following alternative definition of function.

DEFINITION 3.2   A *function* is a nonempty collection of ordered pairs of real numbers no two of which have the same first entry.

In this framework the collection of ordered pairs (x, x^2) defines the function given by the formula $f(x) = x^2$, and the collection of ordered pairs $(x, \sqrt{x})$ for $x \geq 0$ defines the square root function. These sets are sometimes written as

$$\{(x, x^2) | x \text{ is a real number}\}$$

and

$$\{(x, \sqrt{x}) | x \geq 0\}$$

respectively.

Although we have two definitions of function, we prefer to rely only on the first, the one appearing as Definition 3.1.

EXERCISES 3.2

In Exercises 1–12, find the indicated values of the given function.

1. $f(x) = -14$; $f(-2)$, $f(\sqrt{3})$, $f(-x)$

2. $f(x) = x^2 - 3$; $f(5)$, $f(-4)$, $f(x - 2)$

3. $f(t) = t^2 - 3t + 4$; $f(0)$, $f(-1)$, $f(t^2)$

4. $f(t) = \dfrac{t + 4}{t - 1}$; $f(2)$, $f\left(\dfrac{1}{3}\right)$, $f\left(\dfrac{1}{t}\right)$

5. $f(u) = \dfrac{4u^2 - 2u}{u^2 - 0.09}$; $f(10^{-1})$, $f(a)$

6. $g(x) = \dfrac{\dfrac{1}{x} - 1}{\dfrac{1}{x} + 2}$; $g(1)$, $g(-2)$, $g\left(\dfrac{1}{x}\right)$

7. $g(x) = \sqrt{x + 5}$; $g(-1)$, $g(2)$, $g(x^2 - 5)$

8. $h(z) = \sqrt[3]{z - 4}$; $h(12)$, $h(-4)$, $h(z + a)$

9. $h(z) = 2z^{1/3} - z^{3/2}$; $h(64)$, $h(z^6)$

10. $h(z) = \sqrt{\dfrac{2z^2 - 1}{5z^2 - 1}}$; $h(-1)$, $h(1)$, $h(\sqrt{z})$

11. $f(x) = \begin{cases} -x & \text{for } x < 1 \\ 0 & \text{for } x > 3 \end{cases}$; $f(-5)$, $f(\pi)$

12. $f(x) = \begin{cases} \dfrac{x - 1}{x + 2} & \text{for } x \neq -2 \\ 1 & \text{for } x = -2 \end{cases}$; $f(2)$, $f(-2)$

In Exercises 13–20, find the values of y at the given values of x.

13. $y = -2x + 3$; $x = -4$, $x = 0$

14. $y = 3x^2 - 2x + 7$; $x = -2$, $x = -\frac{1}{3}$

15. $y = \dfrac{3x - 4}{2x^2 + 3x - 5}$; $x = 0$, $x = \frac{1}{2}$

16. $y = \dfrac{x^2 - 1}{x^2 + 1}; x = 1, x = -3$

17. $y = \dfrac{1}{x} - \dfrac{1}{x^2}; x = -1, x = \frac{1}{3}$

18. $y = \sqrt{\dfrac{x}{x - 5}}; x = 9, x = 6$

19. $y = \dfrac{\sqrt[3]{x}}{\sqrt[3]{x} - 1}; x = -8, x = 54$

20. $y = 3|x|\sqrt{1 - x}; x = 1, -2$

In Exercises 21–42, find the domain of the given function.

21. $f(x) = 4x^6 + \sqrt{3x^3 - \dfrac{1}{3}x^2 + 1}$

22. $f(x) = \dfrac{1}{x - 7}$

23. $g(x) = \dfrac{3}{4x} - \dfrac{4}{2x - 5}$

24. $g(x) = \dfrac{x + 1}{x^2 - 4}$

25. $g(x) = \dfrac{x + 2}{x^2 + 4}$

26. $g(x) = \dfrac{x - 3}{x^2 - 9}$

27. $h(z) = \dfrac{1}{(z - 1)(z + 6)}$

28. $h(z) = \dfrac{1}{z^2 + 2z + 1}$

29. $h(z) = \dfrac{1}{z^4 + z^2 + 3}$

30. $f(x) = \dfrac{1}{\sqrt{x}}$

31. $g(x) = \dfrac{1}{\sqrt{-x}}$

32. $h(x) = \dfrac{1}{\sqrt{x - 5}}$

33. $y = \sqrt{x^2 - 4}$

34. $y = \dfrac{1}{\sqrt{x^2 - 4}}$

35. $y = \sqrt{\dfrac{x}{x - 1}}$

36. $y = \sqrt{x(x - 3)}$

37. $y = \sqrt[3]{x - 1}$

38. $y = \sqrt{\sqrt{x} - 1}$

39. $y = \sqrt{x - \dfrac{1}{\sqrt{x}}}$

40. $y = \sqrt{\dfrac{x^2 - 1}{x^2 + 1}}$

41. $f(x) = \dfrac{|x|}{x}$

42. $f(x) = \begin{cases} 2x - 3 & \text{for } x \le 0 \\ x + \dfrac{5}{x} & \text{for } x > 1 \end{cases}$

43. Determine which of the following functions have a range that contains -2.

a. $f(x) = \sqrt{x + 1}$
b. $f(x) = \sqrt{1 - x}$

c. $f(x) = \dfrac{x + 2}{x^2 - 4}$ ***d.** $f(x) = \dfrac{x - 2}{x^2 + 4}$

44. Determine which of the following collections of pairs of real numbers define a function according to Definition 3.2.
a. all (x, y) with $x^2 + y^2 = 1$ and $y \ge 0$
b. all (x, y) with $x^2 + y^2 = 1$ and $x \ge 0$
c. all (x, y) with $x = y^3$.
d. all (x, y) with $x = \sqrt{y}$

45. Let $f(x) = (x - 2)^2 + 3(x - 2) - 1$ and $g(x) = x^2 - 3x - 3$. Determine whether or not $f = g$.

46. Let

$$f(x) = \dfrac{|x|}{x} \quad \text{and} \quad g(x) = \begin{cases} -1 & \text{for } x < 0 \\ 1 & \text{for } x > 0 \end{cases}$$

Determine whether or not $f = g$.

47. Let

$$f(x) = \dfrac{1}{x - \sqrt{x^2 - 1}} \quad \text{and} \quad g(x) = x + \sqrt{x^2 - 1}$$

Show that $f = g$.

48. Let $f(x) = \dfrac{2x - 3}{x - 2}$. Show that $f(f(x)) = x$.

49. Let $f(x) = \dfrac{1}{x + 1}$. Show that $f\left(\dfrac{1}{x}\right) = xf(x)$.

50. Let $f(x) = \dfrac{1 + x}{1 - x}$. Show that $f\left(\dfrac{1}{x}\right) = -f(x)$.

51. Let $f(x) = 2.3x^3 + \pi x - 1.7$. Use a calculator to approximate
a. $f(3.5)$ b. $f(-0.2)$ c. $f(13.9)$

52. Let $g(t) = 6.54t^2 - \dfrac{0.55}{t}$. Use a calculator to approximate
a. $g(1.21)$ c. $g(4.97)$
b. $g(6.01)$ d. $g(0.03)$

53. Let $h(z) = \sqrt{z^2 + 3}$. Use a calculator to approximate
a. $h(2.3)$ b. $h\left(\dfrac{4}{3}\right)$ c. $h(\sqrt{2} + 1)$

54. Define a function that expresses the circumference of a circle as a function of
a. the radius
b. the diameter

55. Define a function that expresses the area of a square as a function of
a. the length of a side of the square
b. the length of a diagonal

56. One inch is 2.54 centimeters. Define a function that converts
a. from inches to centimeters
b. from centimeters to inches

3.3 FUNCTIONS AS MODELS

Experimental data collected by scientists usually consist of long lists of numbers. For example, if a ball is dropped into a deep well and the distance traveled by the ball is ascertained each tenth of a second, then the following data might result:

Time elapsed (in seconds):	.1	.2	.3	.4	.5	.6
Depth (in feet):	.16	.64	1.44	2.56	4.00	5.76

Aside from the fact that measurements are at best only approximations to reality, it is unwieldy to handle long lists of numbers. Moreover, it is hard to extrapolate precisely most values not listed in a table. For instance, how far will the ball have traveled in 1.43 seconds or in $\sqrt{2}$ seconds?

For these reasons scientists usually construct "models" for the phenomena they investigate. This means that they formulate physical laws, formulas, or functions that express the dependence of certain quantities on other quantities. One such formula gives the distance an object dropped from rest will fall in t seconds if influenced only by gravity:

$$f(t) = 16t^2 \quad \text{for} \quad t \geq 0 \tag{1}$$

Here t is measured in seconds and $f(t)$ in feet. With such a formula we can confidently compute the values mentioned in the preceding paragraph.

EXAMPLE 1.   Using the formula in (1), determine the distance traveled by the ball during
a. 1.43 seconds
b. $\sqrt{2}$ seconds

Solution.
a. By (1) with $t = 1.43$, the distance traveled in 1.43 seconds is given by

$$f(1.43) = 16(1.43)^2 = 32.7184 \text{ (feet)}$$

b. By (1) with $t = \sqrt{2}$, the distance traveled in $\sqrt{2}$ seconds is given by

$$f(\sqrt{2}) = 16(\sqrt{2})^2 = 16(2) = 32 \text{ (feet)} \quad \square$$

A second model arises from Sir Isaac Newton's famous Second Law of Motion, which states that if an object has a mass of m kilograms, then the force F (in newtons) required to accelerate it at a rate of a (meters per second squared) is the product of the mass and the acceleration. This is expressed by the formula

$$F = ma \qquad (2)$$

Under the assumption that the mass is constant, the force F is a function of the acceleration.

EXAMPLE 2.   Use (2) to determine the force required to accelerate a moped with a mass of 100 kilograms at a rate of 3 meters per second squared.

    Solution.   By (2), with $m = 100$ and $a = 3$, we have

$$F = (100)(3) = 300 \text{ (newtons)} \qquad \square$$

Variation and Proportionality

Certain types of models have occurred so frequently in the sciences that they have prompted some special terminology. For example, if c is a fixed nonzero number and if x and y are related by the formula

$$y = cx \qquad (3)$$

for all x in a given collection of real numbers, then y is said to be **proportional** to (or to be **directly proportional** to or to **vary directly** as) x. In this case c is called the **constant of proportionality**; sometimes the letter k is used in place of c. Thus length measured in centimeters is proportional to length measured in inches, because if x denotes the length of an object in inches and y the length in centimeters, then x and y satisfy the equation

$$y = 2.54x \quad \text{for} \quad x \geq 0$$

Likewise, Hooke's Law states that within certain bounds the force y needed in order to keep a spring stretched x units beyond its natural length is given by the formula

$$y = kx$$

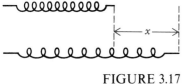

FIGURE 3.17

where k is the so-called **spring constant** of the spring. (See Figure 3.17.) In this case the force is proportional to the distance the spring is stretched.

EXAMPLE 3.   Ohm's Law states that the current i flowing in an electrical circuit is proportional to the voltage V applied to the circuit. Express Ohm's Law by means of a formula.

    Solution.   Using (3) with V and i replacing x and y, respectively, we may write

$$i = cV$$

where c is a constant. $\square$

In passing, we mention that the voltage V in Example 3 can be either positive or negative (corresponding to the two directions the current can flow through the circuit). The constant c is the **conductivity** of the wire, and its reciprocal is the **resistance** of the wire.

EXAMPLE 4.   The volume V of a spherical balloon is proportional to the cube of the radius r. Express this fact by means of a formula involving V and r.

Solution.   We use (3), with V and r^3 substituted for y and x, respectively. This leads us to the formula

$$V = cr^3$$

If appropriate units are used, then $c = \frac{4}{3}\pi$, so that

$$V = \frac{4}{3}\pi r^3 \quad \square$$

EXAMPLE 5.   One of Albert Einstein's most important contributions to physics was his observation that mass and energy are equivalent. According to Einstein's theory, when mass is converted to energy, the amount E of energy released is proportional to the amount m of mass converted. Express this result by means of a formula.

Solution.   Using (3) and realizing that mass must be nonnegative, we have

$$E = km \quad \text{for} \quad m \geq 0 \tag{4}$$

where we have used k for the constant of proportionality. Actually, if appropriate units are used, then $k = c^2$, where c is the speed of light in a vacuum. Thus (4) becomes

$$E = mc^2 \quad \text{for} \quad m \geq 0$$

perhaps the most famous physical formula of all time. $\square$

If c is a nonzero constant and x and y satisfy

$$y = \frac{c}{x} \tag{5}$$

for all x in a given set of real numbers, then y is said to be **inversely proportional** to x. Notice that if $c > 0$, then an increase in x corresponds to a decrease in y. Again, c is the **constant of proportionality**.

EXAMPLE 6.   Boyle's Law states that if the temperature of a gas is constant, then the pressure p of the gas is inversely proportional to the volume V of the gas. Express Boyle's Law by means of a formula.

Solution.   Using formula (5) and realizing that the volume of gas must be positive, we may express Boyle's Law as

$$p = \frac{c}{V} \quad \text{for} \quad V > 0$$

where, as usual, c is a constant.   □

Joint Proportionality

Frequently a physical quantity depends on several other quantities, rather than just one. The terms "proportional" and "inversely proportional" may also be employed in such cases. For example, if c is a nonzero constant and if x, y, and z are related by the formula

$$z = cxy$$

then we say that z is *jointly proportional* (or more simply, *proportional*) to x and y. If

$$z = \frac{cx}{y}$$

then we say that z is *proportional* to x and *inversely proportional* to y.

<u>EXAMPLE 7.</u>   The volume V of a cylinder is jointly proportional to the square of the radius r of the base and to the height h of the cylinder. Express this fact by means of a formula involving r and h.

Solution.   We have

$$V = cr^2h$$

In fact, the constant c of proportionality is π, which yields for formula

$$V = \pi r^2 h$$

for the volume of the cylinder.   □

The terminology of joint proportionality or inverse proportionality extends to any number of variables. In the next example we have a quantity that is jointly proportional to two other quantities and inversely proportional to a third.

<u>EXAMPLE 8.</u>   Newton's Law of Gravitation states that the force of attraction F between two objects is proportional to each of their masses m_1 and m_2 and is inversely proportional to the square of the distance r between them. Express this law by means of a formula.

Solution.   Using G as the constant of proportionality, we have

$$F = \frac{Gm_1 m_2}{r^2}$$

The constant G is known as the ***universal gravitational constant.***   □

EXERCISES 3.3

1. The perimeter p of a square is proportional to the length s of a side. Express this fact by means of a formula. (The constant is 4.)

2. The circumference C of a circle is proportional to the diameter d. Express this fact by means of a formula. (The constant is π.)

3. The surface area S of a cube is proportional to the square of the side length s. Express this fact by means of a formula. (The constant is 6.)

4. The water pressure p at a point is proportional to the depth h of the point below the surface of the water. Express this fact by means of a formula.

5. Planck's quantum theory of light states (in part) that a light ray may be considered as a particle called a photon. The energy E of the photon is proportional to the frequency v of the light ray. Express this result by means of a formula. (The constant of proportionality, known as ***Planck's constant***, is denoted by h.)

6. Joule's Law states that the rate P at which heat is produced in a wire is proportional to the square of the current i flowing in the wire. Express Joule's Law by means of a formula.

7. In the study of motion it is sometimes assumed that the air resistance R on an object is proportional to the square of the velocity v of the object. Express this assumption by means of a formula.

8. Use the result of Example 4 to solve this problem.
    a. Would three spherical lead balls of radius 2 inches produce enough lead for a single spherical lead ball of radius 3 inches? Explain your answer.
    ☰ b. Would the three spherical balls of part (a) produce enough lead for one cube of side length 4.7 inches? Explain your answer.

9. The area A of an equilateral triangle is proportional to the square of the length s of a side.
    a. Express this fact by means of a formula. (The constant is $\sqrt{3}/4$.)

    b. If the area of an equilateral triangle is $9\sqrt{3}$ square inches, determine the length of a side.

10. The volume V of a cube is proportional to the cube of the length s of a side.
    a. Express this fact by means of a formula. (The constant is 1.)
    b. By how much is the volume increased when the length of a side is increased from 3 inches to 4 inches?

11. The length l_f in feet of an object is proportional to the length l_m in meters.
    a. Express this fact by means of a formula. (The constant is $\frac{1250}{381}$.)
    b. Express the fact that the length l_y in yards of an object is proportional to the length l_m in meters by means of a formula. Using (a), determine the constant of proportionality.

12. Distance D_k in kilometers is proportional to distance D_m in miles.
    a. Express this fact by means of a formula. (The constant is 1.609344.)
    ☰ b. Express by means of a formula the fact that the distance in miles is proportional to distance in kilometers. Using a calculator, approximate the value of the constant of proportionality.
    ☰ c. Use (a), with the numerical value for the constant, to determine the approximate number of kilometers in 93×10^6 miles (which is often taken as the average distance between the earth and sun).

13. For each gram of hydrogen that is converted to helium in a star, approximately 0.0277 gram is converted to energy. Use the result of Example 5 to compute the amount of energy released for each gram of hydrogen converted to helium. (*Hint:* The speed of light is approximately 3×10^{10} centimeters per second. Your answer will be in gram centimeters square per second squared, commonly called "ergs.")

14. According to Hubble's Law, the velocity v with which a galaxy is receding from any other galaxy is pro-

portional to the distance d between the galaxies. (Hubble's Law is sometimes viewed as evidence for the "big bang" theory, which holds that the universe originated with a gigantic explosion.)

a. Express Hubble's Law by means of a formula.

b. If a galaxy whose distance from the earth is 10 million light years (that is, 9.46×10^{19} kilometers) is receding from the earth with a velocity of 170 kilometers per second, find the approximate value of the constant of proportionality.

c. How fast would a galaxy recede if it were located 15 million light years from earth?

15. The kinetic energy K of an object of given mass m is proportional to the square of the speed v of the object.

a. Express this fact by means of a formula. (The constant of proportionality in this case is equal to one half the mass of the object, if appropriate units are used.)

b. Suppose two bundles of newspapers, one weighing 10 kilograms and the other 30 kilograms, hit a wall with velocities v_1 and v_2 meters per second, respectively. If both bundles possess the same kinetic energy, what is the relationship between their velocities?

16. To a very good approximation, the time (or period) T it takes a pendulum to make one complete swing is proportional to the square root of the length l of the pendulum.

a. Express this fact by means of a formula.

b. To double the period, how must the length of the pendulum be changed?

17. The wavelength λ of a light ray is inversely proportional to its frequency v (the number of waves per second). Express this fact by means of a formula. (The constant of proportionality in this case is c, the velocity of light.)

18. Kepler's Third Law of Planetary Motion states that the square of the period T of a planet (the time required for the planet to make one revolution about the sun) is proportional to the cube of the average distance a from the planet to the sun.

a. Express Kepler's Third Law by means of a formula.

☐ b. If the period is 365 days and the average distance is 93×10^6 miles, use a calculator to determine the constant c.

☐ c. Use a calculator to determine the constant c in (b) if the period is given in seconds and the average distance in kilometers (see Exercise 12).

19. Coulomb's Law states that the electric force F exerted by one charged particle on another is proportional to the charges q_1 and q_2 on the two particles and is inversely proportional to the square of the distance r between the two particles. Express Coulomb's Law by means of a formula.

20. The *escape velocity* on a planet is the velocity that a spaceship must possess when leaving the surface of the planet in order to escape the gravitational field of the planet without firing any of the ship's rockets. The *escape energy* of a spaceship on a planet is the amount of energy required to impart escape velocity to the spaceship.

a. The escape energy E of a spaceship is proportional to the square of the escape velocity v_e and to the mass m of the spaceship. Express this fact by means of a formula.

b. The escape velocity on a star is defined as for a planet. The escape velocity v_e on a star that is contracting is inversely proportional to the square root of the radius r of the star. Express this fact by means of a formula. (If the radius of the star becomes so small that the escape velocity exceeds the velocity of light, not even light can escape from the star, and the star becomes a "black hole.")

3.4
THE GRAPH
OF A FUNCTION

When a function is described by means of a formula, the formula can tell us much about the function. But pictures are also associated with functions and can themselves give much information about the function. A pictorial representation of a function is called a graph.

DEFINITION 3.3 Let f be a function. Then the set of all points $(x, f(x))$ such that x is in the domain of f is called the **graph** of f (Figure 3.18), and we say that such a point $(x, f(x))$ is **on the graph of** f.

Since no two distinct points on the graph of a function can have the same x coordinate (see Definition 3.2), and since a vertical line is characterized by

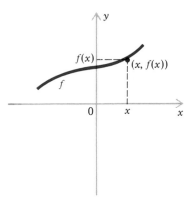

The graph of a function f

FIGURE 3.18

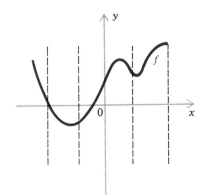

Vertical lines intersect the graph of
a function f at most once

FIGURE 3.19

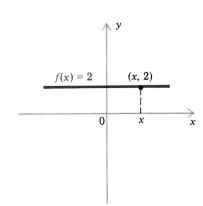

FIGURE 3.20

all its points having the same x coordinate, it follows that no vertical line can intersect the graph of a function more than once (Figure 3.19). You might look quickly at the graphs that follow and see that this is true.

EXAMPLE 1.   Let $f(x) = 2$. Sketch the graph of f.

Solution.   Notice that if x is any real number, then $(x, f(x))$ is the point $(x, 2)$, which means that the y coordinate of each point on the graph of f is 2. Thus the graph is the horizontal line drawn in Figure 3.20.   □

If f is any constant function, say $f(x) = c$, then as in Example 1, the graph of f is a horizontal line that crosses the y axis at the point whose y coordinate is c.

EXAMPLE 2.   Let $g(x) = x$. Sketch the graph of g.

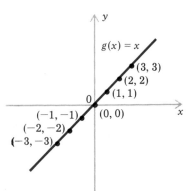

FIGURE 3.21

Solution.   To get an idea of what the graph looks like, let us plot a few of its points with the help of the following table:

x	-3	-2	-1	0	1	2	3
$g(x)$	-3	-2	-1	0	1	2	3

Then we smoothly connect the points plotted. The graph appears to be a straight line bisecting the first and third quadrants (Figure 3.21).   □

EXAMPLE 3.   Let $h(x) = -x$. Sketch the graph of h.

Solution.   Again we draw a smooth line through a few points plotted with the help of the following table:

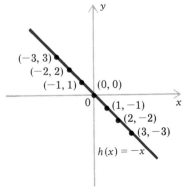

x	-3	-2	-1	0	1	2	3
$h(x)$	3	2	1	0	-1	-2	-3

FIGURE 3.22   The graph appears to be a diagonal line bisecting the second and fourth quadrants (Figure 3.22).   □

EXAMPLE 4.   Sketch the graph of the *absolute value function*:

$$f(x) = |x|$$

Solution.   Since $f(x) = x$ for $x \geq 0$, it follows that the part of the graph of f to the right of the y axis coincides with the corresponding part of the graph of g in Example 2. Since $f(x) = -x$ for $x < 0$, it follows that the part of the graph of f to the left of the y axis coincides with the corresponding part of the graph of h in Example 3. Combining this information, we obtain the graph of f (Figure 3.23).   ☐

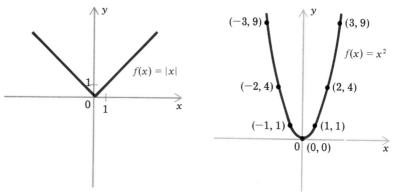

The graph of the absolute value function

FIGURE 3.23   FIGURE 3.24

EXAMPLE 5.   Let $f(x) = x^2$. Sketch the graph of f.

Solution.   As before, we first make a table of values for f:

x	-3	-2	-1	0	1	2	3
$f(x)$	9	4	1	0	1	4	9

Not only do all points in the graph of f appear to be on or above the x axis but also, as x grows, the value $f(x)$ seems to grow ever faster. We obtain the graph sketch in Figure 3.24.   ☐

EXAMPLE 6.   Let $g(x) = x^2 - 4$. Sketch the graph of g.

Solution.   First we fill in the table below:

x	-3	-2	-1	0	1	2	3
$g(x)$	5	0	-3	-4	-3	0	5

Notice that for each value of x, $g(x)$ is exactly 4 units less than the corresponding value for $f(x)$ obtained in the solution of Example 5. Therefore we conclude that the graph of g has the same shape as the graph in

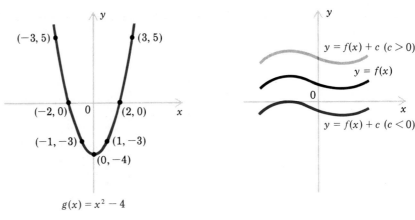

$$g(x) = x^2 - 4$$

FIGURE 3.25

FIGURE 3.26

Example 5 but is shifted down four units. Figure 3.25 reflects this observation. □

We can go a step beyond Example 6. Suppose we know the graph of a function f, and let

$$h(x) = f(x) + c$$

Then the graph of h has the same shape as the graph of f but is raised by c units if $c > 0$ and is lowered $|c|$ units if $c < 0$ (Figure 3.26).

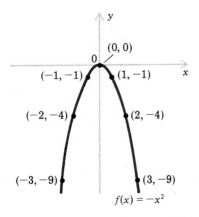

$$f(x) = -x^2$$

FIGURE 3.27

EXAMPLE 7. Let $f(x) = -x^2$. Sketch the graph of f.

Solution. The table we make is related to the one in the solution of Example 5:

x	-3	-2	-1	0	1	2	3
$f(x)$	-9	-4	-1	0	-1	-4	-9

As before, we plot the corresponding points and joint them with a smooth curve (Figure 3.27). □

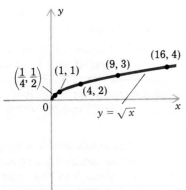

$$y = \sqrt{x}$$

The graph of the square root function

FIGURE 3.28

EXAMPLE 8. Sketch the graph of the function given by $y = \sqrt{x}$.

Solution. We prepare the following table for values of x that are nonnegative and hence in the domain:

x	0	$\frac{1}{4}$	1	4	9	16
y	0	$\frac{1}{2}$	1	2	3	4

Connecting the corresponding points with a smooth curve, we obtain the graph of the square root function (Figure 3.28). □

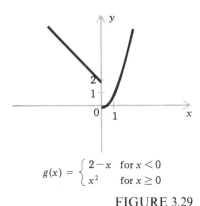

$$g(x) = \begin{cases} 2-x & \text{for } x < 0 \\ x^2 & \text{for } x \geq 0 \end{cases}$$

FIGURE 3.29

EXAMPLE 9.   Let

$$g(x) = \begin{cases} 2 - x & \text{for } x < 0 \\ x^2 & \text{for } x \geq 0 \end{cases}$$

Sketch the graph of g.

Solution.   Since the formula for g is in two parts, one for x in $(-\infty, 0)$ and the other for x in $[0, \infty)$, we will determine the corresponding parts of the graph one at a time. First, $g(x) = 2 - x$ for $x < 0$, so the value of $g(x)$ for any such x is two units greater than the value of $h(x)$ in Example 3 (where $h(x) = -x$). Therefore the part of the graph of g to the left of the y axis has the same shape as the corresponding part of the graph of h but is raised 2 units. Second, $g(x) = x^2$ for $x \geq 0$, so that the part of the graph of g to the right of the y axis coincides with the corresponding part of the graph of f in Example 5 (where $f(x) = x^2$). Combining this information, we obtain the graph of g (Figure 3.29).   □

Symmetry

Symmetry is a quality of art and plays a central role in architecture. One of the most famous examples of symmetry in architecture is the Taj Mahal. The notion of symmetry also relates to graphs of functions in mathematics.

We say that the graph of a function is **symmetric with respect to the y axis** if the portion of the graph to the right of the y axis is mirrored to the left of the y axis. In mathematical terms this means that if (x, y) is on the graph, so is $(-x, y)$ (Figure 3.30). The graphs of the functions in Figures 3.23–3.25 and 3.27 have this property, as you can check. In each case the reason is that

$$f(-x) = f(x) \quad \text{for all } x \text{ in the domain of } f \tag{1}$$

A function that satisfies (1) is called an **even function**. Thus $|x|, x^2, x^2 - 4$, and $-x^2$ are all even functions.

The Taj Mahal, Agra, India.

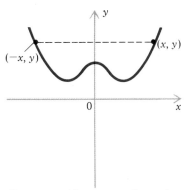

Symmetry with respect to the y axis

FIGURE 3.30

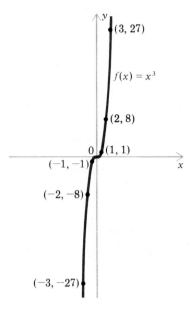

(3, 27)

$f(x) = x^3$

(2, 8)

0 (1, 1)

(−1, −1)

(−2, −8)

(−3, −27)

FIGURE 3.31

Suppose that n is an even integer, so that $n = 2m$ for some integer m. Then

$$(-x)^n = (-x)^{2m} = [(-x)^2]^m = (x^2)^m = x^{2m} = x^n$$

This means that any function that can be described by a formula containing only even powers of x is an even function, and its graph is symmetric with respect to the y axis.

Our next function has a different type of symmetry.

EXAMPLE 10.   Let $f(x) = x^3$. Sketch the graph of f.

x	−3	−2	−1	0	1	2	3
$f(x)$	−27	−8	−1	0	1	8	27

Plotting the corresponding points and connecting them smoothly, we obtain the graph in Figure 3.31.   □

If

$$f(-x) = -f(x) \quad \text{for all } x \text{ in the domain of } f \tag{2}$$

then f is called an **odd function**. The functions x and x^3 both are odd functions, as is any function of the form x^n, where n is odd. The graph of an odd function is said to be **symmetric with respect to the origin**, by which we mean that corresponding pairs of points on opposite sides of the origin are either both on or both not on the graph. Mathematically this means that if (x, y) is on the graph, then so is $(-x, -y)$ (Figure 3.32).

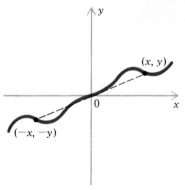

Symmetry with respect to the origin

FIGURE 3.32

Intercepts

Our custom so far has been to plot several points before trying to sketch the graph of a function. This raises the question of which points we should plot. Since we always begin a figure with the coordinates axes, it is reasonable to plot the points of the graph that lie on either of the coordinate axes.

A *y intercept* of the graph of a function f is the y coordinate of any point where the graph of f meets the y axis (Figure 3.33). Since $x = 0$ for any point (x, y) on the y axis, it follows that the graph of a function has a y intercept if and only if 0 is in the domain of f, in which case $f(0)$ is the one and only y intercept. Analogously, an *x intercept* of the graph of a function is the x coordinate of any point where the graph meets the x axis (Figure 3.33). Since $y = 0$ for any point (x, y) on the x axis, it follows that the x intercepts of the graph of f are the values of x (if any) for which $f(x) = 0$. In brief,

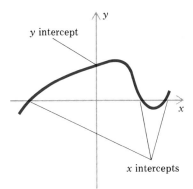

i. $f(0)$ is the y intercept (if 0 is in the domain of f).
ii. x is an x intercept if $f(x) = 0$.

FIGURE 3.33

EXAMPLE 11.   Let $g(x) = x^2 - 4$. Find the y and x intercepts of the graph of g.

Solution.   Since $g(0) = -4$, we know the y intercept is -4. To find the x intercepts we must find those values of x for which $g(x) = 0$, that is, for which $x^2 - 4 = 0$. But these are the values of x for which $x^2 = 4$, which yields $x = 2$ and $x = -2$. Consequently there are two x intercepts, 2 and -2; Figure 3.34 supports this claim.   ☐

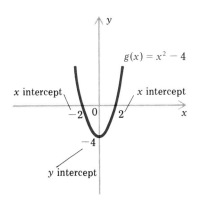

FIGURE 3.34

If we consider the function given by $h(x) = x^2 + 4$, we reach very different conclusions about intercepts. On the one hand, $h(0) = 4$, so the y intercept of h is 4. On the other hand, $h(x) = 0$ if and only if $x^2 + 4 = 0$. But since there are no real solutions of $x^2 + 4 = 0$, the graph of h has no x intercepts (see Figure 3.35). Thus we see that there may or may not be x intercepts for the graph of a given function. Similarly, there may or may not be a y intercept, depending on whether or not 0 is in the domain.

EXAMPLE 12.   Let $f(x) = x^2 + 2x - 3$. Find the x and y intercepts of the graph of f.

Solution.   Since $f(0) = 0^2 + 0 - 3 = -3$, the y intercept is -3. To find the x intercepts we solve the equation $f(x) = 0$ for x. This leads to the equation

$$x^2 + 2x - 3 = 0$$

which, after factorization, becomes

$$(x + 3)(x - 1) = 0$$

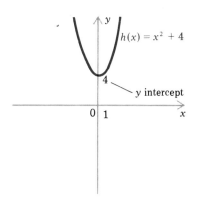

FIGURE 3.35

Thus we see that $f(x) = 0$ if $x = -3$ or $x = 1$. This means that the x intercepts are -3 and 1. The graph of f, which we will learn how to sketch in Section 4.1, is shown in Figure 3.36.   □

EXAMPLE 13.   Let $f(x) = 1/x$. Show that the graph of f has no x or y intercepts.

Solution.   Since division by 0 is undefined, 0 is not in the domain of f, so there is no y intercept. Next, notice that $1/x$ is never 0, so that there are no values of x for which $f(x) = 0$. This means that there are no x intercepts. The graph of f, which we will be able to sketch later (see Section 4.4), is shown in Figure 3.37 and supports the claim that no intercepts exist.   □

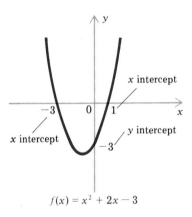

$f(x) = x^2 + 2x - 3$

FIGURE 3.36

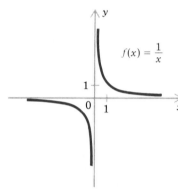

A graph with no intercepts

FIGURE 3.37

EXERCISES 3.4

In Exercises 1–18, sketch the graph of the given function and indicate any intercepts.

1. $f(x) = x + 3$

2. $f(x) = -x - \frac{1}{3}$

3. $f(x) = -x + 0.7$

4. $f(x) = -2 + |x|$

5. $g(x) = 5 - |x|$

6. $g(x) = |x + 2|$

7. $g(x) = |2 - x|$

8. $h(x) = x^2 + 1$

9. $h(x) = 3 - x^2$

10. $y = \dfrac{1}{x} - 3$

11. $y = \dfrac{x + 1}{x} = 1 + \dfrac{1}{x}$

12. $g(x) = -\dfrac{1}{x}$

13. $g(x) = \sqrt{x} - 1$

14. $g(x) = \sqrt{x} + 3$

15. $h(x) = \begin{cases} -x & \text{for } x \le 0 \\ -2x & \text{for } x > 0 \end{cases}$

16. $h(x) = \begin{cases} x^2 & \text{for } x \le 0 \\ x & \text{for } x > 1 \end{cases}$

17. $h(x) = \begin{cases} \dfrac{1}{x} & \text{for } x < -1 \\ 3x & \text{for } x \ge -1 \end{cases}$

18. $h(x) = \begin{cases} |x| & \text{for } x < 1 \\ x^2 & \text{for } x \ge 1 \end{cases}$

In Exercises 19–30, determine all intercepts of the graph of the given function.

19. $f(x) = (x + 1)^2$

20. $f(x) = (x + 1)^2 - 4$

21. $f(x) = x^2 - 4x$

22. $f(x) = x^2 - 6x + 8$

23. $g(x) = x^2 - 6x + 9$

24. $g(x) = x^2 - 6x + 10$

25. $g(x) = \dfrac{1}{x - 1}$

26. $h(x) = \dfrac{1}{x^2}$

27. $h(x) = \dfrac{x + 1}{x + 2}$

28. $f(x) = \sqrt{x + 5}$ **30.** $f(x) = ||x| - 2|$

29. $f(x) = \sqrt{2 - x}$

In Exercises 31–38, sketch the graph and indicate whether the graph is symmetric with respect to the *y* axis, the origin, or neither.

31. $f(x) - 5$

32. $f(x) = 2x$

33. $f(x) = 2x - 1$

34. $g(x) = -|x|$

35. $g(x) = |x| - 4$

36. $g(x) = \dfrac{|x|}{x}$

37. $h(x) = x^2 - \dfrac{1}{2}$

38. $h(x) = \dfrac{1}{x} + 2$

In Exercises 39–48, determine which functions are even, which are odd, and which are neither.

39. $f(x) = x^2 + x$

40. $f(x) = \dfrac{x^2}{x^4 + 2}$

41. $f(x) = x^3 + 4x^5$

42. $f(x) = x - \dfrac{1}{x}$

43. $g(x) = x^2 - \dfrac{1}{x}$

44. $g(x) = \sqrt{x^2 - 1}$

45. $g(x) = \dfrac{x + 3}{x - 5}$

46. $h(x) = \dfrac{x^2 + 3}{x - 5}$

47. $h(x) = \dfrac{x^2 + 1}{x^2 - 1}$

48. $h(x) = \sqrt{x^2 + 3}$

49. Determine which of the graphs in Figure 3.38 are graphs of functions.

50. Are there any nonzero functions that are both even and odd? Give reasons for your answer.

51. Let $f(x) = (x - 1)^2$. By plotting several points, sketch the graph of f. What is the relationship between the graph of f and the graph of $g(x) = x^2$?

52. Let $f(x) = |x + 1|$. By plotting several points, sketch the graph of f. What is the relationship between the graph of f and the graph of $g(x) = |x|$?

53. For any real number x, let $[x]$ denote the largest integer less than or equal to x. Thus $[2] = 2$, $[-\tfrac{7}{5}] = -2$, and $[\pi] = 3$. Let

$$f(x) = [x]$$

Then f is called the **greatest integer function**, or the **staircase function**. Sketch the graph of f.

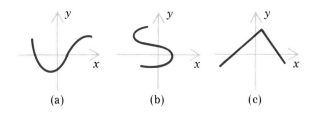

(a)   (b)   (c)

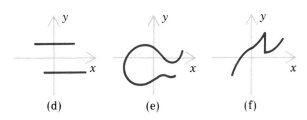

(d)   (e)   (f)

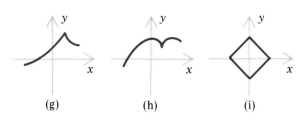

(g)   (h)   (i)

FIGURE 3.38

3.5
LINES
IN THE PLANE

In this section we analyze straight lines in the plane and show that any nonvertical line can be associated with a function. For convenience we will separate the collection of straight lines in the plane into two categories: vertical and nonvertical lines.

Consider the vertical line drawn in Figure 3.39. Notice that the *x* coordinate of each point on the line is -1. Indeed, a point (x, y) is on the line if and only if $x = -1$. As a result, we say that the equation $x = -1$ is an equation of the given vertical line.

More generally, if c is any fixed number, then we call the equation $x = c$ an equation of the vertical line whose x intercept is c.

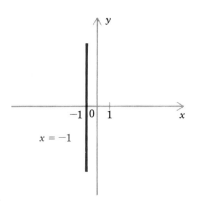

FIGURE 3.39

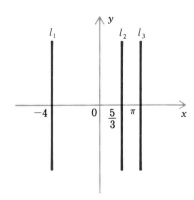

FIGURE 3.40

EXAMPLE 1.   Find equations for the vertical lines l_1, l_2, and l_3 appearing in Figure 3.40.

Solution.   We have the following:

$$l_1 : x = -4 \qquad l_2 : x = \frac{5}{3}, \quad \text{and} \quad l_3 : x = \pi \quad \square$$

This is really all there is to say about vertical lines. Henceforth we will restrict our discussion to nonvertical lines.

Slope of a Nonvertical Line   Let l be any nonvertical line in the plane, and let $P_1(x_1, y_1)$ and $P_2(x_2, y_2)$ be any two distinct points on l. Since l is not vertical, $x_1 \ne x_2$. It turns out that the ratio

$$\frac{y_2 - y_1}{x_2 - x_1}$$

depends only on the line l and *not* on the particular points P_1 and P_2. To see this, observe that if $P_3(x_3, y_3)$ and $P_4(x_4, y_4)$ are any two points on l with P_3 distinct from P_4, then triangles $P_1 Q P_2$ and $P_3 R P_4$ are similar (Figure 3.41), so that

$$\frac{y_2 - y_1}{x_2 - x_1} = \frac{y_4 - y_3}{x_4 - x_3}$$

This common ratio is called the *slope* of l and is usually denoted by m. Thus if $P_1(x_1, y_1)$ and $P_2(x_2, y_2)$ are any two distinct points on l, then the slope of l is given by the following equation:

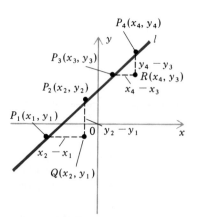

FIGURE 3.41

Slope of a Line
$m = \dfrac{y_2 - y_1}{x_2 - x_1}$

(1)

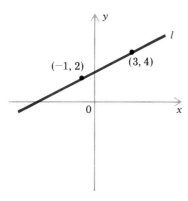

(−1, 2)   (3, 4)

0   x

l

y

FIGURE 3.42

EXAMPLE 2.   Find the slope of the line *l* that contains the points (−1, 2) and (3, 4) (Figure 3.42).

Solution.   If we let (−1, 2) be $P_1(x_1, y_1)$ and (3, 4) be $P_2(x_2, y_2)$, then by (1) the slope *m* of *l* is given by

$$m = \frac{4 - 2}{3 - (-1)} = \frac{2}{4} = \frac{1}{2} \quad \square$$

Caution: We have not defined the slope of a vertical line, nor will we do so. If one tries to apply (1) to a vertical line $x = c$, one obtains the meaningless expression

$$\frac{y_2 - y_1}{c - c}$$

Next we will use the notion of slope to lead us to two types of equations for nonvertical lines—the point–slope equation and the slope–intercept equation.

Point–Slope Equation of a Line

Suppose that we are given the slope *m* and a point $P_1(x_1, y_1)$ on a nonvertical line *l*. We would like to find an equation that is satisfied by any point $P(x, y)$ on *l*. If we apply (1) with $P_2(x_2, y_2)$ replaced by $P(x, y)$, we find that

$$m = \frac{y - y_1}{x - x_1}$$

which is equivalent to the following equation:

Point–Slope Equation of a Line
$y - y_1 = m(x - x_1)$

(2)

A point–slope equation involves the slope *m* and a single point $P_1(x_1, y_1)$ on the line. A point $P(x, y)$ is on the line if and only if *x* and *y* satisfy (2).

EXAMPLE 3.   Find a point–slope equation of the line *l* that has slope 0 and passes through the point (−2, −1). Then sketch the line.

Solution.   Using (2) with $m = 0$, $x_1 = -2$, and $y_1 = -1$, we obtain the point–slope equation

$$y - (-1) = 0(x - (-2))$$

To sketch *l* we first simplify the equation:

$$y + 1 = 0$$
$$y = -1$$

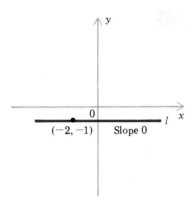

FIGURE 3.43

Consequently l is horizontal and crosses the y axis at the point $(0, -1)$ (Figure 3.43).   □

In Example 3 we found that the given line, which has slope 0, is horizontal. By the same method we deduce that any line having slope 0 and passing through, say (x_1, y_1), has an equation of the form

$$y - y_1 = 0(x - x_1)$$

which simplifies to

$$y - y_1 = 0$$

and then to

$$y = y_1 \qquad (3)$$

It follows from (3) that a line with slope 0 is horizontal.

EXAMPLE 4.   Find a point–slope equation of the line that passes through the point $(3, 2)$ and has the given slope. Then sketch the line.

    a. $\frac{1}{2}$ b. 3   c. -1

Solution.   In each case we use (2):

a. We have $m = \frac{1}{2}, x_1 = 3$, and $y_1 = 2$, so that a point–slope equation of the line is

$$y - 2 = \frac{1}{2}(x - 3)$$

which simplifies to

$$y - 2 = \frac{1}{2}x - \frac{3}{2}$$

or

$$y = \frac{1}{2}x + \frac{1}{2}$$

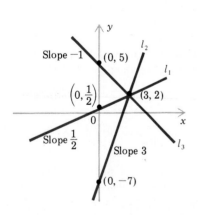

FIGURE 3.44

In order to sketch the graph, we find a second point on the line by setting $x = 0$ and noting that $y = \frac{1}{2}$. Therefore $(0, \frac{1}{2})$ is a second point, and the line, l_1, can be drawn (Figure 3.44).

b. This time $m = 3$, $x_1 = 3$, and $y_1 = 2$, so that

$$y - 2 = 3(x - 3)$$

which simplifies to

$$y - 2 = 3x - 9$$

or
$$y = 3x - 7$$

For the graph we again find a second point on the line by letting $x = 0$ and noting that $y = -7$. Therefore $(0, -7)$ is a second point, and the line, l_2, can be drawn (Figure 3.44).

c. Now we have $m = -1$, $x_1 = 3$, and $y_1 = 2$, so that

$$y - 2 = -1(x - 3)$$

which simplifies to

$$y - 2 = -x + 3$$

or
$$y = -x + 5$$

As before, we let $x = 0$ and find that $y = 5$, so that $(0, 5)$ is a second point on the line. The line, l_3, is drawn in Figure 3.44. ☐

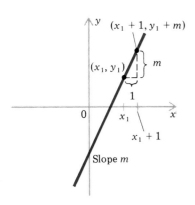

FIGURE 3.45

Let us tie together the results of Examples 3 and 4 and the accompanying Figures 3.43 and 3.44. First, if $m = 0$, then the line is horizontal. Next, if $m > 0$, then the line moves upward from left to right. Finally, if $m < 0$, then the line moves downward from left to right. Moreover, the larger the absolute value of the slope, the steeper the line is.

If we know the slope m and a point (x_1, y_1) on a given line l, then a point–slope equation of l is $y - y_1 = m(x - x_1)$. Thus the point on l whose x coordinate is $x_1 + 1$ satisfies

$$y - y_1 = m(x_1 + 1 - x_1) = m$$

so that
$$y = y_1 + m$$

Therefore an increase of 1 unit in the value of x (from x_1 to $x_1 + 1$) causes a change of m units in the value of y (from y_1 to $y_1 + m$), and the point $(x_1 + 1, y_1 + m)$ also lies on l (Figure 3.45). We will use these ideas to help sketch the line discussed in the next example.

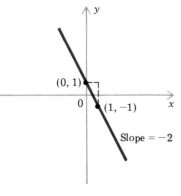

FIGURE 3.46

EXAMPLE 5.   Sketch the line that contains the point $(0, 1)$ and has slope -2.

Solution.   We can sketch the line once we have two points on it. We are given one point on the line, $(0, 1)$. To find a second point we use the fact that the line has slope -2. Thus if we start at $(0, 1)$ and *increase* the value of x by 1, then the value of y must *decrease* by 2. This yields the point $(0 + 1, 1 - 2)$, or $(1, -1)$, on the line. Now we can sketch the line (Figure 3.46). ☐

In order to find an equation of the line *l* that contains two given distinct points, we first use (1) to determine the slope of *l*, and then use a point–slope equation.

EXAMPLE 6.   Find an equation of the line *l* that contains the points $(-1, 2)$ and $(3, 4)$. Then show that $(7, 6)$ is on *l*.

Solution.   By the solution of Example 2 we know that *l* has slope $\frac{1}{2}$. Thus if we apply the point–slope equation of the line with $m = \frac{1}{2}$ and $P_1(x_1, y_1) = (-1, 2)$, we obtain

$$y - 2 = \frac{1}{2}(x - (-1))$$

which is equivalent to

$$y - 2 = \frac{1}{2}(x + 1)$$

To show that $(7, 6)$ is on *l* we let $x = 7$ and $y = 6$ and see that the equation is valid:

$$6 - 2 = \frac{1}{2}(7 + 1) \quad \square$$

Slope–Intercept Equation of a Line

A nonvertical line *l* must cross the *y* axis at some point. The *y* coordinate of that point is called the **y intercept** of *l* and is usually denoted by *b*. Thus $(0, b)$ is a point on *l*. Now suppose that we are given the slope *m* and the *y* intercept *b* of a nonvertical line *l*. If we let $P_1(x_1, y_1) = (0, b)$ and apply the formula for the point–slope equation of a line, we find that $P(x, y)$ is on the line *l* provided that

$$y - b = m(x - 0)$$

which simplifies to give us the following equation:

Slope–Intercept Equation of a Line
$y = mx + b$

(4)

The slope–intercept equation of a line involves the slope *m* and the *y* intercept *b* of the line. As before, (x, y) is on the line if and only if (4) is satisfied.

EXAMPLE 7.   Find the slope–intercept equation of the line that has slope -2 and *y* intercept $\frac{1}{2}$. Then sketch the line.

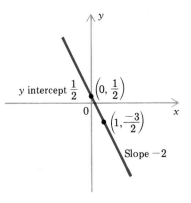

Solution. By (4) with $m = -2$ and $b = \frac{1}{2}$, the slope–intercept equation is

$$y = -2x + \frac{1}{2} \qquad (5)$$

Since the y intercept is $\frac{1}{2}$, we know that $(0, \frac{1}{2})$ is one point on the line. Taking $x = 1$ in equation (5), we find that

$$y = -2(1) + \frac{1}{2} = -\frac{3}{2}$$

FIGURE 3.47   Thus $(1, -\frac{3}{2})$ is a second point on the line, which is sketched in Figure 3.47. □

We have presented two types of equations for a nonvertical line: point–slope and slope–intercept. Other types of equations will appear in Exercises 53, 56, and 58. However, all equations for a given line are equivalent, and information that may be explicit in one equation of the line may be obtained from any other. For example, the y intercept b of a line is explicitly written in the slope–intercept equation $y = mx + b$, but it does not appear explicitly in a point–slope equation $y - y_1 = m(x - x_1)$ unless $x_1 = 0$ and $y_1 = b$. Nevertheless we can determine the y intercept from *any* equation of the line by setting $x = 0$ and solving for y.

<u>EXAMPLE 8.</u>   Find the y intercept of the line with equation $y - 5 = -4(x + 2)$.

Solution.   Setting $x = 0$ and solving for y, we obtain

$$y - 5 = -4(0 + 2) = -8$$
$$y = -8 + 5 = -3$$

Thus the y intercept is -3. □

Parallel and Perpendicular Lines

Among the geometric properties of lines that are easily studied by means of slopes of lines, the two most important are the properties of parallelism and perpendicularity:

> i. Two nonvertical lines are parallel if and only if their slopes are equal.
> ii. Two nonvertical lines are perpendicular if and only if the product of their slopes is -1.

We will not prove the results just stated, which depend only on the definition of slope and basic geometric facts.

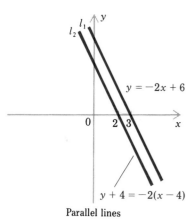

Parallel lines

FIGURE 3.48

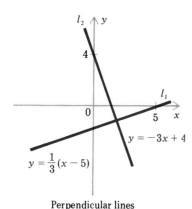

Perpendicular lines

FIGURE 3.49

EXAMPLE 9.   Show that the lines

$$l_1: y = -2x + 6 \quad \text{and} \quad l_2: y + 4 = -2(x - 4)$$

are parallel.

Solution.   The equation of l_1 is the slope–intercept equation of l_1; from it we see that the slope of l_1 is -2. The equation of l_2 is a point–slope equation of l_2; from it we see that the slope of l_2 is also -2. Since the two slopes are equal, the lines are parallel by (i) (Figure 3.48).   □

EXAMPLE 10.   Show that the lines

$$l_1: y = \frac{1}{3}(x - 5) \quad \text{and} \quad l_2: y = -3x + 4$$

are perpendicular.

Solution.   The equation of l_1 is a point–slope equation; from it we see that l_1 has slope $\frac{1}{3}$. The equation of l_2 is the slope–intercept equation of l_2; from it we see that l_2 has slope -3. Since

$$\frac{1}{3}(-3) = -1$$

it follows from (ii) that l_1 and l_2 are perpendicular (Figure 3.49).   □

EXAMPLE 11.   Find an equation of the line l that contains the point $(\frac{1}{2}, \frac{2}{3})$ and is parallel to the line with equation $y = -4x + 7$.

Solution.   The slope of the line with equation $y = -4x + 7$ is -4. It follows from (i) that l also has slope -4. Since we are given a point on l and we now know the slope of l, we use the point–slope equation in (2) with $m = -4$, $x_1 = \frac{1}{2}$, and $y_1 = \frac{2}{3}$ to obtain the equation

$$y - \frac{2}{3} = -4\left(x - \frac{1}{2}\right)$$

for l.   □

Linear Functions   The slope–intercept equation of a nonvertical line is

$$y = mx + b \tag{6}$$

Such an equation yields the function f described by

$$f(x) = mx + b \tag{7}$$

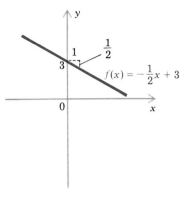

$$f(x) = -\tfrac{1}{2}x + 3$$

FIGURE 3.50

Accordingly, any function of the form given in (6) or (7) is called a **linear function**. Thus the graph of a linear function is a nonvertical line with appropriate slope and y intercept. (Since a vertical line cannot be the graph of a function, a straight line is the graph of a function if and only if it is nonvertical.)

EXAMPLE 12.   Let $f(x) = -\tfrac{1}{2}x + 3$. Sketch the graph of f.

Solution.   We know already that the graph of f is a line. Since the equation is in slope–intercept form, we know the slope is $-\tfrac{1}{2}$ and the y intercept is 3. We place $(0, 3)$ on the graph, and note that an increase of 1 unit in the value of x yields a decrease of $\tfrac{1}{2}$ unit in the y value. From this information we can sketch the graph (Figure 3.50). $\square$

EXERCISES 3.5

In Exercises 1–6, find the slope of the line that contains the points P_1 and P_2.

1. $P_1 = (4, 3), P_2 = (6, 9)$

2. $P_1 = (0, 4), P_2 = (-3, 10)$

3. $P_1 = (-3, -3), P_2 = (-5, 2)$

4. $P_1 = (-5, -4), P_2 = (-9, -7)$

5. $P_1 = (\tfrac{1}{4}, \tfrac{1}{6}), P_2 = (\tfrac{3}{4}, \tfrac{1}{2})$

6. $P_1 = (1.57, 2.86), P_2 = (1.84, 2.56)$

In Exercises 7–12, find a point–slope equation of the line that has slope m and passes through the point P. Then sketch the line.

7. $m = 0, P = (3, -4)$

8. $m = 2, P = (-2, -1)$

9. $m = \tfrac{2}{3}, P = (3, 0)$

10. $m = -4, P = (-\tfrac{1}{2}, 2)$

11. $m = -1, P = (0, \tfrac{3}{2})$

12. $m = \sqrt{2}, P = (\sqrt{2}, 3)$

In Exercises 13–18, find an equation of the line that contains the points P_1 and P_2.

13. $P_1 = (0, 1), P_2 = (2, 5)$

14. $P_1 = (0, -1), P_2 = (3, 8)$

15. $P_1 = (-1, -2), P_2 = (-2, 2)$

16. $P_1 = (2, 3), P_2 = (0, 0)$

17. $P_1 = (2, -1), P_2 = (2, 6)$

18. $P_1 = (4, -3), P_2 = (2, 4)$

In Exercises 19–24, find the slope–intercept equation of the line that has slope m and y intercept b. Then sketch the line.

19. $m = 0, b = -2$

20. $m = 3, b = 4$

21. $m = -\tfrac{4}{5}, b = -1$

22. $m = 1, b = \pi$

23. $m = -\sqrt{3}, b = \tfrac{1}{2}$

24. $m = 0.25, b = 0.4$

In Exercises 25–32, find the slope m and the y intercept b (if any) of the line with the given equation. Then sketch the line.

25. $y = 3$

26. $x = -4$

27. $y = 3x - \tfrac{1}{2}$

28. $3y = x - 4$

29. $2x + 3y = 6$

30. $\tfrac{1}{2}y - 2x = 2$

31. $3.6x - 1.2y = 0$

32. $y - 7 = 3(x - 2)$

In Exercises 33–44, determine whether the two lines having the given equations are parallel, perpendicular, or neither.

33. $x = 0; y = 0$

34. $y = 3; x = -4$

35. $y = x; y = -x$

36. $y = 2x; y = -\tfrac{1}{2}x$

37. $y = 2x; y = -2x$

38. $2x - 3y = 4; y = -\tfrac{3}{2}x$

39. $\tfrac{1}{2}y = x; y = 2(x - 4)$

40. $x = -y; y = 4x + 2$

41. $3x - 5y = 4; 2x - 6y = 3$

42. $4x + 6y = 5; 6x + 9y = 1$

43. $y = 0.01(x - 1); y + 100x = 50$

44. $y - 5 = \frac{1}{2}(x + 3); x - 2y = 6$

In Exercises 45–52, sketch the graph of f.

45. $f(x) = -2$

46. $f(x) = \frac{1}{3}x$

47. $f(x) = -4x$

48. $f(x) = -x + 3$

49. $f(x) = 3(x - 2)$

50. $f(x) = \frac{1}{2}x - 3$

51. $f(x) = -\frac{2}{3}x + 4$

52. $f(x) = -0.5x + 1.5$

53. Let $P_1(x_1, y_2)$ and $P_2(x_2, y_2)$ be two distinct points not on the same vertical line, and let the line l pass through P_1 and P_2. Then $P(x, y)$ is on l provided that

$$(x_2 - x_1)(y - y_1) = (y_2 - y_1)(x - x_1) \qquad (8)$$

This equation is called a **two-point equation** of l.
a. Show that P_1 and P_2 satisfy (8).
b. Use (8) to obtain a point–slope equation of l.

54. Find a two-point equation of the line that passes through each of the following pairs of points. Then sketch the line.
a. $(4, 2)$ and $(3, -3)$
b. $(-1, -2)$ and $(-5, 0)$
c. $(\frac{1}{2}, \frac{3}{2})$ and $(-\frac{5}{2}, \frac{3}{2})$

55. Let l be a nonhorizontal line. The x coordinate of the point at which l crosses the x axis is the **x intercept** of l, and is usually denoted by a. Thus $(a, 0)$ is on l. The x intercept is determined by setting $y = 0$ in an equation of l and then solving for x. Determine the x intercept of the lines having the following equations.
a. $y = 6x - 2$ b. $y - 2 = \frac{1}{2}(x + 1)$ c. $x = -\frac{3}{4}$

56. Let l be a line that is neither horizontal nor vertical, and let its x and y intercepts be a and b, respectively. Then (x, y) is on l provided that

$$\frac{x}{a} + \frac{y}{b} = 1$$

This equation is called the **two-intercept equation** of l. Show that the slope of l is $-b/a$ by writing the equation in the equivalent slope–intercept form.

57. Find the two-intercept equation of the line that has x intercept a and y intercept b. Then sketch the line.
a. $a = -1, b = 2$ c. $a = 0.7, b = 0.3$
b. $a = 3, b = \frac{1}{2}$ d. $a = -2, b = -3$

58. *Any* line l can be described by an equation of the form

$$Ax + By = 1$$

where either $A \neq 0$ or $B \neq 0$. This equation is called a **general linear equation**.
a. Determine the x intercept a (if any) and the y intercept b (if any) of l.
b. Determine the slope (if any) of l.

59. Determine which of the following lines contain the point $(-2, 4)$.
a. $y = 2x + 8$ d. $y - 2 = 2(x + 1)$
b. $y = 2x - 8$ e. $3x + 4y = 22$
c. $y + 2 = -2(x - 1)$ f. $3x + 4y = 10$

60. Determine which of the following lines contain the point $(-3, -1)$.
a. $y = 2x - 5$ d. $y - 1 = \frac{1}{2}(x - 3)$
b. $y = 2x + 5$ e. $-2x + y = -5$
c. $y + 1 = \frac{1}{2}(x + 3)$ f. $2x - y = -5$

61. Suppose that a line l is to pass through $(2, 3)$. What slope must it have in order to pass through $(5, -1)$ as well?

62. Show that a line whose equation is $Ax + By = 0$ passes through the origin. If $B \neq 0$, what is the slope of the line?

63. Find a formula for the linear function whose graph passes through the point $(-1, -4)$ and has slope -2.

64. Find a formula for the linear function whose graph passes through the point $(3, 0)$ and has y intercept $-\frac{1}{3}$.

65. Find a formula for the linear function whose graph has x and y intercepts -1 and 4, respectively.

66. Line l_1 passes through the points $(-2, 1)$ and $(4, -1)$, and line l_2 passes through the points $(5, -3)$ and $(-7, -6)$. Which line has the greater slope?

67. Find an equation of the line that passes through the point $(-2, 5)$ and is parallel to the line with equation $4x + 3y = 1$.

68. Find an equation of the line that passes through the point $(3, -3)$ and is perpendicular to the line with equation $y = 6x - 2$.

69. Find the point of intersection of the lines having equations $2x + y = 0$ and $x - y = -3$. (*Hint:* First solve for y in terms of x in the first equation, and then substitute for y in the second equation.)

70. Find the point of intersection of the lines having equations $2x - 4y = 3$ and $x - y = 1$.

71. Show that the lines having the following equations determine a rectangle: $y = 2x + 1, y = 2x + 5, y = -\frac{1}{2}x + 1, y = -\frac{1}{2}x - 3$.

72. Show that the figure determined by the four points (2, 1), (3, 3) (2, 5), and (1, 3) is a parallelogram by proving that the opposite sides are parallel.

73. Show that the figure determined by the four points (3, 2), (4, 3), (3, 4), and (2, 3) is a square.

74. Show that the figure determined by the three points (1, 1), (5, 8) and (−1, 5) is a right triangle.

75. The Kelvin temperature scale can be obtained from the Celsius scale by adding 273.15 to temperatures given in Celsius.
   a. Write a formula for Kelvin temperatures K in terms of Celsius temperatures C.
   b. Write a formula for Kelvin temperatures K in terms of Fahrenheit temperatures F.
   c. Convert 100° Celsius to degrees Kelvin.
   d. Convert 68°F to degrees Kelvin.

76. Within limits, the length of a copper rod is essentially a linear function of the temperature of the rod. Suppose that at 10°C the length is 15 centimeters and at 40°C the length is 15.04 centimeters.
   a. Find a formula for the length L in centimeters of the rod in terms of the temperature in degrees Celsius.
   b. At what temperature Celsius would the length of the rod be 15.1 centimeters?

77. Suppose a pill is proposed to have the shape of a cylinder of radius $\frac{1}{8}$ inch and height h inches, capped on each end by a hemisphere of radius $\frac{1}{8}$ inch. Show that the volume V of the pill is a linear function of the height h, and find a formula for V in terms of h.

3.6 COMPOSITION OF FUNCTIONS

When we compute the value $h(x)$ of a function h we frequently perform the computation in more than one step. For example, if

$$h(x) = \sqrt{x + 5} \tag{1}$$

then we obtain $h(x)$ for $x \geq -5$ by first computing $x + 5$ and then taking the square root of $x + 5$. Thus

$$h(4) = \sqrt{4 + 5} = \sqrt{9} = 3$$

If we let

$$f(x) = x + 5 \quad \text{and} \quad g(x) = \sqrt{x}$$

then for $x \geq 5$ we have

$$\sqrt{x + 5} = \sqrt{f(x)} = g(f(x))$$

so that

$$h(x) = g(f(x))$$

Similarly, if

$$h(x) = \frac{1}{x - 4} \tag{2}$$

then to compute $h(x)$ for any x such that $x - 4 \neq 0$, that is, for $x \neq 4$, we first compute $x - 4$ and then take its reciprocal. Therefore, if

$$f(x) = x - 4 \quad \text{and} \quad g(x) = \frac{1}{x}$$

then for $x \neq 4$ we have

$$\frac{1}{x - 4} = \frac{1}{f(x)} = g(f(x))$$

so once again

$$h(x) = g(f(x))$$

In general, if f and g are two functions, then we define the ***composite function*** $g \circ f$ to be the function whose rule is given by

$$(g \circ f)(x) = g(f(x)) \tag{3}$$

and whose domain consists of all the numbers x in the domain of f such that $f(x)$ is in the domain of g. (This is just the set of all numbers x for which the right side of (3) is meaningful.) Thus each of the functions h defined in (1) and (2) may be expressed as the composite of two functions f and g.

EXAMPLE 1.   Let $h(x) = \dfrac{1}{(x + 3)^2}$. Express h as the composite $g \circ f$ of two functions f and g.

Solution.   If we let

$$f(x) = x + 3 \quad \text{and} \quad g(x) = \frac{1}{x^2}$$

then for $x \neq -3$ we have

$$h(x) = \frac{1}{(x + 3)^2} = \frac{1}{(f(x))^2} = g(f(x))$$

Thus $h = g \circ f$.   □

In the solution of Example 1, if we had let

$$f(x) = (x + 3)^2 \quad \text{and} \quad g(x) = \frac{1}{x}$$

then it would still have been true that $h = g \circ f$. In general, there are many ways of writing a given function as the composite of two other functions.

EXAMPLE 2. Let $f(x) = \sqrt{x}$ and $g(x) = x^2 - 1$. Find the domain and rule of each of the following composite functions.

a. $g \circ f$ b. $f \circ g$

Solution.

a. The domain of $g \circ f$ consists of all numbers x in the domain of f such that $f(x)$ is in the domain of g. But since the domain of g consists of all numbers, it follows that $f(x)$ is in the domain of g for any number x in the domain of f. Therefore the domain of $g \circ f$ is the domain of f, which is the set of all nonnegative numbers. The rule of $g \circ f$ is given by

$$(g \circ f)(x) = g(f(x)) = g(\sqrt{x}) = (\sqrt{x})^2 - 1 = x - 1 \qquad (4)$$

b. The domain of $f \circ g$ consists of all numbers x in the domain of g such that $g(x)$ is in the domain of f, which consists of all nonnegative numbers. Thus the domain of $f \circ g$ consists of all numbers x such that $x^2 - 1 \geq 0$, which is equivalent to $x^2 \geq 1$. But $x^2 \geq 1$ for x in $(-\infty, -1]$ or $[1, \infty)$. We conclude that the domain of $f \circ g$ consists of the intervals $(-\infty, -1]$ and $[1, \infty)$. The rule of $f \circ g$ is given by

$$(f \circ g)(x) = f(g(x)) = f(x^2 - 1) = \sqrt{x^2 - 1} \qquad \square$$

Caution: Notice that in Example 2 the functions $g \circ f$ and $f \circ g$ are not the same. This is usually the case. Notice also that the right side of (4) is meaningful for all real numbers x, whereas the domain of $g \circ f$ consists only of the nonnegative numbers. Therefore we must be careful when specifying the domain of a composite function.

EXERCISES 3.6

In Exercises 1–6, express h as the composite $g \circ f$ of two functions f and g (neither of which is equal to h).

1. $h(x) = 2(x - 1)$

2. $h(x) = -7\sqrt{x}$

3. $h(x) = \dfrac{3}{x^4}$

4. $h(x) = \dfrac{-5}{x + 4}$

5. $h(t) = \dfrac{1}{2\sqrt{t}}$

6. $h(t) = (t + 1)^{1/3}$

In Exercises 7–20, find the domain and rule of $g \circ f$.

7. $f(x) = x - 1$; $g(x) = 2x^2 + x + 1$

8. $f(x) = 2x - 1$; $g(x) = x^2$

9. $f(x) = \dfrac{1}{2x}$; $g(x) = \sqrt{x}$

10. $f(x) = \sqrt{x}$; $g(x) = \dfrac{1}{2x}$

11. $f(x) = x - 1$; $g(x) = \sqrt{x + 1}$

12. $f(x) = \sqrt{x}$; $g(x) = \dfrac{1}{2x - 4}$

13. $f(x) = \dfrac{1}{2x}$; $g(x) = \dfrac{1}{x^2 - 1}$

14. $f(x) = \dfrac{x^2 - 1}{x^2 + 1}$; $g(x) = \dfrac{1}{x}$

15. $f(x) = \dfrac{1}{x}$; $g(x) = \dfrac{1}{x}$

16. $f(x) = \dfrac{x - 1}{x + 1}$; $g(x) = \dfrac{x + 1}{x - 1}$

17. $f(x) = \dfrac{x-1}{x+1}; g(x) = \dfrac{x+3}{x-2}$

18. $f(x) = \sqrt{x^2 + 1}; g(x) = \sqrt{x^2 - 1}$

19. $f(x) = \sqrt{x^2 + 1}; g(x) = \sqrt{x^2 - 4}$

20. $f(x) = \dfrac{2x}{x-1}; g(x) = \sqrt{2x - 4}$

21. Let $f(x) = x$. Show that $g \circ f = g$ and $f \circ g = g$ for any function g.

22. Let $f(x) = \dfrac{x}{x-1}$. Show that $(f \circ f)(x) = x$ for all $x \neq 1$.

23. Let $f(x) = \dfrac{1}{1-x}$. Show that $(f \circ f \circ f)(x) = x$ for all x except 0 and 1. (By definition, $(f \circ f \circ f)(x) = [f \circ (f \circ f)](x)$.)

24. Suppose the domain of f is $[0, 1)$ and $g(x) = f(x + 10)$. Find the domain of g.

25. Suppose the domain of f is $[0, 1)$ and $g(x) = f(10 - x)$. Find the domain of g.

26. Suppose the domain of f is $(-\pi/2, \pi/2)$ and $g(x) = f(x - \pi/6)$. Find the domain of g.

27. Let $f(x) = mx + b$, where m and b are constants. Let $g(x) = f(x + 1) - f(x)$. Show that g is a constant function, and determine the constant.

28. Let $f(x) = ax^2 + bx + c$, where a, b, and c are constants. Show that if

$$g(x) = f(x + 1) - f(x) \quad \text{and} \quad G(x) = g(x + 1) - g(x)$$

then G is a constant function that is independent of b and c.

29. The volume $V(r)$ of a spherical balloon of radius r is given by

$$V(r) = \frac{4}{3}\pi r^3$$

If the radius of the balloon is increasing with time t according to the formula

$$r(t) = \frac{3}{2}t^2 \quad \text{for} \quad t \geq 0$$

find a formula for the volume of the balloon at any time $t \geq 0$.

30. Suppose a car is traveling 40 miles per hour, so the distance D_m in miles it travels in t hours is given by

$$D_m(t) = 40t$$

If m represents miles and k kilometers, then the conversion from miles to kilometers is given by

$$k = g(m) = 1.609344m$$

Find a formula for the distance D_k in kilometers as a function of time.

3.7
INVERSES
OF FUNCTIONS

We can convert from degrees Celsius to degrees Fahrenheit by means of the formula

$$F = \frac{9}{5}C + 32 \tag{1}$$

By solving this equation for C we obtain a formula for converting from degrees Fahrenheit to degrees Celsius:

$$C = \frac{5}{9}(F - 32) \tag{2}$$

In order to explore further the relationship between (1) and (2), we view (1) and (2) as defining the functions f and g given by

$$f(x) = \frac{9}{5}x + 32 \quad \text{and} \quad g(x) = \frac{5}{9}(x - 32) \tag{3}$$

Notice that if $y = f(x)$, then

$$y = \frac{9}{5}x + 32$$

so that

$$g(y) = g\left(\frac{9}{5}x + 32\right) = \frac{5}{9}\left[\left(\frac{9}{5}x + 32\right) - 32\right] = \frac{5}{9}\left(\frac{9}{5}x\right) = x$$

Thus if $y = f(x)$ then $x = g(y)$. A similar calculation shows that the converse is true: if $x = g(y)$, then $y = f(x)$. We express this relationship between f and g by saying that g is an inverse of f.

DEFINITION 3.4   Let f be a function. Then f *has an inverse* if there is a function g such that

    i. the domain of g is the range of f
    ii. for all x in the domain of f and all y in the range of f,

$$f(x) = y \quad \text{if and only if} \quad g(y) = x \tag{4}$$

Under these conditions g is an *inverse* of f.

Notice that in Definition 3.4, g is a *function*. Its domain is specified by (i) and its rule by (ii). From this it follows that g is unique. To emphasize its connection with f, we write f^{-1} for g and call f^{-1} *the inverse of f*. With g replaced by f^{-1}, (4) becomes

$$f(x) = y \quad \text{if and only if} \quad f^{-1}(y) = x \tag{5}$$

for all x in the domain of f and all y in the range of f. We also find that if f has an inverse, then f^{-1} has f as inverse, so f and f^{-1} are inverses of each other.

Caution: Not every function has an inverse. At the end of this section we will show that, among other functions, x^2 does not have an inverse.

From (5) we obtain two important formulas relating a function f and its inverse f^{-1}:

$$f^{-1}(f(x)) = x \text{ for all } x \text{ in the domain of } f \tag{6}$$
$$f(f^{-1}(y)) = y \text{ for all } y \text{ in the range of } f \tag{7}$$

EXAMPLE 1.   Let $f(x) = 5x^3$, and assume that we already know that f has an inverse and that

$$f^{-1}(y) = \sqrt[3]{\frac{y}{5}}$$

Verify (6) and (7) from the formulas given for f and f^{-1}.

Solution.   For (6) we have

$$f^{-1}(f(x)) = f^{-1}(5x^3) = \sqrt[3]{\frac{5x^3}{5}} = \sqrt[3]{x^3} = x$$

and for (7) we have

$$f(f^{-1}(y)) = f\left(\sqrt[3]{\frac{y}{5}}\right) = 5\left(\sqrt[3]{\frac{y}{5}}\right)^3 = 5\left(\frac{y}{5}\right) = y \quad \square$$

Because we usually write formulas for functions in terms of x (rather than y), we will also usually write the formula for f^{-1} in terms of x. Thus in Example 1 the formula for the inverse f^{-1} would normally be written as

$$f^{-1}(x) = \sqrt[3]{\frac{x}{5}}$$

Finding Formulas for Inverses

In Example 1 we assumed the formula for f^{-1} on faith. Now we will describe a procedure that frequently allows us to determine a formula for the inverse of a function f, provided that f is given by a simple formula. The individual steps in the procedure are:

> Step 1. Write $y = f(x)$.
> Step 2. Solve the equation in Step 1 for x in terms of y.
> Step 3. In the formula for x arising from Step 2, replace x by $f^{-1}(y)$.
> Step 4. Replace each y in the result of Step 3 by x.

Let us see how the method works for the temperature conversion function f appearing in (3).

EXAMPLE 2.   Let $f(x) = \frac{9}{5}x + 32$. Find a formula for f^{-1}.

Solution.   Following the steps listed above, we obtain

Step 1: $$y = \frac{9}{5}x + 32$$

Step 2:
$$\begin{cases} 5y = 9x + 160 \\ 9x = 5y - 160 = 5(y-32) \\ x = \frac{5}{9}(y-32) \end{cases}$$

Step 3: $$f^{-1}(y) = \frac{5}{9}(y-32)$$

Step 4: $$f^{-1}(x) = \frac{5}{9}(x-32) \quad \square$$

Notice that f^{-1} is the function g appearing in (3), so g is shown to be the inverse of f in a second way.

EXAMPLE 3.   Let $f(x) = 2 - x^3$. Find a formula for f^{-1}.

Solution.   Again we use the steps given above:

Step 1: $$y = 2 - x^3$$

Step 2: $$\begin{cases} x^3 = 2 - y \\ x = \sqrt[3]{2 - y} \end{cases}$$

Step 3: $$f^{-1}(y) = \sqrt[3]{2 - y}$$

Step 4: $$f^{-1}(x) = \sqrt[3]{2 - x} \quad \square$$

Because mistakes are easy to make, it is advisable to check either that $f(f^{-1}(x)) = x$ or that $f^{-1}(f(x)) = x$ after obtaining a formula for f^{-1}. Using this advice for the function in Example 3, we have

$$f(f^{-1}(x)) = 2 - (f^{-1}(x))^3 = 2 - (\sqrt[3]{2 - x})^3 = 2 - (2 - x) = x$$

Graphs of Inverses

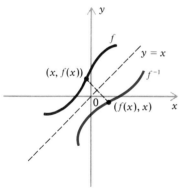

The graphs of f and f^{-1} are symmetric with respect to the line $y = x$.

FIGURE 3.51

To find a method of graphing the inverse of a function f, we begin by observing that if $(1, 3)$ is on the graph of f, then $f(1) = 3$; therefore by the definition of f^{-1}, it follows that $f^{-1}(3) = 1$, which in turn means that $(3, 1)$ is on the graph of f^{-1}. More generally, for any real numbers a and b, if (a, b) is on the graph of f, then (b, a) is on the graph of f^{-1}, and vice versa. In particular, $(x, f(x))$ is on the graph of f for any x in the domain of f, so that by the preceding comments $(f(x), x)$ is on the graph of f^{-1}. Notice that $(x, f(x))$ and $(f(x), x)$ are opposite one another with respect to the line $y = x$ (Figure 3.51). For this reason we say that $(x, f(x))$ and $(f(x), x)$ are **symmetric with respect to the line $y = x$**, and when we pass from one of these points to the other, we say that we **reflect** the point through the line $y = x$. Therefore we can obtain the graph of f^{-1} by reflecting all the points on the graph of f through the line $y = x$, or as we usually say, by reflecting the graph of f through the line $y = x$ (Figure 3.51). The graphs of f and f^{-1} are said to be symmetric with respect to the line $y = x$.

EXAMPLE 4.   Let $f(x) = -2x + 3$. First sketch the graph of f, and then obtain the graph of f^{-1} by reflecting the graph of f through the line $y = x$.

Solution.   The graph of f is the line with slope -2 and y intercept 3. It is sketched in Figure 3.52. Next we draw the line $y = x$, and finally we sketch the graph of f^{-1} by reflecting through the line $y = x$ (Figure 3.52). $\square$

There is a geometric method for telling which functions f have inverses: f has an inverse if and only if no horizontal line intersects the graph of f

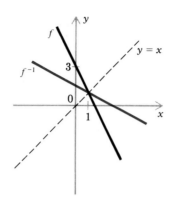

FIGURE 3.52

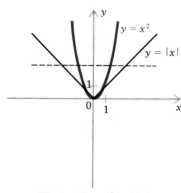

The functions x^2 and $|x|$
do not have inverses

FIGURE 3.53

more than once. As you can easily see from Figure 3.53, this implies that the functions x^2 and $|x|$ do not have inverses. In algebraic terms, a function has an inverse if and only if the following condition holds:

For any x and z in the domain of f

$$\text{if } x \neq z, \quad \text{then} \quad f(x) \neq f(z)$$

In this case f is said to be **one-to-one**.

EXERCISES 3.7

In Exercises 1–22, find a formula for f^{-1}.

1. $f(x) = 2x$

2. $f(x) = 2x + 1$

3. $f(x) = -\dfrac{1}{2}(x - 3)$

4. $f(x) = \pi x - \sqrt{2}$

5. $f(x) = x^3$

6. $f(x) = x^5$

7. $f(x) = 3x^3 - 5$

8. $f(x) = \pi x^2$ for $x \geq 0$

9. $f(x) = -x^5 + 1$

10. $f(x) = \dfrac{1}{x}$

11. $f(x) = \sqrt{x}$

12. $f(x) = \sqrt{x} + 5$

13. $f(x) = \sqrt{x - 3}$

14. $f(x) = 7 + \sqrt{2x - 1}$

15. $f(x) = \dfrac{1}{\sqrt{x}}$

16. $f(x) = \sqrt[3]{x + 5}$

17. $f(x) = \dfrac{1}{\sqrt[3]{4 - 2x}}$

18. $f(x) = \dfrac{-4}{x^3}$

19. $f(x) = \dfrac{x - 1}{x + 1}$

20. $f(x) = \dfrac{-x + 3}{2x - 5}$

21. $f(x) = \dfrac{2x^3 - 1}{x^3 + 3}$

22. $f(x) = \dfrac{x^5}{-x^5 + 3}$

In Exercises 23–32, sketch the graphs of f and f^{-1} in the same coordinate system.

23. $f(x) = x + 1$

24. $f(x) = 3x$

25. $f(x) = 3(x + 1)$

26. $f(x) = 3x + 1$

27. $f(x) = -2x + 4$

28. $f(x) = \sqrt{x}$

29. $f(x) = \dfrac{1}{x}$

30. $f(x) = -\dfrac{1}{x}$

31. $f(x) = \sqrt{x} + 1$

32. $f(x) = \sqrt{x} - 1$

33. Show that each of the following functions is equal to its own inverse:

a. $f(x) = x$

c. $f(x) = -x$

b. $f(x) = \dfrac{1}{x}$

d. $f(x) = \sqrt{1 - x^2}$ for $0 \le x \le 1$

34. Does a constant function have an inverse? Explain your answer.

35. The area A of an equilateral triangle of side length s is given by $A = \dfrac{\sqrt{3}}{4}s^2$.

a. Find a formula for the side length in terms of the area.

b. Determine the value of s for which $A = 16$.

36. The surface area S of a cylinder of height 4, radius r, and closed ends is given by

$$S = 2\pi r^2 + 8\pi r$$

Find a formula for the radius in terms of the surface area.

3.8 GRAPHS OF EQUATIONS

Thus far in Chapter 3 we have concentrated on functions and their graphs. The importance of functions and their graphs cannot be overemphasized; yet equations involving x and y that do not necessarily represent functions play a role in mathematics, and so do their graphs; it is therefore well to study some of the more common ones. The **graph of an equation** in x and y is the collection of all ordered pairs (x, y) of real numbers that satisfy the equation.

We begin with circles. A **circle** of radius r centered at the point (a, b) in the plane is by definition the collection of all points (x, y) whose distance from (a, b) is r. From the distance formula (in Section 3.1), the distance between (a, b) and (x, y) is

$$\sqrt{(x - a)^2 + (y - b)^2}$$

Therefore (x, y) is on the circle if and only if

$$\sqrt{(x - a)^2 + (y - b)^2} = r$$

or, when both sides are squared,

$$(x - a)^2 + (y - b)^2 = r^2 \tag{1}$$

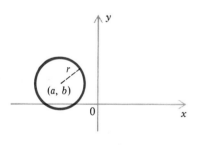

FIGURE 3.54

In summary, the graph of (1) is the circle of radius r centered at (a, b) (Figure 3.54). If the center of the circle is $(0, 0)$, then (1) becomes

$$x^2 + y^2 = r^2 \tag{2}$$

EXAMPLE 1. Find an equation of the circle of radius 1 centered at $(0, 0)$. Then sketch the circle.

Solution. Using (2) with $r = 1$, we obtain the equation

$$x^2 + y^2 = 1$$

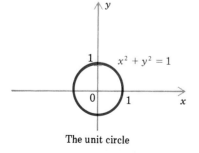

The unit circle

FIGURE 3.55 The circle is sketched in Figure 3.55. ☐

The circle $x^2 + y^2 = 1$ is usually called the **unit circle**; it will play a prominent role in Chapter 6.

EXAMPLE 2.   Find an equation of the circle of radius 4 centered at the point $(2, -3)$. Then sketch the circle.

Solution.   By (1) an equation is given by

$$(x - 2)^2 + (y - (-3))^2 = 4^2$$

which simplifies to

$$(x - 2)^2 + (y + 3)^2 = 16$$

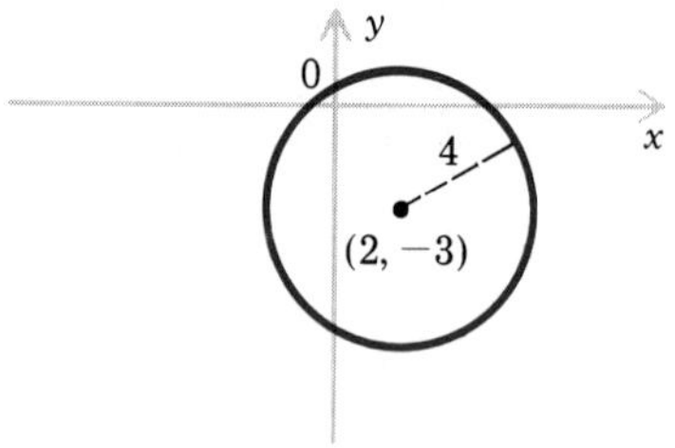

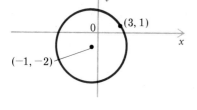

FIGURE 3.56   The circle is sketched in Figure 3.56.   □

EXAMPLE 3.   Find an equation of the circle whose center is $(-1, -2)$ and which passes through the point $(3, 1)$.

Solution.   The radius r of the circle is the distance between the center $(-1, -2)$ and the point $(3, 1)$ on the circle. By the distance formula,

$$r = \sqrt{(3 - (-1))^2 + (1 - (-2))^2} = \sqrt{16 + 9} = \sqrt{25} = 5$$

Thus by (1) the equation we seek is

$$(x - (-1))^2 + (y - (-2))^2 = 5^2$$

which simplifies to

$$(x + 1)^2 + (y + 2)^2 = 25$$

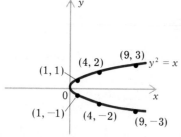

FIGURE 3.57

The circle is sketched in Figure 3.57.   □

Symmetry in Graphs

We begin our discussion of symmetry by sketching the graph of a simple equation.

EXAMPLE 4.   Sketch the graph of $y^2 = x$.

Solution.   First we make a table:

x	0	1	4	9
y	0	1 and -1	2 and -2	3 and -3

Then we plot the corresponding points (x, y) on the graph and connect the points with a smooth curve (Figure 3.58).   □

FIGURE 3.58

The graph of $y^2 = x$ in Figure 3.58 has the feature that it is symmetric with respect to the x axis, that is, the part above the x axis is mirrored below the x axis, and vice versa. Indeed, if the point (x, y) is on the graph, then the point $(x, -y)$, which lies opposite to (x, y) across the x axis, also lies on the graph (Figure 3.58). More generally, we say that the graph of an equation is **symmetric with respect to the x axis** if $(x, -y)$ is on the graph whenever (x, y) is (Figure 3.59).

The definition of symmetry with respect to the y axis given in Section 3.4 for graphs of functions applies equally well to graphs of equations. The graph of an equation is **symmetric with respect to the y axis** if $(-x, y)$ is on the graph whenever (x, y) is (Figure 3.60). For example, the circle $x^2 + (y - 1)^2 = 2$ is symmetric with respect to the y axis (Figure 3.61).

Similarly, the definition of symmetry with respect to the origin given in Section 3.4 for graphs of functions transfers to graphs of equations: The graph of an equation is **symmetric with respect to the origin** if $(-x, -y)$ is on the graph whenever (x, y) is on the graph (Figure 3.62).

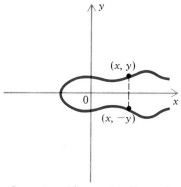

Symmetry with respect to the x axis

FIGURE 3.59

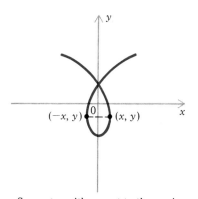

Symmetry with respect to the y axis

FIGURE 3.60

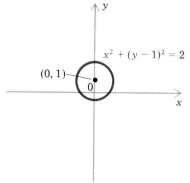

A circle that is symmetric with respect to the y axis.

FIGURE 3.61

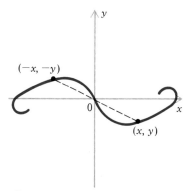

Symmetry with respect to the origin

FIGURE 3.62

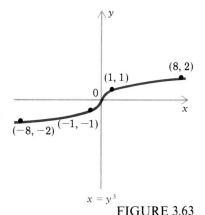

$x = y^3$

FIGURE 3.63

EXAMPLE 5. Sketch the graph of $x = y^3$.

Solution. With the following table

x	-8	-1	0	1	8
y	-2	-1	0	1	2

we are able to plot the corresponding points and connect them with a smooth curve (Figure 3.63). We notice that the graph is symmetric with respect to the origin, since if $x = y^3$, then $-x = -y^3 = (-y)^3$. ☐

The graph of an equation need not be symmetric with respect to either of the axes or the origin (Figure 3.64). In contrast, the graph of an equation can be symmetric with respect to *both* axes as well as to the origin (see Figure 3.55). Finally, we mention that symmetry can facilitate sketching of graphs, because a portion of the graph then determines the remainder of the graph.

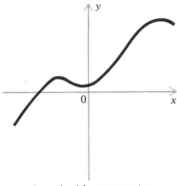

A graph with no symmetry

FIGURE 3.64

Intercepts   As we sketch the graph of an equation, the points where the graph of an equation meets the coordinate axes are of interest. A **y intercept** of the graph of an equation is the *y* coordinate of a point at which the graph meets the *y* axis (Figure 3.65). Since any point on the *y* axis has *x* coordinate 0, the *y* intercepts (if any) can be found by setting $x = 0$ in the equation and then solving for *y*. Similarly, an **x intercept** of the graph of an equation is the *x* coordinate of a point at which the graph meets the *x* axis (Figure 3.65), and the *x* intercepts (if any) can be found by setting $y = 0$ in the equation and then solving for *x*.

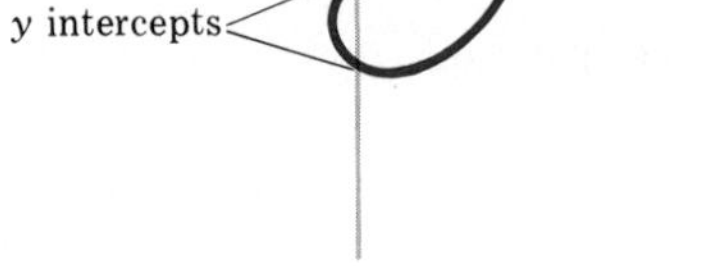

x intercepts

y intercept

y intercepts

FIGURE 3.65

EXAMPLE 6.   Find the intercepts of the graph of the equation $4x^2 + y^2 = 4$.

Solution.   To find the *y* intercepts we set $x = 0$ and solve for *y*:

$$4(0)^2 + y^2 = 4$$
$$y^2 = 4$$
$$y = 2 \quad \text{or} \quad y = -2$$

To find the *x* intercepts we set $y = 0$ and solve for *x*:

$$4x^2 + (0)^2 = 4$$
$$x^2 = 1$$
$$x = 1 \quad \text{or} \quad x = -1$$

Thus the *y* intercepts are 2 and -2, and the *x* intercepts are 1 and -1.   □

EXAMPLE 7. Find the intercepts of the graph of the equation $x^2 - y^2 = 1$.

Solution. To find the y intercepts (if any) we set $x = 0$ and obtain

$$0^2 - y^2 = 1$$
$$y^2 = -1$$

Since this equation has no real solutions, there are no y intercepts. (Thus the graph of $x^2 - y^2 = 1$ does not meet the y axis.) To find the x intercepts we set $y = 0$ and solve for x. We find that

$$x^2 - (0)^2 = 1$$
$$x^2 = 1$$
$$x = 1 \quad \text{or} \quad x = 1$$

Thus the x intercepts are 1 and -1. $\square$

EXAMPLE 8. Show that the graph of $y^2 - 5 = x + \dfrac{10}{x}$ has no intercepts.

Solution. There are no y intercepts because $10/x$ is not defined for $x = 0$, and hence the given equation is meaningless for $x = 0$. To search for x intercepts, we set $y = 0$ and attempt to solve the equation for x:

$$(0)^2 - 5 = x + \frac{10}{x}$$
$$-5x = x^2 + 10$$
$$x^2 + 5x + 10 = 0 \qquad (3)$$

If we attempt to use the quadratic formula, we obtain

$$x = \frac{-5 \pm \sqrt{5^2 - 4(1)(10)}}{2(1)} = \frac{-5 \pm \sqrt{-15}}{2}$$

Since $-15 < 0$, (3) has no real solutions, and consequently the given equation has no x intercepts. $\square$

EXERCISES 3.8

In Exercises 1–8, find an equation of the given circle and then sketch the circle.

1. The circle with radius 3 and center $(0, 0)$.

2. The circle with radius 5 and center $(3, 6)$.

3. The circle with radius 2 and center $(-1, 4)$.

4. The circle with radius $\frac{1}{2}$ and center $(-1, -2)$.

5. The circle with center $(0, 0)$ that passes through $(4, -1)$.

6. The circle with center $(5, 12)$ that passes through $(0, 0)$.

7. The circle with center $(-2, 3)$ that passes through $(1, -1)$.

8. The circle with center $(-3, -1)$ that passes through $(0, 4)$.

In Exercises 9–20, sketch the graph of the given equation and label all intercepts. Determine whether the graph possesses symmetry with respect to either axis or the origin.

9. $y^2 - 6y + 9 = 0$

10. $6x^2 + x - 2 = 0$

11. $x = |y|$

12. $|y| = |x|$

13. $y^2 = x - 1$

14. $x^2 = y^4$

15. $x = -y^3$

16. $x^3 = y^2$

17. $x^2 + y^2 = 4$

18. $(x - 1)^2 + (y + 2)^2 = 4$

19. $x = \sqrt{y}$

20. $x = \begin{cases} \sqrt{-y} & \text{for } y < 0 \\ \sqrt{y} & \text{for } y \geq 0 \end{cases}$

In Exercises 21–32, determine the intercepts (if any) of the graph of the given equation.

21. $x^2 + 4y^2 = 1$

22. $2y^2 - 3x^2 = 1$

23. $x - y^2 = 3$

24. $\frac{1}{x} + \frac{1}{y} = 1$

25. $y^2 = \sqrt{x^2 - 1}$

26. $y^2 - 4 = x + \frac{4}{x}$

27. $y^2 = |x - 5|$

28. $y + \frac{6}{y} = x^2 - 1$

29. $y - \frac{6}{y} = x^2 + 1$

30. $\frac{2}{x} - \frac{4}{x^2} = \pi xy - 5y^2 - 1$

31. $|x - 2| = |y + 1|$

32. $|x - 2| + y^2 = 4$

KEY TERMS

coordinate axes
  x axis
  y axis
coordinates
  x coordinate (abscissa)
  y coordinate (ordinate)
origin
ordered pair
Cartesian coordinate system
quadrant
distance between points
function
  domain
  rule
  range

even
odd
composite
inverse
variable
  dependent variable
  independent variable
proportion
  directly proportional
  inversely proportional
  jointly proportional
  constant of proportionality
circle
  unit circle

graph
  of a function
  of an equation
slope
point–slope equation
slope–intercept equation
linear equation
symmetry
  with respect to the x axis
  with respect to the y axis
  with respect to the origin
intercept
  x intercept
  y intercept

KEY FORMULAS

$$d(P, Q) = \sqrt{(x_2 - x_1)^2 + (y_2 - y_1)^2}$$
$$M(x, y) = \left(\tfrac{1}{2}(x_1 + x_2), \tfrac{1}{2}(y_1 + y_2)\right)$$
$$m = \frac{y_2 - y_1}{x_2 - x_1}$$
$$f^{-1}(f(x)) = x \quad \text{and} \quad f(f^{-1}(y)) = y$$

REVIEW EXERCISES

1. Determine the distance between each of the following pairs of points.
  a. $(-1, -3)$ and $(-4, 3)$
  b. $(\sqrt{2}, \frac{1}{2}\sqrt{2})$ and $(3\sqrt{2}, -2\sqrt{2})$

2. For each of the following functions, determine which of the numbers $-2, 2, -4$, and 4 are in the domain.
  a. $f(x) = \dfrac{x+2}{x-2}$
  b. $f(x) = \dfrac{x+2}{x^2-4}$

In Exercises 3–10, determine the domain of the function.

3. $f(x) = \dfrac{1}{x+3}$

4. $f(x) = \dfrac{x-2}{x+5} - \dfrac{x+1}{x-2}$

5. $g(x) = \dfrac{1}{x^2+4x-7}$

6. $g(x) = \sqrt{x+\pi}$

7. $h(x) = \sqrt{6x-4}$

8. $h(x) = |x-3|$

9. $h(x) = \sqrt{|x|-5}$

10. $f(x) = \begin{cases} x^2 & \text{for } x < -2 \\ 3 & \text{for } -2 < x \le 1 \\ \sqrt{x-4} & \text{for } 5 < x \end{cases}$

11. Show that if $a \ne 0$, then the domain and range of $\sqrt{x+a}$ are distinct.

12. Let $f(x) = \dfrac{3x-4}{2x+5}$. Find
  a. $f(0)$ b. $f(-1)$ c. $f\left(\dfrac{3}{4}\right)$ d. $f\left(-\dfrac{2}{x}\right)$

13. Let $f(x) = x^2 + 3ax - 5$, and suppose that $f(-2) = 1$. Determine a.

14. Let $f(t) = 1 - t^2$. Find $\dfrac{f(b) - f(a)}{b - a}$.

15. Let $f(x) = \dfrac{16}{x^2 - \sqrt{x^4 - 16}}$.
  a. Find the domain of f.
  b. Let $g(x) = x^2 + \sqrt{x^4 - 16}$. Show that $f = g$.

16. Let
$$f(x) = \sqrt{x+1} - \sqrt{x-2}$$
and
$$g(x) = \dfrac{3}{\sqrt{x+1} + \sqrt{x-2}}$$
Determine whether or not $f = g$.

In Exercises 17–26, sketch the graph of the function, noting any intercepts and symmetry.

17. $f(x) = 2x - 5$

18. $f(x) = -\frac{1}{2}x - \frac{3}{2}$

19. $f(x) = x^2 - 6$

20. $f(x) = -2x^2$

21. $f(x) = ||x| - 2|$

22. $f(x) = \dfrac{3}{x}$

23. $g(x) = \sqrt{-x}$

24. $g(x) = \sqrt{x+1}$

25. $g(x) = \begin{cases} x^2 + 4 & \text{for } x \le -1 \\ -4 - x^2 & \text{for } x \ge 1 \end{cases}$

26. $g(x) = \begin{cases} -x & \text{for } x < -1 \\ 2 & \text{for } x = -1 \\ 3x & \text{for } x \ge 0 \end{cases}$

In Exercises 27–34, sketch the graph of the equation, noting any intercepts and symmetry.

27. $(x + \sqrt{2})(y - \sqrt{3}) = 0$

28. $x^2 = y^2$

29. $|x - 4| = |y + 1|$

30. $x^2 + y^2 = 16$

31. $(x - 1)^2 = 4 - (y + 3)^2$

32. $7 + y^2 = x$

33. $y^2 = x + 3$

34. $x^2 + 4x - 12 = 0$

In Exercises 35–40, find $f \circ g$ and $g \circ f$ for the given functions f and g.

35. $f(x) = \dfrac{25}{x^2}$ and $g(x) = 5x$

36. $f(x) = \dfrac{1}{x+2}$ and $g(x) = x - 2$.

37. $f(x) = x^3$ and $g(x) = \sqrt{2x+3}$

38. $f(x) = \dfrac{x+4}{x-3}$ and $g(x) = x^2 + 3$

39. $f(x) = |x|$ and $g(x) = \sqrt{x^2 - 4}$

40. $f(x) = \dfrac{2 - 3x}{4 + x} = g(x)$

41. Let $f(x) = \dfrac{x+1}{x-1}$. Show that $(f \circ f)(x) = x$ for all $x \ne 1$, and thus that f is its own inverse.

42. Let $f(x) = 1/x$ and $g(x) = \dfrac{x^2 - 1}{x^2 + 1}$. Show that

$(g \circ f)(x) = -g(x)$.

43. Let $f(x) = \sqrt{x^2 - 3}$ and $g(x) = \sqrt{x^2 + 9}$. Determine whether or not $f \circ g = g \circ f$. Explain your answer.

44. Let $f(x) = \sqrt{2 + x}$.

C a. Use a calculator to approximate the following numbers.

   i. $f(\sqrt{2})$ ii. $(f \circ f)(\sqrt{2})$ iii. $(f \circ f \circ f)(\sqrt{2})$

   b. Can you guess the value that

$$\underbrace{(f \circ f \circ \cdots \circ f)}_{n \text{ of these}}(\sqrt{2})$$

approaches as n increases without bound?

In Exercises 45–48, find a formula for the inverse of the given function.

45. $f(x) = \dfrac{1}{x + 3}$ **47.** $f(x) = \dfrac{4x - 1}{3x + 2}$

46. $f(x) = \dfrac{5}{x^{1/3} + 4}$ **48.** $f(x) = \sqrt{x - 2}$

49. Determine the distance between the points $(-\sqrt{a}, -\sqrt{b})$ and $(\sqrt{a}, \sqrt{b})$.

50. Find the points on the line $y = 2x + 1$ that are 1 unit from the point $(0, 3)$.

51. Find the points on the line $x - 2y = 3$ that are 5 units from the point $(7, 2)$.

52. Find an equation of all points (x, y) that are 4 units from the point $(-2, -3)$.

53. Find an equation of the circle centered at $(\frac{1}{2}, -\frac{1}{4})$ and passing through $(0, -\frac{1}{4} + \frac{3}{4}\sqrt{7})$.

54. A line l has x intercept -2 and y intercept -4.
   a. Write the slope–intercept equation of l.
   b. Write a point–slope equation of l.

55. Find an equation of the line that is parallel to the line $x + 2y = 3$ and passes through the point $(-1, -3)$.

56. Find an equation of the line that is perpendicular to the line $3x - 2y = 6$ and passes through the point $(4, 6)$.

57. Find an equation of the line consisting of all points (x, y) that are equidistant from the points $(3, 1)$ and $(-1, 4)$.

58. Determine whether or not the points $(-1, -1)$, $(1, 3)$, and $(57, 115)$ lie on a straight line.

59. A converse of the Pythagorean Theorem states that if the lengths a, b, and c of the sides of a triangle satisfy the equation $c^2 = a^2 + b^2$, then the triangle is a right triangle. Use this result to determine whether or not the three points $(2, 1)$, $(1, -2)$, and $(-2, 3)$ form the vertices of a right triangle.

60. Let M be the midpoint of the line segment joining the points $(-2, 1)$ and $(-1, 3)$. Find the distance between M and the origin.

61. Prove that the midpoint of the hypotenuse of a right triangle is equidistant from all three vertices. (*Hint:* Let the two legs lie on the coordinate axes.)

62. Prove that the line joining the midpoints of two sides of a triangle is parallel to the third side. (*Hint:* Set up the coordinate system so that the vertices of the triangle are $P_1(0, 0)$, $P_2(a, 0)$, and $P_3(b, c)$.)

63. Prove that the points $(-2, 2)$, $(-1, -1)$, $(1, 3)$, and $(2, 0)$ are the vertices of a square.

64. A cylindrical can has a height of 5 inches, and a top and bottom. In terms of the radius, the surface area S of the can is given by

$$S = 2\pi(r^2 + 5r) \quad \text{for} \quad r \geq 0$$

Write a formula for the radius in terms of the surface area. (*Hint:* $r^2 + 5r = (r + \frac{5}{2})^2 - \frac{25}{4}$.)

65. The illumination L from a lamp is directly proportional to the intensity I of the light bulb and is inversely proportional to the square of the distance D from the bulb.
   a. Express this fact by means of a formula.
   b. If one moves from 12 feet to 4 feet away from a light source, how is the illumination of the person affected?

66. The safe load L of a horizontal beam supported at both ends is jointly proportional to the width w and the square of the depth d and is inversely proportional to the length l of the beam.
   a. Express this fact by means of a formula.
   b. If the safe load on such a beam 2 inches wide, 8 inches deep, and 10 feet long is 1000 pounds, find the safe load on a beam 4 inches wide, 10 inches deep, and 12 feet long.

4

Polynomial and Rational Functions

A ny polynomial expression defines a function whose domain consists of all real numbers. For example, the polynomial expression $2x^4 - x^3 + 3x^2 + \sqrt{2}$ defines the function f whose rule is given by

$$f(x) = 2x^4 - x^3 + 3x^2 + \sqrt{2}$$

In general, if n is a nonnegative integer and $a_n, a_{n-1}, a_{n-2}, \ldots, a_1$, and a_0 are constants with $a_n \neq 0$, then a function f is defined by the formula

$$f(x) = a_n x^n + a_{n-1} x^{n-1} + a_{n-2} x^{n-2} + \cdots + a_1 x + a_0$$

Such a function is called a ***polynomial function***. This chapter will be mainly devoted to the graphs of polynomial functions and quotients of polynomial functions, which are called ***rational functions***.

Special polynomial and rational functions describe many physical quantities and properties. Examples are the shape of a suspension bridge or a mirror in a telescope, the gravitational force exerted on a satellite by the earth, and the trajectory of a golf ball. These types of functions will help us solve problems such as the following:

> A rancher has 2 miles of fencing and wishes to fence in a rectangular grazing field with an area as large as possible. What should the dimensions of the field be? (See Example 3 of Section 4.2.)

4.1
QUADRATIC
FUNCTIONS

Because they have the lowest degree, constant functions and linear functions are the simplest polynomial functions. Nonzero constant functions have degree 0, and linear functions have degree 1. The next simplest kind of polynomial functions are those of degree 2, called *quadratic functions*. A quadratic function is usually given in the form

$$f(x) = ax^2 + bx + c$$

where a, b, and c are constants with $a \neq 0$. The graph of a quadratic function is called a *parabola*. It has been known since antiquity that this type of curve results when a cone is sliced by a plane (Figure 4.1). The emphasis in this section will be on sketching parabolas.

Let f and g be the quadratic functions defined by

$$f(x) = x^2 \quad \text{and} \quad g(x) = -x^2$$

The graphs of these functions were sketched in Section 3.4 and are reproduced in Figure 4.2a, b. Notice that the graph of x^2 opens upward, whereas the graph of $-x^2$ opens downward. Both graphs are symmetric with respect to the y axis, which is referred to as the *axis* of each parabola. The point $(0, 0)$ is the lowest point on the graph of f and the highest point on the graph of g; it is called the *vertex* of each parabola.

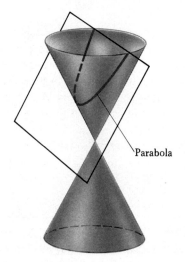

Parabola

FIGURE 4.1

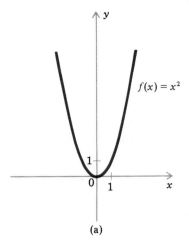

(a)

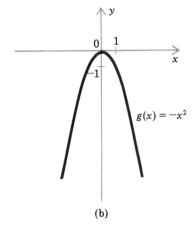

(b)

FIGURE 4.2

If

$$f(x) = ax^2 \quad \text{with} \quad a \neq 0$$

then the graph of f has the same general shape as that of x^2 or $-x^2$, depending on whether $a > 0$ or $a < 0$. However, the graph is flatter if $|a| < 1$ and is more pointed if $|a| > 1$ (Figure 4.3a, b).

FIGURE 4.3

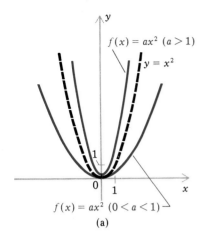

$f(x) = ax^2 \ (a > 1)$

$y = x^2$

$f(x) = ax^2 \ (0 < a < 1)$

(a)

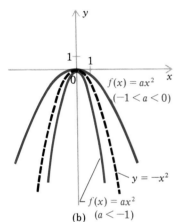

$f(x) = ax^2$
$(-1 < a < 0)$

$y = -x^2$

$f(x) = ax^2$
(b) $(a < -1)$

EXAMPLE 1.

    a. Let $f(x) = 2x^2$. Sketch the graph of f.

    b. Let $g(x) = -\frac{1}{3}x^2$. Sketch the graph of g.

Solution.

    a. The y coordinate of each point on the graph of f is twice the y coordinate of the corresponding point on the graph of x^2 (see Figure 4.2a). This leads us to the sketch in Figure 4.4a, with a few points plotted for assistance.

    b. The y coordinate of each point on the graph of g is $\frac{1}{3}$ the y coordinate of the corresponding point on the graph of $-x^2$ (see Figure 4.2b). This leads us to the sketch in Figure 4.4b, again with a few points plotted for assistance.   ☐

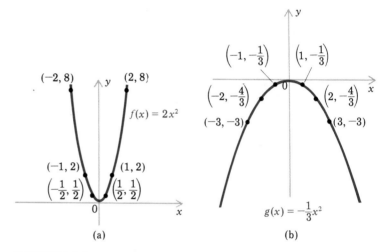

(a)   (b)

FIGURE 4.4

Vertical Shifts in Graphs

Next we let g be a given function, and let

$$f(x) = g(x) + k$$

with k a fixed nonzero number. Since $f(x)$ is obtained from $g(x)$ by adding k, it follows that the graph of f is k units above the graph of g if $k > 0$, and is $-k$ units below the graph of g if $k < 0$ (Figure 4.5). In either case, the graph of f is obtained by shifting the graph of g vertically.

    Now let $g(x) = ax^2$, so that $f(x) = ax^2 + k$. Then the graph of f is obtained from the graph of g simply by making a vertical shift. In particular the vertex $(0, 0)$ of the graph of g is shifted to the point $(0, k)$, which is called the **vertex** of the graph of f. It is immediate that the y intercept of the graph of f is k. In contrast, there are 2, 1, or no x intercepts for the graph of f, depending on the value of k.

EXAMPLE 2.   Let $f(x) = 2x^2 - 2$. Sketch the graph of f, and determine the vertex and the x intercepts of the parabola.

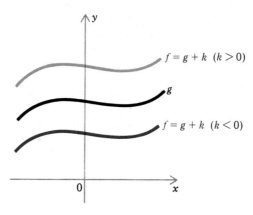

FIGURE 4.5

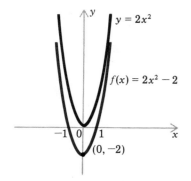

FIGURE 4.6

Solution.   We have $k = -2$, so the graph of f is obtained by shifting the graph of $2x^2$ down 2 units (Figure 4.6). Thus the vertex is $(0, -2)$. For the x intercepts we solve the equation $f(x) = 0$ for x:

$$2x^2 - 2 = 0$$
$$2x^2 = 2$$
$$x^2 = 1$$

Therefore $x = 1$ or $x = -1$

Consequently the x intercepts are 1 and -1.   □

EXAMPLE 3.   Let $y = 2x^2 + \frac{1}{2}$. Sketch the graph of the equation, and determine the vertex and any x intercepts of the parabola.

Solution.   In this case $k = \frac{1}{2}$, so the graph of the equation is obtained by shifting the graph of $2x^2$ up $\frac{1}{2}$ unit (Figure 4.7). Thus the vertex is $(0, \frac{1}{2})$. Concerning the x intercepts, we observe that

$$y = 2x^2 + \frac{1}{2} > 0 \quad \text{for all } x$$

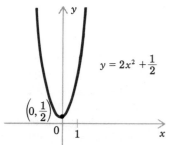

FIGURE 4.7

so there are no x intercepts.   □

As Examples 2 and 3 illustrate, if

$$f(x) = ax^2 + k$$

then there are two x intercepts if $k < 0$, and no x intercepts if $k > 0$.

Horizontal Shifts in Graphs   Now we study horizontal shifts of a graph. Let g be a given function, and let

$$f(x) = g(x - h)$$

where h is a fixed nonzero number. Then $y = f(x)$ if and only if $y = g(x - h)$, so (x, y) is on the graph of f if and only if $(x - h, y)$ is on the graph of g. But observe that the point (x, y) lies h units to the right of the point $(x - h, y)$ if $h > 0$ and $-h$ units to the left if $h < 0$. We conclude that if $h > 0$, then the graph of f is h units to the right of the graph of g, and if $h < 0$, then the graph of f is $-h$ units to the left of the graph of g (Figure 4.8).

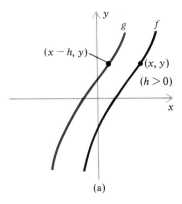
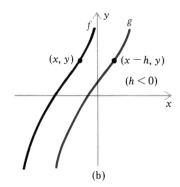

FIGURE 4.8

EXAMPLE 4.   Let $f(x) = (x - 2)^2$. Sketch the graph of f.

Solution.   Here $h = 2 > 0$, so by the comments above, the graph of f is 2 units to the right of the graph of x^2 (Figure 4.9a).   □

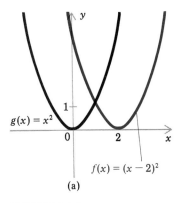
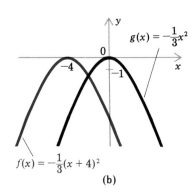

FIGURE 4.9

EXAMPLE 5.   Let $f(x) = -\frac{1}{3}(x + 4)^2$. Sketch the graph of f.

Solution.   Since $x + 4 = x - (-4)$, we have $h = -4 < 0$, so by the comments above, the graph of f is 4 units to the left of the graph of $-\frac{1}{3}x^2$ (see Figure 4.4b). The graph of f is sketched in Figure 4.9b.   □

The Graph of a General Quadratic Function

Suppose that we have an arbitrary quadratic function, given by

$$f(x) = ax^2 + bx + c \quad \text{with} \quad a \neq 0 \tag{1}$$

In order to sketch the graph of f, we complete the square in $ax^2 + bx + c$, as we did in Section 2.3, and obtain

$$f(x) = a\left(x + \frac{b}{2a}\right)^2 + c - \frac{b^2}{4a} \tag{2}$$

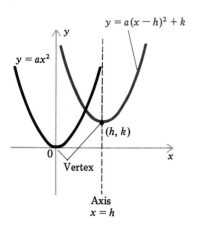

$y = a(x - h)^2 + k$

$y = ax^2$

(h, k)

Vertex

Axis $x = h$

FIGURE 4.10

At first glance it seems no easier to draw the graph of f from (2) than from (1). However, let us simplify the right side of (2) by letting

$$h = -\frac{b}{2a} \quad \text{and} \quad k = c - \frac{b^2}{4a} \tag{3}$$

Then (2) becomes

$$f(x) = a(x - h)^2 + k$$

The graph of $a(x - h)^2 + k$ is obtained from the graph of ax^2 by a vertical shift of $|k|$ units (upward if $k > 0$ and downward if $k < 0$) and a horizontal shift of $|h|$ units (to the right if $h > 0$ and to the left if $h < 0$) (Figure 4.10). For example, if

$$f(x) = -\frac{1}{3}(x + 4)^2 + 2$$

then the graph of f can be obtained by shifting the graph of $-\frac{1}{3}x^2$ upward 2 units and to the left 4 units.

When we shift from the graph of ax^2 to the graph of $a(x - h)^2 + k$, the vertex $(0, 0)$ of ax^2 is shifted to the point (h, k) and the axis $x = 0$ is shifted to the line $x = h$ (Figure 4.10). Thus (h, k) is called the **vertex** of the graph of $a(x - h)^2 + k$, and the line $x = h$ is called the **axis** of the graph. The vertex is the lowest point on the graph if $a > 0$ and is the highest point if $a < 0$.

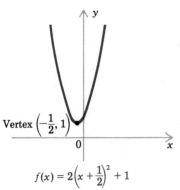

y

Vertex $\left(-\frac{1}{2}, 1\right)$

$f(x) = 2\left(x + \frac{1}{2}\right)^2 + 1$

FIGURE 4.11

EXAMPLE 6.   Let $f(x) = 2\left(x + \frac{1}{2}\right)^2 + 1$. Sketch the graph of f, and locate the vertex of the parabola.

Solution.   The graph of f can be obtained by shifting the graph of $2x^2$ upward 1 unit and to the left $\frac{1}{2}$ unit (Figure 4.11). The vertex is $\left(-\frac{1}{2}, 1\right)$.   □

Let us return to the formula

$$f(x) = ax^2 + bx + c$$

and use the results of the preceding discussion to outline a procedure for sketching the graph of f. As is usually the case, it helps to plot a few points, so our outline includes finding the intercepts.

i. Find the y intercept by calculating $f(0)$. Plot the corresponding point.
ii. Find the x intercepts (if any) by solving $f(x) = 0$ for x. Plot the corresponding points (if any).
iii. Complete the square of the right side to express $f(x)$ in the form

$$f(x) = a(x - h)^2 + k \qquad (4)$$

iv. From the new equation notice that the graph of f is a parabola whose vertex is (h, k) and whose axis is $x = h$. Plot the vertex, (h, k).
v. Obtain the graph of f by shifting the graph of ax^2 vertically $|k|$ units (upward if $k > 0$ and downward if $k < 0$) and horizontally $|h|$ units (to the right if $h > 0$ and to the left if $h < 0$). The graph will contain the points already plotted and will open upward if $a > 0$ and downward if $a < 0$.

From (3) it follows that the vertex and axis of the parabola are given (in terms of the original coefficients a, b, and c) by

$$\text{Vertex:} \quad \left(-\frac{b}{2a}, c - \frac{b^2}{4a}\right)$$

$$\text{Axis:} \quad x = -\frac{b}{2a}$$

But the vertex and axis are much simpler to find after we have completed the square. As a result, when we set out to sketch the graph of a quadratic function, we usually first complete the square, and then sketch the graph by reading off the information we need from the expression for $f(x)$.

EXAMPLE 7. Let $f(x) = x^2 + 2x - 3$. Sketch the graph of f, and locate the vertex of the parabola.

Solution. Since $f(0) = -3$, the y intercept is -3. To find the x intercepts we solve the equation $f(x) = 0$ for x:

$$x^2 + 2x - 3 = 0$$

$$(x + 3)(x - 1) = 0$$

Consequently

$$x = -3 \quad \text{or} \quad x = 1$$

Therefore the x intercepts are -3 and 1. Now we complete the square:

$$\begin{aligned}
f(x) &= (x^2 + 2x) - 3 \\
&= (x^2 + 2x + 1 - 1) - 3 \\
&= (x^2 + 2x + 1) - 1 - 3 \\
&= (x^2 + 2x + 1) - 4 \\
&= (x + 1)^2 - 4
\end{aligned}$$

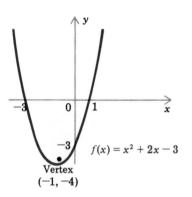

$f(x) = x^2 + 2x - 3$

Vertex
$(-1, -4)$

FIGURE 4.12

In the notation of (4) we have $a = 1$, $h = -1$, and $k = -4$, so the vertex is $(-1, -4)$. Thus we obtain the graph of f by shifting the graph of x^2 downward 4 units and to the left 1 unit (Figure 4.12).   □

EXAMPLE 8.   Sketch the graph of

$$y = -3x^2 + 12x - 8 \tag{5}$$

and locate the vertex of the parabola.

Solution.   Since $y = -8$ for $x = 0$, the y intercept is -8. The x intercepts are obtained by setting $y = 0$ and solving for x by the quadratic formula:

$$-3x^2 + 12x - 8 = 0$$

$$x = \frac{-12 \pm \sqrt{(12)^2 - 4(-3)(-8)}}{2(-3)}$$

$$= \frac{-12 \pm \sqrt{48}}{-6} = \frac{-12 \pm 4\sqrt{3}}{-6} = 2 \pm \frac{2}{3}\sqrt{3}$$

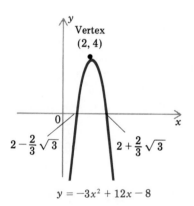

Vertex
$(2, 4)$

$2 - \frac{2}{3}\sqrt{3}$ $2 + \frac{2}{3}\sqrt{3}$

$y = -3x^2 + 12x - 8$

FIGURE 4.13

Therefore the x intercepts are $2 + \frac{2}{3}\sqrt{3}$ and $2 - \frac{2}{3}\sqrt{3}$. Completing the square in (5), we find that

$$\begin{aligned}
y &= -3x^2 + 12x - 8 \\
&= -3(x^2 - 4x) - 8 \\
&= -3(x^2 - 4x + 4 - 4) - 8 \\
&= -3(x^2 - 4x + 4) + 12 - 8 \\
&= -3(x - 2)^2 + 4
\end{aligned}$$

Using the notation of (4), we have $a = -3$, $h = 2$, and $k = 4$, so the vertex is $(2, 4)$. Thus we obtain the graph of the given equation by shifting the graph of $-3x^2$ upward 4 units and to the right 2 units (Figure 4.13).   □

In Section 4.2 we will continue our study of quadratic functions. However, the main point of interest will be the largest, or smallest, value assumed by such a function.

EXERCISES 4.1

In Exercises 1–20, sketch the graph of the given function. In each case, determine the vertex and all intercepts of the parabola.

1. $f(x) = \frac{3}{4}x^2$

2. $f(x) = -3x^2$

3. $f(x) = x^2 + 1$

4. $f(x) = x^2 - \frac{9}{4}$

5. $f(x) = -2x^2 + 8$

6. $f(x) = 1 - 4x^2$

7. $f(x) = -\frac{3}{2}(x - 1)^2$

8. $f(x) = \frac{1}{2}(x + 3)^2$

9. $f(x) = \frac{3}{2}(x + \frac{1}{3})^2 - \frac{1}{3}$

10. $f(x) = 2(x - \frac{1}{2})^2 + \frac{3}{2}$

11. $g(x) = x^2 + 2x$

12. $g(x) = x^2 - 2x$

13. $g(x) = x^2 - 2x + 2$

14. $f(x) = x^2 + 2x + 2$

15. $f(x) = -x^2 + 6x - 5$

16. $f(x) = 2x^2 + 4x + 5$

17. $f(x) = 1 - 2x - 3x^2$

18. $y = \frac{1}{2}x^2 + x + 1$

19. $y = (x + 1)(2x - 1)$

20. $y = (3x + 1)(x - 1)$

In Exercises 21–24, find an equation of the sketched parabola. (*Hint:* Use (4).)

21. The parabola in Figure 4.14a.

22. The parabola in Figure 4.14b.

23. The parabola in Figure 4.14c.

24. The parabola in Figure 4.14d.

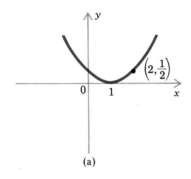(a)
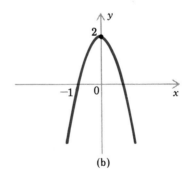(b)
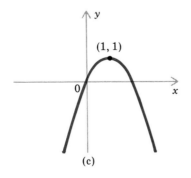(c)
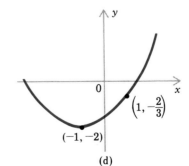(d)

FIGURE 4.14

25. Let $f(x) = -2x^2 + 3$. Find an equation of the function g whose graph is obtained from the graph of f by
   a. shifting upward 1 unit
   b. shifting downward 3 units
   c. shifting to the right $\frac{1}{2}$ unit
   d. shifting to the left 4 units

26. Let $g(x) = -x^2 + 2x - 5$. Find an equation of the function f whose graph is obtained from the graph of g by
   a. shifting upward $\frac{3}{2}$ units and to the left 4 units
   b. shifting downward 2 units and to the right $\frac{1}{2}$ unit

27. For what value of a does the graph of the equation $y = ax^2 + 6x + 3$ pass through the point $(-1, 0)$?

28. For what value of b does the graph of the equation $y = -\frac{1}{2}x^2 + bx + \frac{1}{3}$ pass through the point $(-2, -1)$?

29. For what value of c does the graph of the equation $y = -x^2 + x + c$ pass through the point $(2, 6)$?

30. In each part find the value of c for which the graph of f touches but does not cross the x axis.
   a. $f(x) = \frac{1}{2}x^2 - 2x + c$
   b. $f(x) = x^2 - 3x + c$
   c. $f(x) = 3x^2 + 10x + c$

31. Let $f(x) = ax^2 + c$. Prove that the graph of f has
   a. two x intercepts if a and c have opposite signs
   b. no x intercept if a and c have the same sign

32. Let $f(x) = 2x^2 - 8x + 3$. Determine the points on the graph of f that lie on the horizontal line 18 units above the vertex.

33. A parabolic arch is 16 feet tall at its highest point and spans 36 feet at its base. Find its height 9 feet from the plane on which its axis is located.

4.2
MAXIMUM AND MINIMUM VALUES OF QUADRATIC FUNCTIONS

From Section 4.1 we know that the graph of any quadratic function f is a parabola. If

$$f(x) = ax^2 + bx + c \quad \text{with} \quad a \neq 0 \tag{1}$$

then we also know that the vertex of the parabola is

$$\left(-\frac{b}{2a}, c - \frac{b^2}{4a}\right) \tag{2}$$

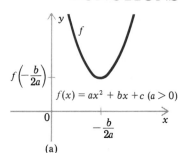

$f(x) = ax^2 + bx + c \ (a > 0)$

(a)

The vertex is the lowest point on the graph if $a > 0$ and is the highest point if $a < 0$. It follows that

> if $a > 0$, then $f\left(-\dfrac{b}{2a}\right) \leq f(x)$ for all x

so $f(-b/2a)$ is called the **minimum value** of f (Figure 4.15a). Analogously,

> if $a < 0$, then $f\left(-\dfrac{b}{2a}\right) \geq f(x)$ for all x

and $f(-b/2a)$ is the **maximum value** of f (Figure 4.15b). Thus every quadratic function has either a minimum or a maximum value, depending on the sign of the leading coefficient (a in (1)).

To compute the minimum or maximum value of such a function f, we can calculate $f(-b/2a)$, either by substituting $-b/2a$ for x in (1) or by calculating $c - b^2/4a$ (see (2)).

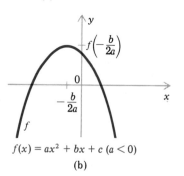

$f(x) = ax^2 + bx + c \ (a < 0)$

(b)

FIGURE 4.15

EXAMPLE 1.   Let $f(x) = 2x^2 - 3x + 1$. Verify that f assumes a minimum value, and compute the minimum value.

Solution.   In this example $f(x)$ has the form of (1) with $a = 2$, $b = -3$, and $c = 1$. Since $a > 0$, it follows that $f(-b/2a)$ is the minimum value. Now

$$-\frac{b}{2a} = -\frac{(-3)}{2(2)} = \frac{3}{4}$$

Consequently to find the minimum value of $f(x)$, we calculate $f(\frac{3}{4})$:

$$f\left(\frac{3}{4}\right) = 2\left(\frac{3}{4}\right)^2 - 3\left(\frac{3}{4}\right) + 1 = \frac{9}{8} - \frac{9}{4} + 1 = -\frac{1}{8}$$

Therefore the minimum value of f is $-\frac{1}{8}$.   □

EXAMPLE 2.   Let $f(x) = -x^2 + \sqrt{2}x + \frac{1}{2}$. Verify that f assumes a maximum value, and compute the maximum value.

Solution.   Here $f(x)$ has the form of (1) with $a = -1$, $b = \sqrt{2}$, and $c = \frac{1}{2}$. Since $a < 0$, $f(-b/2a) = c - b^2/4a$ is the maximum value of f. Next we calculate $c - b^2/4a$:

$$c - \frac{b^2}{4a} = \frac{1}{2} - \frac{(\sqrt{2})^2}{4(-1)} = \frac{1}{2} - \left(-\frac{1}{2}\right) = 1$$

Consequently the maximum value of f is 1.   □

Applications   In applications it is frequently desirable to maximize or minimize a variable quantity, such as area, profit, or cost. If the quantity to be maximized or minimized can be expressed as a quadratic function of another quantity, the discussion in this section provides us with a method for finding the maximum or minimum value.

EXAMPLE 3.   A rancher has 2 miles of fencing and wishes to fence in a rectangular grazing field with an area as large as possible. What should the dimensions of the field be?

Solution.   Since the rancher has 2 miles of fencing, the perimeter of the rectangle will be 2 miles. Let x and z be the dimensions in miles of any rectangle with a perimeter of 2 miles (Figure 4.16). Then

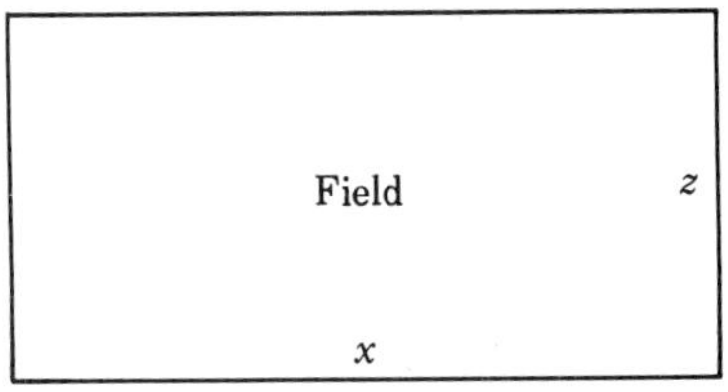

$$2x + 2z = 2$$

so that

$$x + z = 1$$

or equivalently,

$$z = 1 - x$$

FIGURE 4.16   Since $x \geq 0$ and $z \geq 0$, it follows that $0 \leq x \leq 1$. Next we rewrite the area xz of the rectangle solely in terms of x:

$$xz = x(1 - x) = x - x^2 = -x^2 + x$$

If we let

$$f(x) = -x^2 + x \quad \text{for} \quad 0 \le x \le 1$$

then $f(x)$ represents the area of a field one of whose sides has length x. Thus we wish to maximize $f(x)$. Now $f(x)$ has the form of (1) with $a = -1, b = 1$, and $c = 0$. Since $a < 0$, we know that there is a maximum value and that the maximum value occurs for

$$x = -\frac{b}{2a} = -\frac{1}{2(-1)} = \frac{1}{2}$$

But if $x = \frac{1}{2}$, then

$$z = 1 - \frac{1}{2} = \frac{1}{2}$$

Thus the maximum area results if the rectangle is a square with sides of length $\frac{1}{2}$ mile.   □

EXAMPLE 4.   On a rainy day the manager of a store estimates that she can sell 40 umbrellas at $9 each, and that for each nickel decrease in price, one more umbrella can be sold (which corresponds to 20 more umbrellas sold for each dollar decrease in price). Assuming that this is true and that she acquires umbrellas at a cost of $3 each, determine the price at which she should sell umbrellas in order to maximize her profit. How many umbrellas will be sold at that price?

Solution.   We will express all monetary quantities in dollars. Let x be the price of each umbrella and z the number of umbrellas sold. Then the number z of umbrellas sold at price x equals 40 (the number that would be sold at $9) plus 20 times the amount (in dollars) that the price has been decreased from $9. Thus

$$z = 40 + 20 \text{ (decrease from 9)} = 40 + 20(9 - x)$$

or equivalently,

$$z = 220 - 20x$$

Now the revenue R resulting from the sale of the z umbrellas at the price x is given by

$$R = xz = x(220 - 20x) = 220x - 20x^2$$

Since each umbrella costs $3, the cost C of the z umbrellas sold is given by

$$C = 3z = 3(220 - 20x) = 660 - 60x$$

As a result, the profit P is given by

$$P = R - C = (220x - 20x^2) - (660 - 60x)$$
$$= -20x^2 + 280x - 660$$

Thus if we let

$$f(x) = -20x^2 + 280x - 660$$

then we wish to find the maximum value of f. Since $f(x)$ has the form of (1) with $a = -20, b = 280$, and $c = -660$, it follows that the maximum value of f occurs for

$$x = -\frac{b}{2a} = -\frac{280}{2(-20)} = 7$$

Consequently each umbrella should be priced at $7 in order to maximize profit. If $x = 7$, then

$$z = 220 - 20(7) = 80$$

so 80 umbrellas will be sold at the optimal price of $7. □

EXAMPLE 5.   Find the point on the line $y = 3x$ that is closest to the point $(5, 0)$.

Solution.   The distance between a point (x, y) on the line $y = 3x$ and the point $(5, 0)$ is given by

$$\sqrt{(x - 5)^2 + (3x - 0)^2}$$

so we let

$$f(x) = \sqrt{(x - 5)^2 + (3x - 0)^2}$$
$$= \sqrt{(x^2 - 10x + 25) + 9x^2}$$
$$= \sqrt{10x^2 - 10x + 25}$$

We desire to find the minimum value of $f(x)$. However, f is not a quadratic function. But if we let

$$g(x) = 10x^2 - 10x + 25$$

then since $g(x) = (f(x))^2$, the minimum value of f occurs for the same number x as the minimum value of g. Now g is a quadratic function, and $g(x)$ has the form of (1) with $a = 10, b = -10$, and $c = 25$. Since $a > 0$, there is a minimum value, which occurs for

$$x = -\frac{b}{2a} = -\frac{-10}{2(10)} = \frac{1}{2}$$

Therefore $y = 3(\frac{1}{2}) = \frac{3}{2}$, so the closest point is $(\frac{1}{2}, \frac{3}{2})$. □

Notice that in the preceding example, we did not need to find the minimum value of f, but only the number x at which the minimum value occurs.

Problems concerning the motion of an object moving vertically under the sole influence of gravity involve quadratic functions. Recall from (1) in Section 2.4 that the height at time t of an object moving solely under the influence of gravity is given by

$$h(t) = -16t^2 + v_0 t + h_0 \tag{3}$$

where v_0 is the initial velocity and h_0 is the initial height. Since h is a quadratic function, if we know the initial velocity and initial height of an object, we can determine the maximum height.

EXAMPLE 6.   A tape measure is thrown vertically upward from an initial height of 6 feet with an initial velocity of 32 feet per second toward a third floor balcony 21 feet above ground. Does the tape measure reach the balcony?

Solution.   Taking $v_0 = 32$ and $h_0 = 6$ in (3), we have

$$h(t) = -16t^2 + 32t + 6$$

which has the form of (1) with $a = -16$, $b = 32$, and $c = 6$, and with t and h replacing x and f, respectively. Since $a < 0$, we know that h assumes a maximum value for

$$t = -\frac{b}{2a} = -\frac{32}{2(-16)} = 1$$

Thus the tape measure attains its maximum height after 1 second. The fact that

$$h(1) = -16(1)^2 + 32(1) + 6$$
$$= -16 + 32 + 6 = 22$$

implies that the maximum height attained by the tape measure is 22 feet. Therefore the tape measure does reach the balcony. $\square$

In solving maximum and minimum problems, the following approach is frequently effective:

> i. After reading the problem carefully, choose a letter for the quantity to be maximized or minimized. Also choose auxiliary variables for the other quantities appearing in the problem.
> ii. Express the quantity to be maximized or minimized in terms of the auxiliary variables.

iii. Choose one auxiliary variable x to serve as master variable, and use the information given in the problem to express all other auxiliary variables in terms of x.

iv. Use the results of steps (ii) and (iii) to express the quantity to be maximized or minimized in terms of x alone.

v. Use the theory of this section to find the desired maximum or minimum value.

EXERCISES 4.2

In Exercises 1–10, determine whether the given quadratic function assumes a maximum value or a minimum value. Then find the maximum or the minimum value.

1. $f(x) = 3x^2 + 5$

2. $f(x) = x^2 + x - \frac{1}{2}$

3. $f(x) = 20x^2 - 40x$

4. $f(x) = -x^2 + 2x + 3$

5. $g(x) = x - \frac{1}{4}x^2$

6. $g(x) = -5x^2 - x + 1$

7. $g(t) = 16t^2 + 64t + 36$

8. $g(t) = -\frac{1}{4}t^2 + t - 13$

9. $y = t^2 + 3t$

10. $y = \frac{1}{2}x^2 - 3x + \frac{3}{2}$

11. A rancher has 2 miles of fencing and wishes to fence in a rectangular grazing field with an area as large as possible by erecting a fence on only 3 sides, a river forming the other side (Figure 4.17). What should the dimensions of the rectangle be?

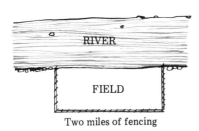

FIGURE 4.17

12. A farmer plans to fence in two adjacent rectangular grazing fields of the same dimensions, one for cows and the other for horses, with a common fence separating the two fields. If the farmer has 900 meters of fencing, how large can the area of each field be?

13. A piece of wire 1 meter long is to be cut into two pieces, each of which is then to be bent into the shape of a square (Figure 4.18). How should the wire be cut

if the sum of the areas of the two squares is to be minimized?

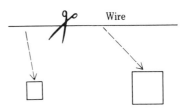

FIGURE 4.18

14. Find the dimensions of the largest rectangle that can be inscribed in the triangle sketched in Figure 4.19.

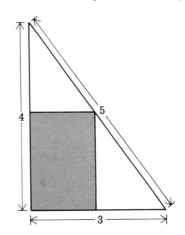

FIGURE 4.19

15. A rectangle is to have a perimeter of 12 inches. Determine the lengths of the sides that will yield the largest area.

16. The base of a given isosceles triangle is 12, and the height is 8. Find the dimensions of the rectangle with maximum area that can be inscribed in the triangle if the base of the rectangle is to lie along the base of the triangle (Figure 4.20).

17. Find the dimensions of the rectangle whose perimeter is 16 and whose diagonal is as small as possible.

18. Find two numbers whose sum is 24 and whose product is as large as possible.

19. Find two numbers whose sum is -16 and whose product is as large as possible.

20. Find two numbers whose difference is 6 and whose product is as small as possible.

21. Find two numbers whose product is as large as possible, under the condition that the sum of one of the numbers and 3 times the other is 48.

22. Find two numbers whose sum is 36 and the sum of whose squares is as small as possible. Then determine the sum of the two squares.

23. Find the point on the line $2x + y = 2$ that is closest to the point $(0, -3)$.

24. Find the point on the line $y = 10 - 3x$ that is closest to the origin.

25. Find the points on the parabola $y = x^2$ that are closest to the point $(0, 1)$. (*Hint:* The square of the distance is a quadratic function of x^2. First find the value of x^2 for which the distance is minimal.)

26. The Hot-Cake Company can sell 100 pounds of pancake mix a day at $1.80 a pound, and it believes that sales will decrease $\frac{1}{5}$ pound for each penny increase in price per pound. For what price would the

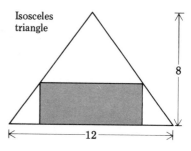

FIGURE 4.20

revenue be maximized? What will the corresponding revenue be?

27. The Fuzzy-Wuzzy Fabric Shop can sell 36 yards of denim per day if it charges $2 per yard, and it figures that each increase of $.25 per yard will reduce sales of denim by 3 yards. Determine the price per yard that will maximize the daily revenue from the sale of denim.

28. A construction company estimates that it can build and sell x houses in a certain development if it sets the price of each home at $100,000 - 1000x$ dollars. If each house costs $60,000 to construct, how many houses must the company sell to maximize its profit? How much must the company charge for each house in order to maximize its profit?

29. Suppose a baseball is thrown vertically upward from a height of 64 feet with an initial velocity of 48 feet per second. Find the maximum height attained by the baseball and the length of time it takes the ball to attain that height.

4.3 GRAPHS OF HIGHER-DEGREE POLYNOMIAL FUNCTIONS

In Section 4.1 we discussed the graph of a quadratic function. The graph is a parabola, and with a minimum of calculation we can determine its vertex, intercepts, and general shape. Now we turn to higher-degree polynomial functions. The higher the degree, the less accurate our graph is apt to be. Nevertheless, there are techniques that allow us to sketch the general shape of many such functions.

First we will analyze the graphs of polynomials of the form ax^n, and then we will study the graphs of other higher-degree polynomial functions.

EXAMPLE 1.   Let $f(x) = x^3$. Sketch the graph of f.

Solution.   We begin by observing that

$$f(-x) = (-x)^3 = (-1)^3x^3 = -x^3 = -f(x)$$

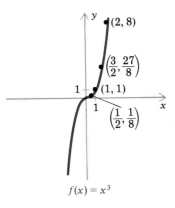

$f(x) = x^3$

FIGURE 4.21

which means that the graph of f is symmetric with respect to the origin. Next, we plot a few points on the graph, using the following table:

x	0	$\frac{1}{2}$	1	$\frac{3}{2}$	2
$f(x)$	0	$\frac{1}{8}$	1	$\frac{27}{8}$	8

Now we connect the points with a smooth curve and then use the symmetry mentioned above to finish the graph (Figure 4.21). □

EXAMPLE 2.   Let $f(x) = x^4$. Sketch the graph of f.

Solution.   Since

$$f(-x) = (-x)^4 = (-1)^4 x^4 = x^4 = f(x)$$

the graph of f is symmetric with respect to the y axis. We plot a few points on the graph, using the following table:

x	0	$\frac{1}{2}$	1	$\frac{3}{2}$	2
$f(x)$	0	$\frac{1}{16}$	1	$\frac{81}{16}$	16

Next we draw a smooth curve through the points we have plotted, and finally we use the symmetry of the graph to obtain the sketch in Figure 4.22. □

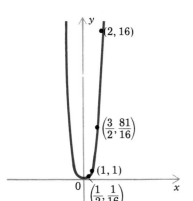

$f(x) = x^4$

FIGURE 4.22

The shapes of the graphs of x^2 and x^4 appear generally similar (Figure 4.23), and in fact the shapes of the graphs of x^2 and x^n are generally similar for any *positive even* integer n. More particularly, they are symmetric with respect to the y axis (since $(-x)^n = x^n$ if n is even), have their minimum value of 0 at 0, and lie above the x axis (because $x^n \geq 0$ if n is even).

A similar pattern emerges with respect to the graph of x^3 and the graph of x^n for any odd integer $n \geq 3$. Here the graphs are symmetric with respect to the origin (since $(-x)^n = -x^n$ if n is odd) and have neither maximum nor minimum values.

If a is a nonzero constant, the graphs of ax^n and x^n are related to each other in the same way as the graphs of ax^2 and x^2 are. The following example illustrates this fact.

EXAMPLE 3.   Sketch the graphs of the following functions.

a. $f(x) = 2x^3$ b. $g(x) = -\frac{1}{3}x^4$

Solution.
a. Since we already have the graph of x^3 in Figure 4.21, to obtain the graph of f we need only multiply the y coordinate of each point on the graph of x^3 by 2, which has the effect of "pulling" the graph of x^3 away from the x axis by a factor of 2 (Figure 4.24).
b. The graph of x^4 is shown in Figure 4.22. To obtain the graph of g, just multiply the y coordinate of each point on the graph of x^4 by $-\frac{1}{3}$,

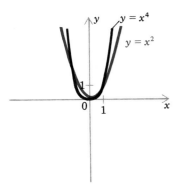

FIGURE 4.23

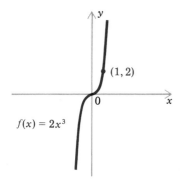

$f(x) = 2x^3$

(1, 2)

FIGURE 4.24

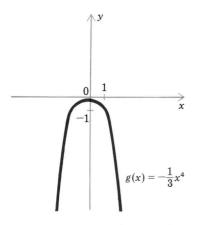

$g(x) = -\frac{1}{3}x^4$

FIGURE 4.25

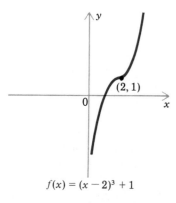

(2, 1)

$f(x) = (x - 2)^3 + 1$

FIGURE 4.26

which has the effect of reflecting the graph of x^4 through the x axis and "pulling" it toward the x axis by a factor of 3 (Figure 4.25).   □

Sometimes the graph of a polynomial can be obtained by a vertical and/or horizontal shift of the graph of a polynomial of the form ax^n.

<u>EXAMPLE 4.</u>   Let $f(x) = (x - 2)^3 + 1$. Sketch the graph of f.

Solution.   The graph of f can be obtained by shifting the graph of x^3 upward 1 unit and to the right 2 units (Figure 4.26).   □

Now we turn to graphs of other polynomial functions of degree greater than 2. We could plot as many points on the graph of such a function as we like by calculating the values of the function for various numbers in the domain, but this would be tedious and, moreover, would not indicate any possible relationship between the graphs of different polynomial functions. In the discussion that follows we will concentrate instead on such features as how a graph looks far away from the y axis and where it is above and where below the x axis.

Let

$$f(x) = a_nx^n + a_{n-1}x^{n-1} + a_{n-2}x^{n-2} + \cdots + a_1x + a_0 \quad \text{with} \quad a_n \neq 0 \quad (1)$$

Far away from the y axis the graph of f has the general shape of the graph of the first term, a_nx^n. The reason is that when x is large, a_nx^n dominates $a_{n-1}x^{n-1} + a_{n-2}x^{n-2} + \cdots + a_1x + a_0$. Thus, far away from the y axis the graphs of $x^3 - x$ and x^3 have the same general shape, and likewise, far away from the y axis the graphs of $2x^4 - 8x^3 + 8x^2$ and $2x^4$ have the same general shape.

To sketch the portions of the graph of the function in (1) nearer the y axis, we are helped enormously if we can factor $f(x)$ as

$$f(x) = a_n(x - c_1)(x - c_2)\cdots(x - c_n) \quad (2)$$

As a result, we will assume henceforth that we can find such a factorization of each polynomial function under consideration. The numbers $c_1, c_2, \ldots, c_n$ appearing in (2) are the x intercepts of f, because $f(x) = 0$ if and only if

$$a_n(x - c_1)(x - c_2)\cdots(x - c_n) = 0$$

which happens if and only if any one of the factors on the left is 0. In other words, $f(x) = 0$ if and only if x is one of the numbers $c_1, c_2, \ldots, c_n$.

To illustrate, we let

$$f(x) = x^3 - x$$

Factoring, we find that

$$x^3 - x = x(x^2 - 1) = x(x - 1)(x + 1)$$

If we write x as $x - 0$, then

$$f(x) = (x - 0)(x - 1)(x + 1)$$

It follows that the x intercepts of f are 0, 1, and -1, so the graph of f intersects the x axis at $(0, 0)$, $(1, 0)$, and $(-1, 0)$.

Knowing the x intercepts allows us to determine the intervals on which $f(x)$ is positive and those on which $f(x)$ is negative. We see how this is done in the following example.

EXAMPLE 5.   Let $f(x) = x^3 - x$. Sketch the graph of f.

Solution.   First of all, the y intercept is 0, since $f(0) = 0$. Next,

$$f(x) = x(x - 1)(x + 1)$$

so the x intercepts are 0, 1, and -1 (as we said above). Now the x intercepts divide the x axis into the subintervals $(-\infty, -1), (-1, 0), (0, 1)$, and $(1, \infty)$. To determine whether $f(x)$ is positive or negative on each of these subintervals, we study the factors x, $x - 1$, and $x + 1$ individually on each subinterval. Figure 4.27 gives the results.

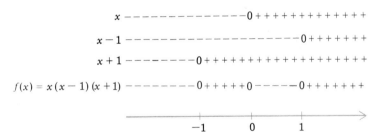

FIGURE 4.27

In constructing Figure 4.27 we used the fact that $x - c < 0$ if $x < c$, and $x - c > 0$ if $x > c$. We also used the fact that $f(x) > 0$ if an even number of the factors of $f(x)$ are negative and the rest are positive, and $f(x) < 0$ if an odd number of the factors are negative and the rest are positive. At this point we know the x intercepts and the intervals on which $f(x)$ is positive or negative. Moreover, because

$$f(-x) = (-x)^3 - (-x) = -x^3 + x = -(x^3 - x) = -f(x)$$

the graph is symmetric with respect to the origin. Finally, far away from the y axis the graph has the general shape of the graph of x^3 (see Figure 4.21). Consequently we can draw a smooth curve for the graph of f (Figure 4.28).   □

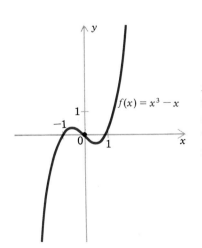

FIGURE 4.28

The fact that the "hilltop" to the left of the y axis in Figure 4.28 occurs for $x = -\sqrt{3}/3$, and the "valley bottom" to the right of the y axis occurs for $x = \sqrt{3}/3$, can be proved by methods of calculus. Even without this specific information we were able to give a reasonably accurate sketch of the graph of f.

Notice that the graph of f in Example 5 *crosses* the x axis at the points corresponding to the x intercepts. Example 6 shows that the graph of a function may touch, but not cross, the x axis at a point corresponding to an x intercept.

EXAMPLE 6.   Let $g(x) = x^3 + 2x^2 + x$. Sketch the graph of g.

Solution.   Since $g(0) = 0$, the y intercept is 0. Next we factor $g(x)$:

$$g(x) = x^3 + 2x^2 + x = x(x^2 + 2x + 1) = x(x + 1)^2$$

Therefore the x intercepts are 0 and -1, and we prepare the chart shown in Figure 4.29, which gives the signs of x and $(x + 1)^2$, as well as the resulting signs of $g(x)$, on the intervals $(-\infty, -1)$, $(-1, 0)$, and $(0, \infty)$.

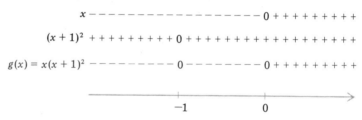

FIGURE 4.29

Using the information in Figure 4.29, along with the fact that far away from the y axis the graph of g has the same general shape as the graph of x^3, we sketch the graph of g in Figure 4.30.   □

Through these examples we have developed a procedure that facilitates drawing a rough sketch of the graph of any polynomial function f that we can factor:

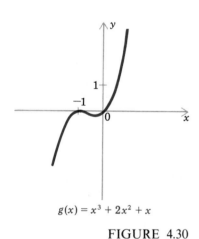

$g(x) = x^3 + 2x^2 + x$

FIGURE 4.30

i. Find the y intercept, $f(0)$.
ii. Factor $f(x)$ to obtain a formula in the form of (2).
iii. Obtain the x intercepts from the factorization in (ii), and note the intervals into which the x intercepts divide the real line.
iv. Prepare a chart that tells on which of the intervals in (iii) the value $f(x)$ is positive and on which negative.
v. Determine the general shape of the graph far away from the y axis (the same as the shape of the graph of $a_n x^n$).
vi. Using the information obtained in (i)–(v), along with any symmetry of the graph, sketch the graph of f.

Notice in Figure 4.30 that the graph of g *touches*, but does not *cross*, the x axis at -1. The reason is that in the factorization of g, $x + 1$ appears *twice*, that is, $(x + 1)^2$ appears in the factorization. Because $(x + 1)^2 \geq 0$ for all x, $g(x)$ does not change sign at -1. In general, if c is an x intercept of a polynomial function f and if $x - c$ appears an even number of times in the factorization of f, then $f(x)$ does not change sign at c and the graph of f merely touches the x axis at c, whereas if $x - c$ appears an odd number of times, then $f(x)$ changes sign at c and the graph of f crosses the x axis at c. We make use of this observation in the solution of the following example.

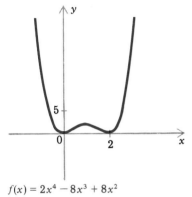

$f(x) = 2x^4 - 8x^3 + 8x^2$

FIGURE 4.31

EXAMPLE 7.   Let $f(x) = 2x^4 - 8x^3 + 8x^2$. Sketch the graph of f.

Solution.   The y intercept is 0, since $f(0) = 0$. For the x intercepts we factor $f(x)$:

$$f(x) = 2x^4 - 8x^3 + 8x^2 = 2x^2(x^2 - 4x + 4) = 2x^2(x - 2)^2$$

Therefore the x intercepts are 0 and 2. Since x appears twice, and similarly $x - 2$ appears twice, the graph touches but does not cross the x axis at $x = 0$ and at $x = 2$. In fact, $f(x) \geq 0$ for all x, so there is no need for a chart in this case. Finally, far from the y axis the graph of f has the general shape of the graph of $2x^4$. Because there is no symmetry, we are now ready to sketch the graph (Figure 4.31).   ☐

Caution: The methods of this section do not give precise, detailed information about the local shape of a graph. For example, if

$$f(x) = x^4 - x^3$$

then $$f(x) = x^3(x - 1)$$

so by using the procedure outlined above we would probably draw the graph appearing in Figure 4.32a. A more precise sketch of the graph appears in Figure 4.32b. Such refinements in the graph can be obtained by methods of calculus, or by calculating a large number of values of f.

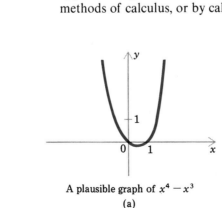

A plausible graph of $x^4 - x^3$

(a)

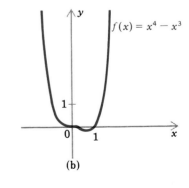

$f(x) = x^4 - x^3$

(b)

FIGURE 4.32

EXERCISES 4.3

In Exercises 1–24, sketch the graph of the given function, and indicate all intercepts.

1. $f(x) = x^3 + 1$

2. $f(x) = 1 - x^4$

3. $f(x) = \frac{1}{4} + 2x^3$

4. $f(x) = 4 - \frac{1}{4}x^4$

5. $f(x) = -2(x + 1)^3$

6. $g(x) = \frac{1}{8}(x - 2)^4 - 2$

7. $g(x) = x^5 + 1$

8. $g(x) = -\frac{1}{81}(x - 3)^6$

9. $g(x) = (x + 1)(x - 1)(x - 2)$

10. $f(x) = x(x + 1)^3$

11. $f(x) = -x^2(x + 1)(x + 2)$

12. $f(x) = 2x^2(x - 4)$

13. $f(x) = x^2(x - 1)^2$

14. $f(x) = x^3(x - 1)^2$

15. $g(x) = (x^2 - 1)(x + 1)^2$

16. $g(x) = (x^2 - 3x + 2)(x^2 - 4)$

17. $g(x) = x^3 - x^2 - 2x$

18. $g(x) = x^3 - 4x^2 + 4x$

19. $f(x) = x^4 - x^2$

20. $f(x) = -x^5 - x^4 + 6x^3$

21. $f(x) = x^5 - x^4$

22. $y = x^5 - x^3$

23. $y = x^4 - 5x^2 + 4$

24. $y = -x^4 + 2x^3 + x^2 - 2x$

25. Find the number c such that the graph of the equation $y = x^4 - x^2 + 3x + c$ passes through (1, 5).

26. Find the number c such that the graph of the equation $y = 3x^3 - x^2 + 2x + c$ has y intercept 7.

27. Let $f(x) = ax^3 - 2x^2 + x - 6$. If -1 is an x intercept of the graph of f, find the value of a.

28. Let $f(x) = x^3 + bx^2 + 3x + 9$. If 3 is an x intercept of the graph of f, find the value of b.

4.4
GRAPHS OF RATIONAL FUNCTIONS

Now that we have studied graphs of polynomial functions at some length, we turn to graphs of quotients of polynomials. As we said in the introduction to Chapter 4, the quotient of two polynomials is called a rational function. In other words, if P and Q are polynomials, then the function f defined by

$$f(x) = \frac{P(x)}{Q(x)}$$

is a rational function. The domain of f consists of all real numbers x such that $Q(x) \neq 0$ (since we cannot divide by 0). Ironically, those numbers that are *not* in the domain of a rational function f frequently play an essential role in sketching the graph of f. Let us illustrate this with the graph of $1/x$.

EXAMPLE 1. Let $f(x) = 1/x$. Sketch the graph of f.

Solution. The domain of f consists of all real numbers x except 0. There are no x or y intercepts, as we showed in Example 13 of Section 3.4. Moreover,

$$f(-x) = \frac{1}{-x} = -\frac{1}{x} = -f(x)$$

and thus the graph of f is symmetric with respect to the origin. Next we construct tables of values of $f(x)$ for small and for large positive values of x:

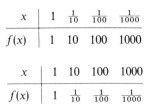

x	1	$\frac{1}{10}$	$\frac{1}{100}$	$\frac{1}{1000}$
$f(x)$	1	10	100	1000

x	1	10	100	1000
$f(x)$	1	$\frac{1}{10}$	$\frac{1}{100}$	$\frac{1}{1000}$

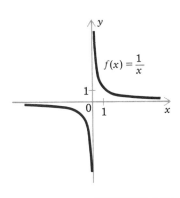

FIGURE 4.33

The first table suggests that as x shrinks toward 0, $f(x)$ grows without bound; accordingly, the part of the graph of f corresponding to small positive values of x is far from the x axis. From the second table we conclude that as x grows without bound, $f(x)$ shrinks toward 0; thus the part of the graph of f corresponding to large positive values of x is close to the x axis. Combining this information and then using the symmetry with respect to the origin, we obtain the graph of f shown in Figure 4.33.   □

Because $1/x$ becomes arbitrarily large if x is positive and approaches 0 (that is, the point $(x, 1/x)$ on the graph gets arbitrarily far from the x axis as x approaches 0 from the right), the vertical line $x = 0$ is called a vertical asymptote of the graph. In general, we say that a vertical line $x = a$ is a **vertical asymptote** of the graph of a function g if either (or both) of the following conditions hold:

1. As x approaches a from the left or from the right (or both), $g(x)$ is positive and becomes arbitrarily large (Figure 4.34a, b).
2. As x approaches a from the left or from the right (or both), $g(x)$ is negative and becomes arbitrarily large in absolute value (Figure 4.34c, d).

The graph in Figure 4.34e satisfies both condition 1 and condition 2.

It can be proved that if P and Q are polynomial functions such that $P(a) \neq 0$ and $Q(a) = 0$, then the line $x = a$ is a vertical asymptote of the

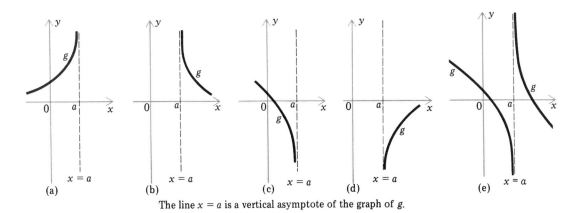

(a)   (b)   (c)   (d)   (e)

The line $x = a$ is a vertical asymptote of the graph of g.

FIGURE 4.34

graph of the rational function P/Q. Thus both the rational functions

$$\frac{x}{(x-2)^3} \quad \text{and} \quad \frac{x^3 + 2x - 1}{(x-2)(x-4)}$$

have the same vertical asymptote $x = 2$, and the second function also has the vertical asymptote $x = 4$.

Recall that $1/x$ approaches 0 if x is positive and becomes arbitrarily large (so that far away from the y axis the graph approaches the horizontal line $y = 0$). We say that the horizontal line $y = 0$ is a horizontal asymptote of the graph. In general, we say that a horizontal line $y = c$ is a **horizontal asymptote** of the graph of a function g if either (or both) of the following conditions hold:

3. If x is positive and becomes arbitrarily large, $g(x)$ gets close to c (Figure 4.35a).
4. If x is negative and becomes arbitrarily large in absolute value, $g(x)$ gets close to c (Figure 4.35b).

If $g(x) = x^2/(x^2 + 1)$ and $c = 1$, then g satisfies both condition 3 and condition 4, and hence the line $y = 1$ is a horizontal asymptote by either condition (Figure 4.35(c)).

Asymptotes play a major role in graphing most rational functions that are not polynomials, as we will presently see.

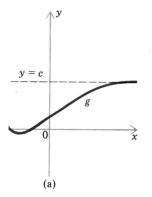

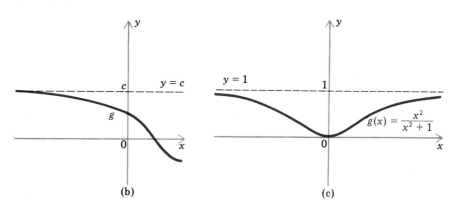

(a)   (b)   (c)

FIGURE 4.35

EXAMPLE 2.   Let $f(x) = 1/x^2$. Sketch the graph of f.

Solution.   The domain consists of all $x \neq 0$, and $1/x^2$ is never 0. Consequently there are no x or y intercepts. Next, observe that

$$f(-x) = \frac{1}{(-x)^2} = \frac{1}{x^2} = f(x)$$

so the graph of f is symmetric with respect to the y axis. Notice that if x is

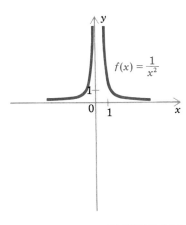

FIGURE 4.36

positive and gets close to 0, so does x^2, and consequently $1/x^2$ becomes very large. Therefore the vertical line $x = 0$ is a vertical asymptote of the graph. Similarly, if x is positive and becomes arbitrarily large, so does x^2, and thus $1/x^2$ approaches 0. Consequently the horizontal line $y = 0$ is a horizontal asymptote of the graph. Combining this information and using the symmetry of the graph with respect to the y axis, we draw the graph of f in Figure 4.36. ☐

When n is any positive integer, the graph of $1/x^n$ may be determined in the same way that the graphs of $1/x$ and $1/x^2$ are. It is only necessary to notice whether n is even or odd. On the one hand, if n is even, then the graph of $1/x^n$ is symmetric with respect to the y axis and resembles the graph in Figure 4.37a. On the other hand, if n is odd, then the graph of $1/x^n$ is symmetric with respect to the origin and resembles the graph in Figure 4.37b.

Finally, if a is a nonzero constant, then the graph of $1/(x - a)^n$ has the same appearance as the graph of $1/x^n$ does, except that the vertical asymptote is $x = a$ (instead of $x = 0$), and the graph is shifted horizontally so as to be centered about the line $x = a$.

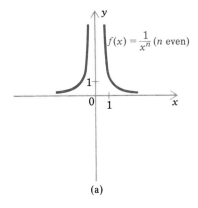

(a)

EXAMPLE 3. Sketch the graphs of the following functions, and identify their vertical asymptotes.

a. $f(x) = \dfrac{1}{(x - 2)^3}$ b. $g(x) = \dfrac{1}{(x + 4)^4}$

Solution.

a. The graph of f has the vertical asymptote $x = 2$, and the graph is as sketched in Figure 4.38a, with y intercept $-\frac{1}{8}$, since

$$f(0) = \frac{1}{(-2)^3} = -\frac{1}{8}$$

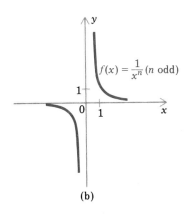

(b)

FIGURE 4.37

FIGURE 4.38

b. Since $x + 4 = x - (-4)$, the vertical asymptote is $x = -4$. The graph is as sketched in Figure 4.38b. The y intercept is $\frac{1}{256}$, since

$$g(0) = \frac{1}{4^4} = \frac{1}{256} \quad \square$$

Graphs of Factored Rational Functions

If we can factor both numerator and denominator of a rational function, we can sometimes obtain a reasonably accurate sketch of the graph with the techniques already at hand. These include finding the domain and any intercepts. locating where the function is positive and where negative, and determining any asymptotes and symmetry of the graph.

<u>EXAMPLE 4.</u>   Let $f(x) = 1/(x - 2)(x + 3)$. Sketch the graph of f.

Solution.   The domain consists of all numbers except 2 and -3. The y intercept is $-\frac{1}{6}$, since

$$f(0) = \frac{1}{(-2)(3)} = -\frac{1}{6}$$

Since the numerator of $f(x)$ cannot be 0, so that $f(x) \neq 0$ for all x in the domain, there are no x intercepts. The vertical asymptotes are determined from the values at which the denominator is 0, which are $x = 2$ and $x = -3$. Next, we prepare the chart in Figure 4.39 to determine the sign of $f(x)$ on the three intervals $(-\infty, -3), (-3, 2)$, and $(2, \infty)$, into which -3 and 2 divide the x axis.

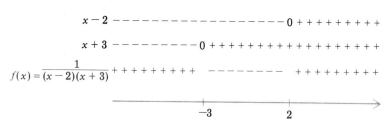

FIGURE 4.39

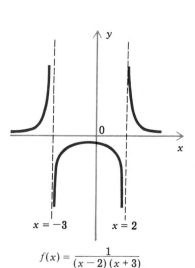

$$f(x) = \frac{1}{(x - 2)(x + 3)}$$

FIGURE 4.40

As our final preparation for sketching the graph, we observe that $f(x)$ approaches 0 if x is positive and becomes large, or if x is negative and becomes large in absolute value. This yields the horizontal asymptote $y = 0$. Because there is no symmetry, we are ready to use the information we have assembled and sketch the graph of f in Figure 4.40. $\square$

That the "hilltop" in Figure 4.40 occurs at $x = -\frac{1}{2}$ can conveniently be found by methods of calculus. Nevertheless, the general appearance of the graph can be determined by the methods we are using.

Now we will see when an arbitrary rational function P/Q has horizontal asymptotes. Let

$$P(x) = a_n x^n + a_{n-1} x^{n-1} + \cdots + a_1 x + a_0 \quad \text{with } a_n \neq 0$$

and

$$Q(x) = b_m x^m + b_{m-1} x^{m-1} + \cdots + b_1 x + b_0 \quad \text{with } b_m \neq 0$$

so that

$$\frac{P(x)}{Q(x)} = \frac{a_n x^n + a_{n-1} x^{n-1} + \cdots + a_1 x + a_0}{b_m x^m + b_{m-1} x^{m-1} + \cdots + b_1 x + b_0}$$

Then

    i. If $n < m$, then the line $y = 0$ is a horizontal asymptote.
    ii. If $n = m$, then the line $y = a_n/b_m$ is a horizontal asymptote.
    iii. If $n > m$, then there is no horizontal asymptote. Instead, far away from the y axis the graph of P/Q has the general appearance of the graph of

$$\frac{a_n}{b_m} x^{n-m}$$

whose graph was described in Section 4.3.

EXAMPLE 5.   Let $f(x) = (1 - x^2)(x + 2)/(x^2 + 2x)$. Sketch the graph of f.

Solution.   The denominator $x^2 + 2x$ can be rewritten as $x(x + 2)$, which implies that the domain of f consists of all numbers except 0 and -2. Since 0 is not in the domain, there is no y intercept. Now we factor the numerator and denominator:

$$f(x) = \frac{(1 - x^2)(x + 2)}{x^2 + 2x} = \frac{(1 - x)(1 + x)(x + 2)}{x(x + 2)}$$

Canceling the factor $x + 2$ in the numerator and denominator yields

$$f(x) = \frac{(1 - x)(1 + x)}{x} \quad \text{for} \quad x \neq 0, -2 \tag{1}$$

From (1) we see that -1 and 1 are x intercepts. Since the denominator is 0 for $x = 0$ but the numerator is not 0 for $x = 0$, the line $x = 0$ is a vertical asymptote. Using (1), we assemble the chart in Figure 4.41 that shows the sign of $f(x)$ on each of the intervals $(-\infty, -1)$, $(-1, 0)$, $(0, 1)$ and $(1, \infty)$ determined by the vertical asymptote and the x intercepts.

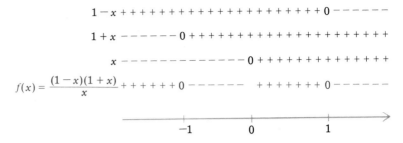

$$1 - x \; + 0 \; - - - - - -$$

$$1 + x \; - - - - - - \; 0 +$$

$$x \; - - - - - - - - - - - - \; 0 + + + + + + + + + + + + + +$$

$$f(x) = \frac{(1-x)(1+x)}{x} \; + + + + + + 0 \; - - - - - - \quad + + + + + + + 0 \; - - - - - -$$

FIGURE 4.41

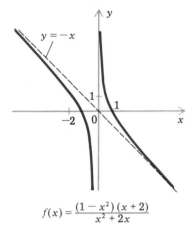

$$f(x) = \frac{(1-x^2)(x+2)}{x^2 + 2x}$$

FIGURE 4.42

By (1),

$$f(x) = \frac{(1-x)(1+x)}{x} = \frac{1-x^2}{x} \tag{2}$$

so that $f(x)$ satisfies possibility (iii). Therefore far from the y axis the graph of f resembles the graph of $-x^{2-1} = -x$, and consequently there is no horizontal asymptote. Finally, from (2) we find that

$$f(-x) = \frac{1-(-x)^2}{-x} = -\frac{1-x^2}{x} = -f(x)$$

so that the graph of f is symmetric with respect to the origin. Now we are ready to draw the graph of f (Figure 4.42). □

The line $y = -x$ in Figure 4.42 is called an *oblique asymptote* of the graph of f, because far away from the y axis the graph of f approaches the line $y = -x$, which is neither horizontal nor vertical. The graph of any rational function for which the degree of the numerator is one more than the degree of the denominator has an oblique asymptote.

In our final example of this section we will sketch the graph of a rational function f that has the factor $x^2 + 1$ that itself cannot be factored. The polynomial $x^2 + 1$ is positive for every value of x, and hence does not create any intercepts or vertical asymptotes in the graph and, in particular, does not affect the sign of $f(x)$.

EXAMPLE 6.   Let $f(x) = (x^2 + 1)/(4x^2 - 1)$. Sketch the graph of f.

Solution.   The y intercept is -1, because $f(0) = -1$. There are no x intercepts, because $x^2 + 1 > 0$ for all x in the domain. Since

$$f(x) = \frac{x^2 + 1}{4x^2 - 1} = \frac{x^2 + 1}{4(x^2 - \frac{1}{4})} = \frac{x^2 + 1}{4(x + \frac{1}{2})(x - \frac{1}{2})}$$

it follows that the lines $x = -\frac{1}{2}$ and $x = \frac{1}{2}$ are vertical asymptotes (and that $-\frac{1}{2}$ and $\frac{1}{2}$ are not in the domain of f). The sign of $f(x)$ on each of the intervals

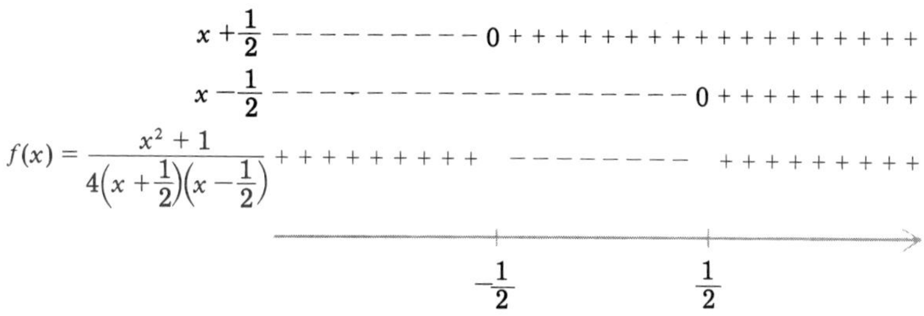

$$x + \frac{1}{2} \quad -------- \quad 0 + + + + + + + + + + + + + + + + + + +$$

$$x - \frac{1}{2} \quad ------------------ \quad 0 + + + + + + + + +$$

$$f(x) = \frac{x^2 + 1}{4\left(x + \frac{1}{2}\right)\left(x - \frac{1}{2}\right)} \quad + + + + + + + + + \quad --------- \quad + + + + + + + + +$$

$$-\frac{1}{2} \qquad\qquad \frac{1}{2}$$

FIGURE 4.43

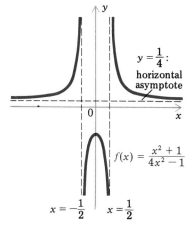

$y = \frac{1}{4}$:
horizontal
asymptote

$f(x) = \frac{x^2 + 1}{4x^2 - 1}$

$x = -\frac{1}{2}$ $x = \frac{1}{2}$

FIGURE 4.44

$\left(-\infty, -\frac{1}{2}\right), \left(-\frac{1}{2}, \frac{1}{2}\right)$, and $\left(\frac{1}{2}, \infty\right)$ determined by the vertical asymptotes is given in the chart in Figure 4.43.

We also find that f satisfies possibility (ii) above, so that since $\frac{1}{4}x^{2-2} = \frac{1}{4}$, it follows that $y = \frac{1}{4}$ is a horizontal asymptote. Finally, $f(-x) = f(x)$, as you can easily verify, so the graph of f is symmetric with respect to the y axis. With all the assembled information we can draw the graph of f (Figure 4.44). □

The following 8 steps, which we have used in the preceding examples, can help in graphing a rational function $f = P/Q$, where P and Q are polynomial functions.

1. If it has not been done already, factor the numerator $P(x)$ and the denominator $Q(x)$ into linear polynomials of the form $x - a$, quadratic polynomials that cannot be factored (such as $x^2 + 1$), and constants.
2. Determine the domain of f, which consists of all x for which $Q(x) \neq 0$. (Notice that a number a is *not* in the domain of f if and only if $x - a$ is a factor of $Q(x)$.)
3. Determine the intercepts (if any). The y intercept is $f(0)$ if 0 is in the domain of f. The x intercepts are the numbers a in the domain of f such that $x - a$ is a factor of the numerator.
4. Cancel any factors common to both numerator and denominator.
5. Determine the vertical asymptotes (if any). The line $x = a$ is a vertical asymptote if a factor of $x - a$ remains in the denominator after all cancellation has been completed in step 4.
6. Determine the sign of $f(x)$ on each of the intervals into which the x intercepts and the vertical asymptotes divide the x axis.
7. Determine the general appearance of the graph far away from the y axis as follows:
   Let

$$f(x) = \frac{a_n x^n + a_{n-1} x^{n-1} + \cdots + a_1 x + a_0}{b_m x^m + b_{m-1} x^{m-1} + \cdots + b_1 x + b_0}$$

   a. If $n < m$, then the line $y = 0$ is a horizontal asymptote.
   b. If $n = m$, then the line $y = a_n/b_n$ is a horizontal asymptote.

c. If $n > m$, then far away from the y axis the graph of f resembles the graph of

$$y = \frac{a_n}{b_m} x^{n-m}$$

In case $n - m = 1$, the latter graph is a straight line.

8. Use the information in steps 1–7, along with any symmetry of the graph, to sketch the graph of f.

EXERCISES 4.4

In Exercises 1–24, sketch the graph of the given function. Indicate all intercepts and asymptotes, as well as the behavior for large values of the variable.

1. $f(x) = -\dfrac{1}{x}$

2. $f(x) = \dfrac{1}{x - 1}$

3. $f(x) = \dfrac{4}{2 - x}$

4. $f(x) = 2 + \dfrac{1}{x^2}$

5. $f(x) = -\dfrac{1}{(x + 1)^2}$

6. $f(x) = \dfrac{1}{x^3} - 1$

7. $g(x) = -\dfrac{1}{4(x + \frac{1}{2})^4}$

8. $g(x) = \dfrac{3}{x^2 - 1}$

9. $g(x) = \dfrac{x}{x - 2}$

10. $g(x) = \dfrac{x - 1}{x + 1}$

11. $g(x) = \dfrac{-x}{x^2 - 4}$

12. $f(x) = -\dfrac{3}{x^2 + x}$

13. $f(x) = \dfrac{2}{(x - 1)(x + 2)}$

14. $f(x) = \dfrac{x(x + 2)}{(x - 1)(x^2 - 4)}$

15. $f(x) = \dfrac{x}{x^2 + 4x + 3}$

16. $f(x) = \dfrac{2x - 1}{x^2 - x - 2}$

17. $f(x) = \dfrac{x^2}{2x^2 - 3x - 2}$

23. $f(x) = \dfrac{3x^2 + 6x + 3}{x^3 + 4x^2 + x + 4}$

24. $f(x) = \dfrac{x + \frac{1}{2}}{4x^3 + 4x^2 + x}$

18. $y = \dfrac{4x^2 + 4x + 1}{x}$

19. $y = \dfrac{4x^2 + 4x + 1}{x^2}$

20. $f(t) = \dfrac{t^2 + 1}{t + 1}$

21. $g(t) = \dfrac{-3t - 1}{t^2 + 4}$

22. $g(t) = \dfrac{t^4 - 1}{t^2 - 1}$

4.5
THE PARABOLA

Conic sections have been of interest to mathematicians since the third century B.C., when the Greek geometer Apollonius wrote eight treatises about conic sections and became known as the "Great Geometer." A *conic section* is formed when a double right circular cone is sliced by a plane. The intersection can be a *parabola*, an *ellipse* (a special case of which is a circle), or a *hyperbola*, and in very special cases it can be a point, a line, or two intersecting lines (called *degenerate* conic sections). Figure 4.45 depicts the various kinds of conic sections.

A conic section is a curve in a plane that cuts a double cone. If we think of the plane as the Cartesian plane, then the conic section can be described as the graph of an equation in x and y. In this and the next two sections we will associate the three main kinds of conic sections—parabola, ellipse, and hyperbola—with special types of equations, and we will sketch the graphs of such equations.

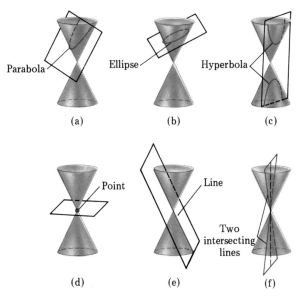

FIGURE 4.45

The graph of an equation of the form $y = ax^2 + bx + c$, with $a \neq 0$, is a parabola, as defined in Section 4.1, and has a graph as in Figure 4.46a, b. If the graph is turned on its side, then it remains a parabola and has an equation of the form

$$x = ay^2 + by + c \quad \text{with} \quad a \neq 0 \tag{1}$$

We sketched the graph of one such equation, $x = y^2$, in Figure 3.58. In general, to sketch the graph of (1) we complete the square on the right side of (1) to obtain an equation of the form

$$x = a(y - k)^2 + h, \quad \text{or equivalently,} \quad x - h = a(y - k)^2 \tag{2}$$

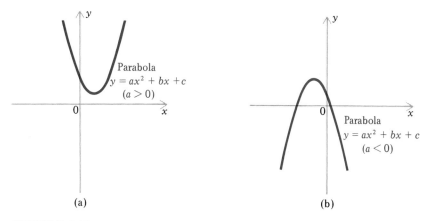

FIGURE 4.46

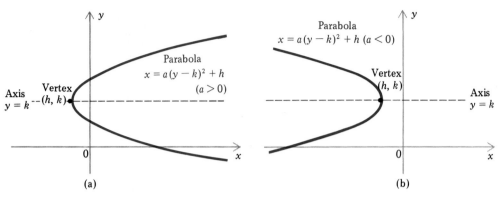

FIGURE 4.47

Then the graph opens to the right of the point (h, k) if $a > 0$ and to the left of (h, k) if $a < 0$ (Figure 4.47a, b). The point (h, k) is the *vertex* of the parabola. Moreover, the graph is symmetric with respect to the line $y = k$, called the *axis* of the parabola.

EXAMPLE 1.   Sketch the parabola having equation

$$x = 2y^2 - 4y + 5$$

and note the intercepts, vertex, and axis of the parabola.

Solution.   The x intercept is 5 because if $y = 0$, then $x = 5$. To determine any y intercepts, we set $x = 0$ in the given equation and solve for y:

$$2y^2 - 4y + 5 = 0 \qquad (3)$$

$$y = \frac{-(-4) \pm \sqrt{(-4)^2 - 4(2)(5)}}{2(2)}$$

$$= \frac{4 \pm \sqrt{16 - 40}}{4} = \frac{4 \pm \sqrt{-24}}{4}$$

Since $-24 < 0$, there is no real number y satisfying equation (3), and hence there are no y intercepts. To help locate the vertex of the parabola, we complete the square:

$$x = 2y^2 - 4y + 5$$
$$= 2(y^2 - 2y) + 5$$
$$= 2(y^2 - 2y + 1 - 1) + 5$$
$$= 2(y^2 - 2y + 1) - 2 + 5$$
$$= 2(y - 1)^2 + 3$$

In the notation of (2) we have $a = 2, h = 3$, and $k = 1$. Therefore the vertex is

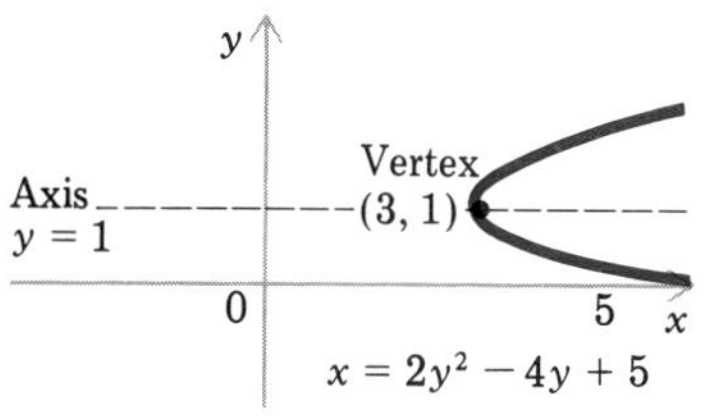

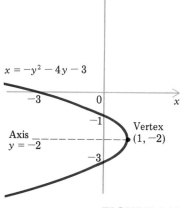

Axis
y = 1

Vertex
(3, 1)

$x = 2y^2 - 4y + 5$

FIGURE 4.48

the point $(3, 1)$, the axis is the line $y = 1$, and the parabola opens to the right because $a = 2 > 0$. The parabola is sketched in Figure 4.48. ☐

EXAMPLE 2.   Sketch the parabola having equation

$$x = -y^2 - 4y - 3$$

and note the intercepts, vertex, and axis of the parabola.

Solution.   The x intercept is -3 because if $y = 0$, then $x = -3$. To find any y intercepts, we set $x = 0$ in the given equation and solve for y:

$$-y^2 - 4y - 3 = 0$$
$$y^2 + 4y + 3 = 0$$
$$(y + 3)(y + 1) = 0$$
$$y = -3 \quad \text{or} \quad y = -1$$

As a result, the y intercepts are -3 and -1. Next we complete the square on the right side of the given equation:

$$x = -y^2 - 4y - 3$$
$$= -(y^2 + 4y) - 3$$
$$= -(y^2 + 4y + 4 - 4) - 3$$
$$= -(y^2 + 4y + 4) + 4 - 3$$
$$= -(y + 2)^2 + 1$$

$x = -y^2 - 4y - 3$

−3

−1

Axis
y = −2

Vertex
(1, −2)

−3

FIGURE 4.49

This time when we use the notation of (2), we find that $a = -1$, $h = 1$, and $k = -2$. Therefore the vertex is $(1, -2)$, and the axis of the parabola is the line $y = -2$. Since $a = -1 < 0$, the parabola opens to the left. The parabola is sketched in Figure 4.49. ☐

The graph of any equation of the form

$$Ax^2 + By^2 + Cx + Dy = E$$

with $A = 0$ and $B \neq 0$, or $A \neq 0$ and $B = 0$ \hfill (4)

is a parabola. By using the technique of completing the square, one can rewrite (4) as either

$$x = a(y - k)^2 + h \tag{5}$$

or

$$y = a(x - h)^2 + k \tag{6}$$

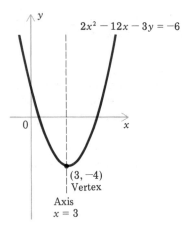

$$2x^2 - 12x - 3y = -6$$

(3, −4)
Vertex

Axis
$x = 3$

FIGURE 4.50

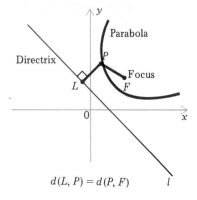

Directrix

Parabola

P

Focus

L

F

$d(L, P) = d(P, F)$ l

FIGURE 4.51

FIGURE 4.52a

In each case the vertex of the parabola is (h, k). For (5) the axis is the horizontal line $y = k$, and for (6) the axis is the vertical line $x = h$. In either case the parabola is symmetric with respect to its axis.

EXAMPLE 3.   Sketch the parabola having equation

$$2x^2 - 12x - 3y = -6$$

noting the vertex and axis of the parabola.

Solution.   The given equation has the form of (4) with $A = 2$, $B = 0$, $C = -12$, $D = -3$, and $E = -6$. Since $B = 0$, we can put the given equation into the form of (6) by completing the square:

$$2x^2 - 12x = 3y - 6$$
$$2(x^2 - 6x) = 3y - 6$$
$$2(x^2 - 6x + 9 - 9) = 3y - 6$$
$$2(x^2 - 6x + 9) - 18 = 3y - 6$$
$$2(x - 3)^2 = 3y + 12$$
$$2(x - 3)^2 = 3(y + 4)$$
$$\frac{2}{3}(x - 3)^2 = y + 4$$
$$y = \frac{2}{3}(x - 3)^2 - 4$$

Thus the vertex is $(3, -4)$, and the axis is the vertical line $x = 3$. The graph is sketched in Figure 4.50.   □

There is an alternative, geometric way of defining the parabola that is geometric in nature. Let l be a fixed line in the plane and F a fixed point not on l. The set of all points P in the plane equidistant from l and F is a parabola (Figure 4.51). The line l is called the ***directrix*** and the point F the ***focus*** of the parabola.

There are many theoretical and practical applications of parabolas. We list several of them here.

1. When an object moves under the sole influence of the earth's gravity and eventually strikes the earth, its path is normally parabolic (or linear if the motion is vertical). Thus the path of a golfball, baseball, or football in flight would be parabolic if there were no air resistance.

2. The shapes of certain mirrors can be obtained by revolving a parabola about its axis. Such mirrors, called ***parabolic mirrors***, are found in automobile headlights and reflecting telescopes. In a headlight all light rays emitted from a light source at the focus are reflected from the mirror along lines parallel to the axis of the parabola. In a reflecting telescope such as the one at Mt. Palomar and various radio telescopes, all incoming

rays from a star, which are essentially parallel to the axis, are reflected by the mirror and then pass through the focus, where the eyepiece of the telescope should be located (see Figure 4.52a).

3. A suspended cable tends to hang in the shape of a parabola if the weight of the cable is small compared to the weight it supports and if the weight is uniformly distributed along the cable. Bridges that are supported by cables in this fashion are called *suspension bridges* (see Figure 4.52b). All the larger bridges in the world, and most of the more famous ones, are suspension bridges; examples are the Golden Gate Bridge, the Brooklyn Bridge, and the Delaware Memorial Bridge.

FIGURE 4.52b

EXERCISES 4.5

In Exercises 1–14, sketch the parabola, noting the vertex and the axis of the parabola.

1. $x = 3y^2$

2. $x + \dfrac{1}{2}y^2 = 0$

3. $x + 5 = -\dfrac{1}{3}(y - 1)^2$

4. $x - 1 = 4(y + 2)^2$

5. $y^2 = x - y$

6. $x = y^2 + 2y - 4$

7. $x = 2y^2 - 2y + 1$

8. $5x - 19 = -y(y + 2)$

9. $y + 4x^2 - 24x + 34 = 0$

10. $y + x^2 - 8x + 22 = 0$

11. $2x - y^2 + 6y - 7 = 0$

12. $3x + 2y^2 + 4y + 8 = 0$

13. $x^2 + 4x = 3y + 5$

14. $(x - y)^2 + (x + y)^2 = 2x^2 + 4y - 2x$

In Exercises 15–20, find an equation of the parabola having the given properties.

15. Vertex $(0, 0)$; axis: y axis; $(2, 5)$ on the parabola

16. Vertex $(0, 0)$; axis: x axis; $(2, 5)$ on the parabola

17. Vertex $(-2, 5)$; axis: $x = -2$; $(0, 7)$ on the parabola

18. Vertex $(1, -4)$; axis: $y = -4$; $(-1, -1)$ on the parabola

19. Vertex $(3, 4)$; $(2, 5)$ and $(4, 5)$ on the parabola

20. Vertex $(-1, 1)$; $(3, 0)$ and $(0, \frac{1}{2})$ on the parabola

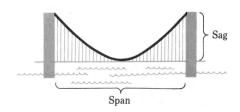

Suspension bridge

FIGURE 4.53

In Exercises 21–24, sketch the graph of the given equation. In each case the graph is a degenerate conic section.

21. $x^2 - 9y^2 + 2x + 18y = 8$

22. $4x^2 + 3y^2 - 24x - 18y + 63 = 0$

23. $36x^2 + 36x + 11 = -18y^2 + 12y$

24. $3x^2 + 6\sqrt{3}x = 2y^2 - 16y + 23$

25. For what value of c will the parabola with equation $cx = (y + 2)^2 - 6$ pass through the point $(-2, 4)$?

*26. For what value of c will the axis of the parabola with equation $2x - 5 = (1/c)y^2 + 4y + 9$ be the line $y = 3$?

27. Determine all points on the parabola with equation $3x - 4y^2 + 2y = 7$ that have x coordinate 3.

28. Determine all points on the parabola with equation $x^2 - 2x - y = -\frac{1}{2}$ that have y coordinate 3.

29. In a suspension bridge the horizontal distance between the supports is the **span** of the bridge, and the vertical distance between the points where the cable is attached to the supports and the lowest point of the cable is the **sag** (Figure 4.53). The span of the George Washington Bridge is 3500 feet, and its sag is 316 feet. Find an equation of the parabola that represents the cable.

4.6
THE ELLIPSE

The graph of an equation of the form

$$Ax^2 + By^2 = E \quad \text{with } A, B, \text{ and } E \text{ of the same sign} \qquad (1)$$

is an ellipse. Dividing each side of the equation in (1) by E allows us to rewrite the equation in the standard form

$$\frac{x^2}{a^2} + \frac{y^2}{b^2} = 1 \quad \text{or} \quad \frac{x^2}{b^2} + \frac{y^2}{a^2} = 1 \quad \text{where } a \geq b > 0 \qquad (2)$$

Ellipses given in standard form are symmetric with respect to both axes as well as the origin, which is called the **center** of the ellipse (Figure 4.54a, b). Setting $y = 0$ in the first equation of (2), and $x = 0$ in the second equation, yields $-a$ and a for two intercepts of the ellipse. The corresponding points of the ellipse are called **vertices** of the ellipse, and the line segment joining the vertices is the **major axis** of the ellipse. The length of the major axis is $2a$ (Figure 4.54a, b). The remaining two intercepts of the ellipse are obtained by

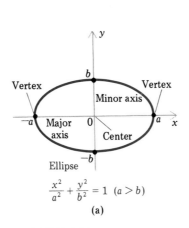

$$\frac{x^2}{a^2} + \frac{y^2}{b^2} = 1 \quad (a > b)$$

(a)

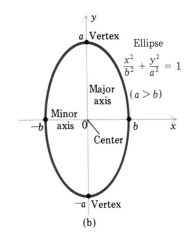

Ellipse

$$\frac{x^2}{b^2} + \frac{y^2}{a^2} = 1$$

$(a > b)$

(b)

FIGURE 4.54

setting $x = 0$ in the first equation of (2) and $y = 0$ in the second, which results in the intercepts $-b$ and b. Although the corresponding points have no special name, the line segment joining those points is called the **minor axis** of the ellipse. The minor axis has length $2b$ (Figure 4.54a, b). Since $a \geq b$ by assumption in (2),

h of the major axis $\geq$ the length of the minor axis

EXAMPLE 1.   Sketch the ellipse having equation

$$\frac{x^2}{9} + \frac{y^2}{4} = 1$$

and note the major and minor axes.

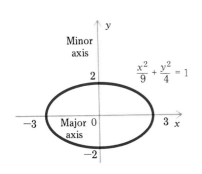

FIGURE 4.55

Solution.   Since the coefficient in the denominator of x^2 is larger than that of y^2, we compare the given equation with the first equation of (2). This yields $a^2 = 9$, so that $a = \sqrt{9} = 3$, and it also yields $b^2 = 4$, so that $b = \sqrt{4} = 2$. It follows that the x intercepts are -3 and 3 and the y intercepts are -2 and 2. The ellipse is sketched in Figure 4.55.   □

EXAMPLE 2.   Sketch the ellipse having equation

$$16x^2 + 2y^2 = 32$$

and note the major and minor axes.

Solution.   In order to use (2), we first divide each side by 32 to obtain

$$\frac{x^2}{2} + \frac{y^2}{16} = 1$$

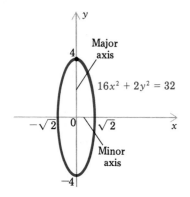

FIGURE 4.56

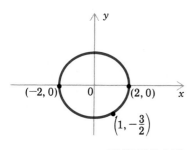

FIGURE 4.57

Since the coefficient in the denominator of y^2 is larger than that of x^2, we compare our new equation with the second equation of (2). We find that $a = \sqrt{16} = 4$ and $b = \sqrt{2}$. Therefore the x intercepts are $-\sqrt{2}$ and $\sqrt{2}$, and the y intercepts are -4 and 4. The ellipse is sketched in Figure 4.56.   ☐

EXAMPLE 3.   Find an equation in standard form for the ellipse whose vertices are $(-2, 0)$ and $(2, 0)$ and which passes through the point $(1, -\frac{3}{2})$ (Figure 4.57).

Solution.   Since the vertices lie on the x axis, the equation we seek has the form

$$\frac{x^2}{a^2} + \frac{y^2}{b^2} = 1$$

The value of a is 2, because the vertices are $(-2, 0)$ and $(2, 0)$. We will be done once we find the numerical value of b^2. To that end we notice that the point $(1, -\frac{3}{2})$ is on the ellipse, so that

$$\frac{1^2}{4} + \frac{(-3/2)^2}{b^2} = 1$$

Solving for b^2, we find that

$$\frac{(-3/2)^2}{b^2} = 1 - \frac{1}{4} = \frac{3}{4}$$

and consequently

$$b^2 = \left(-\frac{3}{2}\right)^2 \cdot \frac{4}{3} = \frac{9}{4} \cdot \frac{4}{3} = 3$$

Therefore the equation we seek is

$$\frac{x^2}{4} + \frac{y^2}{3} = 1   ☐$$

The ellipses in Figures 4.55 and 4.57 are wider than they are tall, whereas the ellipse in Figure 4.56 is taller than it is wide. One can tell immediately from an equation of an ellipse in standard form whether it is wider or taller: if the number under x^2 is larger than the number under y^2, the ellipse is wider; if the number under y^2 is larger than the number under x^2, the ellipse is taller. In the event that the numbers under x^2 and y^2 are the same, the ellipse is a circle, because in that case the equation becomes

$$\frac{x^2}{a^2} + \frac{y^2}{a^2} = 1$$

or equivalently, $x^2 + y^2 = a^2$

Thus a circle is indeed a special case of an ellipse.

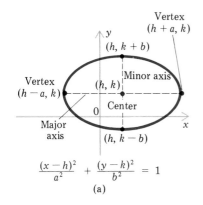

$$\frac{(x-h)^2}{a^2} + \frac{(y-k)^2}{b^2} = 1$$

(a)

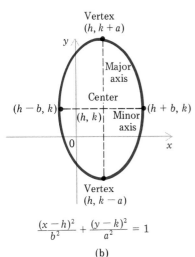

$$\frac{(x-h)^2}{b^2} + \frac{(y-k)^2}{a^2} = 1$$

(b)

FIGURE 4.58

Except in very special cases that will not concern us here, the graph of an equation of the form

$$Ax^2 + By^2 + Cx + Dy = E \quad \text{with } A, B, \text{ and } E \text{ of the same sign} \qquad (3)$$

is an ellipse whose center can be any point in the plane, depending on the values of the constants. By completing squares, one can rearrange (3) so that it has the form

$$\frac{(x-h)^2}{a^2} + \frac{(y-k)^2}{b^2} = 1 \quad \text{or} \quad \frac{(x-h)^2}{b^2} + \frac{(y-k)^2}{a^2} = 1$$

$$\text{where } a \geq b > 0 \qquad (4)$$

The **center** of the ellipse is the point (h, k), and the ellipse is symmetric with respect to the two lines $x = h$ and $y = k$. For an ellipse defined by the first equation in (4), the **vertices** are $(h - a, k)$ and $(h + a, k)$ (Figure 4.58a), and the line segment joining the vertices is the **major axis** of the ellipse, whereas the line segment joining the points $(h, k - b)$ and $(h, k + b)$ is the **minor axis** of the ellipse. Analogously, for an ellipse defined by the second equation in (4), the vertices are $(h, k - a)$ and $(h, k + a)$ (Figure 4.58b), with the line joining them the **major axis**, and the line joining the points $(h - b, k)$ and $(h + b, k)$ the **minor axis** of the ellipse. Again the length of the major axis is $2a$, and the length of the minor axis is $2b$.

EXAMPLE 4.   Sketch the ellipse having equation

$$9x^2 + 54x + y^2 - 4y = -76$$

and note the center and vertices, as well as the major and minor axes.

Solution.   We complete squares and then rearrange so that the right side is 1:

$$(9x^2 + 54x) + (y^2 - 4y) = -76$$
$$9(x^2 + 6x) + (y^2 - 4y) = -76$$
$$9(x^2 + 6x + 9 - 9) + (y^2 - 4y + 4 - 4) = -76$$
$$9(x^2 + 6x + 9) - 81 + (y^2 - 4y + 4) - 4 = -76$$
$$9(x + 3)^2 + (y - 2)^2 = -76 + 81 + 4 = 9$$
$$\frac{(x + 3)^2}{1} + \frac{(y - 2)^2}{9} = 1$$

This equation has the form of the second equation in (4), with $h = -3, k = 2$, $a = 3$, and $b = 1$. Consequently the center of the ellipse is $(-3, 2)$. The vertices are $(-3, 2 - 3)$ and $(-3, 2 + 3)$, that is, $(-3, -1)$ and $(-3, 5)$, and these points serve as the endpoints of the major axis. The minor axis is bounded by the points $(-3 - 1, 2)$ and $(-3 + 1, 2)$, that is, $(-4, 2)$ and

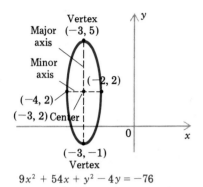

$9x^2 + 54x + y^2 - 4y = -76$

FIGURE 4.59

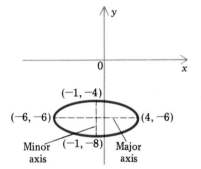

FIGURE 4.60

$(-2, 2)$. Now we are ready to sketch the graph, with all pertinent information on it (Figure 4.59).   □

EXAMPLE 5.   Find an equation of the ellipse whose major axis extends from $(-6, -6)$ to $(4, -6)$ and whose minor axis extends from $(-1, -8)$ to $(-1, -4)$ (Figure 4.60).

Solution.   The center of the ellipse is the midpoint of the major (or minor) axis, which is $(-1, -6)$. Since the major axis is horizontal, the equation we seek has the form

$$\frac{(x - (-1))^2}{a^2} + \frac{(y - (-6))^2}{b^2} = 1$$

or equivalently,

$$\frac{(x + 1)^2}{a^2} + \frac{(y + 6)^2}{b^2} = 1$$

The distance between $(-6, -6)$ and $(4, -6)$ is $4 - (-6) = 10$, and this is the length of the major axis, which is also $2a$. Therefore $2a = 10$, so that $a = 5$. Similarly, the distance between $(-1, -8)$ and $(-1, -4)$ is $-4 - (-8) = 4$, and this is the length of the minor axis, which is also $2b$. Therefore $2b = 4$, so that $b = 2$. Consequently the equation we seek is

$$\frac{(x + 1)^2}{25} + \frac{(y + 6)^2}{4} = 1 \quad □$$

For a geometric description of an ellipse, we let F_1 and F_2 be points in the plane and k a number greater than the distance between F_1 and F_2. The set of all points P in the plane such that

$$d(F_1, P) + d(F_2, P) = k$$

is an ellipse (Figure 4.61a). The points F_1 and F_2 are called the **foci** of the ellipse. In the event that F_1 and F_2 are the same point, the ellipse is a circle (Figure 4.61b).

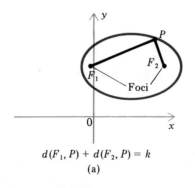

$d(F_1, P) + d(F_2, P) = k$

(a)

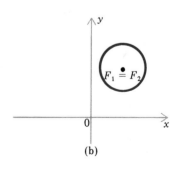

(b)

FIGURE 4.61

We end our discussion of ellipses with a few applications:

1. According to Kepler's laws of planetary motion, a planet moves in an elliptical orbit with the sun at one of the foci. Some comets, such as Halley's comet, also move in elliptical orbits about the sun. Halley's comet returns to the earth's vicinity every 76 years and should be visible from earth in May 1986.

2. An electron in an atom normally moves in an approximately elliptical orbit.

3. The dome of a whispering gallery has the shape of an ellipse that has been revolved about its major axis. Sound emanating from one focus bounces from anywhere on the dome and is then directed to the second focus. It is said that John C. Calhoun used this phenomenon in the Statuary Hall of the Capitol in Washington, D.C., in order to eavesdrop on his adversaries.

4. Ellipses are aesthetically pleasing and are used in the design of buildings and the construction of formal gardens.

EXERCISES 4.6

In Exercises 1–14, sketch the ellipse, noting the center, the major and minor axes, and the vertices.

1. $\dfrac{x^2}{16} + \dfrac{y^2}{4} = 1$

2. $x^2 + \dfrac{y^2}{4} = 1$

3. $\dfrac{x^2}{3} + \dfrac{y^2}{3} = 1$

4. $4x^2 + y^2 = 1$

5. $9x^2 + 4y^2 = 36$

6. $2x^2 + y^2 = 2$

7. $\dfrac{(x-4)^2}{9} + \dfrac{(y+3)^2}{16} = 1$

8. $\dfrac{(x+4)^2}{2} + \dfrac{(y-1)^2}{25} = 1$

9. $(x+2)^2 + 4(y+1)^2 = 16$

10. $4x^2 + y^2 + 6y + 5 = 0$

11. $4x^2 - 8x + y^2 + 4y = 8$

12. $9x^2 + 18x + 4y^2 + 16y = 11$

13. $25x^2 + 16y^2 + 150x - 96y = 31$

14. $4x^2 + 9y^2 - 16x + 54y + 96 = 0$

In Exercises 15–24, find an equation of the ellipse having the given properties.

15. Center: $(0, 0)$; x intercept: 2; y intercept: -3

16. Center: $(0, 0)$; x intercept: -4; y intercept: 1

17. Center: $(0, 0)$; vertex: $(0, 4)$; $(\sqrt{3}, 2)$ on the ellipse

18. Center: $(-1, 2)$; vertex: $(3, 2)$; $(0, 4)$ on the ellipse

19. Center: $(3, 2)$; $(3, 7)$ and $(-4, 2)$ on the ellipse

20. Center: $(-1, -3)$; $(-1, 5)$ and $(3, -3)$ on the ellipse

21. Vertices: $(-2, -3)$ and $(-2, -7)$; length of minor axis: 2

22. Vertices: $(6, -1)$ and $(-10, -1)$; length of minor axis: $\frac{1}{4}$

23. Center: $(-\frac{3}{2}, \frac{1}{2})$; length of major axis: 5; length of minor axis: 1; minor axis horizontal

24. Major axis from $(1, 1)$ to $(1, 10)$; minor axis from $(0, \frac{11}{2})$ to $(2, \frac{11}{2})$

25. Find equations for the two distinct ellipses each of which has center $(0, 4)$, major axis of length 3, and minor axis of length 2.

26. Find equations for the two distinct ellipses each of which has center $(-3, -2)$, major axis of length $\frac{3}{2}$, and minor axis of length $\frac{1}{2}$.

27. Show that if the major and minor axes of an ellipse have the same length, the ellipse is a circle.

28. For what value of c does the ellipse with equation $9(x-1)^2 + 4(y+2)^2 = 36c$ pass through the origin?

29. For what values of h does the ellipse with equation $(x-h)^2/9 + (y+5)^2/4 = 1$ pass through the point $(4, -4)$?

30. Let a and b be positive numbers. The area of the ellipse having an equation in the form of either of the equations in (2) in πab.

 a. Show that the above result implies that the area of the circle with radius r and center at the origin is πr^2.

 b. Find the area of the ellipse in Example 1.

31. a. Let a and b be positive numbers, and let u and v be numbers such that $u^2 + v^2 \neq 0$. Show that the point

$$\left(\frac{au}{\sqrt{u^2 + v^2}}, \frac{bv}{\sqrt{u^2 + v^2}} \right)$$

lies on the ellipse having equation

$$\frac{x^2}{a^2} + \frac{y^2}{b^2} = 1$$

 b. Taking $u = 4$ and $v = -3$ in part (a), find the corresponding point on the ellipse having equation

$$\frac{x^2}{36} + \frac{y^2}{16} = 1$$

32. The *eccentricity* e of an ellipse with equation in the form of (4) is given by

$$e = \sqrt{1 - \frac{b^2}{a^2}}$$

 a. Show that $0 \leq e < 1$.

 b. Show that $e = 0$ if and only if the ellipse is a circle.

 c. Find a formula for b in terms of a and e.

33. Find the eccentricities of the following ellipses.

 a. $\dfrac{x^2}{4} + \dfrac{y^2}{16} = 1$ c. $\dfrac{x^2}{4} + \dfrac{y^2}{9} = 1$

 b. $\dfrac{x^2}{16} + \dfrac{y^2}{4} = 1$ d. $x^2 + y^2 = 1$

34. A planetary orbit is elliptical and is frequently identified by the length $2a$ of the major axis and by the eccentricity e of the orbit. For the earth's orbit around the sun, the numbers $2a$ and e are approximately 299,000,000 (kilometers) and 0.17, respectively.

 a. Use part (c) of Exercise 32 and a calculator to determine an approximate value of b.

 b. Using the given information, find an equation in the form

$$\frac{x^2}{a^2} + \frac{y^2}{b^2} = 1$$

 that approximately represents the earth's orbit around the sun.

4.7
THE HYPERBOLA

The last of the major kinds of conic sections is the hyperbola. The graph of an equation of the form

$$Ax^2 + By^2 = E \quad \text{with } A \text{ and } B \text{ of different signs}$$

is a hyperbola, and by arranging the constants suitably we can write it in one of the two standard forms

$$\frac{x^2}{a^2} - \frac{y^2}{b^2} = 1 \quad \text{or} \quad \frac{y^2}{a^2} - \frac{x^2}{b^2} = 1 \quad \text{with} \quad a > 0 \quad \text{and} \quad b > 0$$

The corresponding hyperbolas are sketched in Figure 4.62a, b. They are symmetric with respect to both axes and the origin, which is called the *center* of the hyperbolas.

Notice that the hyperbola given by

$$\frac{x^2}{a^2} - \frac{y^2}{b^2} = 1 \tag{1}$$

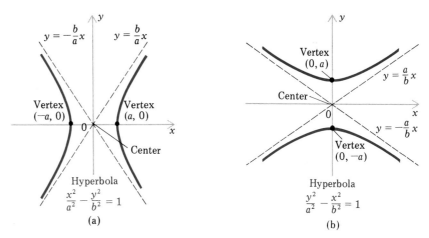

FIGURE 4.62

has no y intercepts but has x intercepts a and $-a$. The points $(a, 0)$ and $(-a, 0)$ are called the **vertices** of the hyperbola, and the line segment determined by the vertices is called the **transverse axis** of the hyperbola. The hyperbola also has two oblique asymptotes, as we now show. Let us first consider the part of the graph in the first quadrant, corresponding to $x > 0$ and $y > 0$. Solving the equation in (1) for y (with $y > 0$), we obtain

$$\frac{y^2}{b^2} = \frac{x^2}{a^2} - 1$$

$$y^2 = b^2\left(\frac{x^2}{a^2} - 1\right)$$

$$y = b\sqrt{\frac{x^2}{a^2} - 1} = b\sqrt{\frac{x^2}{a^2}\left(1 - \frac{a^2}{x^2}\right)} = \frac{bx}{a}\sqrt{1 - \frac{a^2}{x^2}}$$

If x is positive and becomes arbitrarily large, a^2/x^2 gets close to 0, so $1 - (a^2/x^2)$ and hence $\sqrt{1 - (a^2/x^2)}$ get close to 1. Thus for large positive values of x, y is close to bx/a. This means that the line

$$y = \frac{b}{a}x$$

is an oblique asymptote of the part of the graph in the first quadrant.

Now let us consider the portions of the graph in the other quadrants. By the symmetry of the hyperbola with respect to the origin, the same line is an oblique asymptote of the portion of the hyperbola in the third quadrant (Figure 4.62a). It now follows from the symmetry of the graph with respect to the axes that the line

$$y = -\frac{b}{a}x$$

is an oblique asymptote of the part of the graph in the second and fourth quadrants, as is also indicated in Figure 4.62a.

The hyperbola given by

$$\frac{y^2}{a^2} - \frac{x^2}{b^2} = 1 \qquad (2)$$

has no x intercepts but has the two y intercepts a and $-a$. In this case the points $(0, a)$ and $(0, -a)$ are called the **vertices** of the hyperbola and the line segment determined by the vertices is called the **transverse axis**. In addition, the hyperbola has the two oblique asymptotes

$$y = \frac{a}{b}x \quad \text{and} \quad y = -\frac{a}{b}x$$

as indicated in Figure 4.62b.

One way to associate equations (1) and (2) with the corresponding graphs is this: If x comes first (as in (1)), then the graph crosses the x axis; if y comes first (as in (2)), then the graph crosses the y axis. It is also useful to note that, whereas the vertices help in graphing points closest to the center of the hyperbola, the asymptotes help in graphing points on the hyperbola far from the center.

EXAMPLE 1.   Sketch the hyperbola having equation

$$\frac{x^2}{4} - y^2 = 1$$

and note the vertices and asymptotes.

Solution.   Since the given equation is equivalent to

$$\frac{x^2}{4} - \frac{y^2}{1} = 1$$

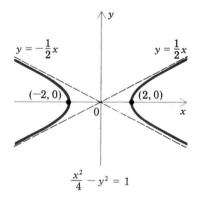

it has the form of equation (1) with $a = 2$ and $b = 1$. Therefore the vertices are $(-2, 0)$ and $(2, 0)$, and the asymptotes are

$$y = \frac{1}{2}x \quad \text{and} \quad y = -\frac{1}{2}x$$

FIGURE 4.63   The hyperbola is sketched in Figure 4.63.   □

EXAMPLE 2.   Sketch the hyperbola having equation

$$16y^2 - 9x^2 = 144$$

and note the vertices and asymptotes.

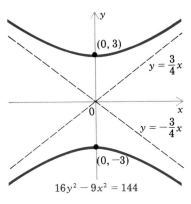

(0, 3)

$y = \frac{3}{4}x$

0

$y = -\frac{3}{4}x$

(0, −3)

$16y^2 - 9x^2 = 144$

FIGURE 4.64

Solution. Dividing both sides of the equation by 144, we obtain

$$\frac{y^2}{9} - \frac{x^2}{16} = 1$$

This equation has the form of (2) with $a = 3$ and $b = 4$. The vertices are $(0, -3)$ and $(0, 3)$, and the asymptotes are

$$y = \frac{3}{4}x \quad \text{and} \quad y = -\frac{3}{4}x$$

The hyperbola is sketched in Figure 4.64. ☐

Except in very special cases, the graph of any equation of the form

$$Ax^2 + By^2 + Cx + Dy = E \quad \text{with } A \text{ and } B \text{ of different signs} \qquad (3)$$

is a hyperbola whose center can be any point in the plane, depending on the values of the constants. After completing squares and rearranging the constants, we can rewrite (3) either as

$$\frac{(x-h)^2}{a^2} - \frac{(y-k)^2}{b^2} = 1 \qquad (4)$$

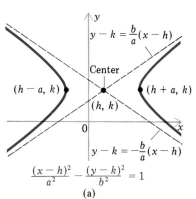

$y - k = \frac{b}{a}(x - h)$

Center

$(h - a, k)$

(h, k)

$(h + a, k)$

0

$y - k = -\frac{b}{a}(x - h)$

$\frac{(x-h)^2}{a^2} - \frac{(y-k)^2}{b^2} = 1$

(a)

whose graph has vertices $(h - a, k)$ and $(h + a, k)$ and asymptotes

$$y - k = \frac{b}{a}(x - h) \quad \text{and} \quad y - k = -\frac{b}{a}(x - h)$$

(Figure 4.65a), or as

$$\frac{(y-k)^2}{a^2} - \frac{(x-h)^2}{b^2} = 1 \qquad (5)$$

whose graph has vertices $(h, k - a)$ and $(h, k + a)$ and asymptotes

$$y - k = \frac{a}{b}(x - h) \quad \text{and} \quad y - k = -\frac{a}{b}(x - h)$$

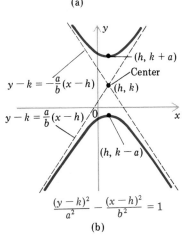

$(h, k + a)$

Center

$y - k = -\frac{a}{b}(x - h)$

(h, k)

$y - k = \frac{a}{b}(x - h)$

0

$(h, k - a)$

$\frac{(y-k)^2}{a^2} - \frac{(x-h)^2}{b^2} = 1$

(b)

FIGURE 4.65

(Figure 4.65b). In either case, the point (h, k) is the **center** of the hyperbola, and the hyperbola is symmetric with respect to the lines $x = h$ and $y = k$.

EXAMPLE 3. Sketch the hyperbola having equation

$$9x^2 - 25y^2 - 18x - 50y = -209$$

and note the center, vertices, and the asymptotes.

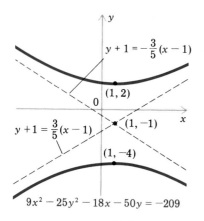

$$9x^2 - 25y^2 - 18x - 50y = -209$$

FIGURE 4.66

Solution.   In order to put the equation into the form of (4) or (5), we complete squares:

$$(9x^2 - 18x) - (25y^2 + 50y) = -209$$
$$9(x^2 - 2x) - 25(y^2 + 2y) = -209$$
$$9(x^2 - 2x + 1 - 1) - 25(y^2 + 2y + 1 - 1) = -209$$
$$9(x^2 - 2x + 1) - 9 - 25(y^2 + 2y + 1) + 25 = -209$$
$$9(x - 1)^2 - 25(y + 1)^2 = -209 + 9 - 25 = -225$$

Now we divide both sides by -225 and obtain

$$\frac{(y + 1)^2}{9} - \frac{(x - 1)^2}{25} = 1 \qquad (6)$$

Equation (6) has the form of (5) with $h = 1$, $k = -1$, $a = 3$, and $b = 5$. As a result, the center is $(1, -1)$ and the vertices are $(1, -1 - 3) = (1, -4)$ and $(1, -1 + 3) = (1, 2)$. The asymptotes are

$$y + 1 = \frac{3}{5}(x - 1) \quad \text{and} \quad y + 1 = -\frac{3}{5}(x - 1)$$

The hyperbola is sketched in Figure 4.66.   □

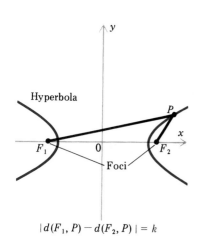

$$|d(F_1, P) - d(F_2, P)| = k$$

FIGURE 4.67

Like the parabola and ellipse, the hyperbola has a geometric description. Let F_1 and F_2 be two distinct points in the plane, and let k be a positive number less than the distance between F_1 and F_2. Then the set of all points P in the plane such that

$$|d(F_1, P) - d(F_2, P)| = k$$

is a hyperbola (Figure 4.67). The points F_1 and F_2 are called the *foci* of the hyperbola.

We mention a few applications of hyperbolas.

1. Comets that do not move in elliptical orbits around the sun almost always move in hyperbolic orbits. (In theory they can also move in parabolic orbits.) Such comets would not repeatedly orbit the earth as Halley's Comet does.
2. Boyle's Law, relating the pressure p and the volume V of a perfect gas at constant temperature, states that $pV = c$, where c is a constant. The graph of p as a function of V is a hyperbola.
3. Hyperbolas can be used to locate the source of a sound heard at three different locations.

Summary   As we mentioned at the outset of Section 4.5, the conic sections are generated by cutting a double cone by various planes (see Figure 4.45). Complementing this relationship between the different conic sections is the fact that our equations for the conic sections can be described in a unified way. More precisely, the graph of any equation of the form

$$Ax^2 + By^2 + Cx + Dy = E$$

is a conic section. The non-degenerate ones can be described as follows:

1. If $A \neq 0$ and $B = 0$, or if $A = 0$ and $B \neq 0$, then the graph is a parabola.
2. If A and B have the same sign, then the graph is an ellipse.
3. If A and B have different signs, then the graph is a hyperbola.

Our short introduction to conic sections gives you some of the basic information about them and their associated equations. Most books on calculus have a more extensive treatment of conic sections.

EXERCISES 4.7

In Exercises 1–14, sketch the hyperbola, noting the center, vertices, and asymptotes.

1. $x^2 - y^2 = 1$

2. $y^2 - x^2 = 4$

3. $\dfrac{x^2}{9} - \dfrac{y^2}{4} = 1$

4. $\dfrac{y^2}{25} - \dfrac{x^2}{16} = 1$

5. $25y^2 - 4x^2 = 100$

6. $4x^2 - y^2 = 4$

7. $\dfrac{(x+2)^2}{16} - \dfrac{(y+1)^2}{25} = 1$

8. $\dfrac{(y-1)^2}{1/4} - x^2 = 1$

9. $4y^2 - x^2 + 8y = 0$

10. $16x^2 - y^2 - 32x - 6y = 57$

11. $9x^2 - 16y^2 - 54x + 225 = 0$

12. $x^2 - 4y^2 - 4x - 8y = 4$

13. $9x^2 - 4y^2 - 18x - 24y = 63$

14. $9x^2 - 4y^2 - 18x - 24y = 26$

In Exercises 15–20, find an equation of the hyperbola with the given properties.

15. Center: $(0, 0)$; x intercept: 2; asymptote: $y = -3x$

16. Center: $(0, 0)$; y intercept: $-\frac{1}{2}$; asymptote: $y = \frac{3}{2}x$

17. Center: $(0, 0)$; vertex: $(0, 4)$; $(-4, -3\sqrt{3})$ on the hyperbola

18. Center: $(-1, 2)$; vertex: $(-1, -3)$; $(-3, -4)$ on the hyperbola

19. Center: $(2, -3)$; $(0, -3)$ and $(6, -3 + 2\sqrt{3})$ on the hyperbola

20. Center: $(7, -6)$; $(7, -1)$ on the hyperbola; asymptote: $y + 6 = \frac{5}{7}(x - 7)$

21. For what values of c does the hyperbola having equation $x^2 + 2x - c^2 y^2 + y = 2$ pass through the point $(1, 2)$?

22. For what values of h does the hyperbola having equation

$$\frac{(x-h)^2}{16} - \frac{(y+1)^2}{36} = 1$$

pass through the point $(\sqrt{10}, 1)$?

23. a. Let a and b be positive numbers, and let u and v be numbers such that $u^2 - v^2 > 0$. Show that the point

$$\left(\frac{au}{\sqrt{u^2 - v^2}}, \frac{bv}{\sqrt{u^2 - v^2}} \right)$$

lies on the hyperbola having equation

$$\frac{x^2}{a^2} - \frac{y^2}{b^2} = 1$$

b. Taking $u = 13$ and $v = -5$ in part (a), find the corresponding point on the hyperbola having equation

$$\frac{x^2}{4} - \frac{y^2}{9} = 1$$

24. A hyperbola having an equation of the form

$$\frac{x^2}{a^2} - \frac{y^2}{b^2} = 1 \quad \text{or} \quad \frac{y^2}{b^2} - \frac{x^2}{a^2} = 1 \qquad (7)$$

with $a = b$ is called an **equilateral hyperbola**. Show that a hyperbola satisfying an equation of the form in (7) is equilateral if and only if its asymptotes are perpendicular to one another.

25. The *eccentricity e* of a hyperbola with an equation in the form of (4) or (5) is given by

$$e = \sqrt{1 + \frac{b^2}{a^2}}$$

Find the eccentricities of the following hyperbolas:

a. $\dfrac{x^2}{4} - \dfrac{y^2}{16} = 1$ c. $\dfrac{x^2}{4} - \dfrac{y^2}{9} = 1$

b. $\dfrac{x^2}{16} - \dfrac{y^2}{4} = 1$ d. $x^2 - y^2 = 1$

In Exercises 26–37, determine the type of conic section represented by the given equation. Determine the vertex of each parabola and the center of each ellipse or hyperbola.

26. $2x^2 - 4x - 3y + 3 = 0$

27. $x^2 + 3y^2 = 3$

28. $9x^2 - 4y^2 - 54x + 32y - 53 = 0$

29. $12y^2 - 12y + 8x + 15 = 0$

30. $x^2 + 6x - y + 11 = 0$

31. $9x^2 + 16y^2 + 72x - 96y + 287 = 0$

32. $9x^2 = 4y^2 - 8y + 3$

33. $x^2 + 6x + 9y^2 - 27 = 0$

34. $3y^2 + 6y - 2x = 3$

35. $25x^2 - 2y^2 = 100x + 8y - 142$

36. $9x^2 + 9y^2 + 6x - 24y = 28$

37. $9x^2 - 4y^2 - 54x + 32y - 19 = 0$

KEY TERMS

polynomial function	ellipse
quadratic function	center
minimum value of a function	vertex
maximum value of a function	major axis
rational function	minor axis
vertical asymptote	focus
horizontal asymptote	hyperbola
oblique asymptote	center
conic section	vertex
parabola	transverse axis
axis	asymptote
vertex	focus
directrix	
focus	

REVIEW EXERCISES

In Exercises 1–4, sketch the graph of the given function, and identify the vertex, axis, and intercepts.

1. $f(x) = \frac{1}{4}x^2 - 4$ **3.** $f(x) = -x^2 - 4x - 7$

2. $f(x) = x^2 + 8x + 15$ **4.** $f(x) = x^2 - 3x + 2$

In Exercises 5–16, sketch the graph of the given function, and identify all intercepts and asymptotes.

5. $f(x) = 2 - x^3$ **7.** $f(x) = x^4 + 2$

6. $f(x) = 4x^3 - x$

8. $g(x) = -\dfrac{1}{(x-3)^2}$

9. $f(x) = -x^2(x-2)$

10. $f(x) = (x+1)(x-2)(x-3)$

11. $f(x) = x^4 - 10x^2 + 9$

12. $g(x) = \dfrac{2x}{x+3}$

13. $g(x) = \dfrac{x+1}{x-1}$

14. $y = \dfrac{(x-1)(x-3)}{(x+2)(x-4)}$

15. $y = \dfrac{x^2 + 3x + 2}{x^3 + 5x^2 + 4x}$

16. $y = \dfrac{x^2}{4x^2 - 1}$

In Exercises 17–22, sketch the graph of the conic section, indicating all pertinent information (center, axes, vertices, asymptotes).

17. $x^2 + y^2 + 2x - 4y + 2 = 0$

18. $6x + y^2 + 6y = 0$

19. $x^2 - y^2 - x - y = 4$

20. $4x^2 + 25y^2 - 8x + 100y + 4 = 0$

21. $12x^2 + 72x + 72 = 9y^2 + 72y$

22. $18y^2 - 36y - 2x + 10 = 0$

23. Let $f(x) = 3x^2 + bx - 4$. One x intercept of the graph of f is 4. Determine the other x intercept.

24. Let $f(x) = 3x^2 + 2x - c$. Determine the value of c for which
   a. the graph contains the point $(-1, 4)$.
   b. the graph has exactly one x intercept.
   c. the x intercepts are 6 units apart.

25. Find the values of h for which the point $(-5, 1)$ lies on the ellipse having equation

$$\dfrac{(x-h)^2}{25} + \dfrac{(y+5)^2}{100} = 1$$

26. Show that if u is any real number, then the point

$$\left(\dfrac{u^2 - 1}{u^2 + 1}, \dfrac{2u}{u^2 + 1} \right)$$

lies on the circle $x^2 + y^2 = 1$.

27. The Student Union plans to organize a group trip to Mexico during spring vacation. The cost per person is set at \$400 for 300 people, and it decreases \$1 for each additional person up to 100 additional people. What is the cost per person that brings the maximum revenue, and how many people must participate at that cost?

28. Find two numbers whose product is as large as possible if the sum of twice one number and 3 times the other is 36.

29. The height at time t of a cannon ball fired at a 30° angle with respect to the ground is given by

$$h = -16t^2 + \dfrac{1}{2}v_0 t$$

where v_0 is the initial, or muzzle, velocity. If the muzzle velocity is 1024 feet per second, determine the maximum height attained by the cannon ball.

30. If the muzzle velocity of the cannon in Exercise 29 is doubled, how is the maximum height affected?

31. A wire 2 feet long is to be cut into 2 pieces, one of which will be bent into the shape of a square and the other into the shape of a circle. Determine the lengths into which the wire must be cut if the sum of the areas of the square and circle is to be minimum.

32. What is the largest possible area of a rectangular tablecloth whose perimeter is 24 feet?

33. Let $f(x) = x^3 + x^2$ for $0 \le x \le 1$. Determine the dimensions of the rectangle of maximum area that has base along the x axis and can be inscribed in the shaded region in Figure 4.68. (*Hint:* The area can be expressed as a quadratic function of x^2. Find the value of x^2 that yields the maximum area.)

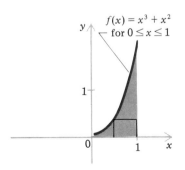

FIGURE 4.68

CUMULATIVE REVIEW I: CHAPTERS 1–4

In Exercises 1–4, simplify the given expression.

1. $\sqrt[3]{-\dfrac{81}{64}}$

3. $\dfrac{(x\sqrt[3]{x}\,y^{2/5})^{15/2}}{x^{1/4}y^2}$

2. $\left(\dfrac{x - x^{-1}}{x - x^{-2}}\right)^{-1}$

4. $(3x - y)^3 - (x + 2y)^3$

In Exercises 5–6, factor the given polynomial.

5. $(x - 3)^2(2x - 3)^4 + 12(x - 3)^3(2x - 3)^2$

6. $8x^3 - 12x^2 - 6x + 9$

In Exercises 7–10, write as one rational expression in simplified form.

7. $\dfrac{\dfrac{4}{x^2} - 2x}{3x + \dfrac{1}{x}}$

9. $\dfrac{x + 4}{x^2 - 4} - \dfrac{x - 4}{(x + 2)^2}$

8. $\dfrac{6x - 3}{2x^3 - x^2 - 2x + 1}$

10. $\dfrac{\sqrt{a + 3} + \sqrt{a}}{\sqrt{a + 3} - \sqrt{a}}$

In Exercises 11–14, find all solutions of the given equation.

11. $|3 - 4x| = 2$

12. $x^2 - \tfrac{3}{2}x - 2 = 0$

13. $\dfrac{2x^2 - 3}{4x - 1} = -2$

14. $\sqrt{x + 2} = 1 + \sqrt{1 - 3x}$

In Exercises 15–18, solve the given inequality and express the solution in terms of intervals.

15. $4x + 3 > 6x - 7$

16. $(4x - 5)(x^2 - 1) \le 0$

17. $\dfrac{x + 4}{2x - 1} \ge -2$

18. $|x - 1| \le |x + 1|$

19. Let l be the line with equation $2x - 3y = -1$.
   a. Show that the point $(-1, -1/3)$ is on l.
   b. Find an equation for the line l_1 that passes through the point $(-1, -1/3)$ and is perpendicular to l.

20. Consider the points $P = (\tfrac{1}{2}, -1)$ and $Q = (-\tfrac{3}{4}, -2)$.
   a. Find the distance between P and Q.

   b. Determine the slope of the line l that contains P and Q.
   c. Find an equation of l.
   d. Find an equation of the line l_1 that is parallel to l and passes through the point $(2, 3)$.

21. Let $f(t) = \sqrt{4 - t^2}$.
   a. Determine the domain of f.
   b. Calculate any values of t for which $f(t) = t$.
   c. Calculate the values of t for which $f(f(t)) = t$.

22. Let $f(x) = \sqrt{x^2 - 16}$ and $g(x) = x^3 + 4$. Find the domain and rule of
   a. $g \circ f$ b. $f \circ g$

23. Let $f(x) = (3 - 2x)/(1 + x)$. Find a formula for f^{-1}.

24. The equation

$$I = I_0\left(\frac{n - 1}{n + 1}\right)^2$$

occurs in the theory of the polarization of light. Solve the equation for n.

In Exercises 25–28, sketch the graph of the given function, indicating any intercepts, as well as vertices and asymptotes when applicable.

25. $f(x) = x^2 - 4x + 7$

26. $f(x) = x^3 - 4x^2 - 4x + 16$

27. $f(x) = x^2(x + 1)^3$

28. $f(x) = \dfrac{x - 2}{x^2 - 1}$

In Exercises 29–32, sketch the conic section, indicating all pertinent information (center, axes, vertices, asymptotes).

29. $\dfrac{(x - 2)^2}{9} + \dfrac{(y + 1)^2}{9} = 1$

30. $y^2 + 4y - 2x + 6 = 0$

31. $y^2 - 2x^2 - 2y - 4x = 17$

32. $x^2 + 4y^2 - 4y = 0$

33. Find an equation of the ellipse whose major axis is bounded by the points $(-1, 1)$ and $(-1, -5)$, and whose minor axis is bounded by the points $(-3, -2)$ and $(1, -2)$.

34. Find an equation of the hyperbola whose center is $(2, 4)$, one of whose vertices is $(2, 7)$, and one of whose asymptotes is the line $y - 4 = \sqrt{3}(x - 2)$.

35. Find the relationship that exists between a and b if
$$\sqrt{a^2 - ab} = \sqrt{3ab - 4b^2}$$

36. a. Let t and u be numbers greater than -1. Show that if $t < u$, then
$$\frac{t}{1 + t} < \frac{u}{1 + u}$$

b. Using part (a), show that for any two numbrs x and y,
$$\frac{|x + y|}{1 + |x + y|} \le \frac{|x|}{1 + |x|} + \frac{|y|}{1 + |y|}$$

37. a. Show that $\dfrac{2}{x} = \dfrac{2}{x + 1} + \dfrac{2}{x(x + 1)}$.

b. The reciprocal $1/n$ of an integer n is called a **unit fraction**. To simplify calculations, the ancient Egyptians expressed fractions as sums of unit fractions. Using (a) with x replaced by n, show that if n is an odd integer greater than 1, then $2/n$ is the sum of two distinct unit fractions. (*Hint:* If n is odd, then $n + 1$ is divisible by 2.)

c. Write $2/13$ as the sum of two unit fractions.

38. Bill Rodgers, the winner of the 1978 Boston Marathon, ran the 26.21875 mile distance in 2 hours, 10 minutes, and 13 seconds.
a. Determine his average speed.
b. Determine the average length of time it took him to run each mile.

39. It has been determined experimentally that the concentration C of a pesticide on orange leaves at any time t (one day or longer after spraying) is inversely proportional to the $3/2$ power of t. Express this by means of a formula.

40. The purchase price P of a given length of wire is proportional to the cross-sectional area x of the wire. The cost L due to power loss while the wire is in use is inversely proportional to x. Express the total cost C, which is the sum of P and L, in terms of x. (There will be two constants of proportionality, but they need not be equal.)

41. An open box is to be made from a rectangular piece of cardboard 24 inches long and 16 inches wide by cutting a square piece x inches on a side from each of the four corners and then bending up the sides. Write a formula for the volume V of the resulting box.

42. The rate R of emission of radiant energy from the surface of a body is directly proportional to the 4th power of the temperature T.
a. Express this fact by means of a formula.
b. The constant of proportionality varies according to the nature of the surface of the body. If R is measured in watts per square meter and the temperature in degrees Kelvin, then for copper the constant is approximately 1.70097×10^{-8}. Suppose the temperature on the surface of a copper wire is $300°$ Kelvin. Use a calculator to determine the rate R of emission of radiant energy. Write your answer in scientific notation.

43. French dressing consists of approximately 72% olive oil, 24% vinegar (or lemon juice), and the remainder salt and pepper. Suppose we have 1 cup of French dressing with 24% vinegar. How much vinegar must be added in order to yield a 25% level of vinegar?

44. Determine how many gallons of a 70% saline solution must be added to 8 gallons of a 20% saline solution in order to obtain a 30% saline solution.

45. A tank can be filled by two pipes together in one hour less than would be required by the larger pipe, and in 4 hours less then would be required by the smaller pipe. How long would it take the smaller pipe to fill the tank? (*Hint:* Let x be the number of hours required by the smaller pipe to fill the tank and y the number for the larger pipe. Express the number of hours required for the two pipes to fill the tank together in terms of x and y.)

46. A garden is 4 yards long and 3 yards wide. There is a crosswalk of uniform width, as in Figure I.1. If the area of the crosswalk is $3\frac{1}{4}$ square yards, how wide is the crosswalk?

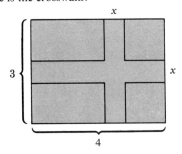

FIGURE I.1

47. When Old Faithful geyser in Yellowstone National Park first erupts, it can send water up some 150 feet.
   a. A water drop attains a maximum height of 150 feet above ground. Assuming no air resistance, determine the velocity of the drop when it leaves the ground.
   b. Using the information of part (a), determine the length of time the drop is in the air.

48. A university plans to rope off a rectangular field for a football game. It has 2000 feet of rope, and it plans to leave a 40-foot entrance free of rope. Find the dimensions of the field with the greatest possible area.

49. The center of gravity of a broad jumper during a given jump traces a curve given by

$$y = -\frac{5}{324}(x^2 - 16x - 260)$$

for all $x \geq 0$ such that $y \geq 0$. Here x and y are measured in feet.

   a. Determine the nonnegative values of x for which $y \geq 0$.
   b. Determine the maximum height of the center of gravity during the jump.

50. Find the value of x in the interval $[0, 2]$ for which the area of the triangle in Figure I.2 is as large as possible.

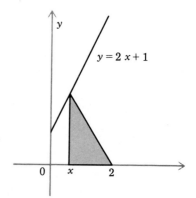

$y = 2x + 1$

FIGURE I.2

5

Exponential and Logarithmic Functions

This chapter is devoted to two kinds of functions that cannot be formed by any of the elementary algebraic operations (addition, subtraction, multiplication, division, and roots). These are the exponential and logarithmic functions. They not only are extremely important in advanced mathematics but also occur with great frequency in the study of such varied topics as radioactivity, noise level, and atmospheric pressure.

For a sample of the applications of exponential and logarithmic functions, consider the following problems:

1. A mummy, known as Whiskey Lil, was discovered in 1955 in Chimney Cave, Lake Winnemucca, Nevada. By carbon dating it was learned that approximately 0.739 of the original C^{14} per gram was still present in 1955. When did Whiskey Lil die?

2. The Alaska Good Friday earthquake of March 28, 1964, measured 8.5 on the Richter scale. Some 19 years later, on May 2, 1983, a massive earthquake measuring 6.5 on the Richter scale rocked Coalinga, California. The Alaska earthquake was how many times as intense as the one in California?

Exponential functions will help us solve the first problem (see Example 6 in Section 5.6), whereas logarithmic functions give us the necessary tools to solve the second problem (see Example 4 in Section 5.5).

Numerical answers to questions such as those mentioned above involve exponential or logarithmic functions, and generally they can only be given approximately in decimal form. For a high degree of accuracy one needs tables or a calculator. We will normally use a calculator and give answers that are accurate to eight places.

**5.1
EXPONENTIAL
FUNCTIONS**

In Section 1.5 we defined the expression a^r for any $a > 0$ and any *rational* number r. However, until now we have not defined expressions such as $2^{\sqrt{3}}$, where the exponent is irrational. In this section we will give a meaning to a^x, in which x can be any number, irrational or rational.

To see how a^x can be defined for an arbitrary irrational number x, let us select a specific number $a > 0$, say $a = 2$. By plotting points from the table

r	-3	-2	-1	0	1	2	3
2^r	$\frac{1}{8}$	$\frac{1}{4}$	$\frac{1}{2}$	1	2	4	8

we obtain Figure 5.1a. The more points $(r, 2^r)$ we plot for other rational values of r, the more the points seem to fall on a smooth curve (Figure 5.1b). In fact, there is exactly one smooth curve that contains all points of the form $(r, 2^r)$, where r is rational (Figure 5.1c). As a result, for any number x we define 2^x to be the y coordinate of the point that lies on the curve directly above x on the x axis. From Figure 5.1c we see that the curve contains the point $(0, 1)$ and slopes upward to the right, which means that

$$\text{if} \quad x < z, \quad \text{then} \quad 2^x < 2^z$$

Notice that if r is close to x, then 2^r is close to 2^x, and moreover, the closer r is to x, the closer 2^r is to 2^x. Thus we can think of 2^x as the "limiting value" of 2^r, where r is rational and r gets closer and closer to x. For example, if $x = \sqrt{3}$, we could let r have the values

$$1, \quad 1.7, \quad 1.73, \quad 1.732, \quad 1.73205, \quad 1.7320508, \ldots$$

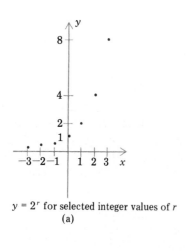

$y = 2^r$ for selected integer values of r
(a)

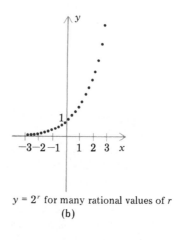

$y = 2^r$ for many rational values of r
(b)

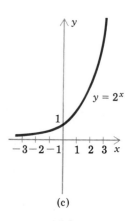

$y = 2^x$

(c)

FIGURE 5.1

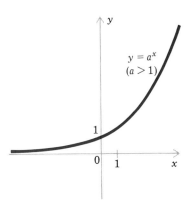

FIGURE 5.2

which are obtained by taking more and more of the decimal expansion of $\sqrt{3}$, so that the successive values of r approach $\sqrt{3}$. The corresponding values of 2^r are

$$2^1, \quad 2^{1.7}, \quad 2^{1.73}, \quad 2^{1.732}, \quad 2^{1.73205}, \quad 2^{1.7320508}, \ldots$$

Evaluating these numbers 2^r with a calculator, we obtain the following table:

r	1	1.7	1.73	1.732	1.73205	1.7320508
2^r	2	3.2490096	3.3172782	3.3218801	3.3219952	3.3219971

It appears from the table that the decimal expansion of $2^{\sqrt{3}}$ begins 3.32199....

The situation is analogous if a is any fixed number with $a > 1$. For rational values of r the points (r, a^r) lie on a unique smooth curve, and for any number x we define a^x to be the y coordinate of the point lying on that curve directly above the number x on the x axis (Figure 5.2). Notice also that the point $(0, 1)$ is on the curve, and the curve slopes upward to the right, which means that

$$\text{for } a > 1: \quad \text{if } x < z, \text{ then } a^x < a^z \tag{1}$$

Moreover, if x is negative with $-x$ very large, then a^x is close to 0, whereas if x is positive and very large, then a^x is very large.

Now if $0 < a < 1$, then the method of defining a^x is the same. The main difference is that the smooth curve connecting the points (r, a^r) for rational values of r slopes downward to the right, which means that

$$\text{for } 0 < a < 1: \quad \text{if } x < z, \text{ then } a^x > a^z \tag{2}$$

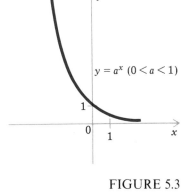

FIGURE 5.3

(Figure 5.3). For example, $\left(\frac{1}{3}\right)^2 > \left(\frac{1}{3}\right)^5$. Moreover, if x is positive and very large, then a^x is close to 0, whereas if x is negative and $-x$ is very large, then a^x is very large.

Finally, if $a = 1$, then $a^r = 1^r = 1$ for any rational value of r, so the points (r, a^r) all lie on the horizontal line $y = 1$ (Figure 5.4). By defining 1^x to be the y coordinate of the point on the line $y = 1$ above x on the x axis, we find that $1^x = 1$ for any real number x.

Led by our discussion above, we define a^x for any fixed $a > 0$ and any number x as follows:

$a^x =$ the y coordinate of the point that lies above the point x on the x axis and lies on the smooth curve that contains all points of the form (r, a^r), where r is a rational number

No matter what the values of a and x are (so long as $a > 0$), a^x is the "limiting value" of a^r, where r is rational and r gets closer and closer to x.

The "definition" we have just presented is intuitive, not a rigorous mathematical definition, which must to be left to more advanced texts.

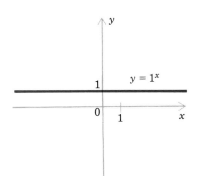

FIGURE 5.4

Nevertheless, we will abide by our "definition," which is consistent with the definition of a^r given earlier for rational values of r.

In the expression a^x the number a is called the **base** and x the **exponent** or **power**. By holding the base fixed and allowing the exponent to vary, we obtain a function.

DEFINITION 5.1   Let $a > 0$. The function f defined by

$$f(x) = a^x$$

is called the **exponential function with base a**.

For any $a > 0$ with $a \neq 1$, the following are properties of the function a^x and its graph:

1. $a^x > 0$ for all x, and the range of a^x consists of *all* positive numbers.
2. The y intercept is 1, and there is no x intercept.
3. The x axis is a horizontal asymptote of the graph of a^x.
4. If $x < z$, then $\begin{cases} a^x < a^z & \text{for } a > 1 \\ a^x > a^z & \text{for } 0 < a < 1 \end{cases}$

We notice that if $a > 1$, then the graph of a^x slopes upward to the right, with the rate of slope becoming larger farther to the right. By contrast, if $0 < a < 1$, then the graph of a^x slopes downward to the right, with the rate of slope becoming smaller farther to the right.

As we did earlier when $a = 2$, we can sketch the graph of the exponential function a^x for any specific positive value of a by computing a few values of a^x for specially chosen rational values of x and then connecting the corresponding points on the graph with a smooth curve. Of course, the more values we compute, the more accurate our sketch is likely to be.

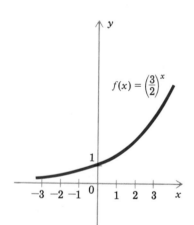

$f(x) = \left(\frac{3}{2}\right)^x$

EXAMPLE 1.   Let $f(x) = \left(\frac{3}{2}\right)^x$. Sketch the graph of f.

Solution.   Using the values in the table

x	-3	-2	-1	0	1	2	3
$f(x)$	$\frac{8}{27}$	$\frac{4}{9}$	$\frac{2}{3}$	1	$\frac{3}{2}$	$\frac{9}{4}$	$\frac{27}{8}$

we plot the corresponding points. Then we sketch a smooth curve through the points (Figure 5.5).   □

FIGURE 5.5

EXAMPLE 2.   Let $g(x) = \left(\frac{1}{3}\right)^x$. Sketch the graph of g.

Solution.   We make a short table of values:

x	-2	-1	0	1	2
$g(x)$	9	3	1	$\frac{1}{3}$	$\frac{1}{9}$

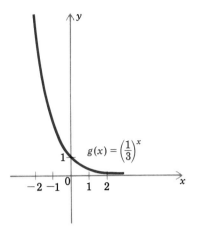

$g(x) = \left(\dfrac{1}{3}\right)^x$

FIGURE 5.6

Then we plot the corresponding points and connect them with a smooth curve (Figure 5.6).   □

The various laws of exponents that apply to a^x for rational values of x remain valid for arbitrary real values of x. Although we will not prove the laws, we list them below:

Laws of Exponents

Let a and b be fixed positive numbers. For any numbers x and y we have

i. $a^x a^y = a^{x+y}$ v. $\left(\dfrac{a}{b}\right)^x = \dfrac{a^x}{b^x}$

ii. $(a^x)^y = a^{xy}$ vi. $\dfrac{a^x}{a^y} = a^{x-y}$

iii. $(ab)^x = a^x b^x$ vii. $a^{-x} = \dfrac{1}{a^x} = \left(\dfrac{1}{a}\right)^x$

iv. $1^x = 1$ viii. $a^0 = 1$

EXAMPLE 3.   Simplify $(4^{\sqrt{6}} \cdot 16^{\sqrt{6}})/(4^{3\sqrt{6}-1})$.

Solution.   By the Laws of Exponents.

$$\frac{4^{\sqrt{6}} \cdot 16^{\sqrt{6}}}{4^{3\sqrt{6}-1}} = \frac{4^{\sqrt{6}}(4^2)^{\sqrt{6}}}{4^{3\sqrt{6}-1}} = \frac{4^{\sqrt{6}} \cdot 4^{2\sqrt{6}}}{4^{3\sqrt{6}-1}} = \frac{4^{\sqrt{6}+2\sqrt{6}}}{4^{3\sqrt{6}-1}}$$

$$= \frac{4^{3\sqrt{6}}}{4^{3\sqrt{6}} \cdot 4^{-1}} = \frac{1}{4^{-1}} = 4 \quad \square$$

From Law (vii) of Exponents,

$$3^{-x} = \left(\frac{1}{3}\right)^x$$

so that the graph of 3^{-x} is the same as the graph of $\left(\frac{1}{3}\right)^x$, which appears in Figure 5.6. We will use this fact in the solution of the following example.

EXAMPLE 4.   Let $f(x) = 3^{1-x}$. Sketch the graph of f.

Solution.   By the Laws of Exponents and our remark above,

$$f(x) = 3^{1-x} = 3^1 3^{-x} = 3\left(\frac{1}{3}\right)^x$$

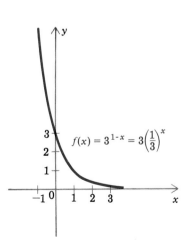

$f(x) = 3^{1-x} = 3\left(\dfrac{1}{3}\right)^x$

FIGURE 5.7

It follows that the graph of f can be obtained by multiplying the y coordinate of each point on the graph of $\left(\frac{1}{3}\right)^x$ (shown in Figure 5.6) by 3. The resulting graph of f appears in Figure 5.7.   □

Another way to solve Example 4 is to observe that

$$3^{1-x} = \left(\tfrac{1}{3}\right)^{-(1-x)} = \left(\tfrac{1}{3}\right)^{x-1}$$

Thus the graph of 3^{1-x} is the same as the graph of $(1/3)^{x-1}$, which can be obtained by shifting the graph of $(1/3)^x$ one unit to the right. A comparison of Figures 5.6 and 5.7 bears this out.

The Exponential Function e^x

Of all the possible bases a for an exponential function, one outshines the others in importance. It is the one which was first called e by the Swiss mathematician Leonhard Euler (1707–1783). This number e is an irrational number whose decimal expansion begins

$$e = 2.71828182845904523536\ldots$$

One mathematical way of obtaining e is as the limit of numbers of the form $(1 + 1/x)^x$ for positive values of x that become arbitrarily large. For a few integral values of x we have

$$\left(1 + \frac{1}{10}\right)^{10} \approx 2.5937425 \qquad \left(1 + \frac{1}{100}\right)^{100} \approx 2.7048138$$

$$\left(1 + \frac{1}{1000}\right)^{1000} \approx 2.7169239 \qquad \left(1 + \frac{1}{10,000}\right)^{10,000} \approx 2.7181459$$

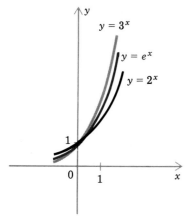

$y = 3^x$

$y = e^x$

$y = 2^x$

FIGURE 5.8

The function e^x is called the **natural exponential function**, or simply the **exponential function**. It is remarkable how often it arises both inside and outside of mathematics (see Section 5.6 for a few applications). Since $2 < e < 3$, the graph of e^x has the same general shape as the graphs of 2^x and 3^x and lies between them (Figure 5.8).

Since e^x occurs so frequently in applications, it is useful to be able to approximate its value for various values of x. A calculator can be used for this purpose. The method of obtaining e^x on a calculator varies, depending on the instrument. But the basic idea is the same. On one such calculator we would evaluate $e^{0.256}$ by pressing first the decimal key that registers a decimal point, then the 2, 5, and 6 keys, and finally the e^x key. Displayed on the calculator would be the approximate value 1.2917527, so that

$$e^{0.256} \approx 1.2917527$$

The symbol "$\approx$" is read "is approximately equal to."

Next we will sketch the graph of a function related to the exponential function and basic to the study of probability.

EXAMPLE 5.   Let $f(x) = e^{-x^2}$. Sketch the graph of f.

Solution.   Because

$$f(-x) = e^{-(-x)^2} = e^{-x^2} = f(x)$$

f is even, and thus its graph is symmetric with respect to the y axis. Next we use a calculator to make a table of approximate values for several positive values of x:

x	0	$\frac{1}{2}$	1	$\frac{3}{2}$	2
$f(x)$	1	0.77880078	0.36787944	0.10539922	0.018315639

After plotting the corresponding points, we connect them with a smooth curve to obtain the graph of f in Figure 5.9. ☐

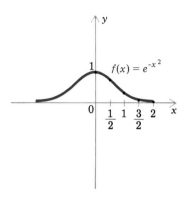

FIGURE 5.9

Exponential Equations

Let $a > 0$ with $a \neq 1$. From (1) and (2) it follows that if $x \neq z$, then $a^x \neq a^z$. This means that for $a > 0$ with $a \neq 1$,

$$\text{if } a^x = a^z, \quad \text{then} \quad x = z \tag{3}$$

Another way of expressing (3) is to say that such a function a^x is one-to-one. As a result, the function a^x has an inverse, a fact that will be of fundamental importance in the next section.

We can use (3) to solve certain equations involving exponential functions.

EXAMPLE 6. Solve the equation $9^x = 3^{x+\sqrt{2}}$ for x.

Solution. In order to use (3), we need the same base on both sides of the equation. Since $9 = 3^2$, the Laws of Exponents imply that

$$9^x = (3^2)^x = 3^{2x}$$

Thus the given equation is equivalent to

$$3^{2x} = 3^{x+\sqrt{2}}$$

From (3) we find that

$$2x = x + \sqrt{2}, \text{ so that } x = \sqrt{2}$$

Therefore the solution of the given equation is $\sqrt{2}$. ☐

EXERCISES 5.1

In Exercises 1–14, sketch the graph of the given function.

1. $f(x) = 4^x$

2. $f(x) = \left(\dfrac{1}{2}\right)^x$

3. $f(x) = -(3^x)$

4. $f(x) = -3(2^x)$

5. $f(x) = 1^{x+2}$

6. $f(t) = 1 + 2^t$

7. $g(t) = 3 - 2^{-t}$

8. $g(x) = e^{-x}$

9. $g(x) = 2^{-2x}$

10. $h(x) = 3^{x+1}$

11. $h(x) = \sqrt{2}(2^{x+1/2})$

12. $g(x) = (0.5)^{2x-1}$

13. $g(x) = \left(\dfrac{1}{2}\right)^{|x|}$

14. $g(x) = \left(\dfrac{1}{2}\right)^{-|x|}$

In Exercises 15–22, simplify the given expression.

15. $\dfrac{\pi^{\sqrt{5}}\pi^{3-\sqrt{5}}}{\pi}$

16. $\dfrac{12^{2e}12^{3-3e}}{12^{-e} \cdot 144}$

17. $\dfrac{5^{\sqrt{6}}25^{3\sqrt{3}}}{125^{\sqrt{12}}}$

18. $(\sqrt{2^{\sqrt{2}}})^2$

19. $(2^{\sqrt{2}})^{-\sqrt{2}}$

20. $\left(\dfrac{1}{7^{\sqrt{5}}}\right)^{2\sqrt{5}}$

21. $(\sqrt[3]{9})^\pi (\sqrt[3]{3})^\pi$

22. $\dfrac{(\sqrt{6})^{\sqrt{3}}(\sqrt{5})^{\sqrt{3}}}{(\sqrt{10})^{\sqrt{3}}}$

In Exercises 23–32, find the base a if the graph of $y = a^x$ contains the given point.

23. $(3, 8)$

24. $(3, 27)$

25. $(1, 7)$

26. $(2, \frac{1}{4})$

27. $(2, \frac{1}{9})$

28. $(\frac{1}{2}, \frac{1}{2})$

29. $(-\frac{1}{2}, 2)$

30. $(3, -\frac{1}{64})$

31. $(\frac{1}{3}, \frac{1}{4})$

32. $(2, a)$

C In Exercises 33–38, use a calculator to approximate the given number.

33. $e^{0.34}$

34. $e^{0.94}$

35. $e^{-1.2}$

36. e^{-3}

37. $\sqrt{e}$

38. $\sqrt[4]{e}$

C In Exercises 39–42, use a calculator to determine the smallest integer x that satisfies the given inequality.

39. $2^{\sqrt{2x}} > 25$

40. $(1.1)^{\sqrt{2x}} > 38$

41. $x^{\sqrt{3}} > 200$

42. $x^{\sqrt{2/3}} > 13.3$

In Exercises 43–48, solve the given equation.

43. $2^{3x} = 2^{5x^2/4}$

44. $(5^x)^{x+1} = 5^{2x+12}$

45. $(\sqrt{3})^{x-1} = 3^{x+\sqrt{2}}$

46. $(6^{1/6})^{x^2} = (\sqrt{6})^{x\sqrt{5}}$

47. $(5^x)^{x+1} = (\sqrt{5})^{2x+10}$

48. $4^{\sqrt{x+1}} = 2^{3x-2}$

49. The **hyperbolic sine** function, whose value at x is denoted by sinh x, and the **hyperbolic cosine** function, whose value at x is denoted by cosh x, are given by

$$\sinh x = \frac{e^x - e^{-x}}{2}$$

$$\cosh x = \frac{e^x + e^{-x}}{2}$$

These functions are important in engineering.
a. Sketch the graph of sinh x.
b. Sketch the graph of cosh x.
c. Show that for every number x,

$$(\cosh x)^2 - (\sinh x)^2 = 1$$

This is the fundamental equation relating the hyperbolic sine and cosine functions.

C **50.** Use a calculator to determine which is larger, 3^π or π^3.

C **51.** Use a calculator to determine which is larger, $(e^e)^e$ or $e^{(e^e)}$.

C **52.** Use a calculator to compute the value of $(1 + 1/x)^x$ for $x = 100,000,000 = 10^8$. The result should be same as the value given by your calculator for e^x with $x = 1$, although the numbers $(1 + 1/10^8)^{10^8}$ and $e^1 = e$ are different.

C **53.** In calculus it is shown that the sums of the form

$$1 + x + \frac{x^2}{2!} + \frac{x^3}{3!} + \cdots + \frac{x^n}{n!} \quad \text{where } n \text{ is a positive integer}$$

approach e^x as n increases without bound.
a. Using a calculator, show that if $x = 1$ and $n = 10$, then the sum agrees with $e^1 = e$ to eight places.
b. Using a calculator, determine to how many places $e^{1.5}$ and the sum agree if $x = 1.5$ and $n = 10$.

54. Let $a > 0$. Use Law (i) of Exponents twice to show that

$$a^{x+y+z} = a^x a^y a^z$$

55. Use (1) of this section to prove (2). (*Hint:* If $0 < a < 1$, then $1/a > 1$.)

56. Let $f(x) = e^{-ae^{-bx}}$, where a and b are positive constants. The graph of f, which is referred to as the *Gompertz growth curve*, is used by actuaries in the preparation of life expectancy tables. If $x < z$, determine whether $f(x) < f(z)$ or $f(x) > f(z)$.

C **57.** The air pressure $p(x)$ at a height of x feet above sea level is given approximately by

$$p(x) = e^{-(4.101 \times 10^{-4})x}$$

where $p(x)$ is given in "atmospheres." Determine the air pressure at a height of
a. 29,028 feet (summit of Mt. Everest)
b. 14,110 feet (summit of Pikes Peak)
c. 1,250 feet (top of Empire State Building)
d. -282 feet (bottom of Death Valley)

C **58.** One test that measures the thyroid condition in a human uses a small dose D of radioactive iodine I^{131}

injected into the blood stream. The amount A remaining in the blood after t days is related to D by the formula

$$A(t) = De^{-0.0086t}$$

Suppose a dose is injected into the bloodstream at noon. Determine the percent in the bloodstream
a. at 1 P.M.
b. one half day later
c. one day later

C **59.** Suppose the charge in coulombs on a capacitor change with time according to the equation

$$C(t) = \frac{1}{500}\left(\frac{1}{50}\right)^{t/4}$$

Use a calculator to find the charge when $t = 1$.

**5.2
LOGARITHMIC
FUNCTIONS**

Let $a > 0$ with $a \neq 1$. As we observed in the preceding section, the range of the function a^x consists of *all* positive numbers. Consequently if $y > 0$, then there is a number x such that $a^x = y$, and in fact this number x is unique. (Indeed, if z also has the property that $a^z = y$, then $a^x = a^z$, so by (3) of Section 5.1, it follows that $x = z$.) A special name is given to x.

DEFINITION 5.2

Let $a > 0$ with $a \neq 1$. For any positive number y, the *logarithm of y to the base a* is the unique number x such that $a^x = y$, and is denoted by $\log_a y$.

According to the definition of $\log_a y$, for any x and any $y > 0$, we have

$$\log_a y = x \quad \text{if and only if} \quad a^x = y \tag{1}$$

The following examples illustrate the use of (1) in passing back and forth between equations involving exponents and equations involving logarithms.

EXAMPLE 1. Convert the following equations involving exponents to equations involving logarithms.
    a. $10^2 = 100$ b. $81^{1/4} = 3$ c. $16 = 3^c$

Solution. In each case we use (1) to transform a statement of the form $a^x = y$ to one of the form $\log_a y = x$:
    a. From $10^2 = 100$ we obtain $\log_{10} 100 = 2$.
    b. From $81^{1/4} = 3$ we obtain $\log_{81} 3 = \frac{1}{4}$.
    c. From $16 = 3^c$ we obtain $\log_3 16 = c$.   □

EXAMPLE 2. Convert the following equations involving logarithms to equations involving exponents.
    a. $\log_3 9 = 2$ b. $\log_2 32 = 5$ c. $\log_c 81 = 7$

Solution.   In each part we use (1) to transform a statement of the form $\log_a y = x$ to one of the form $a^x = y$.

a. From $\log_3 9 = 2$ we obtain $3^2 = 9$.
b. From $\log_2 32 = 5$ we obtain $2^5 = 32$.
c. From $\log_c 81 = 7$ we obtain $c^7 = 81$.   ☐

We can sometimes recognize the numerical value of a logarithm more easily by converting to exponents, as we illustrate now.

EXAMPLE 3.   Evaluate the following expressions.

    a. $\log_2 8$ b. $\log_3 3$ c. $\log_4 256$

Solution.

a. If $\log_2 8 = x$, then by (1), $2^x = 8$. But $2^3 = 8$, so that $x = 3$. Thus $\log_2 8 = 3$.
b. If $\log_3 3 = x$, then by (1), $3^x = 3$. But $3^1 = 3$, so that $x = 1$. Thus $\log_3 3 = 1$.
c. If $\log_4 256 = x$, then by (1), $4^x = 256$. But $4^4 = 256$, so that $x = 4$. Thus $\log_4 256 = 4$.   ☐

EXAMPLE 4.   Evaluate the following expressions.

    a. $\log_{10} 1$ b. $\log_{25} \frac{1}{5}$ c. $\log_4 8$

Solution.

a. If $\log_{10} 1 = x$, then by (1), $10^x = 1$. But $10^0 = 1$, so that $x = 0$. Therefore $\log_{10} 1 = 0$.
b. If $\log_{25} \frac{1}{5} = x$, then by (1), $25^x = \frac{1}{5}$. But $25^{-1/2} = \frac{1}{5}$, so that $x = -\frac{1}{2}$. As a result, $\log_{25} \frac{1}{5} = -\frac{1}{2}$.
c. If $\log_4 8 = x$, then $4^x = 8$. To determine x, we need to write 8 as a power of 4. Since

$$8 = 4 \cdot 2 = 4^1 4^{1/2} = 4^{3/2}$$

it follows that $4^x = 8$ if $4^x = 4^{3/2}$, which means that $x = \frac{3}{2}$. Consequently $\log_4 8 = \frac{3}{2}$.   ☐

Next we solve equations involving logarithms.

EXAMPLE 5.   Solve the following equations for x.

    a. $\log_{10} x = 0$ c. $\log_x \pi = -1$
    b. $\log_e x = 1$ d. $\log_x 81 = 4$

Solution.

a. By (1) the equation $\log_{10} x = 0$ is equivalent to $10^0 = x$, so that $x = 10^0 = 1$. Thus the solution is 1.
b. By (1) the equation $\log_e x = 1$ is equivalent to $e^1 = x$, so that $x = e^1 = e$. Thus the solution is e.
c. By (1) the equation $\log_x \pi = -1$ is equivalent to $x^{-1} = \pi$, or $1/x = \pi$. Thus $x = 1/\pi$, so the solution is $1/\pi$.

d. By (1) the equation $\log_x 81 = 4$ is equivalent to $x^4 = 81$ with $x > 0$. But the only positive value of x for which $x^4 = 81$ is 3. Therefore 3 is the solution. □

By parts (a) and (b) of Example 5, $\log_{10} 1 = 0$ and $\log_e e = 1$, respectively. These are special cases of the following general formulas that hold for any $a > 0$ with $a \neq 1$:

$$\log_a 1 = 0 \tag{2}$$

$$\log_a a = 1 \tag{3}$$

EXAMPLE 6. Solve the following equations for x.

a. $\log_4 \dfrac{x}{5} = 3$ b. $\log_{10}(x - 6) = 2$

Solution.
a. By (1), the equation $\log_4(x/5) = 3$ is equivalent to $4^3 = x/5$, so that

$$x = 5(4^3) = 5(64) = 320$$

Therefore the solution is 320.
b. By (1), the equation $\log_{10}(x - 6) = 2$ is equivalent to $10^2 = x - 6$, so that

$$x = 10^2 + 6 = 106$$

Therefore the solution is 106. □

Having defined $\log_a y$ for any positive number y, we can define the associated function (with x as variable).

DEFINITION 5.3 Let $a > 0$ with $a \neq 1$. The *logarithmic function to the base a* is the function f defined by

$$f(x) = \log_a x \quad \text{for } x > 0$$

Caution: Notice that the domain of $\log_a x$ consists of all $x > 0$ and no other numbers. As a result, we cannot take the logarithm of 0 or any negative number.

From (1) and Definition 5.3 we conclude that the functions $\log_a x$ and a^x are inverses of each other (see Definition 3.4, in Section 3.7). Thus the range of the function $\log_a x$ is identical to the domain of the function a^x, which consists of all real numbers.

The basic formulas

$$f\left(f^{-1}(x)\right) = x \quad \text{and} \quad f^{-1}\left(f(x)\right) = x$$

relating a function and its inverse, yield the formulas

$$a^{\log_a x} = x \quad \text{for } x > 0 \qquad (4)$$

$$\log_a (a^x) = x \quad \text{for all } x \qquad (5)$$

For example,

$$10^{\log_{10}\sqrt{7}} = \sqrt{7} \quad \text{and} \quad \log_e (e^{5.1}) = 5.1$$

Now suppose that x and z are positive numbers and

$$\log_a x = \log_a z$$

Then by (4),

$$x = a^{\log_a x} = a^{\log_a z} = z$$

Thus we have shown that if $\log_a x = \log_a z$, then $x = z$. The converse is obviously true, so we obtain the important result

$$\log_a x = \log_a z \quad \text{if and only if} \quad x = z \qquad (6)$$

We can use (6) to solve certain equations involving logarithms.

EXAMPLE 7.   Solve the equation $\log_2 (x^2 - 7) = \log_2 6x$.

Solution.   By (6) with $a = 2$, the given equation is equivalent to the following equations:

$$x^2 - 7 = 6x$$

$$x^2 - 6x - 7 = 0$$

$$(x - 7)(x + 1) = 0$$

$$x = 7 \quad \text{or} \quad x = -1$$

Check: $\log_2 (7^2 - 7) = \log_2 42$ and $\log_2 (6 \cdot 7) = \log_2 42$
$\log_2 [(-1)^2 - 7]$ is undefined, since $(-1)^2 - 7 = -6$ is negative.

Therefore -1 is an extraneous solution, and consequently the solution of the given equation is 7.   □

Graphs of Logarithmic Functions

Since every logarithmic function is the inverse of an exponential function, we can obtain the graph of a logarithmic function by reflecting the graph of the corresponding exponential function through the line $y = x$.

EXAMPLE 8.   Sketch the graphs of the following functions:
   a. $\log_2 x$ b. $\log_{1/3} x$

Solution.   We sketched the graphs of 2^x and $(\frac{1}{3})^x$ in Figures 5.1c and 5.6, respectively. By reflecting these graphs through the line $y = x$, we obtain the graphs of $\log_2 x$ and $\log_{1/3} x$, respectively, in Figure 5.10a, b.   ☐

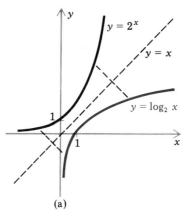

(a)   (b)

FIGURE 5.10

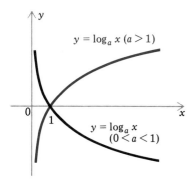

FIGURE 5.11

If $a > 0$ with $a \neq 1$, then the graph of the function $\log_a x$ is as in Figure 5.11, depending on whether $a > 1$ or $0 < a < 1$. The following features of the function $\log_a x$ and its graph are inherited from the corresponding features of the function a^x and its graph.

i. The range of $\log_a x$ consists of all numbers.
ii. The x intercept is 1, and there is no y intercept.
iii. The y axis is a vertical asymptote of the graph.
iv. If $x < z$, then $\begin{cases} \log_a x < \log_a z & \text{for } a > 1 \\ \log_a x > \log_a z & \text{for } 0 < a < 1 \end{cases}$

We also observe that far to the right of the y axis the graph of the function $\log_a x$ becomes very flat, although the graph does not have a horizontal asymptote.

Special Bases   The base of a logarithmic function can be any positive number except 1, which means that $\sqrt{2}$ and $1/\pi$ are perfectly reasonable bases. However, from the viewpoint of mathematical theory and practical applications, the base is almost always greater than 1, and moreover, there are two bases that are far more widely used than any other: e and 10. In calculus and in other branches of mathematics, the logarithm to the base e plays a special role, and is called the ***natural logarithmic function***. Its value at x is normally denoted by $\ln x$, so we would write

$$\ln x \quad \text{instead of} \quad \log_e x$$

(The expression "ln" stands for the Latin phrase "logarithmus naturalis.")

In contrast, in numerical calculation the logarithm to the base 10 has historically reigned supreme, and it is called the **common logarithmic function**. Its value at x is frequently denoted by $\log x$, so that we would write

$$\log x \quad \text{instead of} \quad \log_{10} x$$

The advent of calculators as well as larger computers have made numerical calculations with common logarithms virtually obsolete.

EXERCISES 5.2

In Exercises 1–6, convert the exponential equation to a logarithmic equation.

1. $10^1 = 10$

2. $2^4 = 16$

3. $\left(\frac{1}{2}\right)^{-5} = 32$

4. $343^{1/3} = 7$

5. $x^{-4} = 16$

6. $2^x = 9$

In Exercises 7–12, convert the logarithmic equation to an exponential equation.

7. $\log_{10} 1000 = 3$

8. $\log_{12} 1 = 0$

9. $\log_4 32 = 2.5$

10. $\log_{1.5} x = 3$

11. $\ln x = 2$

12. $\ln e^5 = x$

In Exercises 13–42, evaluate the given logarithm.

13. $\log_{10} 1000$

14. $\log_2 4$

15. $\log_2 16$

16. $\log_2 \frac{1}{4}$

17. $\log_3 \frac{1}{81}$

18. $\log_\pi \pi$

19. $\log_\pi 1$

20. $\log_\pi (1/\pi)$

21. $\log_\pi \pi^{3/2}$

22. $\log_{10} (10^e)$

23. $\ln (e^{-1})$

24. $\log_{10} \sqrt{10}$

25. $\log_8 2$

26. $\log_{16} 2$

27. $\log_{10} 0.01$

28. $\log_{10} 0.0001$

29. $\ln (1/e)$

30. $\log_{1/2} \frac{1}{8}$

31. $\log_{1/3} 81$

32. $\log_4 8$

33. $\log_4 32$

34. $\log_{27} 9$

35. $\log_{27} 81$

36. $\log_9 \frac{1}{27}$

37. $\log_{\sqrt{2}} 8$

38. $\log_{\sqrt{2}} \frac{1}{32}$

39. $\log_{2\sqrt{3}} 144$

40. $\log_2 (2^3 + 2^3)$

41. $\log_4 (8^3 + 8^3)$

42. $\log_{\sqrt{30}} (5^2 + 5)$

In Exercises 43–64, solve the given equation for x.

43. $\ln x = 0$

44. $\log_{10} x = 1$

45. $\log_2 x = 4$

46. $\log_\pi x = -1$

47. $\log_8 x = \frac{1}{3}$

48. $\log_{1/8} x = \frac{1}{3}$

49. $\log_4 x = -2$

50. $\log_{\sqrt{2}} x = 2$

51. $\log_{\sqrt{2}} x = 0$

52. $\log_{1.5} x = -3$

53. $\log_2 (x/3) = 2$

54. $\log_2 \sqrt{x} = 3$

55. $\log_4 (x - 3) = 3$

56. $\log_2 (4x - 7) = 4$

57. $\ln (x^2 - 8) = \ln 2x$

58. $\log_{10} \dfrac{x - 1}{x - 2} = \log_{10} \dfrac{x + 1}{x + 2}$

59. $\log_{10} \dfrac{x^2 + 4}{x} = \log_{10} \dfrac{x^2 + 4x + 4}{x + 1}$

60. $\log_3 |x - 1| = 2$

61. $\log_{10} |x^2 - 21| = 2$

62. $2^{\log_3 27} = x$

63. $3^{\log_2 x} = 27$

64. $4^x = \left(\frac{1}{4}\right)^x$

In Exercises 65–70, solve the given equation for x.

65. $\log_x 16 = 2$

66. $\log_x 64 = 2$

67. $\log_x 125 = 3$

68. $\log_{x-1} 81 = 4$

69. $\log_{x^2} 16 = 2$

70. $\log_{x^2 - 2x} 64 = 2$

In Exercises 71–78, find the base a of the logarithmic function whose graph contains the given point.

71. $(4, 2)$

72. $(64, 3)$

73. $\left(3, \frac{1}{2}\right)$

74. $\left(3, -\frac{1}{2}\right)$

75. $\left(\frac{1}{27}, 3\right)$ **77.** $\left(\frac{1}{2}, \frac{1}{2}\right)$

76. $\left(\frac{1}{27}, -3\right)$ **78.** $\left(\frac{1}{2}, -\frac{1}{2}\right)$

In Exercises 79–86, sketch the graph of the given function.

79. $f(x) = 3\log_2 x$ **81.** $f(x) = \log_2 2x$

80. $f(x) = -\log_2 x$ **82.** $f(x) = \log_2 |x|$

83. $f(x) = \log_4 (2 - x)$ **85.** $f(x) = |\log_{1/3} x|$

84. $g(x) = \log_4 (x - 3)$ **86.** $f(x) = -\log_{1/4} 3x$

87. Is $3\log_{10}\frac{1}{6} > 2\log_{10}\frac{1}{6}$? Explain your answer.

88. Show that if $\log_a b = 0$, then $b = 1$.

5.3 LAWS OF LOGARITHMS AND CHANGE OF BASE

Just as there are laws of exponents, there are also laws of logarithms. The most important ones are presented below. All of them follow from the Laws of Exponents given in Section 5.1 and from the definition of logarithms.

Laws of Logarithms

Let a be a fixed positive number with $a \neq 1$. For any positive numbers x and y,

i. $\log_a (xy) = \log_a x + \log_a y$

ii. $\log_a \frac{1}{x} = -\log_a x$

iii. $\log_a \frac{x}{y} = \log_a x - \log_a y$

iv. $\log_a (x^c) = c\log_a x$ for any number c

v. $\log_a 1 = 0$

vi. $\log_a a = 1$

To prove (i), let $r = \log_a x$ and $s = \log_a y$. It follows from the definition of logarithm that $a^r = x$ and $a^s = y$. Therefore $xy = a^r a^s$. By the Laws of Exponents (Section 5.1), $a^r a^s = a^{r+s}$, so that $xy = a^{r+s}$. By another application of the definition of logarithm, it follows that $\log_a (xy) = r + s$. Thus

$$\log_a (xy) = r + s = \log_a x + \log_a y$$

which is (i). Hints for proving (ii)–(iv) are given in Exercises 86–88. Finally, we observe that (v) and (vi) are restatements, respectively, of (2) and (3) of Section 5.2.

The formula in (i) is frequently referred to as *the* Law of Logarithms. It contains the important feature of logarithms that the logarithm of a *product* of two numbers is the *sum* of the logarithms of the two numbers. In a similar vein, (iii) says that the logarithm of a *quotient* of two numbers is the *difference* of the logarithms of the two numbers.

EXAMPLE 1. Using the facts that $\log_{10} 2 \approx 0.3010$ and $\log_{10} 3 \approx 0.4771$, approximate the numerical value of each of the following logarithms.

a. $\log_{10} 6$ c. $\log_{10} 1.5$

b. $\log_{10} \frac{1}{2}$ d. $\log_{10} \sqrt{2}$

Solution.

a. Since $6 = 2 \cdot 3$, we use (i) to obtain

$$\log_{10} 6 = \log_{10} (2 \cdot 3) = \log_{10} 2 + \log_{10} 3$$
$$\approx 0.3010 + 0.4771 = 0.7781$$

b. We use (ii) to deduce that

$$\log_{10} \frac{1}{2} = -\log_{10} 2 \approx -0.3010$$

c. Since $1.5 = \frac{3}{2}$, we apply (iii), which yields

$$\log_{10} 1.5 = \log_{10} \frac{3}{2} = \log_{10} 3 - \log_{10} 2$$
$$\approx 0.4771 - 0.3010 = 0.1761$$

d. Since $\sqrt{2} = 2^{1/2}$, we can use (iv) to conclude that

$$\log_{10} \sqrt{2} = \log_{10} (2^{1/2}) = \frac{1}{2} \log_{10} 2$$

$$\approx \frac{1}{2}(0.3010) = 0.1505 \quad \square$$

EXAMPLE 2.   Using the facts that $\log_{10} 3 \approx 0.4771$ and $\log_{10} 7 \approx 0.8451$, approximate the numerical value of $\log_{10} \frac{9}{49}$.

Solution.   Since $9 = 3^2$ and $49 = 7^2$, we use (iii) and (iv) to conclude that

$$\log_{10} \frac{9}{49} \overset{\text{(iii)}}{=} \log_{10} 9 - \log_{10} 49 = \log_{10} (3^2) - \log_{10} (7^2)$$

$$\overset{\text{(iv)}}{=} 2 \log_{10} 3 - 2 \log_{10} 7 \approx 2(0.4771) - 2(0.8451)$$
$$= -0.7360 \quad \square$$

Caution: There is no law of logarithms that simplifies either $\log_a (x + y)$ or $(\log_a x)/(\log_a y)$. In particular, $\log_a (x + y)$ is *almost never* equal to $\log_a x + \log_a y$, and $(\log_a x)/(\log_a y)$ is *almost never* equal to $\log_a x - \log_a y$.

The Law of Logarithms (i) can be generalized to more than two variables. For three variables it reads as follows:

$$\log_a xyz = \log_a x + \log_a y + \log_a z \tag{1}$$

EXAMPLE 3.   Using the facts that $\log_{10} 3 \approx 0.4771$ and $\log_{10} 7 \approx 0.8451$, approximate the numerical value of $\log_{10} 210$.

Solution. Since $210 = 3 \cdot 7 \cdot 10$, we use (1) to find that

$$
\begin{aligned}
\log_{10} 210 &= \log_{10} (3 \cdot 7 \cdot 10) \\
&= \log_{10} 3 + \log_{10} 7 + \log_{10} 10 \\
&\approx 0.4771 + 0.8451 + 1 \\
&= 2.3222
\end{aligned}
$$

Therefore $\log_{10} 210 \approx 2.3222$. $\square$

By combining (i) and (iii) we can obtain

$$
\log_a \frac{xy}{z} = \log_a x + \log_a y - \log_a z \tag{2}
$$

EXAMPLE 4. Express $\log_2 \frac{9}{2} + \log_2 \frac{8}{5} - \log_2 \frac{9}{8}$ as a single logarithm, and simplify your answer.

Solution. By (2),

$$
\log_2 \frac{9}{2} + \log_2 \frac{8}{5} - \log_2 \frac{9}{8} = \log_2 \left(\frac{\frac{9}{2} \cdot \frac{8}{5}}{\frac{9}{8}} \right) = \log_2 \left(\frac{9}{2} \cdot \frac{8}{5} \cdot \frac{8}{9} \right)
$$

$$
= \log_2 \frac{32}{5} \quad \square
$$

It is possible to simplify the answer to Example 4 further, although the simplified version is not a single logarithm:

$$
\log_2 \frac{32}{5} = \log_2 32 - \log_2 5 = \log_2 (2^5) - \log_2 5 = 5 - \log_2 5
$$

Solving Equations Involving Logarithms

In the next two examples we will return to solving equations involving logarithms.

EXAMPLE 5. Solve the equation $3 \log_{10} x = \log_{10} 8$ for x.

Solution. We use (iv) above, and then (6) of Section 5.2:

$$
3 \log_{10} x = \log_{10} 8
$$

$$
\log_{10} (x^3) = \log_{10} 8
$$

$$
x^3 = 8
$$

$$
x = 2
$$

Check: $3 \log_{10} 2 = \log_{10} (2^3) = \log_{10} 8$

Consequently the solution is 2. $\square$

EXAMPLE 6.   Solve   the   equation $\log_8 (x + 9) - \log_8 (x - 1) = \frac{1}{3}$ for x.

Solution.   This time we use (iii) and the definition of the logarithm to the base 8:

$$\log_8 (x + 9) - \log_8 (x - 1) = \frac{1}{3}$$

$$\log_8 \frac{x + 9}{x - 1} = \frac{1}{3}$$

$$\frac{x + 9}{x - 1} = 8^{1/3} = 2$$

$$x + 9 = 2(x - 1)$$

$$x + 9 = 2x - 2$$

$$x = 11$$

Check: $\log_8 (11 + 9) - \log_8 (11 - 1) = \log_8 20 - \log_8 10$

$$= \log_8 \frac{20}{10} = \log_8 2$$

$$= \log_8 (8^{1/3}) = \frac{1}{3}$$

Thus the solution of the given equation is 11.   □

Change of Base   It is sometimes necessary to convert from logarithms in one base to logarithms in another base, that is, to write $\log_b x$ in terms of $\log_a x$ (where values of $\log_a x$ are presumably known). Since calculators usually have keys only for natural logarithms and common logarithms, this conversion may even be necessary when one wishes to use a calculator to make computations involving logarithms.

Suppose a and b are positive numbers different from 1 and we wish to express $\log_b x$ in terms of $\log_a x$. Let

$$r = \log_b x \quad \text{so that} \quad x = b^r$$

Since $a \neq 1$ by hypothesis, we can take logarithms to the base a of both sides of the latter equation and obtain

$$\log_a x = \log_a (b^r)$$

But by (iv) in the Laws of Logarithms,

$$\log_a (b^r) = r \log_a b \tag{3}$$

Since $b \neq 1$ by hypothesis, it follows that $\log_a b \neq 0$. Therefore we may divide both sides of (3) by $\log_a b$ to obtain

$$r = \frac{\log_a x}{\log_a b}$$

Substituting $\log_b x$ for r, we obtain the following formula.

> **Change of Base Formula**
>
> $$\log_b x = \frac{\log_a x}{\log_a b} \quad \text{for } x > 0 \tag{4}$$

As we observed earlier, most calculators have a key for the common logarithm of a given number. Now if $a = 10$, then (4) becomes

$$\log_b x = \frac{\log_{10} x}{\log_{10} b} \quad \text{for } x > 0 \tag{5}$$

Since calculators can compute $\log_{10} x$ and $\log_{10} b$ quickly and accurately, a good approximate value for $\log_b x$ can be obtained by calculator. To find the common logarithm of a number by calculator, first key in the number and then press the key for the common logarithm. The value will be displayed after a short pause. Some calculators also have keys for natural logarithms. The procedure for calculating them is the same.

<u>**EXAMPLE 7.**</u>   Use a calculator to approximate the value of $\log_2 5$.

   Solution.   By (5),

$$\log_2 5 = \frac{\log_{10} 5}{\log_{10} 2} \tag{6}$$

Using (6), we find by calculator that

$$\log_2 5 \approx 2.3219281 \quad \square$$

If $x = a$, then the change of base formula (4) becomes

$$\log_b a = \frac{\log_a a}{\log_a b} \tag{7}$$

Since $\log_a a = 1$, (7) simplifies to

$$\log_b a = \frac{1}{\log_a b} \tag{8}$$

a formula that is occasionally of use.

EXAMPLE 8.   Using the fact that $\log_{10} e \approx 0.4343$, approximate $\ln 10$.

Solution.   By (8),

$$\ln 10 = \log_e 10 = \frac{1}{\log_{10} e} \approx \frac{1}{0.4343} \approx 2.303 \quad \square$$

The invention of logarithms is generally credited to John Napier (1550–1617), a Scotsman of noble heritage for whom mathematics was a hobby. His invention arose through his desire to associate ratios of pairs of numbers in a geometric progression with ratios of pairs of numbers in an arithmetic progression. Thus the name *logarithmus*, which means "ratio number" and has been anglicized to "logarithm." What resulted from Napier's creation was an association between quotients and differences, namely Law of Logarithms (iii). The base of Napier's logarithms happened to be related to e, and this is the reason why $\log_e x$ (that is, $\ln x$) is sometimes called a Napierian logarithm. Not only did Napier invent his special logarithms, but he made an extensive table of seven-place logarithms that was published in 1614.

EXERCISES 5.3

In Exercises 1–14, approximate the numerical value of the given logarithm. Use the facts that $\log_{10} 2 \approx 0.3010$, $\log_{10} 3 \approx 0.4771$, and $\log_{10} 5 \approx 0.6990$.

1. $\log_{10} 6$

2. $\log_{10} 30$

3. $\log_{10} 12$

4. $\log_{10} 75$

5. $\log_{10} \frac{1}{3}$

6. $\log_{10} \frac{1}{8}$

7. $\log_{10} \frac{10}{3}$

8. $\log_{10} \frac{2}{25}$

9. $\log_{10} \frac{9}{16}$

10. $\log_{10} 7.5$

11. $\log_{10} 0.2$

12. $\log_{10} \sqrt{5}$

13. $\log_{10} \sqrt{30}$

14. $\log_{10} \sqrt[3]{2}$

In Exercises 15–28, approximate the numerical value of the given logarithm. Use the facts that $\ln 2 \approx 0.6931$, $\ln 7 \approx 1.9459$, and $\ln 10 \approx 2.3026$.

15. $\ln 20$

16. $\ln 70$

17. $\ln \frac{1}{7}$

18. $\ln 0.1$

19. $\ln 49$

20. $\ln 7000$

21. $\ln 5$

22. $\ln 35$

23. $\ln 24.5$

24. $\ln \sqrt{2}$

25. $\ln \sqrt[3]{14}$

26. $\ln \sqrt[4]{40}$

27. $\ln 14^{2/3}$

28. $\ln 40^{0.8}$

In Exercises 29–32, find the numerical value of the given expression.

29. $\log_3 (27^5)$

30. $\log_4 [(\frac{1}{32})^6]$

31. $\log_{16} (8^{7/9})$

32. $\log_{27} (9^{0.2})$

In Exercises 33–36, approximate the numerical value of the given logarithm. Use the fact that $\log_{10} 4 \approx 0.6021$.

33. $\log_{10} 40$

34. $\log_{10} 400,000,000$

35. $\log_{10} \frac{1}{4}$

36. $\log_{10} 0.00000025$

In Exercises 37–42, approximate the numerical value of the given logarithm. Assume that $\log_{10} a = 2.7318$.

37. $\log_{10} (1000a)$

38. $\log_{10} (a \times 10^6)$

39. $\log_{10} (a \times 10^{-9})$

40. $\log_{10} \dfrac{a}{100}$

41. $\log_{10} (a^{10})$

42. $\log_{10} (a^{-5})$

In Exercises 43–48, use a calculator to approximate the numerical value of the given logarithm.

43. $\log_{10} 5$

44. $\log_{10} 217$

45. $\log_{10} 0.00849$

46. $\ln 3.06$

47. $\ln \frac{1}{53}$

48. $\ln 17,329.6$

In Exercises 49–54, rewrite the given expression as a single logarithm.

49. $\log_2 3 + \log_2 \frac{4}{3} + \log_2 \frac{5}{4}$

50. $\ln (3/e) + \ln \frac{1}{6} + \ln 2$

51. $\log_2 (xy^2) + \log_2 (y^{-3}) + \log_2 (x^5)$

*52. $\ln 40 + \log_2 8$

*53. $\log_{10} 2 + 3\log_{\sqrt{10}} x - \log_{10}(2y)$

*54. $\log_2 (9x) - \log_4 (7x^2) - \log_8 (3/x^2)$

In Exercises 55–60, express the given logarithm in terms of $\log_a x$, $\log_a y$, and $\log_a z$.

55. $\log_a \dfrac{xy}{z}$

56. $\log_a (x^2 yz^3)$

57. $\log_a (z^3 \sqrt{xy})$

58. $\log_a \dfrac{1}{xyz}$

59. $\log_a \sqrt{\dfrac{xy^5}{z^3}}$

60. $\log_a \sqrt[4]{\dfrac{x\sqrt{y}}{y^{1/3}}}$

In Exercises 61–66, solve the given equation for x.

61. $2\log_{10} x = \log_{10} 16$

62. $4\log_2 x = \log_2 81$

63. $-2\log_{10} x = \log_{10} 4$

64. $\log_{10} (4x - 1) - \log_{10} \left(\dfrac{x}{5}\right) = 1$

65. $\log_2 (x + 1) + \log_2 (x - 1) = 3$

66. $\ln (2x - 1) - \ln (3x + 1) = -1$

In Exercises 67–72, use the change of base formulas to obtain the desired results.

67. Given that $\log_{10} 7 \approx 0.8450$ and $\log_{10} 2 \approx 0.3010$, approximate $\log_2 7$.

68. Given that $\log_{10} 5 \approx 0.6990$ and $\log_{10} e \approx 0.4343$, approximate $\ln 5$.

69. Given that $\log_2 6 \approx 2.5850$ and $\log_2 3 \approx 1.5850$, approximate $\log_3 6$.

70. Given that $\ln 7 \approx 1.9459$ and $\log_5 e \approx 0.6213$, approximate $\log_5 7$.

71. Given that $\log_2 3 \approx 1.5850$, approximate $\log_3 2$.

72. Given that $\log_{10} 2 \approx 0.3010$, approximate $\log_2 10$.

C In Exercises 73–80, use a calculator and the change of base formulas to obtain the desired results.

73. $\log_2 5$

74. $\log_2 0.48$

75. $\log_7 14.3$

76. $\log_{13} \dfrac{419}{73}$

77. $\log_{1/3} \dfrac{13}{7}$

78. $\log_{9/10} 0.00346$

79. $\log_{.4} 5432$

80. $\log_{.02} 0.00749$

81. a. Show that $\log_2 x = \log_4 x^2$.
   b. Show that $\log_a x = \log_{a^2} x^2$ for any positive numbers a and x such that $a \neq 1$.

82. Prove that $\ln (x + \sqrt{x^2 - 1}) = -\ln (x - \sqrt{x^2 - 1})$.

83. Prove that $\log_a (1/x) = \log_{1/a} x$.

84. Show that if x satisfies $\dfrac{\ln x}{x} < \dfrac{\ln 2}{2}$, then $x^2 < 2^x$.

85. Let $b = e^{-1/10^7}$. Show that $\log_b x = -10^7 \ln x$ for any $x > 0$. (The "logarithm" that John Napier defined in the seventeenth century was actually $\log_b x + 10^7 \ln 10^7$. Thus, although his logarithm was not the natural logarithm, it was very closely related to it.)

86. Prove Law (ii) by using Law (i) with y replaced by $1/x$.

87. Prove Law (iii) by using Laws (i) and (ii).

88. Prove Law (iv). (*Hint:* Let $r = \log_a x$. Using first the definition of the logarithm to the base a, next a law of exponents, and then the definition of the logarithm to the base a again, show in turn that $x = a^r$, $x^c = a^{cr}$, and $cr = \log_a (x^c)$, and complete the proof.)

5.4
NUMERICAL COMPUTATIONS WITH COMMON LOGARITHMS

Before calculators were invented, computations were frequently performed either with a slide rule or with the aid of common logarithms and a table of (approximate) logarithmic values. Common logarithms were used because our number system is based on the number 10. Even though calculators have made these methods obsolete, we will discuss the use of the logarithmic table in computation. There are two reasons for this: First, the method illustrates the basic properties of logarithms. Second, some of the techniques we will use apply to other tables as well. Since all logarithms in this section have base 10, we will write $\log x$ instead of $\log_{10} x$ throughout the section.

Recall from Section 1.3 that any positive number x can be written in scientific notation as

$$x = b \times 10^n \tag{1}$$

where n is an integer and $1 \le b < 10$. For example,

$$4296 = 4.296 \times 10^3$$

$$0.000783 = 7.83 \times 10^{-4}$$

Taking logarithms of both sides of (1) and then using the Laws of Logarithms, we find that

$$\log x = \log (b \times 10^n) = \log b + \log (10^n)$$
$$= (\log b) + n \log 10 = (\log b) + n$$

so that

$$\log x = n + \log b \tag{2}$$

Thus the common logarithm of *any* positive number x is the sum of an integer n and the common logarithm $\log b$ of a number b between 1 and 10. The number n is called the ***characteristic*** of $\log x$, and the number $\log b$ is the ***mantissa*** of $\log x$.* Since $1 \le b < 10$, we have

$$0 = \log 1 \le \log b < \log 10 = 1$$

so the mantissa of $\log x$ is always between 0 and 1.

Table C in the Appendix gives the common logarithms of various numbers between 1 and 10 in steps of 0.01. A portion of one page of the table is reproduced here:

x	0	1	2	3	4	5	6	7	8	9
3.5	.5441	.5453	.5465	.5478	.5490	.5502	.5514	.5527	.5539	.5551
3.6	.5563	.5575	.5587	.5599	.5611	.5623	.5635	.5647	.5658	.5670
3.7	.5682	.5694	.5705	.5717	.5729	.5740	.5752	.5763	.5775	.5786
3.8	.5798	.5809	.5821	.5832	.5843	.5855	.5866	.5877	.5888	.5899
3.9	.5911	.5922	.5933	.5944	.5955	.5966	.5977	.5988	.5999	.6010

To use the table to find the logarithm of a number such as 3.81, which is between 1 and 10, first locate the row labeled 3.8 and then the column labeled 1 at the top. The entry belonging to both row and column is .5809 (without

---

* The word "mantissa" is of Latin origin, meaning "an addition." The mantissa is "added on" to the characteristic, which is the integral, or main, part of the logarithm.

the customary 0 before the decimal in order to keep the table as compact as possible), so that

$$\log 3.81 \approx 0.5809$$

Caution: Most of the entries in Table C are not exact values but only approximations to the exact values, rounded off to four decimal places, which means that the maximum error in any entry is 0.00005. Thus the entry of .5809 for log 3.81 implies that the true value of log 3.81 lies between 0.58085 and 0.58095.

Before finding numerical values of common logarithms, we emphasize that there is no need for the common logarithm table to have logarithms of numbers that are not between 1 and 10, since the use of scientific notation reduces the problem of finding the logarithm of any positive number to that of finding the logarithm of a number between 1 and 10 (see (1) and (2)).

EXAMPLE 1.   Use Table C to approximate the following numbers.
   a. log 143   b. log 0.00756

Solution.
   a. First we write 143 in scientific notation as

$$143 = 1.43 \times 10^2$$

Next we use (2) to deduce that

$$\log 143 = 2 + \log 1.43$$

Finally we see from Table C that log 1.43 $\approx$ 0.1553, so that

$$\log 143 \approx 2 + 0.1553 = 2.1553$$

   b. Proceeding as in (a), we find that

$$0.00756 = 7.56 \times 10^{-3}$$
$$\log 0.00756 = -3 + \log 7.56$$

From Table C we notice that log 7.56 $\approx$ 0.8785, so that

$$\log 0.00756 \approx -3 + 0.8785 \quad \square$$

Notice that in the final answer of (b) we did not combine the characteristic -3 with the mantissa 0.8785. When the characteristic is negative, the logarithm is normally easier to use with the characteristic and mantissa kept separate.

Now we turn the process around and find a number whose common logarithm is given.

EXAMPLE 2.   Use Table C to find a number whose common logarithm is approximately 3.9253.

Solution.   First observe that

$$3.9253 = 3 + 0.9253$$

This means that if we find a number b whose logarithm is given in Table C by .9253, then $b \times 10^3$ is the number we seek, since in that case

$$\log(b \times 10^3) = 3 + \log b \approx 3 + 0.9253 = 3.9253$$

Looking in Table C for the entry .9253, we find it in the row labeled 8.4 and the column labeled 2 at the top. It follows that

$$\log 8.42 \approx 0.9253$$

Therefore 8.42×10^3, which of course is 8420, is a number whose common logarithm is approximately 3.9253.   □

In the next example, where we again will find a number whose common logarithm is given, we will have to be careful to write the given number in the form $n + \log b$, where n is an integer and $0 \le \log b < 1$.

EXAMPLE 3.   Use Table C to find a number whose common logarithm is approximately -1.4123.

Solution.   Although $-1.4123 = -1 - 0.4123$, the expression $-1 - 0.4123$ does not have the form $c + \log b$, where $0 \le \log b < 1$ (since -0.4123 is not between 0 and 1). But notice that

$$-1.4123 = -2 + 0.5877$$

and $-2 + 0.5877$ has the correct form because $0 < 0.5877 < 1$. From Table C we see that

$$\log 3.87 \approx 0.5877$$

Thus 3.87×10^{-2}, which is the same as 0.0387, is a number whose common logarithm is approximately -1.4123.   □

If x is any number, then the positive number y with $\log y = x$ is called the *antilogarithm* of x. For instance, the solution of Example 2 tells us that the antilogarithm of 3.9253 is approximately 8420, and the solution of Example 3 tells us that the antilogarithm of -1.4123 is approximately 0.0387.

To obtain an alternative description of an antilogarithm, we recall from (1) in Section 5.2 that

$$\log y = x \quad \text{if and only if} \quad 10^x = y$$

Since $\log y = x$ means that y is the antilogarithm of x, it follows that the antilogarithm of x is 10^x.

Linear Interpolation

By Table C,

$$\log 268 = \log (2.68 \times 10^2) = 2 + \log 2.68 \approx 2 + 0.4281 = 2.4281$$

and

$$\log 269 = \log (2.69 \times 10^2) = 2 + \log 2.69 \approx 2 + 0.4298 = 2.4298$$

But suppose that we desire to compute log 268.4, which lies between log 268 and log 269. The process starts in the same way:

$$\log 268.4 = \log (2.684 \times 10^2) = 2 + \log 2.684$$

But 2.684 has three digits to the right of the decimal, whereas Table C contains entries for numbers with only 1 or 2 digits to the right of the decimal. The way we go about finding a value for log 268.4 is as follows.
From Table C,

$$\log 2.68 \approx 0.4281$$

$$\log 2.684 = ??$$

$$\log 2.69 \approx 0.4298$$

Since $2.684 - 2.68 = 0.004$ and $2.69 - 2.68 = 0.01$, evidently 2.684 is $0.004/0.01 = \frac{4}{10}$ of the distance from 2.68 to 2.69. Therefore we take log 2.684 to be $\frac{4}{10}$ of the distance from log 2.68 to log 2.69, that is, from 0.4281 to 0.4298. Since

$$0.4298 - 0.4281 = 0.0017 \quad \text{and} \quad \frac{4}{10}(0.0017) = 0.00068 \approx 0.0007$$

(rounded off to four decimal places, as are the entries in Table C), we take

$$\log 2.684 \approx (\log 2.68) + 0.0007 = 0.4281 + 0.0007 = 0.4288$$

Consequently

$$\log 268.4 = 2 + \log 2.684 \approx 2 + 0.4288 = 2.4288$$

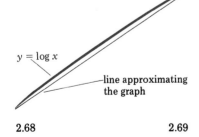

$y = \log x$

line approximating the graph

2.68   2.69

FIGURE 5.12

This method of approximating a logarithm is called **_linear interpolation_**, because in effect the method treats the logarithmic function as a linear function on intervals of length 0.01 (Figure 5.12). On such intervals the graph of log x is very nearly linear, so the estimate is reasonable.

EXAMPLE 4.   Use linear interpolation to find an approximate value of log 2817.

Solution.   Notice that

$$\log 2817 = \log (2.817 \times 10^3) = 3 + \log 2.817$$

From Table C,

$$\log 2.81 \approx 0.4487$$

$$\log 2.817 = \text{??}$$

$$\log 2.82 \approx 0.4502$$

Since $2.817 - 2.81 = 0.007$ and $2.82 - 2.81 = 0.01$, evidently 2.817 is $0.007/0.01 = \frac{7}{10}$ of the distance from 2.81 to 2.82. Therefore we take $\log 2.817$ to be $\frac{7}{10}$ of the distance from $\log 2.81$ to $\log 2.82$, that is, from 0.4487 to 0.4502. Since

$$0.4502 - 0.4487 = 0.0015 \quad \text{and} \quad \frac{7}{10}(0.0015) = 0.00105 \approx 0.0011$$

(rounded off to four decimal places), we take

$$\log 2.817 \approx (\log 2.81) + 0.0011 \approx 0.4487 + 0.0011 = 0.4498$$

Consequently

$$\log 2817 = 3 + \log 2.817 \approx 3 + 0.4498 = 3.4498 \quad \square$$

Linear interpolation can also be used to approximate antilogarithms of numbers not appearing in Table C. For example, suppose we wish to find the antilogarithm of 0.7961. Observe that by the table,

$$\log 6.25 \approx 0.7959$$

$$\log \text{??} = 0.7961$$

$$\log 6.26 = 0.7966$$

Since $0.7961 - 0.7959 = 0.0002$ and $0.7966 - 0.7959 = 0.0007$, it follows that 0.7961 is $0.0002/0.0007 = \frac{2}{7}$ of the distance from 0.7959 to 0.7966. Therefore we take the antilogarithm of 0.7961 to be $\frac{2}{7}$ of the distance from the antilogarithm of 0.7959 to the antilogarithm of 0.7966, that is, from 6.25 to 6.26. Now

$$6.26 - 6.25 = 0.01 \quad \text{and} \quad \frac{2}{7}(0.01) \approx 0.003$$

(rounded off to three decimal places), so the value of the antilogarithm of 0.7961 is approximately

$$6.25 + 0.003 = 6.253$$

Notice that logarithms are given in Table C for numbers with two decimal places, whereas the answer just given with the help of linear interpolation was rounded off to three decimal places. In general the third decimal place will be more reliable than a fourth or higher decimal place, and as a result, we will limit ourselves to three decimal places when we approximate antilogarithms with linear interpolation.

EXAMPLE 5.   Use linear interpolation to find an approximate value of the antilogarithm of 0.3764.

Solution.   By Table C,

$$\log 2.37 \approx 0.3747$$

$$\log \text{??} = 0.3764$$

$$\log 2.38 \approx 0.3766$$

Since $0.3764 - 0.3747 = 0.0017$ and $0.3766 - 0.3747 = 0.0019$, it follows that 0.3764 is $0.0017/0.0019 = \frac{17}{19}$ of the distance from 0.3747 to 0.3766. Therefore we take the antilogarithm of 0.3764 to be $\frac{17}{19}$ of the distance from the antilogarithm of 0.3747 to the antilogarithm of 0.3766, that is, from 2.37 to 2.38. Since

$$\frac{17}{19}(2.38 - 2.37) = \frac{17}{19}(0.01) \approx 0.009$$

(rounded off to three decimal places), the antilogarithm we seek is approximately

$$2.37 + 0.009 = 2.379 \quad \square$$

We complete the section with a computation done with the help of Table C. Before calculators were available, calculations were frequently done this way.

EXAMPLE 6.   Use common logarithms to approximate $(8.53)^{7.27}$.

Solution.   First we take the logarithm of the given expression, and then rearrange the result with the help of the Laws of Logarithms and Table C:

$$\log\left[(8.53)^{7.27}\right] = (7.27)\log 8.53 \approx (7.27)(0.9309) \approx 6.7676$$

In order to obtain the antilogarithm of $6 + 0.7676$, we observe from Table C that

$$\log 5.85 \approx 0.7672$$

$$\log \text{??} = 0.7676$$

$$\log 5.86 \approx 0.7679$$

Therefore 0.7676 is $0.0004/0.0007 = \frac{4}{7}$ of the distance from 0.7672 to 0.7679.

Consequently we take the antilogarithm of 0.7676 to be $\frac{4}{7}$ of the distance from 5.85 to 5.86, that is, approximately 5.856. Thus the antilogarithm of $6 + 0.7676$ is approximately 5.856×10^6, or 5,856,000. We conclude that

$$(8.53)^{7.27} \approx 5,856,000 \quad \square$$

One calculator almost instantly produces the value 5,861,374.311 for the answer to Example 6. Thus modern technology, in the form of calculators, can save much time and effort in computations and, moreover, can provide accuracy vastly superior to that available to people even as recently as 40 years ago.

EXERCISES 5.4

In Exercises 1–20, use Table C to approximate the common logarithm of the given number. Use linear interpolation if needed.

1. 1.01
2. 3.87
3. 520
4. 1,750,000
5. 0.0981
6. 0.00145
7. 5.23×10^4
8. $(7.7)^3$
9. $\sqrt{1.41}$
10. $\sqrt[3]{429}$
11. $\dfrac{1}{693}$
12. $\dfrac{1}{0.00872}$
13. 3.281
14. 29.35
15. 15,340,000
16. 0.1927
17. 0.001437
18. 4.793×10^{-2}
19. $(23.29)^3$
20. $\sqrt{49.75}$

In Exercises 21–36, use Table C to approximate the antilogarithm of the given number. Use linear interpolation if needed.

21. 0.8756
22. 0.5514
23. 4.9983
24. 5.0043
25. 8.8000
26. −1.5331
27. −2.2882
28. −6.3546
29. 0.9752
30. 1.6081
31. 2.1913
32. 6.5499
33. −1.6814
34. −2.6979
35. −0.3933
36. −3.7208

In Exercises 37–44, use common logarithms to approximate the given expression.

37. $\dfrac{2.7 \times 90.73}{3.48}$
38. $\dfrac{5.1 \times 1.021 \times 3.2}{1.015 \times 97.37}$
39. $(7.612)^{2/3}$
40. $\sqrt[4]{3} \cdot \sqrt[3]{4}$
41. $(4.512)^{1/2} \times (923.4)^3$
42. $\dfrac{(3.5)^{1/3}}{\sqrt{62,100}}$
43. $4^{-4.4}$
44. $\dfrac{(28100)^{3.2} \times (54,300)^{1.1}}{(0.00316)^2 \times \sqrt{2}}$

45. The equation

$$\log w = 1.9532 + 3.135 \log s$$

has been used to relate the weight w (in kilograms) to the sitting height s (in meters) of people. Using this equation and Table C or a calculator, approximate
a. the weight of a person whose sitting height is 1 meter.
b. the weight of a person whose sitting height is $\frac{1}{2}$ meter.
c. the sitting height of a person whose weight is 36 kilograms.
d. the sitting height of a person whose weight is 20 kilograms.

46. For small values of $|h|$, the number $\log(1 + h)$ may be approximated by using the formula

$$\log(1 + h) \approx 0.4343\left(h - \frac{1}{2}h^2\right)$$

Use this equation to approximate
a. log 1.1   b. log 1.01
(Compare your answers with the values given by a calculator or Table C.)

5.5 APPLICATIONS OF COMMON LOGARITHMS

This section is devoted to mathematical and scientific applications of common logarithms. The mathematical application concerns solutions of equations involving exponentials, and the scientific applications are represented by loudness of sound and magnitude of earthquakes, both of which employ common logarithms in their formulation. As before, we will write $\log x$ for $\log_{10} x$. Our computations will be made by calculator. They could also be made by using Table C.

Solutions of Equations

For the examples below, we will use the fact that

$$x = y \quad \text{if and only if} \quad \log_a x = \log_a y$$

When we pass from an equation of the form $x = y$ to the equivalent equation $\log_a x = \log_a y$, we say that we **take logarithms of both sides** of the equation $x = y$.

EXAMPLE 1.   Solve the equation $4^x = 17$ for x.

   Solution.   We take common logarithms of both sides and then use Law of Logarithms (iv):

$$\log (4^x) = \log 17$$

$$x \log 4 = \log 17$$

$$x = \frac{\log 17}{\log 4}$$

By calculator we find that

$$x \approx 2.0437314 \quad \square$$

EXAMPLE 2.   Solve the equation $2^{x+1} = 3^{x^2-1}$ for x.

   Solution.   Taking common logarithms of both sides and then using (iv) of the Laws of Logarithms, we obtain

$$\log (2^{x+1}) = \log (3^{x^2-1})$$

$$(x + 1) \log 2 = (x^2 - 1) \log 3$$

$$(x + 1) \log 2 = (x + 1)[(x - 1) \log 3]$$

From this last equation we conclude that either $x = -1$ or

$$\log 2 = (x - 1) \log 3$$

Solving for $x - 1$ in the latter equation, we obtain

$$x - 1 = \frac{\log 2}{\log 3} \quad \text{so that} \quad x = \frac{\log 2}{\log 3} + 1$$

Thus the solutions of the given equation are -1 and $[(\log 2)/(\log 3)] + 1$ (which by calculator we find is approximately 1.6309298). $\square$

Loudness of Sound

We turn now to physical applications of common logarithms, represented by loudness of sound and magnitude of earthquakes. In each of these applications the primary reason for using common logarithms is to restrict the numbers involved to a more manageable range. For instance, if a certain quantity q can take any value from, say, 10^{-100} (which is miniscule) to 10^{100} (which is astronomical), then the value of $\log q$ lies between -100 and 100.

The *intensity* of a sound wave is the amount of energy that the wave carries through a unit area in a unit time, or equivalently, the amount of power per unit area. The unit we will use for the intensity of sound is one watt per square meter. The *threshold of audibility* of a sound wave is the minimal intensity the sound wave can have and still be audible to the normal human ear. The threshold of audibility of a sound wave varies according to the pitch, or frequency, of the wave. To simplify the discussion that follows, let us assume that each sound wave under consideration has the same frequency — 100 hertz (cycles per second), producing a sound about an octave and a half below middle C on the piano.

Let I_0 be the intensity of such a sound wave at the threshold of audibility, approximately 10^{-12} watts per square meter. If x denotes the intensity of any sound wave of the same frequency, then the *noise level* (or *loudness*) $L(x)$ of the sound wave is defined by

$$L(x) = 10 \log \frac{x}{I_0} \tag{1}$$

The units for $L(x)$ are *decibels*, in honor of Alexander Graham Bell (1847–1922). Notice that if $x = I_0$, then (1) becomes

$$L(I_0) = 10 \log \frac{I_0}{I_0} = 10 \log 1 = 0$$

so that at the threshold of audibility the noise level is just 0.

EXAMPLE 3.   Suppose a sound wave from a loud conversation is 25,000 times as intense as a sound wave from a whisper, whose noise level is 22 decibels. Determine the noise level of the loud conversation.

Solution.   Let x and z, respectively, represent the intensities of sound waves from the loud conversation and the whisper. We wish to determine $L(x)$. By hypothesis,

$$\frac{x}{z} = 25{,}000, \quad \text{so that} \quad x = 25{,}000z$$

Using (1) and then the Law of Logarithms, we obtain

$$L(x) = L(25{,}000z) = 10 \log \frac{25{,}000z}{I_0}$$

$$= 10 \log 25{,}000 + 10 \log \frac{z}{I_0}$$

Since $L(z) = 22$ by hypothesis, (1) implies that

$$10 \log \frac{z}{I_0} = L(z) = 22$$

Using a calculator, we find that

$$L(x) = 10 \log 25{,}000 + 22 \approx 10(4.39794) + 22 \approx 66$$

Consequently the noise level of the loud conversation is approximately 66 decibels. ☐

Earthquakes

During the past three decades several formulas have been proposed for converting seismographic readings into a unified scale that would represent the magnitude of an earthquake. The scale most commonly used is the **Richter scale**, devised by Charles F. Richter, an American geologist. In order to describe measurement on the Richter scale, we first introduce a reference, or **zero-level**, earthquake, which by definition is any earthquake whose largest seismic wave would measure 0.001 millimeter (1 micron) on a standard seismograph located 100 kilometers from the epicenter of the earthquake. The **magnitude** M of a given earthquake can be obtained by the formula

$$M = \log \frac{a}{a_0} = \log a - \log a_0 \qquad (2)$$

where a is the amplitude measured on a standard seismograph of the largest seismic wave of the given earthquake, and a_0 is the amplitude on the same seismograph of the largest seismic wave of a zero-level earthquake with the same epicenter. Values of a_0 for various distances from the epicenter have been calculated, so one only needs to measure the number a in order to be able to assess the magnitude of a given earthquake.

Notice that the stronger the earthquake is, the greater the ratio a/a_0 of amplitudes of seismic waves is. Let us call a/a_0 the **intensity** of an earthquake.

EXAMPLE 4.   The Alaska Good Friday earthquake of March 28, 1964, measured 8.5 on the Richter scale. Some 19 years later, on May 2, 1983, a massive earthquake measuring 6.5 on the Richter scale rocked Coalinga, California. The Alaska earthquake was how many times as intense as the one in California?

Solution.   Let a_1 and a_2, respectively, represent the amplitudes of the maximum seismic waves of the Alaska and California earthquakes at, say 1000 kilometers from their respective epicenters. We must determine $(a_1/a_0)/(a_2/a_0)$, which equals a_1/a_2. By hypothesis,

$$8.5 = \log \frac{a_1}{a_0} \quad \text{and} \quad 6.5 = \log \frac{a_2}{a_0}$$

The effect of the Good Friday earthquake on a main street in Anchorage, Alaska. Part of the street was left some 20 feet below the rest of it.

Therefore by the Laws of Logarithms,

$$2 = 8.5 - 6.5 = \log \frac{a_1}{a_0} - \log \frac{a_2}{a_0} = \log \frac{a_1/a_0}{a_2/a_0} = \log \frac{a_1}{a_2}$$

It follows that

$$\frac{a_1}{a_2} = 10^2 = 100$$

so that $a_1 = 100\, a_2$. This means that the Alaska earthquake was some 100 times as intense as the California earthquake. □

Theoretically, the magnitude of an earthquake can be any real number. In practice, however, a standard seismograph can only record those earthquakes whose magnitudes exceed 2. On the other end of the scale, no magnitude has ever exceeded 9.0, and only half a dozen times has one been 8.5 or higher.

EXERCISES 5.5

In Exercises 1–12, solve the given equation. Leave your answer in terms of common logarithms.

1. $2^x = 5$

2. $3^{-x} = 17$

3. $5^{2x} = 51$

4. $7^{\sqrt{2}x} = 3$

5. $4^{x+2} = 5$

6. $(\frac{1}{4})^{x+2} = 5$

7. $5^{2x+3} = 3^{5x-2}$

8. $8^{4x-1} = 6^{1-x}$

9. $3^x = 2^{(x^2)}$

10. $3^x = 2^{(x^3)}$

11. $4^{x+2} = 5^{x^2-4}$

12. $2^{x-1} = 10^{x^2+x-2}$

13. Using (1), show that if $L(x_1)$ decibels corresponds to intensity x_1 and $L(x_2)$ corresponds to intensity x_2, then

$$L(x_2) - L(x_1) = 10 \log \frac{x_2}{x_1}$$

14. Determine the number of decibels that corresponds to each of the following intensities (in watts per square meter).
   a. 10^{-11} (refrigerator motor)
   b. 10^{-5} (conversation)
   c. 10^{-4} (electric typewriter)

15. Using the value of 10^{-12} for I_0, find the intensity of a sound at 50 decibels.

16. What is the ratio of the intensity of a given sound to that of one that is 100 decibels higher?

17. What is the difference in the noise levels of two sounds one of which is 100 times as intense as the other?

18. Suppose the cannon fired at the time of a football touchdown produces a sound intensity 1000 times as strong as a cheerleader's shouts. Determine how many more decibels the noise level from the cannon produces than the cheerleader's cry.

19. The human ear can just barely distinguish between two sounds if one is 0.6 decibels higher than the other. What is the ratio of the intensity of one sound to that of another sound which is lower than the first and is just barely distinguishable from the first sound?

20. Suppose soundproofing a room results in a noise reduction of 30 decibels in the noise from a busy highway. What percentage of the original intensity is the reduced intensity?

21. The magnitude of the Good Friday Alaska earthquake of 1964 is sometimes given as 8.4. What is the ratio of the maximum amplitude of an earthquake of magnitude 8.4 to that of an earthquake of magnitude 8.5?

22. Of seven major earthquakes to hit Iran in recent times, two have measured approximately 6.9 on the Richter scale (those occurring in April 1972 and in March 1977). Find the ratio of the amplitudes of the largest waves of these earthquakes to that of the San Francisco earthquake of 1906, which measured 8.3 in magnitude.

23. The strongest earthquakes ever recorded occurred off the coast of Ecuador and Columbia in 1906, and in Japan in 1933. Each had magnitude 8.9. Find the ratio of the amplitude of the largest wave of such a quake to the corresponding amplitude of a zero-level quake.

24. The Alaska Good Friday earthquake of March 28, 1964, measured 8.5 on the Richter scale. Find the intensity a/a_0.

25. Suppose a seismograph is located exactly 100 kilometers from the epicenter of an earthquake. Determine the magnitude of the earthquake if the largest amplitude of the seismic waves registered on the seismograph is
    a. 1 micron   b. 1 millimeter   c. 1 centimeter

26. If an earthquake has magnitude 2, find the amplitude of its largest wave 100 kilometers from the epicenter.

27. Suppose the amplitude of the maximal seismic wave is doubled. By how much is the magnitude of the earthquake increased?

28. In chemistry the **hydrogen potential** (denoted pH) of a solution is defined by

$$pH = \log \frac{1}{[H^+]} = -\log [H^+]$$

where H^+ is the concentration of hydrogen ions in the solution in moles per liter. The pH of a solution ranges from 0 to 14. If the pH of a solution is less than 7, the solution is acidic and turns litmus paper red. In contrast, if the pH is greater than 7, it is alkaline and turns litmus paper blue. Distilled water has a pH of 7 and is neither acid nor alkaline. Determine the hydrogen ion concentration of pure water.

29. Find the pH of a soft drink whose hydrogen ion concentration is roughly 0.004 mole per liter. Is the drink highly acidic, or highly alkaline?

30. Orange juice has a pH of approximately 3.5, and tomato juice has a pH of approximately 4.2. What would be the ratio of hydrogen ion concentration in orange juice to that in tomato juice?

31. Wheat grows best in soil that has a pH of between 6 and 7.5, whereas oats grow best in soil that has a pH of between 5 and 6.2. Suppose the hydrogen ion concentration in the soil in a field is found to be 4.47×10^{-7} mole per liter. Is the soil better suited to wheat or to oats?

5.6 EXPONENTIAL GROWTH AND DECAY

In physical and mathematical applications the exponential function that is most often used is the exponential function with base e, that is, the function e^x. In particular, many physical quantities, such as the amount of a specific radioactive substance, depend on time according to an equation of the form

$$f(t) = Ae^{ct} \qquad (1)$$

where A and c are constants. If we substitute 0 for t in (1), we have

$$f(0) = Ae^{c \cdot 0} = Ae^0 = A \cdot 1 = A$$

so that $A = f(0)$. Thus (1) becomes

$$f(t) = f(0)e^{ct} \qquad (2)$$

Since $e^{ct} > 0$ for all t, it follows from (2) that $f(t)$ has the same sign as $f(0)$. In applications $f(t)$ is almost always positive. As a result, we will assume throughout this section that $f(t) > 0$ for all t.

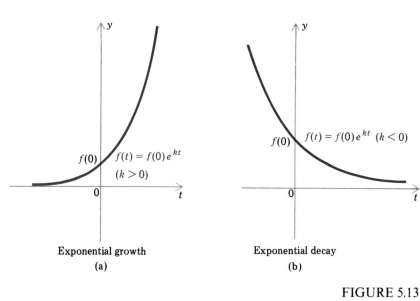

Exponential growth
(a)

Exponential decay
(b)

FIGURE 5.13

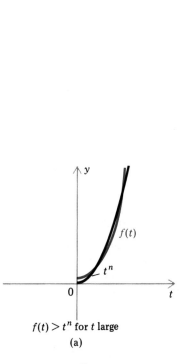

$f(t) > t^n$ for t large
(a)

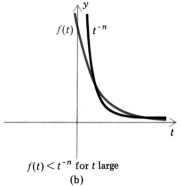

$f(t) < t^{-n}$ for t large
(b)

FIGURE 5.14

If $c > 0$, then customarily one lets $c = k$, so that

$$f(t) = f(0)e^{kt} \quad \text{with } k > 0 \tag{3}$$

In this case, $f(t)$ increases with t (Figure 5.13a), and we say that f *grows exponentially*. Analogously, if $c < 0$, then customarily one lets $c = -k$, so that

$$f(t) = f(0)e^{-kt} \quad \text{with } k > 0 \tag{4}$$

In this case, $f(t)$ decreases with t (Figure 5.13b), and we say that f *decays exponentially*. If f grows exponentially, then for any positive integer n, $f(t) > t^n$ for all sufficiently large t (Figure 5.14a), and we say that $f(t)$ grows faster than any function of the form t^n. Similarly, if f decays exponentially, then for any positive integer n, $f(t) < t^{-n}$ for all sufficiently large t (Figure 5.14b), and we say that $f(t)$ decreases faster than any function of the form t^{-n}.

Solving Exponential Equations by Natural Logarithms

In the examples of exponential growth and decay that we will consider, it will be necessary to solve equations containing expressions like e^{kt}. Since

$$\ln(e^{kt}) = kt \tag{5}$$

the use of natural logarithms will simplify the solutions. This is illustrated in the next two examples.

EXAMPLE 1.   Solve the equation $e^{5k} = 2$ for k.

Solution.   We take the natural logarithm of each side, use (5), and then solve for k:

$$\ln (e^{5k}) = \ln 2$$

$$5k = \ln 2$$

$$k = \frac{1}{5} \ln 2$$

By calculator we find that $k \approx 0.13862944$.   □

EXAMPLE 2.   Solve the equation $e^{(t/5)\ln 2} = 3$ for t.

Solution.   We take the natural logarithm of each side, use (5), and solve for t:

$$\ln (e^{(t/5)\ln 2}) = \ln 3$$

$$\frac{t}{5} \ln 2 = \ln 3$$

$$t = 5 \frac{\ln 3}{\ln 2}$$

By calculator we find that $t \approx 7.9248125$.   □

Population Growth   When the number of organisms in a group is very large, the graph of the function representing the number in the group at any given time is very nearly a smooth curve. If the group has plenty of food and space and no enemies (such as disease or predators), the curve can be essentially the graph of an exponential function of the form of (3). This means that the population of the group can grow exponentially. Statistics show that various types of bacteria, insects, and rodents, and even certain human populations, have in the past grown exponentially for awhile—until conditions such as lack of food, overcrowding, and war changed the growth pattern.

EXAMPLE 3.   Experiment shows that under optimal conditions and at a temperature of 29°C, the population of a colony of 1000 rice weevils can grow exponentially to about 2139 weevils in a week. Assuming that there are 1000 initially and that their population grows exponentially, find a formula for the number of weevils at time t.

Solution.   Let $f(t)$ denote the number of weevils at time t, where t is measured in weeks. We take $t = 0$ to represent the initial time, which means that $f(0) = 1000$, the initial number of weevils. Then (3) becomes

$$f(t) = 1000e^{kt}$$

and we need to determine the numerical value of k. By assumption there are

2139 weevils after 1 week, so that $f(1) = 2139$. Thus

$$2139 = f(1) = 1000e^{k(1)} = 1000e^k$$

or equivalently,

$$e^k = \frac{2139}{1000} = 2.139$$

Taking natural logarithms of both sides yields

$$k = \ln 2.139$$

Therefore

$$f(t) = 1000e^{(\ln 2.139)t}$$

which can be written in the alternative form

$$f(t) = 1000(2.139^t) \quad \square$$

EXAMPLE 4.   From the data given in Example 3, determine the length of time in weeks for the population of weevils to double.

   Solution.   Since the initial population of weevils is 1000, what we seek is the value of t for which

$$f(t) = 2000$$

By the solution of Example 3, this means that we wish to find the value of t for which

$$2000 = 1000e^{(\ln 2.139)t}$$

or equivalently,

$$2 = e^{(\ln 2.139)t}$$

Taking natural logarithms of both sides, we obtain

$$\ln 2 = (\ln 2.139)t$$

so that

$$t = \frac{\ln 2}{\ln 2.139}$$

By calculator we find that

$$t \approx 0.91162981$$

Consequently the population of weevils will double after approximately 0.91162981 weeks, which is just over 6 days and 9 hours.   □

The length of time it takes a population growing exponentially to double is called the ***doubling time*** of the population. Thus under optimal conditions the doubling time for rice weevils is approximately 6 days and 9 hours.

Carbon Dating   Certain isotopes, such as carbon 14 (C^{14}) and uranium 238 (U^{238}), are radioactive, which means that they are unstable and in time decay, changing into other substances by the emission of various particles from their nuclei. For example, C^{14} decays into nitrogen 14 by emitting an electron (a beta particle), and U^{238} decays into thorium 234 by emitting a helium 4 nucleus (an alpha particle).

If $f(t)$ is the amount of a given radioactive element in a substance at time t, then by experiment it is known that f satisfies (4) for a suitable constant k depending on the isotope. Notice that the amount $f(t)$ decreases as t increases, and in fact, $f(t)$ eventually approaches 0, no matter how much of the element is initially present. The ***half-life*** of a radioactive element is the length of time necessary for half of any given amount of it to decay.

EXAMPLE 5.   The half-life of C^{14} is approximately 5730 years. Using this figure, show that the amount $f(t)$ of C^{14} remaining after an elapsed time of t years is given by

$$f(t) = f(0)e^{-(1/5730)(\ln 2)t} \tag{6}$$

Solution.   We need to determine the constant k so that

$$f(t) = f(0)e^{-kt} \quad \text{for } t > 0 \tag{7}$$

By assumption, half the original C^{14} remains after 5730 years, so that

$$f(5730) = \frac{1}{2}f(0)$$

However, by substituting 5730 for t in (7), we obtain

$$f(5730) = f(0)e^{-5730k}$$

From the last two equations it follows that

$$\frac{1}{2}f(0) = f(0)e^{-5730k}$$

or equivalently,

$$\frac{1}{2} = e^{-5730k}$$

Taking natural logarithms of both sides, we obtain

$$\ln \frac{1}{2} = -5730k$$

Therefore

$$k = -\frac{1}{5730} \ln \frac{1}{2} = \left(-\frac{1}{5730} \right)(-\ln 2) = \frac{1}{5730} \ln 2$$

Consequently the formula we seek is

$$f(t) = f(0)e^{-(1/5730)(\ln 2)t} \qquad \square$$

An ancient Egyptian mummy in its wood coffin. Carbon dating is often used in determining the age of mummies.

Every living object contains C^{14}. Because of the radioactivity of C^{14}, you might think that the amount of C^{14} in living objects would be ever decreasing. But to offset the decay of C^{14} in any living object, there is an increase of C^{14} in the atmosphere due to cosmic radiation. This C^{14} enters the carbon cycle of living plants and animals, with the effect that the amount per gram of C^{14} in living organisms remains constant. However, when an organism dies, it abruptly ceases to assimilate any more C^{14}, and consequently the amount of C^{14} per gram in the organism starts decreasing. This decrease in C^{14} after death is the basis for the process known as ***carbon dating*** of once-living organisms.

EXAMPLE 6.   A mummy, known as Whiskey Lil, was discovered in 1955 in Chimney Cave, Lake Winnemucca, Nevada. By carbon dating it was learned that approximately 0.739 of the original C^{14} per gram was still present in 1955. When did Whiskey Lil die?

Solution.   We let $t = 0$ at the time of death and let t_0 be the time elapsed between the death of Whiskey Lil and 1955. First we will determine the value of t_0, and then we will calculate when she died. By hypothesis,

$$f(t_0) = 0.739f(0)$$

and by (6) with t replaced by t_0,

$$f(t_0) = f(0)e^{-(1/5730)(\ln 2)t_0}$$

Therefore

$$0.739f(0) = f(0)e^{-(1/5730)(\ln 2)t_0}$$

or equivalently,

$$0.739 = e^{-(1/5730)(\ln 2)t_0}$$

Taking logarithms of both sides, we find that

$$\ln 0.739 = -\left(\frac{1}{5730} \ln 2\right) t_0$$

Solving for t_0, we obtain

$$t_0 = -\frac{5730 \ln 0.739}{\ln 2}$$

By calculator, we conclude that

$$t_0 \approx 2500.3069$$

Therefore Whiskey Lil died about 2500 years before 1955 A.D., which means about 545 B.C.   □

Since the mid-1950s, when Willard Libby proposed dating objects by analysis of electron emissions of C^{14}, carbon dating has been intimately connected with the questions of when the Americas were settled and what kinds of tools early cultures included. Bones, skulls, artifacts— archaeological finds of all kinds—were subjected to carbon dating analysis. Until recently the best estimates pointed to the first American settlements as having taken place around 11,000 B.C., but new evidence suggests, again with the help of carbon dating, that the first settlements were perhaps as much as 16,000 years earlier. (See Exercise 25.)

Yet there are limitations to the use of carbon dating. First, the outer limits of dates that can be ascertained by carbon dating are approximately 1500 A.D. and 50,000 B.C. Second, carbon dating gives approximate, rather than precise, dates; the error can be more than 10%.

EXERCISES 5.6

In Exercises 1–10, solve the given equation for k or t, whichever is appropriate. Leave your answer in terms of natural logarithms.

1. $e^{3k} = 4$

2. $e^{-4k} = \dfrac{1}{5}$

3. $7e^{-1.2k} = 3.5$

4. $3e^{(t/2)\ln 2} = 4$

5. $1.7e^{-(t/3)\ln 2} = 0.85$

6. $0.005e^{-(t/1.5)\ln 3} = 0.0004$

7. $e^{2t} - 3e^t + 2 = 0$
(*Hint:* This is a quadratic equation in e^t.)

8. $e^{4t} - e^{2t} - 6 = 0$

9. $e^t + e^{-t} = 2$

10. $2^{kt} = e^{-t}$

11. Suppose the number of organisms of a species is given by (3). Show that the doubling time is $(\ln 2)/k$: (*Hint:* Compute $f((\ln 2)/k)$.)

12. With the data given in Example 3, determine how long it would take for the population of 1000 rice weevils to reach
a. 3000   b. 4000

13. If the temperature is lowered to 23°C, then an initial population of 1000 rice weevils could ideally grow to approximately 1537 during a week's time.
a. Determine a formula for the number of rice weevils at a future time t.
b. Determine the doubling time of the weevils.

14. If the temperature is increased to 33.5°C, then the doubling time of the rice beetle expands to approximately 5.78 weeks (about 40.5 days).
    a. Find the numerical value of k in the population formula.
    b. Determine how long it would take for a population of 217 to increase to 864.

15. Experiment has shown that under ideal conditions and a constant temperature of 28.5°C the population of a certain type of flour beetle grows according to the formula

$$f(t) = f(0)e^{0.71t}$$

where t is measured in weeks. Does it take more than a week for the population to double? Explain your answer.

16. Using the information from Exercise 15, determine the length of time it would take for a population of 5000 flour beetles to grow to 17,000.

17. It is known that a population of one type of grain beetle can grow from 1000 to 1994 in a week if the temperature is set at 32.3°C, whereas it can grow from 2500 to 4465 in a week at a temperature of 29.1°C. Determine at which temperature the doubling time is less.

18. Suppose the number of bacteria in a culture grows exponentially and increases from 1 million to 3 million in 6 hours.
    a. Find a formula for the number $f(t)$ of bacteria at time t.
    b. How long would it take for the number of bacteria to quadruple?

19. The world population in 1962 was approximately 3.15 billion, and in 1971 it was approximately 3.706 billion. Assuming that the world population was growing exponentially during the interim and continued to do so, determine what the (approximate) world population would be in 1990.

20. In 1930 the world's population was 2 billion, and in 1976 the population passed the 4 billion mark. Assuming that the population grew exponentially during the interim, determine the constant k.

21. Suppose a radioactive substance decays according to (4). Show that the half-life of the substance is $(\ln 2)/k$. (See Exercise 11.)

22. What percent of the original C^{14} content in a skull would remain now if the animal lived around
    a. 1500 A.D.?   b. 50,000 B.C.?

23. In the late 1960s a wrench made from bone was unearthed in southeast Arizona. The date of the bone was calculated to be approximately 11,200 B.C. What percentage of the original C^{14} remained?

24. About 1960, awls used in basketmaking and formed from antlers were found in an archaeological site in western Tennessee. At that time approximately 32.86% of the original C^{14} remained. How old was the antler?

25. During the beginning of the 1970s a toothed bone scraper found in northern Yukon, Canada, gave new evidence of human settlements in North America more ancient than had been supposed. The C^{14} content remaining in the scraper was calculated to be approximately 3% of the original C^{14}. What would the maximum age of the scraper have been?

26. Some trees that were destroyed during the Fourth Ice Age now contain only 27% of the amount of C^{14} per gram that living trees of the same type now contain. When did the Fourth Ice Age occur?

27. The recently discovered Dead Sea Scrolls were determined by carbon dating to have been written about 33 A.D. What percentage of the original amount of C^{14} presently remains in the Scrolls?

28. Suppose you are told that a bone found in an archaelogical site is more than 17,000 years old. When you submit the bone to carbon dating analysis, you determine that approximately 20% of the original C^{14} remains. Are your findings compatible with the claim of the bone's age? Explain why or why not.

29. Determine the percentage of C^{14} in a skull that decays during a century. (Your answer should show that the same percentage of C^{14} is lost, no matter what century is chosen.)

30. The half-life of strontium 90 (Sr^{90}) is 19.9 years. How long is required for a given amount of Sr^{90} to decay to
    a. 25% of its original amount?
    b. 20% of its original amount?

31. The half-life of uranium 238 (U^{238}) is approximately 4.5 billion years. How long is required for 1% of a given quantity of U^{238} to decay?

32. Suppose the half-life of a radioactive isotope is h years. Show that the amount $f(t)$ of the isotope after t years is given by

$$f(t) = f(0)2^{-t/h}$$

33. Halley's Law states that the barometric pressure $p(x)$ in inches of mercury at x miles above sea level is given by

$$p(x) = 29.92e^{-0.2x} \quad \text{for} \quad x \geq 0$$

Find the barometric pressure
a. at sea level
b. 1 mile above sea level
c. atop Mt. Whitney (14,495 feet above sea level)

34. The Bouguer–Lambert Law states that as a light beam passes through water, the intensity of the beam x meters from the surface of the water is given by

$$f(x) = ce^{-1.4x}$$

where c is the intensity at the surface. Determine the ratio of the intensity 10 meters below the surface to the intensity at the surface—that is, find $f(10)/c$. (From your answer it will be apparent why most flora cannot survive at depths greater than 10 meters.)

35. Let $A(t)$ denote the area in square centimeters of the unhealed portion of a wound t days after it is sustained, and let $A(0)$ represent the area of the original wound. Then according to the Law of Healing, $A(t)$ is given by

$$A(t) = A(0)e^{-0.15t} \quad \text{for} \quad t \geq 0$$

Suppose that just after a collision, a basketball player is told he cannot play until 90% of a leg wound is healed. Would he be able to play in a tournament in two weeks?

KEY TERMS

exponential function	logarithmic function
base	base
exponent	natural logarithm
natural exponential function	common logarithm
exponential growth	characteristic
doubling time	mantissa
exponential decay	antilogarithm
half-life	linear interpolation

KEY FORMULAS

Laws of Exponents

$$a^x a^y = a^{x+y}$$

$$(a^x)^y = a^{xy}$$

$$(ab)^x = a^x b^x$$

$$\left(\frac{a}{b}\right)^x = \frac{a^x}{b^x}$$

$$\frac{a^x}{a^y} = a^{x-y}$$

$$a^{-x} = \frac{1}{a^x} = \left(\frac{1}{a}\right)^x$$

$$a^0 = 1$$

$$1^x = 1$$

Laws of Logarithms

$$a^{\log_a x} = x$$

$$\log_a (a^x) = x$$

$$\log_a (xy) = \log_a x + \log_a y$$

$$\log_a \frac{1}{x} = -\log_a x$$

$$\log_a \frac{x}{y} = \log_a x - \log_a y$$

$$\log_a (x^c) = c \log_a x$$

$$\log_b x = \frac{\log_a x}{\log_a b}$$

$$\log_a 1 = 0$$

$$\log_a a = 1$$

REVIEW EXERCISES

In Exercises 1–4, sketch the graph of the given function.

1. $f(x) = 3^x$

2. $f(x) = (\frac{1}{2})^{x-1}$

3. $f(x) = \frac{1}{2} \log_2 x$

4. $f(x) = |\log_2 x|$

In Exercises 5–8, simplify the given expression.

5. $(\sqrt{2^{\sqrt{2}}})^{\sqrt{2}}$

6. $\dfrac{5^{1-\sqrt{2}} \cdot 25^{1/\sqrt{2}}}{5^{2+\sqrt{2}}}$

7. $\ln 12 - \ln 2 + 2 \ln e^3$

8. $\log_{10} 5 - \log_{10} \dfrac{50}{7}$

In Exercises 9–14, solve the given equation.

9. $7^{x-1} = (\sqrt{7})^{-2x^2}$

10. $2^{x^3} = 4^x$

11. $\log_2 x = -3$

12. $\log_x 5 = 2$

13. $\log_{10} (3 - 5x) - \log_{10} (1 - x) = 1$

14. $3 \log_2 \sqrt{x} = 2 \log_2 8$

In Exercises 15–20, solve the given equation. Leave your answer in terms of logarithms.

15. $2^{-x} = 7$

16. $4^{x-4} = 5^{3x+5}$

17. $2^x = 3^{(x^4)}$

18. $e^{2x} = \dfrac{1}{e^{(x^2)}}$

19. $e^{5k} = 4.37$

20. $e^{-3t \ln 2} = 0.37$

In Exercises 21–24, evaluate the given expression.

21. $\log_2 \frac{1}{16}$

22. $\log_{10} (10^{-5})$

23. $e^{\ln 5}$

24. $10^{\log_{10} 1.23}$

In Exercises 25–28, approximate the numerical value of the given logarithms. Use the approximations $\log_{10} 6 \approx 0.7782$, $\log_{10} 7 \approx 0.8451$, and $\log_{10} 11 \approx 1.0414$.

25. $\log_{10} 420$

26. $\log_{10} \frac{77}{36}$

27. $\log_{10} 0.0077$

28. $\log_{10} (6^{-1} \times 11^{10})$

In Exercises 29–32, use Table C to approximate the common logarithm of the given number. Use linear interpolation if needed.

29. 4.13

30. 0.00956

31. 6.384×10^8

32. 0.1977

In Exercises 33–36, use Table C to approximate the antilogarithm of the given number. Use linear interpolation if needed.

33. 0.5514

34. -3.7372

35. 0.6752

36. -1.1300

37. Given that $\log_{10} 8 \approx 0.9031$ and $\log_{10} e \approx 0.4343$, approximate $\ln 8$.

38. Given that $\ln 5 \approx 1.6094$ and $\ln 2 \approx 0.6931$, approximate $\log_2 5$.

39. Given that $\ln 10 \approx 2.3026$, approximate $\log_{10} e$.

40. Solve the equation $y = \ln (x - \sqrt{x^2 - 1})$ for x.

41. Suppose $y = a + b \ln x$, where a and b are constants. Express x in terms of y.

42. Show that $\log_2 x = 2 \log_4 x$.

43. Show that $\log_2 x = 3 \log_8 x$.

44. Let $f(x) = e^{cx}$, where c is a constant. Suppose that the graph of f passes throught the point $(\frac{1}{3}, e^2)$. Determine c.

45. Let $g(x) = ae^{cx}$, where a and c are constants. Suppose that the graph of g passes through the points $(0, \frac{1}{5})$ and $(-1, \frac{1}{5}e^2)$. Determine a and c.

46. The noise level of one sound is 20 decibels higher than that of a second sound. Find the ratio of the intensity of the first sound to that of the second.

47. If an earthquake measures 6.1 on the Richter scale, what will one that is 100 times as intense measure?

48. It has been shown experimentally that under optimal conditions, the population of a colony of brown rats could grow exponentially from 50 rats to approximately 11,070 during the course of a year.
   a. Determine the doubling time.
   b. How long would it take for the population to increase from 120 to 340 rats?

49. The total amount of timber in a young forest increases exponentially.
   a. If the annual growth rate k is 3%, how long does it take for the total amount of timber to double?
   b. How much sooner would the total amount of timber double if the rate of growth were 4%?

50. The radioactive isotope potassium 40 (K^{40}) has a half-life of approximately 91.3 billion years. If a rock formed 4 billion years ago (shortly after the earth came into existence) initially contained 1 gram of K^{40}, how much of it would the rock contain today?

51. Radioactive argon 39 (Ar^{39}) has a half-life of 4 minutes. How long would it take for 99% of a given amount of Ar^{39} to decay?

6

The Trigonometric Functions

Next we turn to trigonometry and trigonometric functions. Trigonometry is related to the study of triangles, which were studied long ago by the Babylonians and ancient Greeks. In fact, the word trigonometry is derived from the Greek word for "the measurement of triangles." Today trigonometry and trigonometric functions are indispensable tools not only in mathematics, but also in many practical applications, especially those involving oscillations and rotations.

As the first of three chapters on trigonometry, Chapter 6 is devoted mainly to defining the trigonometric functions and studying their properties, with a special emphasis on their graphs. The last section of the chapter involves inverses of trigonometric functions.

Of the many ways the concepts in this chapter can be applied, we will discuss two in detail:

> In the first application, one can study the lengths of the sides and measures of the angles in an appropriate triangle in order to determine the height of the tallest known tree in the world, a mammoth redwood tree located in Humboldt Redwood State Park, California (see Example 1 of Section 6.4). For a second application, trigonometric functions discussed in this chapter can help in understanding sound waves. For example, they can be used to analyze the vibrations of the strings on a violin or guitar (see Section 6.8).

6.1 ANGLE MEASUREMENT

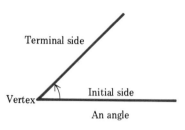

Before we can define the trigonometric functions, we must discuss angles and methods of measuring them. An *angle* in the plane is formed by two lines or line segments, called the *sides* of the angle, that have a common endpoint, called the *vertex* of the angle (Figure 6.1). We call one side the *initial side* of the angle and the other the *terminal side*, and we identify the direction that indicates rotation from the initial side to the terminal side. In a diagram this is indicated by a curved arrow from the initial side to the terminal side (Figure 6.1).

In this chapter angles will generally be located in a plane equipped with a Cartesian coordinate system. Normally the vertex of such an angle will be the origin and the initial side will lie along the positive x axis (Figure 6.2). In this case we say that the angle is in *standard position* in the coordinate system.

There are two common units in which we measure angles: degrees and radians. We will employ both of these units. First we will study degrees, then radians.

FIGURE 6.1

Degrees

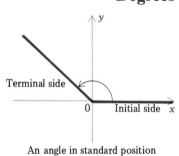

One unit for measuring angles is the *degree*, chosen in such a way that an angle formed by one complete revolution in the counterclockwise direction contains 360 degrees (written 360°), as Figure 6.3a indicates. Sometimes an angle of 360° is called a *round* angle. An angle of 180°, called a *straight* angle, and an angle of 90°, called a *right* angle, are shown in Figure 6.3b, c. If the angle is formed by rotating the initial side counterclockwise, then the angle measurement is *positive*; if the angle is formed by rotating in the clockwise direction, then the angle measurement is *negative*. Figure 6.3d depicts angles of −90° and −180°. We sometimes say that an angle is positive or negative if its degree measurement is positive or negative, respectively.

An angle in standard position

FIGURE 6.2

EXAMPLE 1. Draw angles of 45°, −30°, and 135°.

Solution. Since 45 is one-half of 90, we draw an angle of 45° by mentally bisecting the 90° angle shown in Figure 6.3c. The resulting angle is

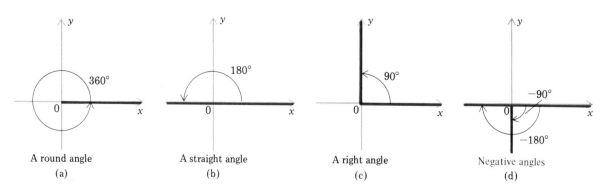

A round angle (a)   A straight angle (b)   A right angle (c)   Negative angles (d)

FIGURE 6.3

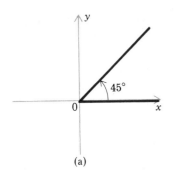

45°

(a)

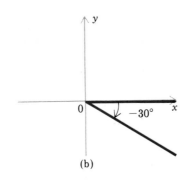

−30°

(b)

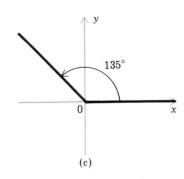

135°

(c)

FIGURE 6.4

shown in Figure 6.4a. We obtain a $-30°$ angle in a similar way, by mentally trisecting the $-90°$ angle in Figure 6.3d. The resulting angle is shown in Figure 6.4b. To obtain an angle of $135°$ we observe that $135 = 90 + 45$, and then we mentally combine the angles of $90°$ and $45°$ in Figures 6.3c and 6.4a. The resulting angle is shown in Figure 6.4c.   □

Since a $45°$ angle is formed by bisecting a $90°$ angle, it follows from the congruence of the triangles in Figure 6.5a that the terminal side of a $45°$ angle in standard position lies along the line $y = x$. Similarly, the terminal side of a $135°$ angle bisects the second quadrant, so it lies along the line $y = -x$ (Figure 6.5b).

Because angle measurement is positive or negative according to whether the rotation of the initial side is counterclockwise or clockwise, it follows that if we have drawn an angle in standard position, then we can draw the negative of the angle by rotating in the opposite direction (clockwise rather than counterclockwise, or counterclockwise rather than clockwise). In Figure 6.4a, b we have drawn angles of $45°$ and $-30°$, respectively. By rotating in the opposite direction, we obtain the angles of $-45°$ and $30°$ drawn in Figure 6.6a, b.

Now we turn to larger angles.

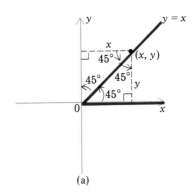

(a)

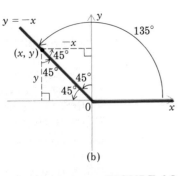

(b)

FIGURE 6.5

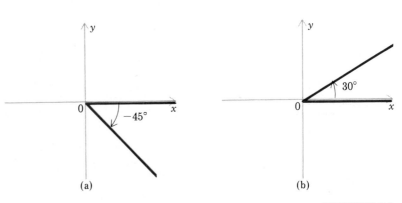

(a)

(b)

FIGURE 6.6

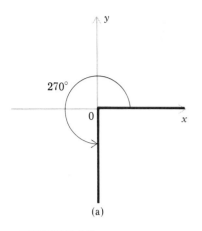

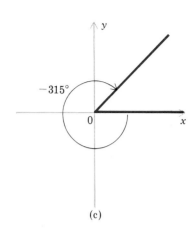

FIGURE 6.7

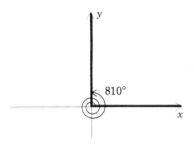

FIGURE 6.8

EXAMPLE 2.   Draw angles of 270° and −315°.

Solution.   Since 270 = 3 · 90, we mentally combine three angles of 90° to form the 270° angle in Figure 6.7a. Next we observe that −315 = −270 − 45 = −270 + (−45). We already have drawn a 270° angle, so our discussion above implies that an angle of −270° is as in Figure 6.7b. Combining this angle with the angle of −45° given in Figure 6.6a yields the −315° angle in Figure 6.7c.   □

Before we present the next example, let us recall that a 360° angle corresponds to one complete revolution, so the initial and terminal sides of the angle coincide. Moreover, the sides of a 360° angle lie along the positive x axis if the angle is in standard position. This observation will help us draw angles of more than 360°.

EXAMPLE 3.   Draw angles of 810° and −390°.

Solution.   Since 810 = 2 · 360 + 90, an angle of 810° is formed by rotating the positive x axis counterclockwise through two complete revolutions and then continuing the rotation through an additional 90° (Figure 6.8). Since −390 = −360 − 30, an angle of −390° is formed by rotating the positive x axis clockwise through one revolution and then continuing the rotation through an additional 30° (Figure 6.9).   □

From Figures 6.3c and 6.8 it is apparent that angles in standard position of 90° and 810° have the same terminal sides. Such angles are called ***coterminal angles***. The only difference between two coterminal angles is that one of them can be formed by rotating the initial side through more complete revolutions than the other. Since each revolution corresponds to 360°, it follows that the degree measures of two coterminal angles differ by an integral multiple of 360°. For example, the coterminal angles of 90° and 810° have a difference of 720° (which is twice 360°), whereas the coterminal angles of −30° and −390° (see Figures 6.4b and 6.9) have a difference of 360°.

FIGURE 6.9

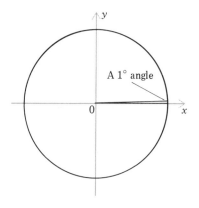

FIGURE 6.10

As one can see from Figure 6.10, an angle of $1°$ is essentially imperceptible in a small circle. However, suppose a circle is as large as the silhouette of the earth, some 24,881 miles in circumference. In this case, $1°$ subtends an arc of $\frac{24,881}{360}$ (approximately 69) miles, or about the length of Death Valley in California. In light of this it might not be surprising that degrees have been subdivided. Indeed, each degree is divided into 60 minutes, and each minute is divided into 60 seconds. A minute is denoted by the symbol $'$, and a second by the symbol $''$. Thus an angle of 34 degrees, 23 minutes, and 59 seconds is written $34°23'59''$. For accurate location of stars and satellites in the sky, astronomers use not only degrees but also minutes and seconds. Longitude and latitude of points on earth are also described in terms of degrees and minutes (and seconds if greater precision is required).

Radians

The measure of angles most frequently used in mathematics is the radian, a unit much larger than the degree. In order to set the stage for the definition of radian, we first draw the circle $x^2 + y^2 = 1$ of radius 1, called the **unit circle**. Any positive angle in standard position intercepts an arc on the unit circle, and the length of the arc intercepted is proportional to the size of the angle. The radian is chosen so that the **radian measure** of a positive angle equals the length of the arc it intercepts on the unit circle (Figure 6.11).

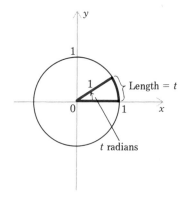

FIGURE 6.11

Since a straight angle intercepts half of the unit circle (Figure 6.12), the radian measure of a straight angle is half the circumference of the unit circle, or $\frac{1}{2}(2\pi \cdot \text{radius}) = \frac{1}{2}(2\pi \cdot 1) = \pi$. Thus a straight angle contains π radians. Since such as angle also contains $180°$, we have

$$180° = \pi \text{ radians}$$

From this it follows that for any number θ,

$$\theta° = \left(\frac{\theta}{180} \cdot 180\right)° = \left(\frac{\theta}{180} \cdot \pi\right) \text{ radians} = \left(\frac{\pi}{180} \cdot \theta\right) \text{ radians}$$

where θ (the Greek letter *theta*) represents the degree measure of an angle. Likewise,

$$t \text{ radians} = \left(\frac{t}{\pi} \cdot \pi\right) \text{ radians} = \left(\frac{t}{\pi} \cdot 180\right)° = \left(\frac{180}{\pi} \cdot t\right)°$$

where t represents the radian measure of an angle. The preceding two sets of equations yield the formulas

$$\theta° = \left(\frac{\pi}{180} \cdot \theta\right) \text{ radians} \tag{1}$$

and

$$t \text{ radians} = \left(\frac{180}{\pi} \cdot t\right)° \tag{2}$$

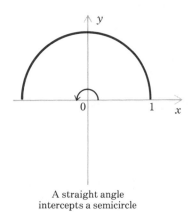

A straight angle
intercepts a semicircle

FIGURE 6.12

In particular,

$$1 \text{ radian} = \left(\frac{180}{\pi}\right)^\circ \approx 57°17'45''$$

Formulas (1) and (2) provide the conversion from degrees to radians and from radians to degrees.

EXAMPLE 4.   Convert the following to radian measure.
  a. 30°   b. 60°   c. 291°

Solution.   In each case we use (1):

a. $30° = \left(\dfrac{\pi}{180} \cdot 30\right) \text{ radians} = \dfrac{\pi}{6} \text{ radians}$

b. $60° = \left(\dfrac{\pi}{180} \cdot 60\right) \text{ radians} = \dfrac{\pi}{3} \text{ radians}$

c. $291° = \left(\dfrac{\pi}{180} \cdot 291\right) \text{ radians} = \dfrac{97\pi}{60} \text{ radians}$ □

EXAMPLE 5.   Convert the following to degree measure.

a. $\dfrac{\pi}{4}$ radians   b. $\dfrac{\pi}{12}$ radians   c. $\dfrac{7\pi}{10}$ radians

Solution.   In each case we use (2):

a. $\dfrac{\pi}{4} \text{ radians} = \left(\dfrac{180}{\pi} \cdot \dfrac{\pi}{4}\right)^\circ = 45°$

b. $\dfrac{\pi}{12} \text{ radians} = \left(\dfrac{180}{\pi} \cdot \dfrac{\pi}{12}\right)^\circ = 15°$

c. $\dfrac{7\pi}{10} \text{ radians} = \left(\dfrac{180}{\pi} \cdot \dfrac{7\pi}{10}\right)^\circ = 126°$ □

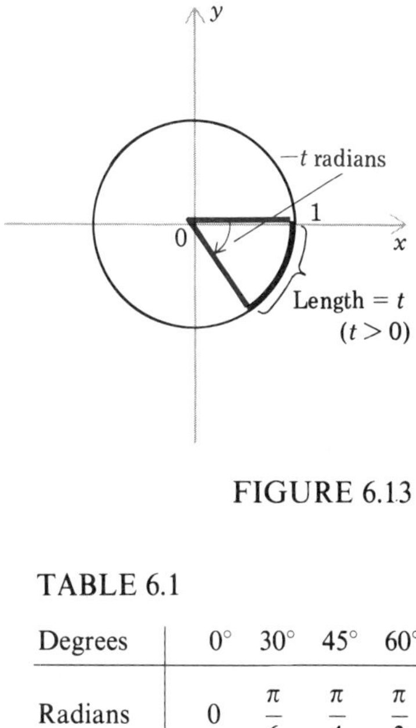

FIGURE 6.13   Table 6.1 gives the conversion from degrees to radians for several angles that occur frequently.

TABLE 6.1

Degrees	0°	30°	45°	60°	90°	120°	135°	150°	180°	210°	225°	240°	270°	300°	315°	330°	360°
Radians	0	$\dfrac{\pi}{6}$	$\dfrac{\pi}{4}$	$\dfrac{\pi}{3}$	$\dfrac{\pi}{2}$	$\dfrac{2\pi}{3}$	$\dfrac{3\pi}{4}$	$\dfrac{5\pi}{6}$	π	$\dfrac{7\pi}{6}$	$\dfrac{5\pi}{4}$	$\dfrac{4\pi}{3}$	$\dfrac{3\pi}{2}$	$\dfrac{5\pi}{3}$	$\dfrac{7\pi}{4}$	$\dfrac{11\pi}{6}$	2π
					Right angle				Straight angle								Round angle

If an angle in standard position has negative measure in degrees, then the radian measure is also negative; it is the negative of the length of the arc subtended on the circle (Figure 6.13).

EXAMPLE 6.   Convert −1110° to radian measure.

Solution.   By the preceding remark, along with (1), we have

$$-1110° = -\left(\frac{\pi}{180}\right)(1110) \text{ radians} = -\frac{37\pi}{6} \text{ radians} \quad \square$$

Radian Measure for Circles of Arbitrary Radius

If $t > 0$, then an angle of t radians subtends an arc of length t on the unit circle. If the unit circle is replaced by the circle $x^2 + y^2 = r^2$, with an arbitrary radius $r > 0$, then t can be related to the length of the arc on the new circle subtended by an angle of t radians. Since the circumference of the new circle $x^2 + y^2 = r^2$ is $2\pi r$, it follows that the length s of the arc subtended by an angle of t radians is given by the formula

$$s = \left(\frac{t}{2\pi}\right)(2\pi r) = rt$$

or as the formula often appears (with θ substituted for t),

$$s = r\theta \tag{3}$$

In other words, the length of the circular arc subtended by an angle whose vertex is the center of the circle is the product of the radius and the radian measure of the angle. If $\theta > 2\pi$, then the arc subtended is to be interpreted as consisting of more than one complete circle, and the value of s will be greater than the circumference of the circle. For example, if $r = 3$ and $\theta = 5\pi/2$, then

$$s = 3\left(\frac{5\pi}{2}\right) = \frac{15\pi}{2}$$

EXAMPLE 7.   Find the length s of the arc on a circle of radius 6 subtended by the central angle of the given radian measure.

a. $\dfrac{\pi}{6}$ b. $\dfrac{7\pi}{4}$ c. $\dfrac{43\pi}{3}$

Solution.   Each part follows from (3).

a. $s = 6\left(\dfrac{\pi}{6}\right) = \pi$

b. $s = 6\left(\dfrac{7\pi}{4}\right) = \dfrac{21\pi}{2}$

c. $s = 6\left(\dfrac{43\pi}{3}\right) = 86\pi \quad \square$

Now that we have discussed angle measurement and have introduced radians, we are prepared to define the trigonometric functions. We will take up that topic in the next section.

EXERCISES 6.1

In Exercises 1–16, draw the angle in standard position having the given degree measure.

1. $60°$

2. $120°$

3. $150°$

4. $210°$

5. $225°$

6. $300°$

7. $-75°$

8. $-120°$

9. $-150°$

10. $-210°$

11. $-240°$

12. $-330°$

13. $510°$

14. $750°$

15. $-450°$

16. $855°$

In Exercises 17–32, draw the angle in standard position with the given radian measure.

17. $\dfrac{\pi}{6}$

18. $\dfrac{3\pi}{4}$

19. $\dfrac{7\pi}{6}$

20. $\dfrac{5\pi}{3}$

21. 2π

22. $\dfrac{7\pi}{3}$

23. $\dfrac{9\pi}{2}$

24. $-\dfrac{\pi}{4}$

25. $-\dfrac{\pi}{3}$

26. $-\dfrac{\pi}{2}$

27. $-\dfrac{5\pi}{6}$

28. $-\dfrac{3\pi}{2}$

29. $-\dfrac{7\pi}{4}$

30. $-\dfrac{17\pi}{6}$

31. $-\dfrac{11\pi}{3}$

32. -3π

In Exercises 33–46, convert from degrees to radians.

33. $0°$

34. $15°$

35. $-45°$

36. $-210°$

37. $390°$

38. $450°$

39. $-180°$

40. $-300°$

41. $1470°$

42. $20°$

43. $-72°$

44. $1°$

45. $4°30'$

46. $15°15'$

In Exercises 47–58, convert from radians to degrees.

47. $\dfrac{9\pi}{4}$ radians

48. $\dfrac{7\pi}{3}$ radians

49. 6π radians

50. $\dfrac{17\pi}{2}$ radians

51. 11π radians

52. $-\dfrac{2\pi}{3}$ radians

53. $-\dfrac{13\pi}{4}$ radians

54. $-\dfrac{16\pi}{3}$ radians

55. $\dfrac{7\pi}{18}$ radians

56. $\dfrac{\pi}{90}$ radians

57. 1.8 radians

58. -0.6 radians

In Exercises 59–68, determine the quadrant in which the terminal side of the angle in standard position with the given measure is located.

59. $\dfrac{4\pi}{7}$ radians

60. $\dfrac{29\pi}{5}$ radians

61. 3.15π radians

62. $-\dfrac{19\pi}{5}$ radians

63. $-\dfrac{37\pi}{6}$ radians

64. -4.57π radians

65. 2 radians

66. $721°$

67. $-809°$

68. $-1261°$

69. Identify those angles that the coterminal with an angle of $2\pi/3$ radians and those coterminal with an angle of $4\pi/3$ radians. Assume all angles are in standard position.

a. $\dfrac{16\pi}{3}$

b. $\dfrac{20\pi}{3}$

c. $-\dfrac{4\pi}{3}$

d. $-\dfrac{8\pi}{3}$

e. $-\dfrac{20\pi}{3}$

f. $-\dfrac{26\pi}{3}$

g. $240°$

h. $-240°$

i. $840°$

j. $-840°$

70. Find the length of the arc on a circle of radius 8 subtended by a central angle of the given radian measure.

a. $\dfrac{4\pi}{3}$

b. $\dfrac{17\pi}{6}$

c. $\dfrac{2\pi}{5}$

71. Find the length of the arc on a circle of the given radius subtended by a central angle of $\pi/8$ radians.

a. $r = 1$

b. $r = \frac{1}{3}$

c. $r = 4$

72. Find the radius of the circle on which a central angle of the given radian measure subtends an arc of length 5.

a. $\dfrac{2\pi}{3}$ b. $\dfrac{5\pi}{4}$ c. $\dfrac{13\pi}{6}$

73. An eighth of a right angle, sometimes called a *point*, has seen use in navigation. Give the degree and the radian measures of a point.

74. Determine the number of radians in the angle formed by the hands of a clock at the following times.
a. 7 A.M.   b. 4 P.M.
c. 1:30 P.M.   d. 11:15 A.M.

75. The earth rotates 360° in a 24-hour period. Determine through how many radians it rotates in
a. 1 hour   b. 1 minute   c. 1 second

76. By definition, one *nautical mile* is equal to the length of arc an angle of 1′ subtends on a great circle of the earth.
a. Assuming that the radius of the earth is 3960 miles, approximate the length of one nautical mile in ordinary (that is, *statute*) miles.
b. The distance between San Francisco and Honolulu is 2091 nautical miles. Approximate the number of statute miles separating the two cities.

77. An airplane is flying 900 feet directly above the summit of Pike's Peak. If the angle subtended by the airplane appears to an observer on the summit to be 5° (Figure 6.14), use (3) to approximate the length of the airplane.

FIGURE 6.14

78. Miami, Florida, and Pittsburgh, Pennsylvania, have approximately the same longitudes. Using the latitudes of Miami as 25°46′37″ and of Pittsburgh as 42°26′19″ and taking the radius of the earth to be 3960 miles, determine the approximate distance between Miami and Pittsburgh.

6.2
TRIGONOMETRIC FUNCTIONS OF ACUTE ANGLES

Having discussed angles measured in degrees and radians in Section 6.1, we are ready to define six basic trigonometric functions. For the present, when we write θ for an angle, it will not matter whether the angle is measured in degrees or in radians.

Let θ be an acute angle, as in Figure 6.15a, and form an associated right triangle as in Figure 6.15b, with legs of length a and b, and hypotenuse of length c. Using the customary abbreviations, we define the functions *sine*, *cosine*, *tangent*, *cotangent*, *secant*, and *cosecant* at θ as follows:

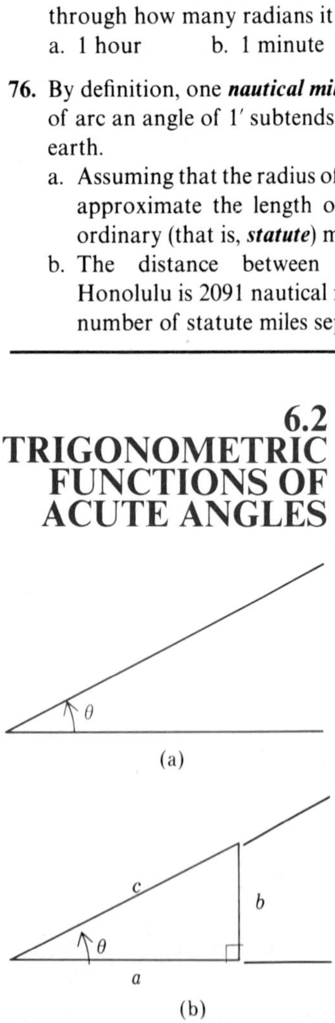

(a)

(b)

FIGURE 6.15

$$\sin \theta = \frac{b}{c} \quad \tan \theta = \frac{b}{a} \quad \sec \theta = \frac{c}{a}$$

$$\cos \theta = \frac{a}{c} \quad \cot \theta = \frac{a}{b} \quad \csc \theta = \frac{c}{b}$$

(1)

The angle θ appearing in the formulas above can lie in many right triangles (see Figure 6.16a). However, all such triangles are similar, so the ratios of the lengths of any two corresponding sides are identical. Thus for any two triangles as in Figure 6.16b, we have

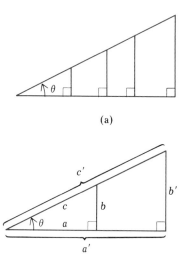

(a)

(b)

FIGURE 6.16

$$\frac{b}{c} = \frac{b'}{c'}$$

Consequently $\sin\theta$, $\cos\theta$, $\tan\theta$, $\cot\theta$, $\sec\theta$, and $\csc\theta$ do not depend on the lengths of the sides of the right triangle, but *only* on the angle θ. Collectively the six functions defined for acute angles are referred to as the **trigonometric functions**. Observe that each of the trigonometric values in (1) is nonnegative.

The following formulas are suggested by Figure 6.17 and should aid in calculating values of the trigonometric functions:

$$\sin\theta = \frac{\text{length of opposite side}}{\text{length of hypotenuse}}$$

$$\cos\theta = \frac{\text{length of adjacent side}}{\text{length of hypotenuse}}$$

$$\tan\theta = \frac{\text{length of opposite side}}{\text{length of adjacent side}}$$

$$\cot\theta = \frac{\text{length of adjacent side}}{\text{length of opposite side}}$$

$$\sec\theta = \frac{\text{length of hypotenuse}}{\text{length of adjacent side}}$$

$$\csc\theta = \frac{\text{length of hypotenuse}}{\text{length of opposite side}}$$

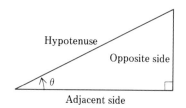

FIGURE 6.17

Special Values of the Trigonometric Functions

Next we find some values of the trigonometric functions for specific values of θ. In so doing we will repeatedly refer to the following results from geometry (see Figure 6.18):

$$a^2 + b^2 = c^2 \qquad (2)$$

$$\theta + \phi = 90°$$

The first of these formulas is the Pythagorean Theorem; the second can be proved by recalling that the sum of the angles of any triangle is 180° and that a right triangle has a 90° angle.

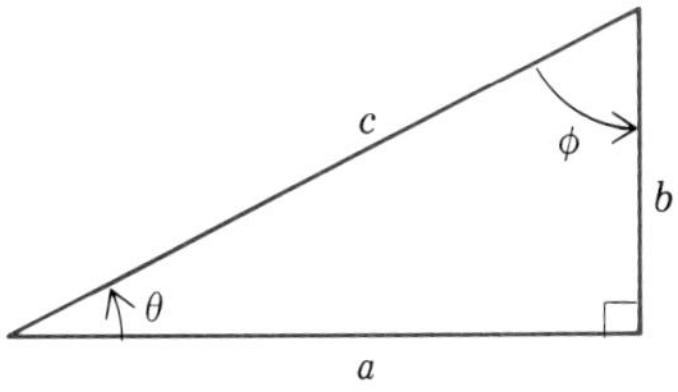

First we will compute the values of the trigonometric functions for $\theta = 45°$, which means that the other acute angle in the triangle is also 45°, so the triangle is isosceles.

FIGURE 6.18

EXAMPLE 1. Find $\sin 45°$.

Solution. Since the value of $\sin 45°$ is independent of the lengths of sides of the triangle, assume for simplicity that the legs each have length 1

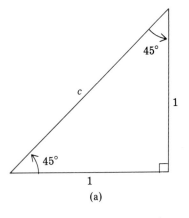

(a)

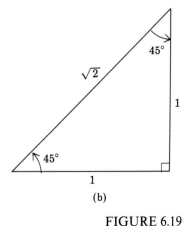

(b)

FIGURE 6.19

(Figure 6.19a). Then by (2) we find that

$$c^2 = 1^2 + 1^2 = 2$$

so that $c = \sqrt{2}$. It then follows from the definition of the sine function and the lengths shown in Figure 6.19b that $\sin 45° = 1/\sqrt{2} = \frac{1}{2}\sqrt{2}$. □

The values of the other trigonometric functions at $45°$ can also be found by using the defining formulas in (1), along with the lengths of the sides in Figure 6.19b. All these values are listed below:

$$\sin 45° = \frac{1}{2}\sqrt{2} \qquad \tan 45° = 1 \qquad \sec 45° = \sqrt{2}$$

$$\cos 45° = \frac{1}{2}\sqrt{2} \qquad \cot 45° = 1 \qquad \csc 45° = \sqrt{2}$$

Our next goal is to determine $\sin 60°$.

EXAMPLE 2.   Find $\sin 60°$.

Solution.   First we draw the triangle in Figure 6.20a, with hypotenuse of length 2. Next we place a congruent triangle beside it (Figure 6.20b).

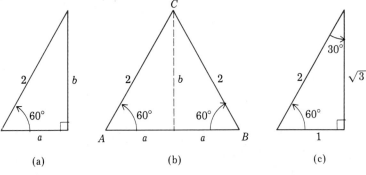

(a)   (b)   (c)

FIGURE 6.20

Because two right triangles are placed side by side, the resulting figure ABC is in fact a triangle. Moreover, since it has three angles of $60°$, it is equilateral, each side having length 2. Therefore $2a = 2$, and $a = 1$. Consequently by the Pythagorean Theorem,

$$b^2 = 2^2 - a^2 = 4 - 1 = 3$$

so that $b = \sqrt{3}$. Thus we have the triangle shown in Figure 6.20c, so the definition of $\sin \theta$ yields

$$\sin 60° = \frac{\sqrt{3}}{2} = \frac{1}{2}\sqrt{3} \quad □$$

Using the triangle in Figure 6.20c we also find that

$$\sin 30° = \frac{1}{2}$$

In fact, from Figure 6.20c we can assemble Table 6.2a, which gives the values for all trigonometric functions at $\theta = 30°$ and $\theta = 60°$.

TABLE 6.2a

θ	$\sin \theta$	$\cos \theta$	$\tan \theta$	$\cot \theta$	$\sec \theta$	$\csc \theta$
30°	$\frac{1}{2}$	$\frac{1}{2}\sqrt{3}$	$\frac{1}{3}\sqrt{3}$	$\sqrt{3}$	$\frac{2}{3}\sqrt{3}$	2
60°	$\frac{1}{2}\sqrt{3}$	$\frac{1}{2}$	$\sqrt{3}$	$\frac{1}{3}\sqrt{3}$	2	$\frac{2}{3}\sqrt{3}$

Angles of 30°, 45°, and 60° contain $\pi/6$, $\pi/4$, and $\pi/3$ radians, respectively. Therefore we can assemble Table 6.2b for the values of the trigonometric functions for these angles measured in radians.

TABLE 6.2b

θ	$\sin \theta$	$\cos \theta$	$\tan \theta$	$\cot \theta$	$\sec \theta$	$\csc \theta$
$\frac{\pi}{6}$	$\frac{1}{2}$	$\frac{1}{2}\sqrt{3}$	$\frac{1}{3}\sqrt{3}$	$\sqrt{3}$	$\frac{2}{3}\sqrt{3}$	2
$\frac{\pi}{4}$	$\frac{1}{2}\sqrt{2}$	$\frac{1}{2}\sqrt{2}$	1	1	$\sqrt{2}$	$\sqrt{2}$
$\frac{\pi}{3}$	$\frac{1}{2}\sqrt{3}$	$\frac{1}{2}$	$\sqrt{3}$	$\frac{1}{3}\sqrt{3}$	2	$\frac{2}{3}\sqrt{3}$

In general, if we know the lengths of any two sides of the right triangle in Figure 6.18, then by using the Pythagorean Theorem we can determine the length of the third side and consequently the values of the six trigonometric functions—even if we do not know the value of θ. In our next example we show how to determine them.

EXAMPLE 3.   Suppose θ is an acute angle and $\tan \theta = \frac{2}{3}$. Find the values of the other five trigonometric functions at θ.

Solution.   Since

$$\tan \theta = \frac{\text{length of opposite side}}{\text{length of adjacent side}} = \frac{2}{3}$$

we obtain the triangle in Figure 6.21a, with the lengths of the two legs as indicated. The Pythagorean Theorem tells us that $c^2 = 2^2 + 3^2 = 13$, so

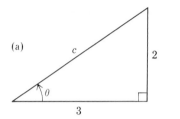

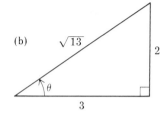

FIGURE 6.21

$c = \sqrt{13}$. Thus the triangle is as in Figure 6.21b, and we can compute the required values:

$$\sin \theta = \frac{2}{\sqrt{13}} = \frac{2}{13}\sqrt{13} \qquad \cos \theta = \frac{3}{\sqrt{13}} = \frac{3}{13}\sqrt{3}$$

$$\cot \theta = \frac{3}{2} \qquad \sec \theta = \frac{\sqrt{13}}{3} \qquad \csc \theta = \frac{\sqrt{13}}{2} \quad \square$$

In the solution of Example 3 we could have selected another size triangle; the only stipulation is that the *ratio* of the lengths of the opposite and adjacent sides has to be $\frac{2}{3}$.

Relations among the Trigonometric Functions

Of the many relationships among the trigonometric functions, we note the following:

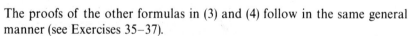

$$\tan \theta = \frac{\sin \theta}{\cos \theta} \qquad \cot \theta = \frac{\cos \theta}{\sin \theta} \qquad \cot \theta = \frac{1}{\tan \theta} \tag{3}$$

$$\sec \theta = \frac{1}{\cos \theta} \qquad \csc \theta = \frac{1}{\sin \theta} \tag{4}$$

Each of these formulas can be proved from the formulas in (1). For example, the first formula in (3) can be proved from (1) by observing that

$$\frac{\sin \theta}{\cos \theta} = \frac{b/c}{a/c} = \frac{b}{a} = \tan \theta$$

The proofs of the other formulas in (3) and (4) follow in the same general manner (see Exercises 35–37).

By applying the Pythagorean Theorem to the triangle in Figure 6.22 and then dividing by c^2, we obtain the fundamental formula relating the sine and cosine:

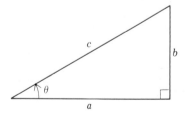

FIGURE 6.22

$$b^2 + a^2 = c^2$$

$$\left(\frac{b}{c}\right)^2 + \left(\frac{a}{c}\right)^2 = 1$$

$$\sin^2 \theta + \cos^2 \theta = 1 \tag{5}$$

Because of the intimate relationship between (5) and the Pythagorean Theorem, (5) is frequently called a ***Pythagorean Identity***. If we divide each side of (5) by $\cos^2 \theta$, we obtain

$$\frac{\sin^2 \theta}{\cos^2 \theta} + \frac{\cos^2 \theta}{\cos^2 \theta} = \frac{1}{\cos^2 \theta}$$

which by (3) and (4) simplifies to

$$\tan^2 \theta + 1 = \sec^2 \theta \qquad (6)$$

Similarly, dividing both sides of (5) by $\sin^2 \theta$ leads to the formula

$$1 + \cot^2 \theta = \csc^2 \theta \qquad (7)$$

Exercise 38 asks for proof of this formula. Together (5)–(7) are the Pythagorean Identities; they are valid for any acute angle θ.

   The preceding formulas provide a second method of finding the values of all the trigonometric functions of an acute angle when the value of one of the functions is known. Furthermore, it will not be necessary to draw a triangle.

   <u>**EXAMPLE 4.**</u>   Suppose θ is an acute angle and $\cos \theta = \frac{1}{4}$. Find the values of the other five trigonometric functions at θ.

   Solution.   From (5) we have

$$\sin^2 \theta + \left(\frac{1}{4}\right)^2 = 1$$

so that

$$\sin^2 \theta = 1 - \frac{1}{16} = \frac{15}{16}$$

Therefore

$$\sin \theta = \sqrt{\frac{15}{16}} = \frac{1}{4}\sqrt{15}$$

Then we use (3) and (4) to deduce that

$$\tan \theta = \frac{\sin \theta}{\cos \theta} = \frac{\frac{1}{4}\sqrt{15}}{1/4} = \sqrt{15}$$

$$\cot \theta = \frac{1}{\tan \theta} = \frac{1}{\sqrt{15}} = \frac{1}{15}\sqrt{15}$$

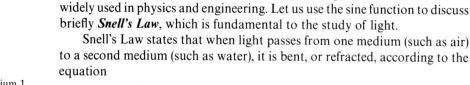

$$\sec \theta = \frac{1}{\cos \theta} = \frac{1}{1/4} = 4$$

$$\csc \theta = \frac{1}{\sin \theta} = \frac{1}{\frac{1}{4}\sqrt{15}} = \frac{4}{15}\sqrt{15} \quad \square$$

Snell's Law

As indicated in the introduction to the chapter, trigonometric functions are widely used in physics and engineering. Let us use the sine function to discuss briefly **Snell's Law**, which is fundamental to the study of light.

Snell's Law states that when light passes from one medium (such as air) to a second medium (such as water), it is bent, or refracted, according to the equation

$$\sin i = \mu \sin r$$

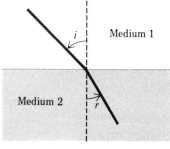

Medium 1

Medium 2

FIGURE 6.23

where i is the angle of incidence, r is the angle of refraction, and μ is the index of refraction (Figure 6.23). The index of refraction depends on the color (or frequency) of the light and on the two media through which the light passes, but not on the angles of incidence or refraction. The index of refraction for a light ray in the visible spectrum passing from air into water varies from approximately 1.330 for red to 1.342 for violet.

EXERCISES 6.2

In Exercises 1–6, evaluate the expression.

1. $2 \sin 45° - 3 \cos 30°$

2. $5 \cos 60° + \sec 30° - 1$

3. $\tan 45° - \cot 45°$

4. $\sin \dfrac{\pi}{4} + \cos \dfrac{\pi}{4} + \tan \dfrac{\pi}{4}$

5. $\csc \dfrac{\pi}{6} + \csc \dfrac{\pi}{3}$ **6.** $\cot \dfrac{\pi}{3} - \sec \dfrac{\pi}{6}$

7. Determine which of the following are equal.
   a. $\tan 45°$ b. $\csc 60°$ c. $\csc 30°$
   d. $\frac{1}{2} \sec 60°$ e. $\sin 30°$ f. $\cos 30°$
   g. $\cos 60°$ h. $2 \sin 30°$

8. Determine which of the following are equal.

   a. $\sin \dfrac{\pi}{3}$ b. $\cos \dfrac{\pi}{3}$ c. $\cos \dfrac{\pi}{6}$ d. $2 \tan \dfrac{\pi}{6}$

   e. $\dfrac{1}{2} \tan \dfrac{\pi}{6}$ f. $\dfrac{1}{2} \tan \dfrac{\pi}{3}$ g. $\sec \dfrac{\pi}{6}$ h. $\csc \dfrac{\pi}{3}$

In Exercises 9–20, use the given values along with Figure 6.22 to determine the values of the six trigonometric functions at θ.

9. $a = 6, b = 1$ **15.** $a = 1.5, b = 1.5$

10. $a = 2, b = 4$ **16.** $a = 1, b = \sqrt{2}$

11. $a = 4, c = 5$ **17.** $a = 7, c = 25$

12. $a = 1, c = 6$ **18.** $a = 3, c = 6$

13. $b = 1.2, c = 1.3$ **19.** $b = \frac{1}{2}, c = 3$

14. $b = 5, c = 13$ **20.** $b = 0.3, c = 0.8$

In Exercises 21–32, use the given value to determine the values of the remaining five trigonometric functions at θ.

21. $\sin \theta = \dfrac{2}{5}$ **25.** $\tan \theta = 2$

22. $\sin \theta = 0.6$ **26.** $\tan \theta = \dfrac{1}{\pi}$

23. $\cos \theta = \dfrac{3}{4}$ **27.** $\cot \theta = 1.3$

24. $\cos \theta = 0.2$ **28.** $\cot \theta = \sqrt{2}$

                                              29. $\sec \theta = 1.1$

30. $\sec \theta = 4$

31. $\csc \theta = \dfrac{5}{3}$

32. $\csc \theta = 1.2$

33. For what values of θ in $(0, \pi/2)$ is $\tan \theta > 1$? [*Hint:* From (1), $\tan \theta = b/a$. Notice that $a = b$ if $\theta = \pi/4$. Now determine those θ in $(0, \pi/2)$ for which $b > a$.]

34. For what values of θ in $(0, \pi/2)$ is $\csc \theta < 2$?

35. a. Show that $\cot \theta = \dfrac{\cos \theta}{\sin \theta}$ for $0 < \theta < \pi/2$.

   b. Show that $\cot \theta = \dfrac{1}{\tan \theta}$ for $0 < \theta < \pi/2$.

36. Show that $\sec \theta = \dfrac{1}{\cos \theta}$ for $0 < \theta < \pi/2$.

37. Show that $\csc \theta = \dfrac{1}{\sin \theta}$ for $0 < \theta < \pi/2$.

38. Show that $1 + \cot^2 \theta = \csc^2 \theta$ for $0 < \theta < \pi/2$.

39. Show that $\sin \theta = \cos \left(\dfrac{\pi}{2} - \theta \right)$ for $0 < \theta < \pi/2$. (*Hint:* If one of the acute angles in a right triangle is θ, then the other is $\dfrac{\pi}{2} - \theta$.)

40. Show that $\cos \theta = \sin \left(\dfrac{\pi}{2} - \theta \right)$ for $0 < \theta < \pi/2$.

41. Show that $\tan \theta = \cot \left(\dfrac{\pi}{2} - \theta \right)$ for $0 < \theta < \pi/2$.

42. Show that $\sec \theta = \csc \left(\dfrac{\pi}{2} - \theta \right)$ for $0 < \theta < \pi/2$.

43. Can the angle of incidence of a ray passing from air to water be $45°$ and the angle of refraction be $30°$? Explain your answer.

**6.3
COMPUTING
VALUES OF
TRIGONOMETRIC
FUNCTIONS**

In Section 6.2 we determined the values of the trigonometric functions at $30°$, $45°$, and $60°$. However, the values of the trigonometric functions for other acute angles are generally impossible to calculate. In such cases the best we can do is to approximate the values by using either a calculator or trigonometric tables.

**Computing
Trigonometric
Values by
Calculator**

Many calculators can approximate the values of the trigonometric functions when either degrees or radians are used. However, the procedure used in setting the calculator for degrees or radians varies from manufacturer to manufacturer. One must be aware of the functioning of the keys on any given calculator. After that, calculating trigonometric values proceeds something like this: To evaluate $\sin 40°$, first push the key representing degrees, so the calculator will compute with respect to degrees, then key in 40, and finally press the key marked "sin." The display on one calculator would then read .64278761, meaning that

$$\sin 40° \approx 0.64278761$$

To approximate $\cos 40°$ or $\tan 40°$, we would proceed the same way, but press the key marked "cos" or "tan" instead of the one marked "sin." Although there generally are no keys representing cotangent, secant, or cosecant, we can still approximate their values on a calculator. For example, to approximate $\csc 40°$ we would approximate $\sin 40°$ as above, and then press the "$1/x$" key to obtain

$$\csc 40° = \frac{1}{\sin 40°} \approx 1.5557238$$

Evaluating an expression like sin 18°49′ on a calculator usually necessitates converting the minutes portion, 49′, into a decimal fraction of a degree, utilizing the fact that

$$1' = \left(\frac{1}{60}\right)^\circ$$

First we find that

$$49' = \left[49\left(\frac{1}{60}\right)\right]^\circ \approx (0.81666667)^\circ$$

so that $\qquad\qquad 18°49' \approx (18.816667)^\circ$

where two decimal places were lost in the last expression because the display on this particular calculator has only eight places. After we have 18°49′ in the calculator in decimal form, we press the key marked "sin" and find that

$$\sin 18°49' \approx 0.32254105$$

If the angle measure includes seconds as well as minutes, we use the fact that

$$1'' = \left(\frac{1}{60}\right)' = \left(\frac{1}{3600}\right)^\circ$$

to help write the decimal expansion of the angle.

EXAMPLE 1.   Use a calculator to approximate sin 64°13′28″.

Solution.   From our remarks above,

$$64°13'28'' = 64° + \left[13\left(\frac{1}{60}\right)\right]^\circ + \left[28\left(\frac{1}{3600}\right)\right]^\circ \approx (64.224444)^\circ$$

The calculator now yields

$$\sin 64°13'28'' \approx 0.90050437 \quad \square$$

EXAMPLE 2.   Use a calculator to approximate the value of $\cos\dfrac{\pi}{5}$.

Solution.   First we set the calculator to compute in radians. Then we key in π (most calculators that compute trigonometric functions have a key for π), divide by 5, and then press the cosine key. We obtain

$$\cos\frac{\pi}{5} \approx 0.80901699 \quad \square$$

Computing Trigonometric Values by Tables

In the centuries preceding the advent of the calculator, tables giving approximate values of trigonometric functions were essential to computations involving triangles. Although tables for the values of the trigonometric functions given in radians existed, they were rarely used. Far more important were the tables of the values of the trigonometric functions given in degrees. Table E, in the back of the book, provides such values. A portion of Table E is displayed here as Table 6.3.

TABLE 6.3

θ degrees	$\sin \theta$	$\cos \theta$	$\tan \theta$	$\cot \theta$	$\sec \theta$	$\csc \theta$	
21°00′	.3584	.9336	.3839	2.605	1.071	2.790	**69°00′**
10	.3611	.9325	.3872	2.583	1.072	2.769	50
20	.3638	.9315	.3906	2.560	1.074	2.749	40
30	.3665	.9304	.3939	2.539	1.075	2.729	30
40	.3692	.9293	.3973	2.517	1.076	2.709	20
50	.3719	.9283	.4006	2.496	1.077	2.689	10
22°00′	.3746	.9272	.4040	2.475	1.079	2.669	**68°00′**
10	.3773	.9261	.4074	2.455	1.080	2.650	50
20	.3800	.9250	.4108	2.434	1.081	2.632	40
30	.3827	.9239	.4142	2.414	1.082	2.613	30
40	.3854	.9228	.4176	2.394	1.084	2.595	20
50	.3881	.9216	.4210	2.375	1.085	2.577	10
23°00′	.3907	.9205	.4245	2.356	1.086	2.559	**67°00′**
⋮	⋮	⋮	⋮	⋮	⋮	⋮	⋮
28°00′	.4695	.8829	.5317	1.881	1.133	2.130	**62°00′**
	$\cos \theta$	$\sin \theta$	$\cot \theta$	$\tan \theta$	$\csc \theta$	$\sec \theta$	θ degrees

To find the value of a trigonometric function for θ between $0°$ and $45°$, we locate the degree at the left of the table and the trigonometric function at the top of the columns. Similarly, to find the value for θ between $45°$ and $90°$, we locate the degree at the right and the trigonometric function at the bottom of the columns. The following examples will illustrate the procedure.

EXAMPLE 3.   Use Table 6.3 to approximate $\sin 21°50′$.

   Solution.   Since $21°50′$ is between $0°$ and $45°$, we use the degrees column at the left and the labels at the top of the columns. First we locate $21°00′$ in the left column, and in the batch it heads we find 50. The trigonometric values for $21°50′$ lie on the same row as the 50. Since we wish $\sin 21°50′$, we use the column headed by $\sin \theta$. The entry that appears in the row $21°50′$ is .3719, so

$$\sin 21°50′ \approx 0.3719 \qquad \square$$

EXAMPLE 4.   Use Table 6.3 to approximate tan 67°20′.

Solution.   Since 67°20′ is between 45° and 90°, we use the degrees column at the right and the labels at the bottom of the columns. First we locate 67°00′ in the right column, and in the batch above it we find 20. The trigonometric values for 67°20′ lie on the same row as the 20. Since we wish tan 67°20′, we use the column designated by tan θ on the bottom. The entry that appears in the row 67°20′ is 2.394, so

$$\tan 67°20′ \approx 2.394 \quad \square$$

Table E lists approximate trigonometric values for angles in multiples of 10′. For example, it gives tan 38°10′ $\approx$ 0.7860 and tan 38°20′ $\approx$ 0.7907. But suppose we wish to approximate tan 38°16′. Although 38°16′ does not appear in the table, we can approximate tan 38°16′ by first observing that 38°16′ is $\frac{6}{10}$ of the distance from 38°10′ to 38°20′, and then assuming that tan 38°16′ will be approximately $\frac{6}{10}$ of the distance from tan 38°10′ to tan 38°20′. This method of approximating (or interpolating) values not listed in Table E is called ***linear interpolation***, because in effect the method treats the trigonometric function as a linear function between successive values listed in the table.

EXAMPLE 5.   Use Table E and interpolation to approximate the value of tan 38°16′.

Solution.   By Table E,

$$\tan 38°10′ \approx 0.7860$$
$$\tan 38°16′ \approx ??$$
$$\tan 38°20′ \approx 0.7907$$

Since 0.7907 − 0.7860 = 0.0047, and since 38°16′ is $\frac{6}{10}$ of the distance from 38°10′ to 38°20′, it follows that

$$\tan 38°16′ \approx 0.7860 + \frac{6}{10}(0.0047) \approx 0.7888 \quad \square$$

Later it will be necessary to find (or approximate) an acute angle that has a given trigonometric value. If the given trigonometric value is listed in Table E, then we simply take the corresponding angle in the table as the approximation. For example, to approximate an acute angle θ such that sin θ = 0.1708, we locate .1708 in the column of Table E labeled sin θ at the top. Then we read the angle on the left of the page and conclude that

$$\sin 9°50′ \approx 0.1708$$

However, if the given trigonometric value does not appear in the table, then we may use linear interpolation, as in the following example.

EXAMPLE 6.   Use Table E and interpolation to approximate the value of θ for which $\sin \theta = 0.9200$.

Solution.   By Table E,

$$\sin 66°50' \approx 0.9194$$

$$\sin \ ?? = 0.9200$$

$$\sin 67°00' \approx 0.9205$$

Since $0.9205 - 0.9194 = 0.0011$ and $0.9200 - 0.9194 = 0.0006$, it follows that 0.9200 is $0.0006/0.0011 = \frac{6}{11}$ of the distance from 0.9194 to 0.9205. Therefore we take θ to be $\frac{6}{11}$ of the distance from $66°50'$ to $67°00'$, which means that

$$\theta \approx 66°55' \quad \square$$

It is interesting to note that tables for trigonometric values have existed since biblical times. In fact, the oldest known trigonometric table is a Babylonian clay tablet that was written sometime between 1900 B.C. and 1600 B.C.; it is essentially a table of values of the secant for angles from 31° to 45°. A later table by Ptolemy of Alexandria in about 150 A.D. appeared in the *Almagest*,* the most influential book on astronomy until the time of Copernicus. This table contained the sines of angles from 0° to 90° in steps of 15'. Later a disciple of Copernicus by the name of Georg Rhaeticus (1514–1576) compiled a 15-place sine table in steps of 10″. Rhaeticus was also the first to define the trigonometric functions as ratios of the lengths of the sides of a right triangle.

* *Almagest* is a hybrid word derived from the Latin *magiste* (greatest) and the Arabic article *al*.

EXERCISES 6.3

In Exercises 1–12, use a calculator to approximate the value of the given expression.

1. $\sin 8°$

2. $\sin 74°$

3. $\cos 27°$

4. $\cos 83°$

5. $\tan 14°$

6. $\tan 61°$

7. $\cot 41°$

8. $\cot 87°$

9. $\sec 1°$

10. $\sec 43°$

11. $\csc 54°$

12. $\csc 5°$

In Exercises 13–24, use a calculator to approximate the value of the given expression.

13. $\sin 38°20'$

14. $\sin 64°30'$

15. $\cos 16°10'$

16. $\cos 44°44'$

17. $\tan 57°17'$

18. $\sec 85°59'$

19. $\sin 63°21'49''$

20. $\cos 41°35'23''$

21. $\tan 88°59'59''$

22. $\cot 20°20'20''$

23. $\sec 1'37''$

24. $\csc 18°12'6''$

25. Use a calculator to approximate the value of each of the following expressions.
  a. $\sin 0.1$
  b. $\sin 0.0001$
  c. $\sin 10^{-6}$
  d. $\cos 0.1$
  e. $\cos 0.001$
  f. $\cos 0.00001$

26. Use a calculator to approximate the value of each of the following expressions.
  a. $\tan (89.9)°$
  b. $\tan (89.99)°$
  c. $\tan (89.999)°$
  d. $\tan (89.9999)°$

🔲 **27.** Try to evaluate tan 90° by calculator. Justify the calculator's response.

In Exercises 28–39, use Table E to approximate the value of the given expression.

28. sin 25°10′   **34.** cot 14°10′

29. sin 41°40′   **35.** cot 88°30′

30. cos 54°50′   **36.** sec 1°20′

31. cos 78°20′   **37.** sec 80°40′

32. tan 45°40′   **38.** csc 16°10′

33. tan 84°50′   **39.** csc 29°50′

In Exercises 40–51, use Table E and interpolation to approximate the given expression.

40. sin 32°41′   **46.** cot 62°16′

41. sin 67°36′   **47.** cot 14°4′

42. cos 18°55′   **48.** sec 25°29′

43. cos 79°11′   **49.** sec 53°7′

44. tan 44°44′   **50.** csc 21°1′

45. tan 87°33′   **51.** csc 79°17′

In Exercises 52–57, use Table E to find an approximate value of θ, with $0° < \theta < 90°$, for which the given equation is valid.

52. $\sin \theta = 0.2613$ **55.** $\cot \theta = 0.1876$

53. $\cos \theta = 0.8359$ **56.** $\sec \theta = 2.873$

54. $\tan \theta = 7.132$ **57.** $\csc \theta = 5.555$

58. Using Exercise 39 in Section 6.2, but without using a calculator or a table, compute
   a. $\sin^2 16° + \sin^2 74°$
   b. $\sin^2 10° + \sin^2 20° + \sin^2 30° + \sin^2 40° + \sin^2 50° + \sin^2 60° + \sin^2 70° + \sin^2 80°$

59. During a 24-hour period the ground under a Foucault pendulum located at θ latitude will rotate $(360 \sin \theta)°$, so that the plane in which the pendulum swings appears to rotate through $(360 \sin \theta)°$. Through how many degrees does a Foucault pendulum in New York City (located at latitude 40°40′) appear to rotate during a 24-hour period?

🔲 **60.** In 1729 the English astronomer James Bradley found a way of determining the speed of the earth around the sun. He noticed that to view a star straight overhead by telescope, he had to tilt the telescope 20.49″. He concluded that the speed s in kilometers per second of the earth was given by

$$s = c \tan 20.49″$$

where c is the speed of light (approximately 299,792.5 kilometers per second). Use a calculator to determine s.

🔲 **61.** Recall from Section 6.2 that Snell's Law states that

$$\sin i = \mu \sin r$$

where i is the angle of incidence, r the angle of refraction, and μ the index of refraction. A yellow sodium light wave has an index of refraction of approximately 1.333. Using a calculator or Table E, determine the angle of refraction if the angle of incidence is
   a. 20°   b. 53°   c. 88°

🔲 **62.** Suppose a ramp c feet long is inclined at an angle of θ degrees. In foot-pounds the work W required to move an object weighing p pounds up the ramp is given by the formula

$$W = pc \sin \theta$$

Determine the work involved in pushing a 60-pound crate of apples up a 20-foot long ramp whose angle of inclination is 18°.

🔲 **63.** The perimeter of a regular polygon of n sides inscribed in a circle of radius r is given by

$$p_n(r) = 2nr \sin \left(\frac{180°}{n}\right)$$

Suppose the radius is 8 inches. Find the perimeter of a regular polygon with the given number of sides.
   a. 5   b. 20   c. 180   d. 360

🔲 **64.** Suppose a circular curve in a highway has a radius of r feet and is banked at an angle θ (Figure 6.24). Assume further that the road is icy, so there is no friction. Then in miles per hour the maximum speed v that a car can travel on the curve without skidding satisfies the formula

$$v^2 = \frac{1800}{121} r \tan \theta$$

   a. Suppose the curve has a radius of 1000 feet and is

banked at an angle of 10°. Find the maximum speed on the curve under icy conditions.

b. If the curve is regraded so that it is banked at an angle of 12°, how is the maximum speed on the curve under icy conditions affected?

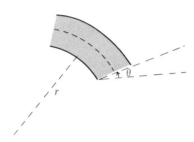

FIGURE 6.24

C **65.** Suppose a projectile is shot from ground level at an angle θ with respect to the horizontal and with an initial speed of v feet per second. If the ground is level,

air resistance is negligible, and distances are measured in feet, then the maximum height H attained by the projectile and the range R of the projectile are given by

$$H = \frac{v^2 \sin^2 \theta}{64} \quad \text{and} \quad R = \frac{v^2 \sin 2\theta}{32}$$

(See Figure 6.25). Suppose a golf ball is hit at an angle of 30° with respect to the horizontal and has an initial speed of 100 feet per second. Find the maximum height and range.

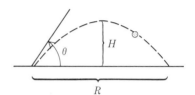

FIGURE 6.25

6.4 SOLVING RIGHT TRIANGLES

An almanac lists a 362-foot tall redwood located in Humboldt Redwood State Park, California, as the world's tallest tree. How was the tree's height first measured? We don't know, but this section contains methods of calculating such heights. More specifically, we will learn how to determine the lengths of the sides and the measures of the angles of a given right triangle when certain of them are known, a process called *solving the triangle*. Later we will learn how to solve triangles that are not necessarily right triangles.

To compute the height of the redwood tree mentioned, assume that the tree stands essentially upright. Then the tree could be represented by one side of a right triangle with one vertex at a point A some distance away from the tree (Figure 6.26a). By measuring the distance between A and the tree, and the angle α in Figure 6.26a, one can then determine the height of the tree (that is, the length a of the appropriate leg of the triangle); one can even solve the triangle completely. In order to prepare for such calculations, we will give some preliminary notation and information.

Let the lengths of the legs of a given right triangle be denoted by a and b, the length of the hypotenuse by c, and the angles opposite a, b, and c by α, β, and γ, respectively (Figure 6.26b). In this section the angle γ will always be the right angle of the triangle, containing 90°, and therefore will need no further mention.

When we solve a right triangle, we will repeatedly use, as we did in Section 6.2, two results from geometry:

Trigonometry can be used in measuring the height of a redwood tree.

$$a^2 + b^2 = c^2 \tag{1}$$

$$\alpha + \beta = 90° \tag{2}$$

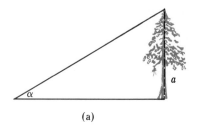

(a)

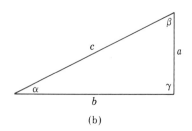

(b)

FIGURE 6.26

We will also use the following formulas for the values of the six basic trigonometric functions:

$$\sin \alpha = \frac{a}{c} \quad \cot \alpha = \frac{b}{a} \quad \sin \beta = \frac{b}{c} \quad \cot \beta = \frac{a}{b}$$

$$\cos \alpha = \frac{b}{c} \quad \sec \alpha = \frac{c}{b} \quad \cos \beta = \frac{a}{c} \quad \sec \beta = \frac{c}{a}$$

$$\tan \alpha = \frac{a}{b} \quad \csc \alpha = \frac{c}{a} \quad \tan \beta = \frac{b}{a} \quad \csc \beta = \frac{c}{b}$$

Using these results, we can solve a right triangle provided that either

(a) one acute angle and the length of one side are known, or
(b) the lengths of two sides are known.

In particular, we can solve the tree problem if we are armed with a tape measure and either a compass or a transit.

EXAMPLE 1.   Suppose we select a spot *A* on the ground from which we can see the top of the redwood tree mentioned above. By tape measure we find that the distance between *A* and the center of the tree is 150 feet, and by a large compass (or a transit) we find that $\alpha = 67.5°$. With this information, determine the height of the tree.

Solution.   We must calculate *a* in Figure 6.27. But

$$\tan 67.5° = \frac{a}{150} \quad \text{so that} \quad a = 150 \tan 67.5°$$

By calculator we find that

$$a \approx 362.1$$

Consequently the height of the tree is approximately 362 feet, as the almanac says.   □

The angle α, projecting upward from the horizontal to the top of the tree, is called the ***angle of elevation*** of the tree from the point *A*. It plays an important role in the determination of heights of trees and structures.

Once we know the tree's height, the horizontal distance from *A* to the center of the tree, and the angle of elevation, we can easily finish solving the corresponding triangle. We do this in Example 2.

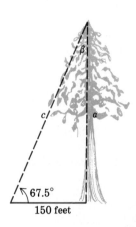

67.5°

150 feet

FIGURE 6.27

EXAMPLE 2.   Solve the right triangle shown in Figure 6.27.

Solution.   We already know that $b = 150$, $a \approx 362.1$, and $\alpha = 67.5°$. As a result, we need only determine *c* and β. By (1),

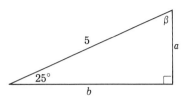

FIGURE 6.28

$$c = \sqrt{a^2 + b^2} \approx \sqrt{(362.1)^2 + (150)^2} \approx 391.9$$

and by (2),

$$\beta = 90° - 67.5° = 22.5° \quad \square$$

In Example 2 the length of one leg and one acute angle were given. In the next two examples we will solve a triangle when other information is provided.

EXAMPLE 3. Solve the right triangle shown in Figure 6.28.

Solution. From the figure, $c = 5$ and $\alpha = 25°$. We must calculate a, b, and β. From (2),

$$\beta = 90° - 25° = 65°$$

Now we use the known values of α and c to determine a and b:

$$\sin 25° = \frac{a}{5}, \quad \text{so that} \quad a = 5 \sin 25°$$

and

$$\cos 25° = \frac{b}{5}, \quad \text{so that} \quad b = 5 \cos 25°$$

By calculator we find that

$$a \approx 2.113 \quad \text{and} \quad b \approx 4.532 \quad \square$$

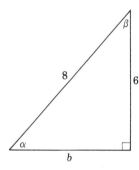

FIGURE 6.29

EXAMPLE 4. Solve the right triangle shown in Figure 6.29.

Solution. This time $a = 6$ and $c = 8$. We need to find b, α, and β. By (1),

$$b = \sqrt{8^2 - 6^2} = \sqrt{28} \approx 5.292$$

For α, we notice that

$$\sin \alpha = \frac{6}{8} = 0.75$$

so that by calculator we have

$$\alpha \approx 48.59° \ (\approx 48°35')$$

Consequently, it follows from (2) that

$$\beta = 90° - \alpha \approx 90° - 48.59° = 41.41° \ (\approx 41°25') \quad \square$$

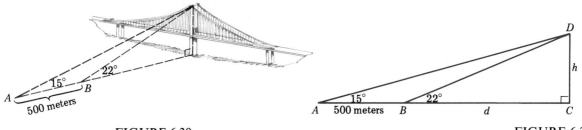

FIGURE 6.30   FIGURE 6.31

We end with a more complicated example involving heights.

EXAMPLE 5.   In order to estimate the height of a bridge, two sightings are made, one each at points A and B, which are 500 meters apart (Figure 6.30). How tall is the bridge?

   Solution.   Figure 6.31 contains the information we use to solve the problem, including the letter h for the height we need to determine. From triangle ACD we obtain

$$\tan 15° = \frac{h}{500 + d}, \quad \text{so that} \quad h = (500 + d)\tan 15° \qquad (3)$$

From triangle BCD we obtain

$$\cot 22° = \frac{d}{h}, \quad \text{so that} \quad d = h\cot 22°$$

Substituting for d in the second equation of (3) yields

$$h = (500 + h\cot 22°)\tan 15° = 500\tan 15° + h\cot 22° \tan 15°$$

Rearranging, we find that

$$h - h\cot 22° \tan 15° = 500\tan 15°$$

or

$$h(1 - \cot 22° \tan 15°) = 500\tan 15°$$

Thus

$$h = \frac{500\tan 15°}{1 - \cot 22° \tan 15°}$$

By calculator we find that

$$h \approx 397.8$$

Consequently the bridge is approximately 397.8 meters tall.   □

EXERCISES 6.4

In Exercises 1–12, solve the right triangle with the given data pertaining to Figure 6.32.

1. $\beta = 15°, c = 4$

2. $\beta = 15°, a = 4$

3. $\beta = 40°, b = 2$

4. $\alpha = 69.2°, b = 5$

5. $\alpha = 38.4°, a = 3.2$

6. $\alpha = 83.7°, c = 11.6$

7. $a = 15, c = 17$

8. $a = 11, b = 60$

9. $b = 3, c = \sqrt{10}$

10. $a = 2.1, b = 2$

11. $a = \frac{1}{3}, b = \frac{2}{5}$

12. $b = \frac{5}{7}, c = \frac{10}{3}$

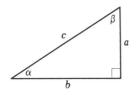

FIGURE 6.32

13. Find the distance from point A to point B on the opposite side of the pond represented in Figure 6.33.

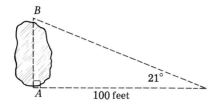

FIGURE 6.33

14. Find the height h of the flagpole atop the building pictured in Figure 6.34.

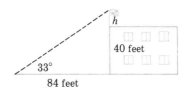

FIGURE 6.34

15. The top of a 12-foot ladder resting against a wall is 9 feet above the ground. Find the angle of elevation of the ladder.

16. A police car is located 80 feet from a street having a speed limit of 30 miles per hour (44 feet per second). Ten seconds after a truck passes the intersection, the line connecting the truck and the police car makes an angle of 10° with the street (Figure 6.35).
    a. Has the truck been speeding?
    b. If the truck had been traveling exactly 30 miles per hour, determine the angle the line connecting the truck and the police car would have made with the street 10 seconds after the truck passed the intersection.

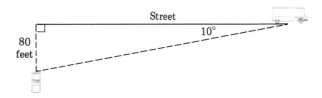

FIGURE 6.35

17. A jet is flying at 400 miles per hour. One minute after the jet passes directly over an airport, its angle of elevation from the airport is 35°. Determine the height of the jet.

18. A pleasure boat travels 10 miles due east from port and then 5 miles due south before running aground on a reef. The Coast Guard plans to send a rescue vessel along a straight line from port to the distressed boat. What angle should that straight line make with the line that points straight east from the port?

19. A surveyor takes a measurement 5 feet above ground and 100 feet from an apartment building (Figure 6.36). He determines that the angle of elevation of the top of the building is 36°. How tall is the building?

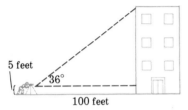

FIGURE 6.36

20. Points *A* and *B* lie directly opposite each other on a river and are invisible from each other because of an island in the river. Using the information in Figure 6.37, determine the distance between *A* and *B*.

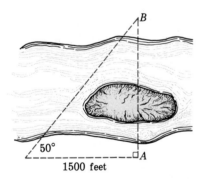

FIGURE 6.37

21. You are standing 4000 feet from the launch pad at Cape Kennedy. A few seconds after launch, the angle of elevation of the tail of the space shuttle is 50° and the angle of elevation of the nose is 52.05°. What is the length of the rocket?

22. The Leaning Tower of Pisa is now approximately 177 feet tall and leans at an angle of approximately 5.2° with respect to the vertical. Determine the original height of the tower.

23. Suppose it is decided to build a bridge 10 miles long from the south rim to the north rim of the Grand Canyon. Using the fact that the north rim is 1200 feet higher than the south rim, find the angle of elevation of the bridge.

24. The sun is approximately 93,000,000 miles from the earth and 865,000 miles in diameter, whereas the moon is approximately 240,000 miles from the earth and 2160 miles in diameter. Using this information, determine which appears larger from the earth. (*Hint:* Determine the angle each subtends in the eye of an observer.)

25. Originally the Great Pyramid of Cheops had a square base approximately 754 feet on a side. In order to determine its height, a visitor 800 feet away from the midpoint of one side noted that the angle of elevation of the pyramid from the visitor's feet was 22.27° (Figure 6.38). How tall was the pyramid?

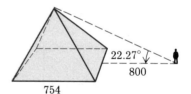

FIGURE 6.38

**6.5
TRIGONOMETRIC
FUNCTIONS OF
ARBITRARY
ANGLES**

Until now we have discussed the trigonometric functions only for acute angles. However, since many applications involve angles that are not acute, it is desirable to define the trigonometric functions for arbitrary angles, acute or not. As in Section 6.2, the definitions in this section are valid whether the angles are measured in degrees or radians.

Let θ be any angle in standard position, and let $P(a, b)$ be an arbitrary point on the terminal side distinct from the origin (Figure 6.39). Furthermore, let $c = \sqrt{a^2 + b^2}$, the distance between the origin and P. Then the six trigonometric functions are defined by formulas identical to those in Section 6.2:

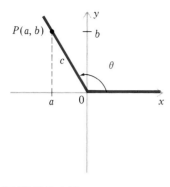

FIGURE 6.39

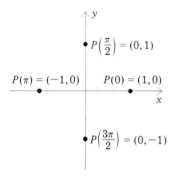

FIGURE 6.40

$$\sin \theta = \frac{b}{c} \qquad \tan \theta = \frac{b}{a} \qquad \sec \theta = \frac{c}{a}$$

$$\cos \theta = \frac{a}{c} \qquad \cot \theta = \frac{a}{b} \qquad \csc \theta = \frac{c}{b}$$

(1)

Moreover, the basic formulas we derived in Section 6.2 remain valid for any angle θ and can be proved the same way as for acute angles:

$$\tan \theta = \frac{\sin \theta}{\cos \theta} \qquad \cot \theta = \frac{\cos \theta}{\sin \theta} \qquad \cot \theta = \frac{1}{\tan \theta}$$

$$\sec \theta = \frac{1}{\cos \theta} \qquad \csc \theta = \frac{1}{\sin \theta}$$

$$\sin^2 \theta + \cos^2 \theta = 1 \qquad \tan^2 \theta + 1 = \sec^2 \theta \qquad 1 + \cot^2 \theta = \csc^2 \theta$$

Notice that $\sin \theta$ and $\cos \theta$ are defined for any angle θ. However, the remaining functions are not defined whenever their respective denominators equal 0. For example, if we use radians to measure angles, then we have $P(0) = (1, 0)$, $P(\pi/2) = (0, 1)$, $P(\pi) = (-1, 0)$, and $P(3\pi/2) = (0, -1)$ (Figure 6.40). Thus, from the definitions of the trigonometric functions, we obtain the values in Table 6.4.

TABLE 6.4

θ (in radians)	θ (in degrees)	$\sin \theta$	$\cos \theta$	$\tan \theta$	$\cot \theta$	$\sec \theta$	$\csc \theta$
0	0°	0	1	0	not defined	1	not defined
$\dfrac{\pi}{2}$	90°	1	0	not defined	0	not defined	1
π	180°	0	-1	0	not defined	-1	not defined
$\dfrac{3}{2}\pi$	270°	-1	0	not defined	0	not defined	-1

Since either or both of the coordinates a and b in Figure 6.39 may be negative, it follows from (1) that each of the trigonometric functions can have negative values as well as positive values. Table 6.5 summarizes the signs of the various trigonometric functions, depending on the quadrant in which the terminal side of the angle lies.

TABLE 6.5

Quadrant in which the terminal side of θ lies	$\sin \theta, \csc \theta$	$\cos \theta, \sec \theta$	$\tan \theta, \cot \theta$
first	+	+	+
second	+	−	−
third	−	−	+
fourth	−	+	−

Since $|a| \leq c$ and $|b| \leq c$, it follows that

$$|\sin \theta| \leq 1 \quad \text{and} \quad |\cos \theta| \leq 1$$

Thus $\sin \theta$ and $\cos \theta$ are numbers between -1 and 1. The absolute value of other trigonometric functions can be arbitrarily large.

The Reference Angle

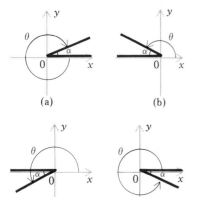

(a)   (b)

(c)   (d)

Reference angles

FIGURE 6.41

If an angle θ is in standard position but is not acute, then there are several ways of computing the trigonometric values at θ. One way involves an associated acute angle that we will henceforth denote by the letter α. This new angle is formed by the terminal side of the given angle and the positive or negative x axis, whichever makes the associated angle acute (Figure 6.41 a–d). The associated acute angle α is called the ***reference angle*** for the given angle. Since the positive and negative x axes correspond to angles of $(180n)°$, or $n\pi$ radians, where n is an integer, we can find the reference angle α of a given angle θ in standard position by determining the acute angle α such that

(in degrees) either $\theta = (180n)° + \alpha$ or $\theta = (180n)° - \alpha$

(in radians) either $\theta = n\pi + \alpha$ or $\theta = n\pi - \alpha$

where n is an integer.

EXAMPLE 1.   Find the reference angle α for each of the following angles.

a. $120°$ b. $-135°$ c. $\dfrac{5\pi}{4}$ d. $\dfrac{11\pi}{6}$

Solution.
a. Since $120° = 180° - 60°$, the reference angle is $60°$ (Figure 6.42a).
b. Since $-135° = -180° + 45°$, the reference angle is $45°$ (Figure 6.42b).
c. Since $\dfrac{5\pi}{4} = \pi + \dfrac{\pi}{4}$, the reference angle is $\dfrac{\pi}{4}$ (Figure 6.42c).
d. Since $\dfrac{11\pi}{6} = 2\pi - \dfrac{\pi}{6}$, the reference angle is $\dfrac{\pi}{6}$ (Figure 6.42d).   □

If a given angle θ has reference angle α, then

$$|\sin \theta| = \sin \alpha \tag{2}$$

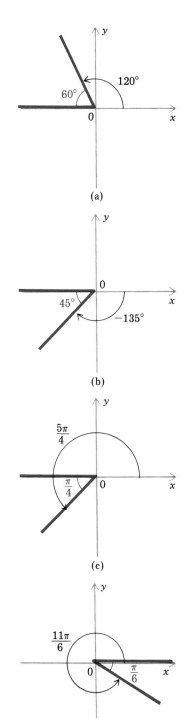

(a)

(b)

(c)

(d)

FIGURE 6.42

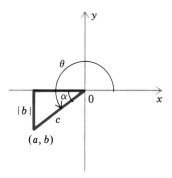

FIGURE 6.43

To prove (2) we observe that if (a, b) is any point on the terminal side of the given angle θ (Figure 6.43), then

$$\sin \alpha = \frac{|b|}{c} = \left|\frac{b}{c}\right| = |\sin \theta|$$

Formula (2) provides us with a procedure for determining the value of $\sin \theta$ for an arbitrary angle θ:

1. Determine the reference angle α.
2. Compute $\sin \alpha$.
3. Determine the sign of $\sin \theta$ from Table 6.5.
4. If $\sin \theta \geq 0$, then $\sin \theta = \sin \alpha$.
   If $\sin \theta < 0$, then $\sin \theta = -\sin \alpha$.

A similar procedure applies to any other trigonometric function.

EXAMPLE 2.   Compute $\sin 120°$ and $\cos 120°$.

Solution.   By the solution of Example 1(a) we know that the reference angle is $60°$. By Table 6.5 we know that $\sin 120° > 0$ and $\cos 120° < 0$. Using our knowledge of the values of the trigonometric functions at $60°$, we find that

$$\sin 120° = \sin 60° = \frac{\sqrt{3}}{2} \quad \text{and} \quad \cos 120° = -\cos 60° = -\frac{1}{2} \quad \square$$

EXAMPLE 3.   Compute the values of the six trigonometric functions at $11\pi/6$.

Solution.   By the solution of Example 1(d) we know that the reference angle is $\frac{\pi}{6}$. By Table 6.5 we know that $\cos\frac{11\pi}{6}$ and $\sec\frac{11\pi}{6}$ are positive, whereas $\sin\frac{11\pi}{6}$, $\tan\frac{11\pi}{6}$, $\cot\frac{11\pi}{6}$, and $\csc\frac{11\pi}{6}$ are negative. Using our knowledge of the values of the trigonometric functions at $\frac{\pi}{6}$, we conclude that

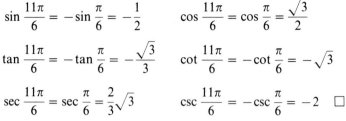

$$\sin \frac{11\pi}{6} = -\sin \frac{\pi}{6} = -\frac{1}{2} \qquad \cos \frac{11\pi}{6} = \cos \frac{\pi}{6} = \frac{\sqrt{3}}{2}$$

$$\tan \frac{11\pi}{6} = -\tan \frac{\pi}{6} = -\frac{\sqrt{3}}{3} \qquad \cot \frac{11\pi}{6} = -\cot \frac{\pi}{6} = -\sqrt{3}$$

$$\sec \frac{11\pi}{6} = \sec \frac{\pi}{6} = \frac{2}{3}\sqrt{3} \qquad \csc \frac{11\pi}{6} = -\csc \frac{\pi}{6} = -2 \quad \square$$

Suppose one trigonometric value at an angle θ is given, along with the quadrant in which the terminal side of θ is located. Then the reference angle and an associated triangle provide the information needed to compute the values of the other trigonometric functions at θ.

EXAMPLE 4. Suppose $\sin \theta = -\frac{2}{7}$ and $\pi < \theta < \frac{3\pi}{2}$. Find the values of the remaining trigonometric functions at θ.

Solution. Let α be the reference angle for θ. Since the terminal side of θ lies in the third quadrant and $\sin \theta = -\frac{2}{7}$, we draw the triangle in Figure 6.44, with all lengths positive. By the Pythagorean Theorem,

$$7^2 = 2^2 + d^2, \quad \text{so that} \quad d = \sqrt{7^2 - 2^2} = \sqrt{45} = 3\sqrt{5}$$

FIGURE 6.44

From Table 6.5 we know that $\cos \theta < 0$, $\tan \theta > 0$, $\cot \theta > 0$, $\sec \theta < 0$, and $\csc \theta < 0$. Using the triangle in Figure 6.44, we conclude that

$$\cos \theta = -\cos \alpha = -\frac{d}{7} = -\frac{3\sqrt{5}}{7} \qquad \cot \theta = \cot \alpha = \frac{d}{2} = \frac{3\sqrt{5}}{2}$$

$$\tan \theta = \tan \alpha = \frac{2}{d} = \frac{2}{3\sqrt{5}} = \frac{2}{15}\sqrt{5} \qquad \sec \theta = -\sec \alpha = -\frac{7}{d} = -\frac{7\sqrt{5}}{15}$$

$$\csc \theta = -\csc \alpha = -\frac{7}{2} \quad \square$$

We have one more observation concerning the tangent function. Let l be a nonvertical line that passes through the origin and makes an angle θ with the positive x axis. If (x, y) is any point on l except the origin, then

$$\tan \theta = \frac{y}{x}$$

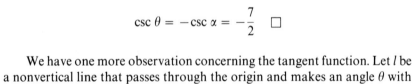

FIGURE 6.45 (Figure 6.45). Since l passes through the origin, it follows from the definition of the slope m of l that

$$m = \frac{y - 0}{x - 0} = \frac{y}{x}$$

Therefore

$$\boxed{\tan \theta = \text{the slope of } l}$$

EXERCISES 6.5

In Exercises 1–8, determine all the trigonometric functions that are defined for the given angle.

1. $360°$

2. $450°$

3. $-270°$

4. $-1080°$

5. $\dfrac{3\pi}{2}$

6. 10π

7. $-\dfrac{\pi}{2}$

8. $-\pi$

In Exercises 9–12, find the reference angle of the given angle.

9. $150°$

10. $-120°$

11. $\dfrac{9\pi}{4}$

12. $-\dfrac{\pi}{3}$

In Exercises 13–24, use the reference angle to find the values of the six trigonometric functions at the given angle.

13. $135°$

14. $210°$

15. $\dfrac{5\pi}{3}$

16. $\dfrac{11\pi}{6}$

17. $-120°$

18. $-225°$

19. $-\dfrac{3\pi}{4}$

20. $\dfrac{8\pi}{3}$

21. $750°$

22. $-945°$

23. $\dfrac{41\pi}{6}$

24. $-\dfrac{20\pi}{3}$

In Exercises 25–30, find the values of the remaining trigonometric functions at θ.

25. $\sin\theta = -\dfrac{3}{5}$ and $\pi < \theta < \dfrac{3\pi}{2}$

26. $\cos\theta = \dfrac{4\sqrt{3}}{7}$ and $-\dfrac{\pi}{2} < \theta < 0$

27. $\tan\theta = \sqrt{2}$ and $3\pi < \theta < \dfrac{7\pi}{2}$

28. $\cot\theta = -\sqrt{3}$ and $\dfrac{5\pi}{2} < \theta < 3\pi$

29. $\sec\theta = -\sqrt{3}$ and $-\dfrac{11\pi}{2} < \theta < -5\pi$

30. $\csc\theta = \dfrac{7}{4}$ and $4\pi < \theta < \dfrac{9\pi}{2}$

In Exercises 31–38, determine which statements are true and which are false. Justify your answer.

31. $\sin\dfrac{\pi}{4} \overset{?}{=} \dfrac{1}{2}\sin\dfrac{\pi}{2}$

32. $\cos 2\pi \overset{?}{=} (\cos\pi)^2$

33. $\cos 2\pi \overset{?}{=} 2\cos\pi$

34. $\sin\dfrac{\pi}{6} + \sin\dfrac{\pi}{3} \overset{?}{=} \sin\dfrac{\pi}{2}$

35. $\sin\dfrac{\pi}{6} + \cos\dfrac{\pi}{3} \overset{?}{=} \sin\dfrac{\pi}{2}$

36. $\sin\dfrac{\pi}{6} + \left(\cos\dfrac{3\pi}{4}\right)^2 \overset{?}{=} \cos 4\pi$

37. $\dfrac{2}{3}\sin\dfrac{7\pi}{3}\cos\dfrac{11\pi}{6} \overset{?}{=} \left(\sin\dfrac{3\pi}{4}\right)^2$

38. $\sin^2 n\pi - \cos^2 n\pi \overset{?}{=} -1$

39. Find a formula for the distance between two points $(\cos s, \sin s)$ and $(\cos t, \sin t)$ on the unit circle. Simplify your answer.

6.6
RELATIONS
INVOLVING
TRIGONOMETRIC
FUNCTIONS

In the preceding sections we found values of the six trigonometric functions by means of triangles, calculators, or tables. Now we turn to various relationships among the trigonometric functions.

Although all the basic ideas appearing in this section apply equally well to angles given in degrees and those given in radians, we will use radian measure throughout the section except in certain examples.

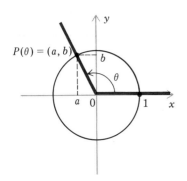

FIGURE 6.46

For any real number θ let $P(\theta) = (a, b)$ be the point on the unit circle that lies on the terminal side of the angle of θ radians in standard position (Figure 6.46). (For the various real numbers θ, the association of the point $P(\theta)$ with θ is sometimes called the *wrapping function*.) Since $a^2 + b^2 = 1$, the formulas in (1) of Section 6.5 become

$$\sin \theta = b \qquad \tan \theta = \frac{b}{a} \qquad \sec \theta = \frac{1}{a}$$

$$\cos \theta = a \qquad \cot \theta = \frac{a}{b} \qquad \csc \theta = \frac{1}{b}$$

Not only do certain properties of the trigonometric functions become easier to visualize when we use the circle and the point $P(\theta)$, but also the description of the sine function by means of a circle sheds light on the etymology of "sine."*

Since there are 2π radians in one complete revolution, it follows that $P(\theta) = P(\theta + 2\pi)$ (Figure 6.47). Thus if $P(\theta) = (a, b)$, then $P(\theta + 2\pi) = (a, b)$ also. We conclude that

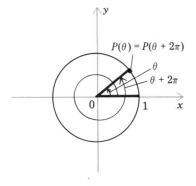

FIGURE 6.47

$$\sin (\theta + 2\pi) = \sin \theta \qquad \tan (\theta + 2\pi) = \tan \theta \qquad \sec (\theta + 2\pi) = \sec \theta$$
$$\cos (\theta + 2\pi) = \cos \theta \qquad \cot (\theta + 2\pi) = \cot \theta \qquad \csc (\theta + 2\pi) = \csc \theta \qquad (1)$$

More generally, for any integer n we have $P(\theta + 2n\pi) = P(\theta)$, so that

$$\sin (\theta + 2n\pi) = \sin \theta \quad \text{and} \quad \cos (\theta + 2n\pi) = \cos \theta \qquad (2)$$

with similar formulas for the other trigonometric functions.

From (1) and the fact that $\sin 0 = 0$, $\cos 0 = 1$, $\sin \pi = 0$, and $\cos \pi = -1$, we conclude that for any integer n,

$$\sin n\pi = 0 \quad \text{and} \quad \begin{cases} \cos 2n\pi = 1 \\ \cos (2n + 1)\pi = -1 \end{cases}$$

Periodic Functions

It is a direct consequence of (1) that the trigonometric functions repeat their values every 2π units. Functions that repeat their values regularly have a special name.

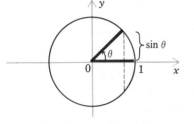

FIGURE 6.48

* Notice from Figure 6.48 that if $0 \le \theta \le \pi$, then $\sin \theta$ is the length of a half-chord, a fact that seems to have influenced the etymology of the word "sine." In the fifth century A.D. the Indian mathematician Aryabbatta called the sine of an angle by the name "ardha-jya," meaning "half-chord." Soon the word condensed to "jya," or "chord." The Arabs translated the word into their vowelless written language as "jb." Later, as the Arabs once again began adding vowels to their language, "jb" somehow became "jaib," which already meant "bosom." Around the twelfth century, when mathematics was translated into Latin, the Arabic "jaib" was translated into "sinus," the Latin equivalent for "bosom," from which our word "sine" comes. It was in 1626 that the abbreviations "sin" and "cos" were introduced, by Albert Girard of the Netherlands.

DEFINITION 6.1   A function f is ***periodic*** if there is a positive number p such that $f(x + p) = f(x)$ for all x. If there is a smallest such number p, then p is called the ***period*** of f.

The formulas in (1) tell us that the trigonometric functions are periodic. It will follow later that the sine, cosine, secant, and cosecant have period 2π, whereas the tangent and the cotangent have period π.

None of the functions encountered in earlier chapters (except the constant functions) is periodic—neither the polynomials nor the rational functions nor the exponential functions nor the logarithmic functions. One of the reasons that the trigonometric functions are so important is their periodic nature. They are indispensable in the mathematical description of such periodic natural phenomena as tides, alternating current, and the oscillation of springs.

Symmetry with Respect to the *x* Axis

Many of the properties of the trigonometric functions can be related to the symmetry of the unit circle with respect to the x and y axes and the origin. First we observe that $P(-\theta)$ and $P(\theta)$ are symmetric with respect to the x axis (Figure 6.49), so that if $P(\theta) = (a, b)$, then $P(-\theta) = (a, -b)$. It follows that

$$\sin(-\theta) = -\sin\theta \qquad \tan(-\theta) = -\tan\theta \qquad \sec(-\theta) = \sec\theta$$
$$\cos(-\theta) = \cos\theta \qquad \cot(-\theta) = -\cot\theta \qquad \csc(-\theta) = -\csc\theta \tag{3}$$

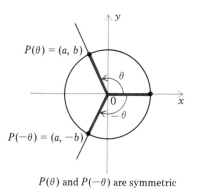

$P(\theta)$ and $P(-\theta)$ are symmetric with respect to the x axis

FIGURE 6.49

EXAMPLE 1.   Use (1)–(3) and values already known for the trigonometric functions to compute the values of the following expressions.

  a. $\sin\dfrac{13\pi}{6}$ b. $\tan\left(-\dfrac{\pi}{4}\right)$ c. $\sec(-60°)$

Solution.

  a. Since $\dfrac{13\pi}{6} = 2\pi + \dfrac{\pi}{6}$, it follows from (1) that

$$\sin\frac{13\pi}{6} = \sin\frac{\pi}{6} = \frac{1}{2}$$

  b. By (3),

$$\tan\left(-\frac{\pi}{4}\right) = -\tan\frac{\pi}{4} = -1$$

  c. By the version of (3) for degrees,
$$\sec(-60°) = \sec 60° = 2 \quad \square$$

Symmetry with Respect to the *y* Axis

Next we observe that $P((\pi/2) + \theta)$ and $P((\pi/2) - \theta)$ are symmetric with respect to the y axis (Figure 6.50), so that if $P((\pi/2) - \theta) = (a, b)$, then $P((\pi/2) + \theta) = (-a, b)$. Therefore

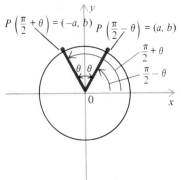

$P\left(\frac{\pi}{2}+\theta\right)$ and $P\left(\frac{\pi}{2}-\theta\right)$ are symmetric with respect to the y axis

FIGURE 6.50

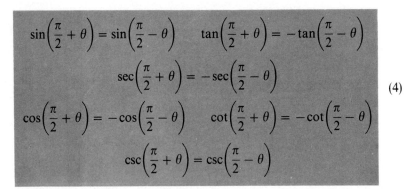

$$\sin\left(\frac{\pi}{2}+\theta\right)=\sin\left(\frac{\pi}{2}-\theta\right) \qquad \tan\left(\frac{\pi}{2}+\theta\right)=-\tan\left(\frac{\pi}{2}-\theta\right)$$

$$\sec\left(\frac{\pi}{2}+\theta\right)=-\sec\left(\frac{\pi}{2}-\theta\right)$$

$$\cos\left(\frac{\pi}{2}+\theta\right)=-\cos\left(\frac{\pi}{2}-\theta\right) \qquad \cot\left(\frac{\pi}{2}+\theta\right)=-\cot\left(\frac{\pi}{2}-\theta\right)$$

$$\csc\left(\frac{\pi}{2}+\theta\right)=\csc\left(\frac{\pi}{2}-\theta\right)$$

(4)

EXAMPLE 2.   Use (4) and values already known for the trigonometric functions to compute the values of the following expressions.

a. $\cos\dfrac{2\pi}{3}$ b. $\csc\dfrac{3\pi}{4}$ c. $\tan 150°$

Solution.

a. Since $\dfrac{2\pi}{3}=\dfrac{\pi}{2}+\dfrac{\pi}{6}$, it follows from (4) that

$$\cos\frac{2\pi}{3}=\cos\left(\frac{\pi}{2}+\frac{\pi}{6}\right)=-\cos\left(\frac{\pi}{2}-\frac{\pi}{6}\right)=-\cos\frac{\pi}{3}=-\frac{1}{2}$$

b. Since $\dfrac{3\pi}{4}=\dfrac{\pi}{2}+\dfrac{\pi}{4}$, it follows from (4) that

$$\csc\frac{3\pi}{4}=\csc\left(\frac{\pi}{2}+\frac{\pi}{4}\right)=\csc\left(\frac{\pi}{2}-\frac{\pi}{4}\right)=\csc\frac{\pi}{4}=\sqrt{2}$$

c. Since $150°=90°+60°$, it follows from the version of (4) for degrees that

$$\tan 150°=\tan(90°+60°)=-\tan(90°-60°)$$
$$=-\tan 30°=-\frac{1}{3}\sqrt{3} \quad \square$$

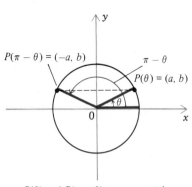

$P(\theta)$ and $P(\pi-\theta)$ are symmetric with respect to the y axis

FIGURE 6.51

The symmetry of the unit circle with respect to the y axis yields another set of formulas. Specifically, the points $P(\pi-\theta)$ and $P(\theta)$ are symmetric with respect to the y axis (Figure 6.51), so that if $P(\theta)=(a, b)$, then $P(\pi-\theta)=(-a, b)$. It follows that

$$\sin(\pi-\theta)=\sin\theta \qquad \tan(\pi-\theta)=-\tan\theta \qquad \sec(\pi-\theta)=-\sec\theta$$

$$\cos(\pi-\theta)=-\cos\theta \qquad \cot(\pi-\theta)=-\cot\theta \qquad \csc(\pi-\theta)=\csc\theta$$

(5)

EXAMPLE 3.   Use (5) and values already known for the trigonometric functions to compute the values of the following expressions.

a. $\sin \dfrac{5\pi}{6}$ b. $\sec \dfrac{2\pi}{3}$

Solution.

a. Since $\dfrac{5\pi}{6} = \pi - \dfrac{\pi}{6}$, it follows from (5) that

$$\sin \frac{5\pi}{6} = \sin\left(\pi - \frac{\pi}{6}\right) = \sin \frac{\pi}{6} = \frac{1}{2}$$

b. Since $\dfrac{2\pi}{3} = \pi - \dfrac{\pi}{3}$, it follows from (5) that

$$\sec \frac{2\pi}{3} = \sec\left(\pi - \frac{\pi}{3}\right) = -\sec \frac{\pi}{3} = -2 \quad \square$$

Symmetry with Respect to the Origin

Next we observe that $P(\pi + \theta)$ and $P(\theta)$ are symmetric with respect to the origin (Figure 6.52), so that if $P(\theta) = (a, b)$, then $P(\pi + \theta) = (-a, -b)$. Consequently

$$\sin(\pi + \theta) = -\sin \theta \qquad \tan(\pi + \theta) = \tan \theta \qquad \sec(\pi + \theta) = -\sec \theta$$
$$\cos(\pi + \theta) = -\cos \theta \qquad \cot(\pi + \theta) = \cot \theta \qquad \csc(\pi + \theta) = -\csc \theta$$
$$(6)$$

The formulas for the tangent and cotangent in (6) give credence to the fact that the tangent and cotangent functions have period π, as we have already mentioned.

EXAMPLE 4.   Use (6) and values already known for the trigonometric functions to compute the values of the following expressions.

a. $\cot \dfrac{7\pi}{6}$ b. $\csc\left(-\dfrac{2\pi}{3}\right)$

Solution.

a. Since $\dfrac{7\pi}{6} = \pi + \dfrac{\pi}{6}$, it follows from (6) that

$$\cot \frac{7\pi}{6} = \cot\left(\pi + \frac{\pi}{6}\right) = \cot \frac{\pi}{6} = \sqrt{3}$$

b. Since $\dfrac{\pi}{3} = \pi + \left(-\dfrac{2\pi}{3}\right)$, it follows from (6) that

$$\csc\left(-\frac{2\pi}{3}\right) = -\csc\left(\pi - \frac{2\pi}{3}\right) = -\csc \frac{\pi}{3} = -\frac{2\sqrt{3}}{3} \quad \square$$

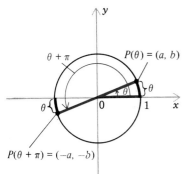

$P(\theta) = (a, b)$

$\theta + \pi$

$P(\theta + \pi) = (-a, -b)$

$P(\theta)$ and $P(\theta + \pi)$ are symmetric with respect to the origin

FIGURE 6.52

EXAMPLE 5.   It is known that $\tan\dfrac{3\pi}{8} = \dfrac{\sqrt{2}}{2-\sqrt{2}}$. Find the value of $\tan\dfrac{11\pi}{8}$.

Solution.   Since $\dfrac{11\pi}{8} = \pi + \dfrac{3\pi}{8}$, it follows from (6) that

$$\tan\frac{11\pi}{8} = \tan\left(\pi + \frac{3\pi}{8}\right) = \tan\frac{3\pi}{8} = \frac{\sqrt{2}}{2-\sqrt{2}} \quad \square$$

Symmetry with Respect to the Line $y = x$

Finally we notice that $P\big((\pi/2)-\theta\big)$ and $P(\theta)$ are symmetric with respect to the line $y = x$ (Figure 6.53). Therefore if $P(\theta) = (a, b)$, then $P\big((\pi/2)-\theta\big) = (b, a)$. The following relations between pairs of trigonometric functions result:

$$\sin\left(\frac{\pi}{2}-\theta\right) = \cos\theta \quad \tan\left(\frac{\pi}{2}-\theta\right) = \cot\theta \quad \sec\left(\frac{\pi}{2}-\theta\right) = \csc\theta$$
$$\cos\left(\frac{\pi}{2}-\theta\right) = \sin\theta \quad \cot\left(\frac{\pi}{2}-\theta\right) = \tan\theta \quad \csc\left(\frac{\pi}{2}-\theta\right) = \sec\theta$$

(7)

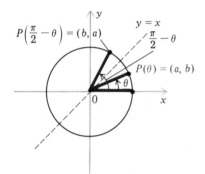

$P(\theta)$ and $P\left(\dfrac{\pi}{2}-\theta\right)$ are symmetric with respect to the line $y = x$

FIGURE 6.53

EXAMPLE 6.   It is known that $\cos\dfrac{\pi}{12} = \dfrac{\sqrt{3}}{4} - \dfrac{1}{2}$. Find $\sin\dfrac{5\pi}{12}$.

Solution.   Since $\dfrac{5\pi}{12} = \dfrac{\pi}{2} - \dfrac{\pi}{12}$, it follows from the first formula in (7) that

$$\sin\frac{5\pi}{12} = \sin\left(\frac{\pi}{2} - \frac{\pi}{12}\right) = \cos\frac{\pi}{12} = \frac{\sqrt{3}}{4} - \frac{1}{2} \quad \square$$

We will use the information gained in this section to graph the trigonometric functions in the next section.

EXERCISES 6.6

In Exercises 1–12, use (1)–(7) and values already known for the trigonometric functions to compute the value of the given expression.

1. $\sin\dfrac{13\pi}{3}$

2. $\tan\dfrac{33\pi}{4}$

3. $\cos\left(-\dfrac{\pi}{3}\right)$

4. $\cot\left(-\dfrac{\pi}{6}\right)$

5. $\sec\dfrac{3\pi}{4}$

6. $\csc\dfrac{2\pi}{3}$

7. $\sin\dfrac{3\pi}{4}$

8. $\tan\dfrac{2\pi}{3}$

9. $\cos\dfrac{5\pi}{4}$

10. $\tan\dfrac{7\pi}{6}$

11. $\sec\left(-\dfrac{25\pi}{6}\right)$

12. $\csc\dfrac{13\pi}{4}$

In Exercises 13–20, use the versions of (1)–(7) for degrees and values already known for the trigonometric functions to compute the value of the given expression.

13. $\tan(-60°)$ **17.** $\cos 390°$

14. $\sec 120°$ **18.** $\tan 405°$

15. $\csc 150°$ **19.** $\cot 780°$

16. $\sin(-210°)$ **20.** $\sec 945°$

In Exercises 21–30, assume that $\sin \theta = \dfrac{1}{3}$ and $\dfrac{\pi}{2} < \theta < \pi$. Determine the value of the given expression.

21. $\sin(\theta + 2\pi)$ **27.** $\sin\left(\dfrac{\pi}{2} + \theta\right)$

22. $\sin(\theta + 8\pi)$

23. $\sin(-\theta)$ **28.** $\sin\left(\dfrac{3\pi}{2} - \theta\right)$

24. $\sin(\theta + \pi)$

25. $\sin(\pi - \theta)$ **29.** $\cos\left(\dfrac{\pi}{2} - \theta\right)$

26. $\sin\left(\dfrac{\pi}{2} - \theta\right)$ **30.** $\cos\left(\dfrac{\pi}{2} + \theta\right)$

In Exercises 31–38, assume that $\cos \theta = -\dfrac{2}{5}$ and $\pi < \theta < \dfrac{3\pi}{2}$. Determine the value of the given expression.

31. $\cos(-\theta)$

32. $\cos(\theta + \pi)$ **36.** $\cos\left(\dfrac{\pi}{2} + \theta\right)$

33. $\cos(\theta - 4\pi)$ **37.** $\sin\left(\dfrac{\pi}{2} - \theta\right)$

34. $\cos\left(\dfrac{\pi}{2} - \theta\right)$ **38.** $\sin(\theta + 3\pi)$

35. $\cos\left(\theta - \dfrac{\pi}{2}\right)$

In Exercises 39–44, assume that $\tan \theta = -2$ and $\dfrac{3\pi}{2} < \theta < 2\pi$. Determine the value of the given expression.

39. $\tan(-\theta)$ **42.** $\tan\left(\theta + \dfrac{\pi}{2}\right)$

40. $\tan(\theta + 9\pi)$

41. $\tan\left(\dfrac{\pi}{2} - \theta\right)$ **43.** $\tan(7\pi - \theta)$

44. $\tan\left(\dfrac{3\pi}{2} + \theta\right)$

In Exercises 45–50, use the fact that $\sin \pi/8 = \dfrac{1}{2}\sqrt{2 - \sqrt{2}}$ to determine the value of the given expression.

45. $\sin\left(-\dfrac{\pi}{8}\right)$ **48.** $\cos\dfrac{3\pi}{8}$

46. $\sin\dfrac{7\pi}{8}$ **49.** $\cos\left(-\dfrac{21\pi}{8}\right)$

47. $\sin\dfrac{9\pi}{8}$ **50.** $\cos\dfrac{27\pi}{8}$

In Exercises 51–54, use the fact that $\tan \pi/8 = \sqrt{2} - 1$ to determine the value of the given expression.

51. $\tan\left(-\dfrac{\pi}{8}\right)$ **53.** $\cot\dfrac{\pi}{8}$

52. $\tan\dfrac{7\pi}{8}$ **54.** $\cot\left(-\dfrac{9\pi}{8}\right)$

In Exercises 55–62, determine the period of the function.

55. $\sin 2x$ **58.** $|\cos x|$ **61.** $|\tan x|$

56. $\cos\dfrac{\pi}{4}x$ **59.** $\tan\left(x + \dfrac{\pi}{2}\right)$ **62.** $\tan x - \sin x$

57. $\cos\sqrt{2}x$ **60.** $\csc\left(\dfrac{\pi}{3} - x\right)$

6.7
GRAPHS OF THE TRIGONOMETRIC FUNCTIONS

Now we are ready to draw the graphs of the trigonometric functions. Throughout this section angles will be measured in radians, and we will use x instead of θ for the angle.

The Graphs of the Sine and Cosine Functions

We begin with the graph of the sine function. We *could* sketch the graph of the sine function on $[0, \pi/2]$ and then use several formulas from Section 6.6 to extend the graph to the real line. However, an easier approach is to use special values of $\sin x$ on $[0, 2\pi]$ to graph $\sin x$ on $[0, 2\pi]$ and then to apply the periodicity of the sine function to complete the graph.

We prepare Table 6.6 with values of sin x:

TABLE 6.6

x	0	$\frac{\pi}{6}$	$\frac{\pi}{4}$	$\frac{\pi}{3}$	$\frac{\pi}{2}$	$\frac{2\pi}{3}$	$\frac{3\pi}{4}$	$\frac{5\pi}{6}$	π	$\frac{7\pi}{6}$	$\frac{5\pi}{4}$	$\frac{4\pi}{3}$	$\frac{3\pi}{2}$	$\frac{5\pi}{3}$	$\frac{7\pi}{4}$	$\frac{11\pi}{6}$	2π
$\sin x$	0	$\frac{1}{2}$	$\frac{\sqrt{2}}{2}$	$\frac{\sqrt{3}}{2}$	1	$\frac{\sqrt{3}}{2}$	$\frac{\sqrt{2}}{2}$	$\frac{1}{2}$	0	$-\frac{1}{2}$	$-\frac{\sqrt{2}}{2}$	$-\frac{\sqrt{3}}{2}$	-1	$-\frac{\sqrt{3}}{2}$	$-\frac{\sqrt{2}}{2}$	$-\frac{1}{2}$	0

As usual, we plot the points obtained by the table and connect them smoothly. Finally we use the fact that $\sin(x + 2\pi) = \sin x$ to obtain the remainder of the graph (Figure 6.54).

We similarly assemble Table 6.7 with values of cos x:

TABLE 6.7

x	0	$\frac{\pi}{6}$	$\frac{\pi}{4}$	$\frac{\pi}{3}$	$\frac{\pi}{2}$	$\frac{2\pi}{3}$	$\frac{3\pi}{4}$	$\frac{5\pi}{6}$	π	$\frac{7\pi}{6}$	$\frac{5\pi}{4}$	$\frac{4\pi}{3}$	$\frac{3\pi}{2}$	$\frac{5\pi}{3}$	$\frac{7\pi}{4}$	$\frac{11\pi}{6}$	2π
$\cos x$	1	$\frac{\sqrt{3}}{2}$	$\frac{\sqrt{2}}{2}$	$\frac{1}{2}$	0	$-\frac{1}{2}$	$-\frac{\sqrt{2}}{2}$	$-\frac{\sqrt{3}}{2}$	-1	$-\frac{\sqrt{3}}{2}$	$-\frac{\sqrt{2}}{2}$	$-\frac{1}{2}$	0	$\frac{1}{2}$	$\frac{\sqrt{2}}{2}$	$\frac{\sqrt{3}}{2}$	1

Then we plot the corresponding points, connect them, and use the fact that $\cos(x + 2\pi) = \cos x$ to complete the graph (Figure 6.55). Observe from Figures 6.54 and 6.55 that the period of sin x and cos x is precisely 2π, a fact alluded to in Section 6.6.

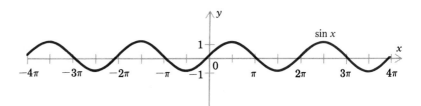

FIGURE 6.54

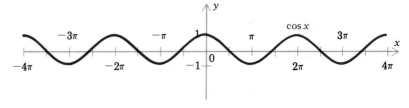

FIGURE 6.55

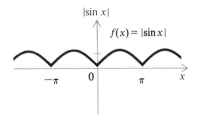

FIGURE 6.56

A knowledge of the graphs of the sine and cosine functions enables us to sketch the graphs of closely related functions.

EXAMPLE 1.   Let $f(x) = |\sin x|$. Sketch the graph of f.

Solution.   Notice that if $\sin x \geq 0$, then $f(x) = \sin x$, whereas if $\sin x < 0$, then $f(x) = -\sin x$. Therefore we obtain the graph of f from the graph of $\sin x$ by leaving unaltered the portions that lie above the x axis and by reflecting through the x axis the portions that lie below the x axis (Figure 6.56).   □

Before we sketch the next trigonometric graph, we recall that if

$$g(x) = f(x) + c \quad \text{and} \quad h(x) = f(x + a)$$

then the graph of g is the graph of f shifted upward c units if $c > 0$ and downward $-c$ units if $c < 0$. Similarly, the graph of h is the graph of f shifted to the left a units if $a > 0$ and to the right $-a$ units if $a < 0$ (see Section 4.1).

EXAMPLE 2.   Sketch the graphs of the following functions, and determine their periods.

a. $f(x) = \sin x + 3$ b. $g(x) = \cos\left(x - \dfrac{4\pi}{3}\right)$

FIGURE 6.57

Solution.

a. The graph of f is the graph of $\sin x$, shifted upward 3 units (Figure 6.57). Since $\sin x$ has period 2π, we find that

$$f(x + 2\pi) = \sin(x + 2\pi) + 3 = \sin x + 3 = f(x)$$

so that the period of f is also 2π, as Figure 6.57 confirms.

b. The graph of g is the graph of $\cos x$, shifted to the right $4\pi/3$ units (Figure 6.58). The fact that $\cos x$ has period 2π implies that

$$g(x + 2\pi) = \cos\left[(x + 2\pi) - \frac{4\pi}{3}\right]$$

$$= \cos\left[\left(x - \frac{4\pi}{3}\right) + 2\pi\right] = \cos\left(x - \frac{4\pi}{3}\right) = g(x)$$

so that the period of g is also 2π, as Figure 6.58 suggests.   □

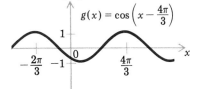

FIGURE 6.58

As Example 2 illustrates, any function of the form

$$\sin x + c \quad \text{or} \quad \cos x + c$$

or any function of the form

$$\sin(x + a) \quad \text{or} \quad \cos(x + a)$$

has period 2π. Thus when the graphs of the sine and cosine functions are shifted vertically or horizontally, the period remains the same—namely, 2π.

EXAMPLE 3. Let $f(x) = \cos\left(x - \dfrac{4\pi}{3}\right) - \dfrac{3}{2}$. Sketch the graph of f.

Solution. The graph of f is obtained from the graph of $\cos x$ by shifting it to the right $4\pi/3$ units and then downward $\frac{3}{2}$ units, which yields the graph in Figure 6.59. ☐

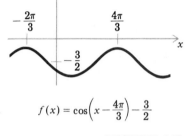

$f(x)$

$f(x) = \cos\left(x - \dfrac{4\pi}{3}\right) - \dfrac{3}{2}$

FIGURE 6.59

In studying the wavy patterns of the graphs of the sine and cosine functions, you might have observed that one is merely the other shifted to the right or the left (see Figures 6.54 and 6.55). We will now confirm this observation. Indeed, combining the formulas

$$\cos(-x) = \cos x \quad \text{and} \quad \cos\left(\frac{\pi}{2} - x\right) = \sin x$$

obtained from (3) and (7) of Section 6.6, we find that

$$\sin x = \cos\left(\frac{\pi}{2} - x\right) = \cos\left[-\left(\frac{\pi}{2} - x\right)\right] = \cos\left(x - \frac{\pi}{2}\right)$$

Consequently the graph of the sine function can be obtained by shifting the graph of the cosine function $\pi/2$ units to the right.

The Graphs of the Remaining Trigonometric Functions

To sketch the graph of the tangent function, we first prepare the values of $\tan x$ as shown in Table 6.8.

TABLE 6.8

x	$-\dfrac{\pi}{3}$	$-\dfrac{\pi}{4}$	$-\dfrac{\pi}{6}$	0	$\dfrac{\pi}{6}$	$\dfrac{\pi}{4}$	$\dfrac{\pi}{3}$
$\tan x$	$-\sqrt{3}$	-1	$-\dfrac{\sqrt{3}}{3}$	0	$\dfrac{\sqrt{3}}{3}$	1	$\sqrt{3}$

Then we connect the corresponding points with a smooth curve, as in Figure 6.60a. Since $\pi/2$ is not in the domain of the tangent function, we investigate the values of $\tan x$ for which x is close to and less than $\pi/2$. Recall that

$$\tan x = \frac{\sin x}{\cos x}$$

Now if x is close to $\pi/2$ and $0 < x < \pi/2$, then $\sin x$ is a positive number close to 1 and $\cos x$ is a positive number close to 0. Thus $\tan x$ is a large positive

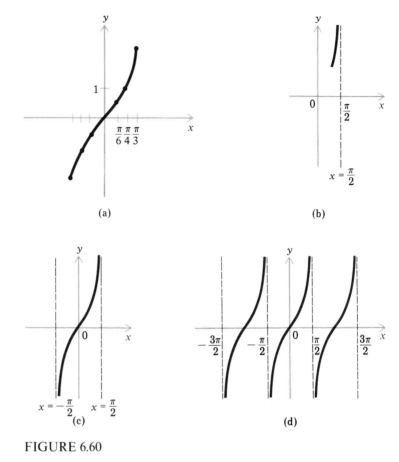

(a)   (b)

(c)   (d)

FIGURE 6.60

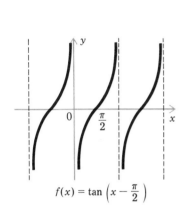

$f(x) = \tan\left(x - \dfrac{\pi}{2}\right)$

FIGURE 6.61

number. The closer x comes to $\pi/2$, the closer cos x comes to 0 and sin x to 1, and hence the larger tan x becomes. This is reflected in the graph drawn in Figure 6.60b. It follows that the vertical line $x = \pi/2$ is a vertical asymptote of the graph of tan x. Since $\tan(-x) = -\tan x$, the vertical line $x = -\pi/2$ is also a vertical asymptote of the graph, so we obtain Figure 6.60c. Finally, from the fact that $\tan(x + \pi) = \tan x$ we conclude that the graph of tan x is as in Figure 6.60d. Notice that the period of tan x is π, as we suggested in Section 6.6.

EXAMPLE 4.   Sketch the graphs of the following functions.

a. $f(x) = \tan\left(x - \dfrac{\pi}{2}\right)$ b. $g(x) = \tan\left(x + \dfrac{\pi}{4}\right) - 1$

Solution.

a. The graph of f is obtained from the graph of tan x by shifting it to the right $\pi/2$ units. Thus we obtain the graph in Figure 6.61.

b. The graph of g is obtained from the graph of tan x by shifting it to the left $\pi/4$ units and then downward 1 unit. This yields the graph in Figure 6.62.   □

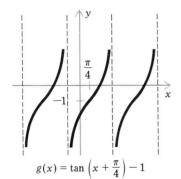

$g(x) = \tan\left(x + \dfrac{\pi}{4}\right) - 1$

FIGURE 6.62

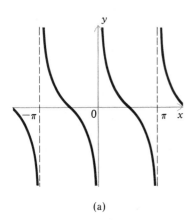
(a)

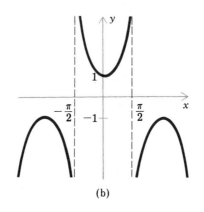
(b)

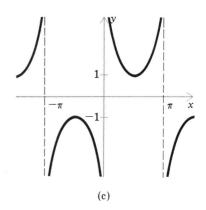
(c)

FIGURE 6.63

We complete this section by drawing the graphs of the remaining trigonometric functions: cot x, sec x, and csc x (Figure 6.63a–c). Notice that any line of the form $x = n\pi$ is a vertical asymptote of the graphs of cot x and csc x, and any line of the form $x = \dfrac{\pi}{2} + n\pi$ is a vertical asymptote of the graph of sec x.

In the next section we will discuss graphs of other functions related to trigonometric functions.

EXERCISES 6.7

In Exercises 1–12, sketch the graph of the function.

1. $\sin x + 2$

2. $\sin x - \dfrac{1}{2}$

3. $\cos x + \dfrac{3}{2}$

4. $\cos x - 1$

5. $\sin\left(x - \dfrac{\pi}{2}\right)$

6. $\sin(x + \pi)$

7. $\cos\left(x + \dfrac{3\pi}{4}\right)$

8. $\cos\left(x - \dfrac{4\pi}{3}\right)$

9. $\sin(-x)$

10. $\sin\left(\dfrac{3\pi}{2} - x\right)$

11. $\cos(-x)$

12. $\cos(\pi - x)$

In Exercises 13–20, sketch the graph of the function.

13. $\sin(x + \pi) - 2$

14. $\sin\left(x - \dfrac{\pi}{3}\right) - \dfrac{1}{3}$

15. $\sin\left(x + \dfrac{\pi}{4}\right) + 1$

16. $\cos\left(x - \dfrac{\pi}{2}\right) - 1$

17. $\cos\left(x + \dfrac{3\pi}{2}\right) + \dfrac{2}{3}$

18. $\cos\left(x + \dfrac{2\pi}{3}\right) + 3$

19. $|\cos x|$

20. $\left|(\sin x) + \dfrac{1}{2}\right|$

In Exercises 21–30, sketch the graph of the function.

21. $\tan\left(x + \dfrac{\pi}{2}\right)$

22. $\tan\left(x - \dfrac{\pi}{3}\right)$

23. $\cot(x - \pi)$

24. $\cot\left(x + \dfrac{\pi}{4}\right)$

25. $\sec\left(x - \dfrac{\pi}{6}\right)$

26. $1 + \sec x$

27. $\csc(x + \pi)$

28. $2 + \csc x$

29. $\tan\left(x - \dfrac{\pi}{3}\right) - 2$

30. $\sec\left(x + \dfrac{\pi}{4}\right) + 3$

In Section 6.7 we analyzed the graphs of the sine and cosine functions, along with horizontal and vertical shifts of these graphs. In this section we will discuss the graphs of *sinusoidal functions*, which are functions of the form $a \sin b(x + c)$ or $a \cos b(x + c)$. These functions and their graphs are important in physical applications.

Vertical Expansions and Contractions

Let $a \neq 0$. For any real number x,

$$-|a| \leq a \sin x \leq |a|$$

Moreover, all numbers between $-|a|$ and $|a|$ are values of $a \sin x$. The number $|a|$ is called the *amplitude* of $a \sin x$. Notice that $a \sin x$ has a greater amplitude than $\sin x$ does if $|a| > 1$ and has a smaller amplitude than $\sin x$ does if $0 < |a| < 1$.

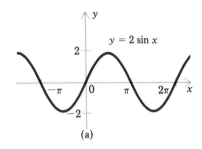

(a)

EXAMPLE 1.   Sketch the graphs of the following functions.

a. $2 \sin x$ b. $\dfrac{1}{3} \sin x$ c. $-\dfrac{1}{3} \sin x$

Solution.

a. To obtain the graph of $2 \sin x$, we multiply the y coordinate of each point on the graph of $\sin x$ by 2 (Figure 6.64a).
b. To obtain the graph of $\frac{1}{3} \sin x$, we multiply the y coordinate of each point on the graph of $\sin x$ by $\frac{1}{3}$ (Figure 6.64b).
c. To obtain the graph of $-\frac{1}{3} \sin x$, we reflect the graph of $\frac{1}{3} \sin x$ through the x axis (Figure 6.64c).   ☐

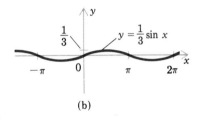

(b)

Similar remarks apply to the function $a \cos x$. In particular, the *amplitude* of $a \cos x$ is $|a|$.

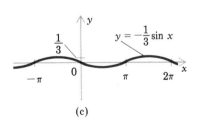

(c)

FIGURE 6.64

EXAMPLE 2.   Sketch the graphs of the following functions.

a. $\dfrac{3}{2} \cos x$ b. $\dfrac{1}{2} \cos x$ c. $-\dfrac{1}{2} \cos x$

Solution.

a. For the graph of $\frac{3}{2} \cos x$ we multiply the y coordinate of each point on the graph of $\cos x$ by $\frac{3}{2}$ (Figure 6.65a).
b. Here we multiply the y coordinate of each point on the graph of $\cos x$ by $\frac{1}{2}$ (Figure 6.65b).
c. For the graph of $-\frac{1}{2} \cos x$ we reflect the graph of $\frac{1}{2} \cos x$ through the x axis (Figure 6.65c).   ☐

Notice that the graph of $2 \sin x$ in Figure 6.64a is obtained by "pulling" the graph of $\sin x$ away from the x axis so that it extends between the lines

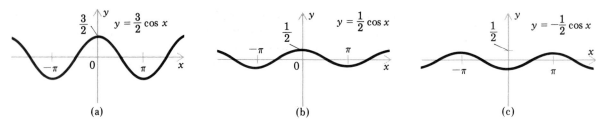

FIGURE 6.65

$y = -2$ and $y = 2$. We say that the graph has been expanded vertically. More generally, the graph of $a \sin x$ is a *vertical expansion* of the graph of $\sin x$ if $a > 1$. Thus the graph of $2 \sin x$ is a vertical expansion of the graph of $\sin x$. In contrast, from Figure 6.64b we see that the graph of $\frac{1}{3} \sin x$ is obtained from the graph of $\sin x$ by compressing it so that it fits between the lines $y = -\frac{1}{3}$ and $y = \frac{1}{3}$. We say that the graph has been contracted vertically. More generally, the graph of $a \sin x$ is a *vertical contraction* of the graph of $\sin x$ if $0 < a < 1$. As a result, the graph of $\frac{1}{3} \sin x$ is a vertical contraction of the graph of $\sin x$. Finally, if $a < 0$, then the graph of $a \sin x$ is obtained by a vertical expansion or contraction of the graph of $\sin x$, plus a reflection through the x axis, as part (c) of Example 1 suggests. Analogous remarks pertain to the graph of $a \cos x$.

Horizontal Expansions and Contractions

Let b be a fixed positive number. Then it can be shown that the function $\sin bx$ is periodic and its values are repeated every $2\pi/b$ units. After all,

$$\sin b\left(x + \frac{2\pi}{b} \right) = \sin(bx + 2\pi) = \sin bx$$

In fact, it turns out that the period of $\sin bx$ is exactly $2\pi/b$, so that the graph of $\sin bx$ is determined by its graph on $[0, 2\pi/b]$. Comparing the period $2\pi/b$ of $\sin bx$ with the period 2π of $\sin x$, we notice that if $b > 1$, then the graph of $\sin x$ must be contracted horizontally in order to obtain the graph of $\sin bx$. In contrast, if $0 < b < 1$, then the graph of $\sin x$ must be expanded horizontally to obtain the graph of $\sin bx$. In either case the graph lies between the horizontal lines $y = -1$ and $y = 1$, just as the graph of $\sin x$ does. Similar comments apply to the graph of $\cos bx$ for $b > 0$.

EXAMPLE 3.   Sketch the graphs of the following functions.

a. $\sin 2x$ b. $\sin \dfrac{2}{3} x$ c. $\cos \pi x$

Solution.

a. In this case $b = 2$, so the period is $2\pi/2 = \pi$. Thus the graph of $\sin 2x$ is obtained by contracting the graph of $\sin x$ horizontally so that the

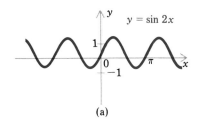

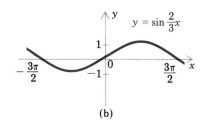

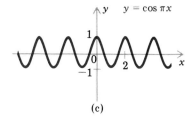

(a)   (b)   (c)

FIGURE 6.66

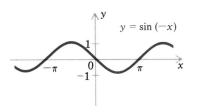

FIGURE 6.67

portion between $x = 0$ and $x = 2\pi$ fits into the region between $x = 0$ and $x = \pi$ (Figure 6.66a).

b. This time $b = \frac{2}{3}$, so the period is $2\pi/(\frac{2}{3}) = 3\pi$. It follows that the graph of $\sin \frac{2}{3}x$ is obtained by expanding the graph of $\sin x$ horizontally so that the portion between $x = 0$ and $x = 2\pi$ fits into the region between $x = 0$ and $x = 3\pi$ (Figure 6.66b).

c. Now $b = \pi$, so the period is $2\pi/\pi = 2$. Therefore the graph of $\cos \pi x$ is obtained by contracting the graph of $\cos x$ horizontally so that the portion between $x = 0$ and $x = 2\pi$ fits into the region between $x = 0$ and $x = 2$ (Figure 6.66c).   □

Since $\sin(-x) = -\sin x$, the graph of $\sin(-x)$ is just the reflection of the graph of $\sin x$ through the x axis (Figure 6.67). Therefore, to obtain the graph of $\sin bx$ when $b < 0$, one draws the graph of $\sin(|b|x)$ and then reflects it through the x axis. The period is $2\pi/|b|$.

EXAMPLE 4.   Sketch the graph of $\sin\left(-\frac{2}{3}x\right)$.

Solution.   We recall the graph of $\sin \frac{2}{3}x$ from Figure 6.66b and reflect it through the x axis to obtain the graph we desire (Figure 6.68).   □

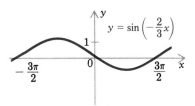

FIGURE 6.68

Since $\cos(-x) = \cos x$, it follows that $\cos bx = \cos(-bx)$ for any constant b, so the graphs of $\cos bx$ and $\cos(-bx)$ are the same. Thus the graph of, say, $\cos(-\pi x)$ is the same as the graph of $\cos \pi x$.

Horizontal Expansions and Contractions Plus Shifts

Next we will discuss the graphs of $\sin b(x + c)$ and $\cos b(x + c)$, where $b \neq 0$ and c may be any number. The graph of $\sin b(x + c)$ is the same as the graph of $\sin bx$ shifted to the left c units if $c > 0$, and to the right $-c$ units if $c < 0$. In particular, $\sin b(x + c)$ and $\sin bx$ have the same period. Similarly, the graph of $\cos b(x + c)$ is the same as the graph of $\cos bx$ shifted $|c|$ units to the left or right. The number c is called the **phase angle** of $\sin b(x + c)$ and of $\cos b(x + c)$.

EXAMPLE 5.   Sketch the graph of $\sin\left[-2\left(x - \dfrac{\pi}{3}\right)\right]$.

Solution.   First we sketch the graph of $\sin 2x$ (see Figure 6.66a again). Then we reflect that graph through the x axis to obtain the graph of $\sin(-2x)$ (Figure 6.69a). Finally, we shift the resulting graph $\pi/3$ units to the right to obtain the graph of the given function (Figure 6.69b).   □

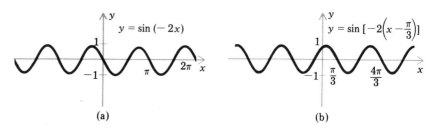

(a)   (b)

FIGURE 6.69

EXAMPLE 6.   Sketch the graph of $\cos\left(-\dfrac{1}{2}x - \dfrac{\pi}{8}\right)$.

Solution.   Notice that

$$\cos\left(-\frac{1}{2}x - \frac{\pi}{8}\right) = \cos\left[-\frac{1}{2}\left(x + \frac{\pi}{4}\right)\right]$$

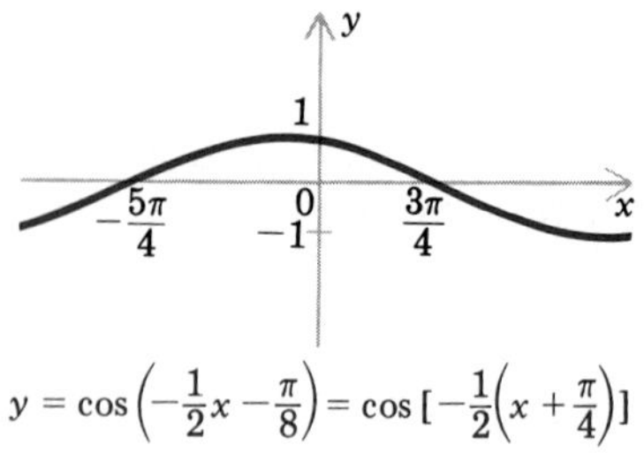

and that the right side of this equation has the form $\cos b(x + c)$. Thus we draw the graph of $\cos(-\tfrac{1}{2}x)$, which is the same as that of $\cos \tfrac{1}{2}x$, and shift the graph $\pi/4$ units to the left (Figure 6.70).   □

$y = \cos\left(-\dfrac{1}{2}x - \dfrac{\pi}{8}\right) = \cos\left[-\dfrac{1}{2}\left(x + \dfrac{\pi}{4}\right)\right]$

FIGURE 6.70

General Sinusoidal Graphs

Finally, we discuss graphs of the general sinusoidal functions, which are obtained from $\sin x$ and $\cos x$ by horizontal shifts as well as horizontal and vertical contractions and expansions. Such altered functions can be written in the form $a \sin b(x + c)$ or $a \cos b(x + c)$.

EXAMPLE 7.   Sketch the graphs of the following functions.

a. $3 \sin\left(2x - \dfrac{4\pi}{3}\right)$ b. $\dfrac{1}{2} \sin 2\left(x - \dfrac{2\pi}{3}\right)$

Solution.

a. When written in the form $a \sin b(x + c)$, the function will become $3 \sin 2(x - 2\pi/3)$. First we sketch the graph of $\sin 2x$ (see Figure 6.66a again) and shift it $2\pi/3$ units to the right to obtain the graph of $\sin 2(x - 2\pi/3)$. Then we expand the latter graph vertically by a factor of 3 to obtain the graph of the given function (Figure 6.71a).

b. If instead of expanding the graph in Figure 6.71a we contract it vertically by a factor of $\frac{1}{2}$, we obtain the desired graph (Figure 6.71b). ☐

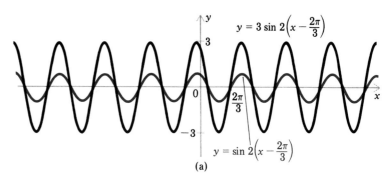

(a)

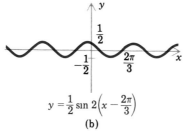

(b)

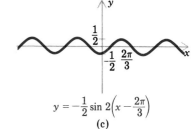

(c)

FIGURE 6.71

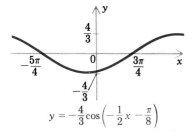

$$y = -\frac{4}{3} \cos\left(-\frac{1}{2}x - \frac{\pi}{8}\right)$$

FIGURE 6.72

In those cases for which $a < 0$ rather than $a > 0$, we first determine the graph of $|a| \sin b(x + c)$ and then reflect it through the x axis in order to obtain the graph of $a \sin b(x + c)$. Thus the graph of $-\frac{1}{2} \sin 2(x - 2\pi/3)$ is as in Figure 6.71c.

The graph of $a \cos b(x + c)$ is obtained from the graph of $\cos b(x + c)$ in exactly the same manner. Thus to obtain the graph of $-\frac{4}{3} \cos(-\frac{1}{2}x - \pi/8)$, we would expand the graph of $\cos(-\frac{1}{2}x - \pi/8)$ in Figure 6.70 vertically by the factor $\frac{4}{3}$ and then reflect the graph through the x axis. The result appears in Figure 6.72.

We call $|a|$ the **amplitude** of $a \sin b(x + c)$ and of $a \cos b(x + c)$.

Simple Harmonic Motion

Many objects in the physical world vibrate, that is, move back and forth in a regular motion. Examples include the heart, waves in an ocean, the strings on a guitar, the vocal cords of a singer, the pendulum on a grandfather clock, a piston in the engine of a car, and the blade of a pneumatic hammer. Because

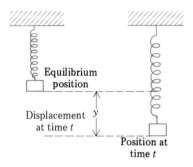

Equilibrium position

Displacement y at time t

Position at time t

FIGURE 6.73

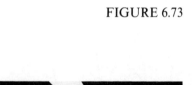

any musical tone (in fact, any sound whatever) is produced by vibrations, the term **harmonic motion** is applied to any vibration, whether it produces a sound or not.

The displacement of a vibrating object from its equilibrium position is frequently given by an equation of the form

$$y = a \sin b(t + c)$$

where t represents time and y the displacement. Any harmonic motion that can be so described is called **simple harmonic motion**. The motion of a bob attached to a spring (Figure 6.73) would be simple harmonic motion if there were no air resistance (for example, if the spring were located in a vacuum).

Another example of simple harmonic motion is provided by a vibrating string, such as a string on a musical instrument. Suppose the string lies along the x axis. Under certain conditions, when the string is plucked or bowed, the disturbance that arises is propagated along the string in a way that can be described by the equation

$$y = a \sin 2\pi \left(\frac{x}{\lambda} - \frac{t}{T} \right) \tag{1}$$

In this equation, y represents the displacement at time t of the point on the string corresponding to x; the letters a, T, and λ have physical interpretations described as follows:

First, the number $|a|$ is the **amplitude** of the vibration. It represents the maximum displacement of any point on the string from the x axis and is related to the loudness of the sound produced by the vibration.

Second, the number T is the **period** of the vibration. If x is held constant in (1), then the resulting function of t is sinusoidal with period T. Physically, T is the length of time it takes a given point on the string to make one complete vibration and is related to the pitch of the sound produced.

Finally, the number λ is called the **wavelength** of the vibration. If we hold t constant in (1), then the resulting function of x is sinusoidal with period λ. Physically, λ is the length between two successive points on the string that experience the same displacement at any given time t (Figure 6.74).

Notice that if we compare the graph of y as a function of x for successive values of t that are close together (Figure 6.74), we see that the points of maximal displacement move along the string. For this reason the disturbance is sometimes called a "traveling wave."

In Section 7.4 we will explain this topic further.

←Wavelength→

FIGURE 6.74

EXERCISES 6.8

In Exercises 1–12, sketch the graph of the given function. Note the amplitude, period, and phase angle of the function.

1. $3 \sin x$

2. $-\frac{2}{3} \sin x$

3. $-\cos x$

4. $3 \cos x$

5. $\sin \pi x$

6. $\sin\left(-\tfrac{1}{2}x\right)$

7. $\cos 2x$

8. $\cos\left(-\tfrac{2}{3}x\right)$

9. $2\sin(x - \pi/3)$

10. $-\sin(-x - 3\pi/4)$

11. $3\cos 2(x + \pi/3)$

12. $\tfrac{3}{2}\cos 4\pi\left(x - \tfrac{1}{2}\right)$

In Exercises 13–15, write down a function with the given amplitude, period, and phase angle, and then sketch the graph of the function.

13. amplitude $= 4$, period $= 2\pi/3$, and phase angle $= \pi$

14. amplitude $= \tfrac{4}{3}$, period $= 3\pi$, and phase angle $= 5\pi/3$

15. amplitude $= \tfrac{3}{2}$, period $= \pi$, and phase angle $= \pi/3$

*16. Assume that as a result of the tides, the depth d of the ocean above a certain barrier reef depends on time x according to the equation

$$d = d_0 + a \sin b\left(x - \frac{\pi}{2}\right)$$

where d_0 is the average depth, a is the difference between the depth at high tide and the average depth, and b is related to the tidal period. Assume that the tides have a period of 12.5 hours, and that the water is 10 feet above the reef at high tide but only 2 feet above the reef at low tide. If a certain boat needs at least 8 feet of water to cross the reef, how long during one tidal period is it safe for the boat to cross the reef?

6.9 INVERSE TRIGONOMETRIC FUNCTIONS

Each of the six basic trigonometric functions is periodic; consequently none has an inverse (see Section 3.7). However, it is possible to restrict the domain of each of these functions to a suitable interval so that the resulting function has an inverse. These inverse functions are called **inverse trigonometric functions**; we will presently define and discuss them.

The Arcsine Function

If we restrict the domain of the sine function to $[-\pi/2, \pi/2]$, then the value of $\sin x$ increases with x, so distinct values of x yield distinct values of $\sin x$. (See Figure 6.75.) Thus, by the comments at the end of Section 3.7, the restricted sine function has an inverse. Since the range of the restricted function is $[-1, 1]$, the domain of the inverse is $[-1, 1]$. The inverse is called the **arcsine function** (or the inverse sine function). Its value at x is usually denoted by arcsin x (sometimes by Arcsin x or $\sin^{-1} x$). Thus

$$\arcsin x = y \quad \text{if and only if} \quad \sin y = x$$
$$\text{for } -1 \le x \le 1 \quad \text{and} \quad -\pi/2 \le y \le \pi/2 \qquad (1)$$

By definition, if $-1 \le x \le 1$, then arcsin x is the unique number y between $-\pi/2$ and $\pi/2$ whose sine is x.

Caution: Even though we write $\sin^2 x$ for $(\sin x)^2$, we *never* write $\sin^{-1} x$ for $(\sin x)^{-1}$. To avoid any possible confusion we will use the notation arcsin x instead of $\sin^{-1} x$ for the arcsine function.

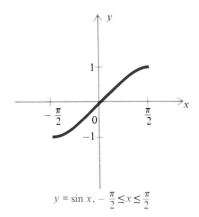

$y = \sin x, \ -\frac{\pi}{2} \le x \le \frac{\pi}{2}$

FIGURE 6.75

EXAMPLE 1.   Evaluate the following expressions.

a. arcsin 0

b. $\arcsin \dfrac{1}{2}$

c. arcsin 1

d. $\arcsin\left(-\dfrac{\sqrt{3}}{2}\right)$

Solution.

a. By (1), arcsin $0 = y$ for the value of y in $[-\pi/2, \pi/2]$ such that $\sin y = 0$. But $\sin y = 0$ if $y = 0$. Thus arcsin $0 = 0$.

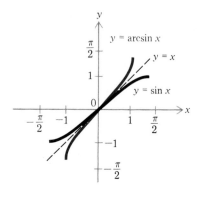

$y = \arcsin x$

$y = x$

$y = \sin x$

FIGURE 6.76

b. By (1), $\arcsin \frac{1}{2} = y$ for the value of y in $[-\pi/2, \pi/2]$ such that $\sin y = \frac{1}{2}$. But $\sin y = \frac{1}{2}$ if $y = \pi/6$. Thus $\arcsin \frac{1}{2} = \pi/6$.
c. By (1), $\arcsin 1 = y$ for the value of y in $[-\pi/2, \pi/2]$ such that $\sin y = 1$. But $\sin y = 1$ if $y = \pi/2$, so $\arcsin 1 = \pi/2$.
d. By (1), $\arcsin (-\sqrt{3}/2) = y$ for the value of y in $[-\pi/2, \pi/2]$ such that $\sin y = -\sqrt{3}/2$. But $\sin y = -\sqrt{3}/2$ if $y = -\pi/3$, so $\arcsin (-\sqrt{3}/2) = -\pi/3$. $\square$

Since the graph of an inverse function can always be obtained from the graph of the original function by reflection through the line $y = x$, the graph of the arcsine function can be obtained from the graph of the restricted sine function in just this manner (Figure 6.76).

The fundamental formulas relating inverses have the following versions for the restricted sine function and the arcsine function:

$$\arcsin (\sin x) = x \quad \text{for} \quad -\frac{\pi}{2} \le x \le \frac{\pi}{2} \tag{2}$$

$$\sin (\arcsin x) = x \quad \text{for} \quad -1 \le x \le 1 \tag{3}$$

For example,

$$\arcsin\left(\sin \frac{\pi}{12} \right) = \frac{\pi}{12} \quad \text{and} \quad \sin\left(\arcsin \frac{1}{3} \right) = \frac{1}{3}$$

Caution: Formula (2) is not valid for values of x outside the interval $[-\pi/2, \pi/2]$. However, for such values of x we can still calculate $\arcsin (\sin x)$ in the following way. Find the number y in $[-\pi/2, \pi/2]$ such that

$$\sin x = \sin y$$

Then by (2),

$$\arcsin(\sin y) = y$$

so that

$$\arcsin(\sin x) = \arcsin(\sin y) = y$$

EXAMPLE 2.   Evaluate $\arcsin (\sin 21\pi/5)$.

Solution.   Notice that $21\pi/5$ is not in $[-\pi/2, \pi/2]$, so we cannot evaluate $\arcsin (\sin 21\pi/5)$ by using (2) directly. But since

$$\sin \frac{21\pi}{5} = \sin\left(4\pi + \frac{\pi}{5} \right) = \sin \frac{\pi}{5}$$

and $-\pi/2 < \pi/5 < \pi/2$, it follows from (2) with $x = \pi/5$ that

$$\arcsin\left(\sin\frac{21\pi}{5}\right) = \arcsin\left(\sin\frac{\pi}{5}\right) = \frac{\pi}{5} \quad \square$$

A well-chosen triangle will assist us in evaluating the expression given in the next example.

EXAMPLE 3. Evaluate $\tan\left(\arcsin\frac{2}{3}\right)$.

Solution. We will evaluate $\tan\left(\arcsin\frac{2}{3}\right)$ by evaluating $\tan y$, where y is the number in $[-\pi/2, \pi/2]$ such that $\arcsin\frac{2}{3} = y$, that is, $\sin y = \frac{2}{3}$. Since $\sin y > 0$, it follows that $0 < y < \pi/2$, and therefore $\tan y > 0$. By applying the Pythagorean Theorem to the triangle drawn in Figure 6.77, which is drawn so that $\sin y = \frac{2}{3}$, we find that

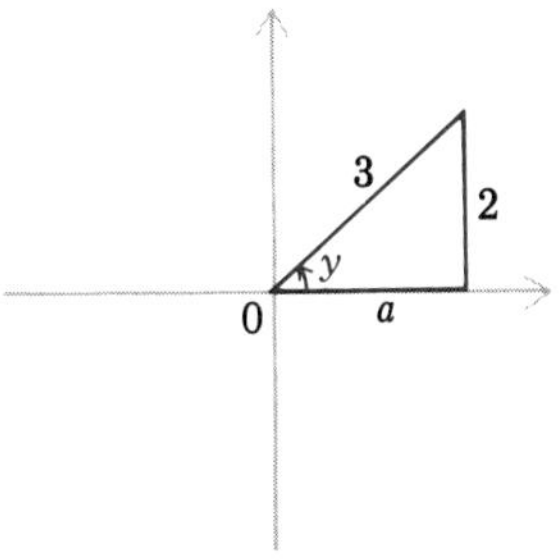

$$a = \sqrt{3^2 - 2^2} = \sqrt{9 - 4} = \sqrt{5}$$

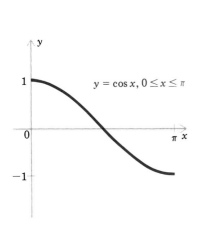

FIGURE 6.77

Consequently

$$\tan y = \frac{2}{\sqrt{5}} = \frac{2}{5}\sqrt{5}$$

so that

$$\tan\left(\arcsin\frac{2}{3}\right) = \tan y = \frac{2}{5}\sqrt{5} \quad \square$$

A notable feature of Example 3 is that we did *not* need to evaluate $\arcsin\frac{2}{3}$ in order to evaluate $\tan\left(\arcsin\frac{2}{3}\right)$.

The Arccosine Function

On the interval $[0, \pi]$ the value of $\cos x$ decreases as x increases, so that the restricted cosine function has an inverse (Figure 6.78). Since the restricted function has range $[-1, 1]$, it follows that the domain of the inverse is also $[-1, 1]$. The inverse is called the **arccosine function** (or inverse cosine function). Its value at x is denoted by $\arccos x$ (sometimes by Arccos x or $\cos^{-1} x$). Thus

> $\arccos x = y$ if and only if $\cos y = x$
> $\qquad\qquad$ for $-1 \leq x \leq 1$ and $0 \leq y \leq \pi$ (4)

By definition, if $-1 \leq x \leq 1$, then $\arccos x$ is the unique number y in $[0, \pi]$ such that $\cos y = x$.

EXAMPLE 4. Evaluate the following expressions.

a. $\arccos 1$ 　　　　b. $\arccos\left(-\frac{1}{2}\right)$

Solution.
a. By (4), $\arccos 1 = y$ for the value of y in $[0, \pi]$ such that $\cos y = 1$. But $\cos y = 1$ if $y = 0$, so that $\arccos 1 = 0$.

$y = \cos x,\ 0 \leq x \leq \pi$

FIGURE 6.78

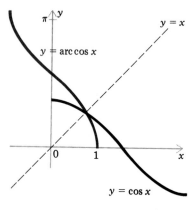

$y = \arccos x$

$y = x$

$y = \cos x$

FIGURE 6.79

b. By (4), $\arccos(-\tfrac{1}{2}) = y$ for the value of y in $[0, \pi]$ such that $\cos y = -\tfrac{1}{2}$. But $\cos y = -\tfrac{1}{2}$ if $y = 2\pi/3$, so $\arccos(-\tfrac{1}{2}) = 2\pi/3$. $\quad\square$

The graph of the arccosine function, which is obtained by reflecting the restricted cosine function through the line $y = x$, appears in Figure 6.79.

The fundamental relations between the restricted cosine function and the arccosine function are:

$$\arccos(\cos x) = x \quad \text{for} \quad 0 \le x \le \pi \qquad (5)$$

$$\cos(\arccos x) = x \quad \text{for} \quad -1 \le x \le 1 \qquad (6)$$

In particular,

$$\arccos\left[\cos\frac{2\pi}{3}\right] = \frac{2\pi}{3} \quad \text{and} \quad \cos\left(\arccos\frac{1}{\sqrt{5}}\right) = \frac{1}{\sqrt{5}}$$

If x lies outside $[0, \pi]$ we can evaluate $\arccos(\cos x)$ by finding the value of y in $[0, \pi]$ for which $\cos y = \cos x$. Then by (5),

$$\arccos(\cos x) = \arccos(\cos y) = y$$

EXAMPLE 5.   Evaluate $\arccos\left(\cos\dfrac{20\pi}{9}\right)$.

Solution.   We cannot evaluate $\arccos(\cos 20\pi/9)$ by invoking (5) with $x = 20\pi/9$ because $20\pi/9$ is not in $[0, \pi]$. But since $20\pi/9 = 2\pi/9 + 2\pi$ and $0 < 2\pi/9 < \pi$, we have

$$\cos\frac{20\pi}{9} = \cos\left(2\pi + \frac{2\pi}{9}\right) = \cos\frac{2\pi}{9}$$

Consequently by (5) with $x = 2\pi/9$, we conclude that

$$\arccos\left(\cos\frac{20\pi}{9}\right) = \arccos\left(\cos\frac{2\pi}{9}\right) = \frac{2\pi}{9} \quad\square$$

EXAMPLE 6.   Evaluate $\sin[\arccos(-\tfrac{4}{7})]$.

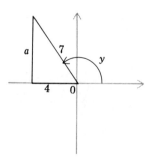

FIGURE 6.80

Solution.   We will evaluate $\sin[\arccos(-\tfrac{4}{7})]$ by evaluating $\sin y$, where y is the number in $[0, \pi]$ such that $\arccos(-\tfrac{4}{7}) = y$, that is, $\cos y = -\tfrac{4}{7}$. Since $\cos y < 0$, it follows that $\pi/2 < y < \pi$, so that $\sin y > 0$. By applying the Pythagorean Theorem to the triangle in Figure 6.80, for which $\cos y = -\tfrac{4}{7}$, we find that

$$a = \sqrt{7^2 - 4^2} = \sqrt{49 - 16} = \sqrt{33}$$

Therefore

$$\sin y = \frac{\sqrt{33}}{7}$$

so that

$$\sin\left[\arccos\left(-\frac{4}{7}\right)\right] = \sin y = \frac{\sqrt{33}}{7} \quad \square$$

The Arctangent Function

On the interval $(-\pi/2, \pi/2)$, the value of $\tan x$ increases as x increases, so that the restricted tangent function has an inverse (Figure 6.81). Since the restricted tangent function has range $(-\infty, \infty)$, we conclude that the domain of the inverse is $(-\infty, \infty)$, that is, all the real numbers. The inverse function is called the **arctangent function** (or inverse tangent function). Its value at x is denoted by arctan x (sometimes Arctan x or $\tan^{-1} x$). Thus

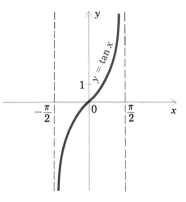

FIGURE 6.81

arctan $x = y$ if and only if $\tan y = x$

$$\text{for any } x \text{ and for } -\frac{\pi}{2} < y < \frac{\pi}{2} \quad (7)$$

For any number x, arctan x is the unique number y in $(-\pi/2, \pi/2)$ such that $\tan y = x$.

EXAMPLE 7.   Evaluate the following expressions.

   a. arctan 1   b. arctan $(-\sqrt{3})$

Solution.

a. By (7), arctan $1 = y$ for the value of y in $(-\pi/2, \pi/2)$ such that $\tan y = 1$. But $\tan \pi/4 = 1$, so that arctan $1 = \pi/4$.

b. By (7), arctan $(-\sqrt{3}) = y$ for the value of y in $(-\pi/2, \pi/2)$ such that $\tan y = -\sqrt{3}$. But $\tan(-\pi/3) = -\sqrt{3}$, so that arctan $(-\sqrt{3}) = -\pi/3$. $\square$

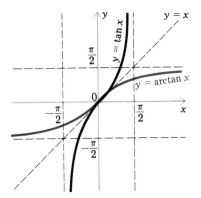

FIGURE 6.82

The graph of the arctangent function, which is obtained by reflecting the graph of the restricted tangent function through the line $y = x$, is shown in Figure 6.82.

We also have

$$\arctan(\tan x) = x \quad \text{for} \quad -\frac{\pi}{2} < x < \frac{\pi}{2}. \quad (8)$$

$$\tan(\arctan x) = x \quad \text{for all } x \quad (9)$$

EXAMPLE 8.   Evaluate sec (arctan $\frac{3}{5}$).

Solution.   We will evaluate sec (arctan $\frac{3}{5}$) by evaluating sec y for the value of y in $(-\pi/2, \pi/2)$ such that arctan $\frac{3}{5} = y$, that is, $\tan y = \frac{3}{5}$. Since $\tan y > 0$, we know that $0 < y < \pi/2$, so that sec $y > 0$. Applying the Pythagorean Theorem to the triangle in Figure 6.83, for which $\tan y = \frac{3}{5}$, we find that

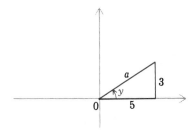

FIGURE 6.83

$$a = \sqrt{3^2 + 5^2} = \sqrt{9 + 25} = \sqrt{34}$$

Therefore

$$\sec y = \frac{\sqrt{34}}{5}$$

so that

$$\sec\left(\arctan\frac{3}{5}\right) = \sec y = \frac{\sqrt{34}}{5} \quad \square$$

EXAMPLE 9.   Express sec (arctan x) in terms of x, without the use of trigonometric functions.

Solution.   First we write sec (arctan x) as sec y, where y = arctan x. But y = arctan x means that tan $y = x$ and y is in $(-\pi/2, \pi/2)$. Now x can be either positive or negative. However, either of the triangles that arise (see Figure 6.84) yields tan $y = x$. For each triangle, the Pythagorean Theorem tells us that

$$a = \sqrt{1^2 + x^2} = \sqrt{1 + x^2}$$

Consequently

$$\sec(\arctan x) = \sec y = \frac{a}{1} = a = \sqrt{1 + x^2} \quad \square$$

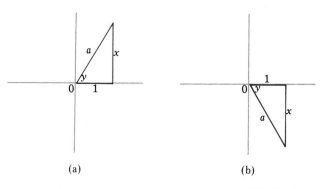

(a)   (b)

FIGURE 6.84

The Remaining Inverse Trigonometric Functions   There are also inverses of suitable restrictions of the cotangent, secant, and cosecant functions. They are not used nearly as frequently as the three we have presented. For that reason we merely give their definitions.

> arccot $x = y$ if and only if   cot $y = x$
>
> for any x and for $0 < y < \pi$
>
> arcsec $x = y$ if and only if   sec $y = x$
>
> for $|x| \geq 1$ and for $0 \leq y \leq \pi$ except $y = \frac{\pi}{2}$

$$\text{arccsc } x = y \quad \text{if and only if} \quad \text{csc } y = x$$

$$\text{for } |x| \ge 1 \text{ and for } 0 < |y| \le \frac{\pi}{2}$$

Approximating Values of the Inverse Trigonometric Functions

If angles are measured in degrees rather than radians, then it is convenient to express the values of the inverse trigonometric functions in terms of degrees. In that case the values of the inverse sine and the inverse tangent functions would lie between $-90°$ and $90°$, rather than between $-\pi/2$ and $\pi/2$. Similarly, the values of the inverse cosine function would lie between $0°$ and $180°$. We can use either a calculator or Table E to approximate the values of the inverse trigonometric functions in terms of degrees. The following example illustrates both methods.

<u>**EXAMPLE 10.**</u> Approximate arcsin 0.3456 by using
    a. a calculator   b. Table E

Solution.

a. One needs to proceed according to the format of one's calculator. On ours we first set the calculator for degree measure, then key in .3456, and finally press the "inv" and "sin" keys. Then the display reads 20.218426, which means that

$$\text{arcsin } 0.3456 \approx (20.218426)° \approx 20°13'$$

b. We look down the sin x column of Table E until we come to numbers approximating 0.3456, and obtain the following:

$$\sin 20°10' \approx 0.3448$$

$$\sin \text{ ??} = 0.3456$$

$$\sin 20°20' \approx 0.3475$$

Since $0.3456 - 0.3448 = 0.0008$ and $0.3475 - 0.3448 = 0.0027$, it follows that 0.3456 is $0.0008/0.0027 = \frac{8}{27}$ of the distance from sin 20°10′ to sin 20°20′. Therefore we take arcsin 0.3456 to be $\frac{8}{27}$ of the distance from 20°10′ to 20°20′, which means that

$$\text{arcsin } 0.3456 \approx 20°13' \quad \square$$

The process of interpolation we followed in part (b) is exactly the one we used in solving Example 6 of Section 6.3, where we approximated the value of θ for which $\sin \theta = 0.9200$. You can see now why there is a resemblance: In Section 6.3 we were merely finding an approximate value for arcsin 0.9200.

EXERCISES 6.9

In Exercises 1–6, find the value of the given expression.

1. $\arcsin(-1)$

2. $\arcsin \sqrt{3}/2$

3. $\arccos 0$

4. $\arccos(-1)$

5. $\arctan 0$

6. $\arctan \sqrt{3}/3$

In Exercises 7–18, find the value of the given expression.

7. $\operatorname{arccot} 0$

8. $\operatorname{arccot} \sqrt{3}$

9. $\operatorname{arccot}(-1)$

10. $\operatorname{arccot}(-\sqrt{3}/3)$

11. $\operatorname{arcsec} 1$

12. $\operatorname{arcsec} \sqrt{2}$

13. $\operatorname{arcsec}(-2)$

14. $\operatorname{arcsec}(-\frac{2}{3}\sqrt{3})$

15. $\operatorname{arccsc} 1$

16. $\operatorname{arccsc} \frac{2}{3}\sqrt{3}$

17. $\operatorname{arccsc}(-1)$

18. $\operatorname{arccsc}(-2)$

In Exercises 19–30, use a calculator (or Table E) to estimate the value of the given expression.

19. $\arcsin 0.3$

20. $\arcsin(-0.84)$

21. $\arccos 0.54$

22. $\arccos(-0.17)$

23. $\arctan 0.11$

24. $\arctan(-7.81)$

25. $\operatorname{arccot} 0.94$

26. $\operatorname{arccot}(-55.5)$

27. $\operatorname{arcsec} 2.4$

28. $\operatorname{arcsec}(-1.5)$

29. $\operatorname{arccsc}(-4.7)$

30. $\operatorname{arccsc} 6.17$

In Exercises 31–42, determine the value of the given expression.

31. $\arcsin(\sin \pi/6)$

32. $\arcsin(\sin \frac{1}{5})$

33. $\arcsin(\sin 5\pi/3)$

34. $\arcsin[\sin(-7\pi/6)]$

35. $\arccos(\cos 2\pi/3)$

36. $\arccos(\cos \pi)$

37. $\arccos(\cos 19\pi/4)$

38. $\arccos[\cos(-8\pi/3)]$

39. $\arctan(\tan \pi/13)$

40. $\arctan[\tan(-5\pi/6)]$

41. $\operatorname{arccot}[\cot(-19\pi/6)]$

42. $\operatorname{arcsec}(\sec 23\pi/4)$

In Exercises 43–46, determine the value of the given expression.

43. $\arcsin(\cos \pi/3)$

44. $\arccos(\sin \pi/4)$

45. $\arctan(\cos \pi/2)$

46. $\arctan(\cos 8\pi)$

In Exercises 47–60, determine the value of the given expression.

47. $\sin(\arcsin \frac{1}{4})$

48. $\sin(\arccos \frac{1}{2})$

49. $\sin[\arctan(-\sqrt{3}/3)]$

50. $\cos(\arcsin 1)$

51. $\cos[\arctan(-1)]$

52. $\tan[\arccos(-1)]$

53. $\tan[\arcsin(-\sqrt{2}/2)]$

54. $\sin(\arccos \frac{1}{6})$

55. $\cos[\arctan(-\frac{4}{3})]$

56. $\cos(\operatorname{arccsc} \frac{9}{2})$

57. $\tan[\arcsin(-\frac{4}{5})]$

58. $\tan[\operatorname{arcsec}(-3)]$

59. $\sec[\arcsin(-\frac{1}{3})]$

60. $\csc[\arctan(-14)]$

In Exercises 61–64, determine the value of the given expression.

61. $\arcsin[\tan(\arccos \sqrt{2}/2)]$

62. $\arccos\{\sin[\arctan(-1)]\}$

63. $\arctan\{\frac{2}{3}\sin[\arccos(-\frac{1}{2})]\}$

64. $\arccos[\frac{1}{2}\tan(\arcsin \sqrt{3}/2)]$

In Exercises 65–70, verify the given equation.

65. $\sin(\arccos x) = \sqrt{1-x^2}$ (*Hint:* See Example 9, but here let $\arccos x = y$, so that $\cos y = x$. Then express $\sin y$ in terms of x.)

66. $\sin(\arctan x) = \dfrac{x}{\sqrt{1+x^2}}$

67. $\sec(\arctan x) = \sqrt{1+x^2}$

68. $\tan(\arcsin x) = \dfrac{x}{\sqrt{1-x^2}}$

69. $\cos(\arcsin x^2) = \sqrt{1-x^4}$

70. $\tan\left(\arccos \dfrac{1}{\sqrt{x}}\right) = \sqrt{x-1}$

In Exercises 71–73, sketch the graph of the function.

71. $|\arcsin x|$

72. $-2 + \arccos x$

73. $\arctan |x|$

In Exercises 74–78, verify the equation.

74. $\arcsin(-x) = -\arcsin x$ (*Hint:* Let $\arcsin(-x) = z$, and show that $\arcsin x = -z$.)

75. $\arctan(-x) = -\arctan x$

76. $\arcsin x = \arctan \dfrac{x}{\sqrt{1 - x^2}}$

77. $\arcsin x + \arccos x = \pi/2$ (*Hint:* Let $\arcsin x = z$, and show that $\arccos x = \pi/2 - z$.)

78. $\arctan x + \text{arccot } x = \pi/2$

79. C. L. Dodgson (better known as Lewis Carroll) proved that if a, b, and c are constants such that $bc = 1 + a^2$, then

$$\arctan \frac{1}{a + b} + \arctan \frac{1}{a + c} = \arctan \frac{1}{a}$$

Taking $a = b = 1$ and $c = 2$, show that

$$\arctan \frac{1}{2} + \arctan \frac{1}{3} = \frac{\pi}{4}$$

KEY TERMS

angle	trigonometric functions	reference angle
initial side	sine function	periodic function
terminal side	cosine function	sinusoidal function
vertex	tangent function	inverse trigonometric function
standard position	cotangent function	arcsine function
coterminal angles	secant function	arccosine function
degree measure	cosecant function	arctangent function
radian measure	Pythagorean Identity	

KEY FORMULAS

$$\theta° = \left(\frac{\pi}{180} \cdot \theta \right) \text{ radians}$$

$$\theta \text{ radians} = \left(\frac{180}{\pi} \cdot \theta \right)°$$

$$\sin \theta = \frac{b}{c} \qquad \cot \theta = \frac{\cos \theta}{\sin \theta} = \frac{a}{b}$$

$$\cos \theta = \frac{a}{c} \qquad \sec \theta = \frac{1}{\cos \theta} = \frac{c}{a}$$

$$\tan \theta = \frac{\sin \theta}{\cos \theta} = \frac{b}{a} \qquad \csc \theta = \frac{1}{\sin \theta} = \frac{c}{b}$$

$$\sin^2 \theta + \cos^2 \theta = 1$$

$$\tan^2 \theta + 1 = \sec^2 \theta$$

$$1 + \cot^2 \theta = \csc^2 \theta$$

For any integer n:

$$\sin(\theta + 2n\pi) = \sin \theta \qquad \cot(\theta + n\pi) = \cot \theta$$

$$\cos(\theta + 2n\pi) = \cos \theta \qquad \sec(\theta + 2n\pi) = \sec \theta$$

$$\tan(\theta + n\pi) = \tan \theta \qquad \csc(\theta + 2n\pi) = \csc \theta$$

$\arcsin x = y$ if and only if $\sin y = x$

$\arccos x = y$ if and only if $\cos y = x$

$\arctan x = y$ if and only if $\tan y = x$

REVIEW EXERCISES

In Exercises 1–6, draw the angle in standard position having the given measure.

1. $315°$

2. $-390°$

3. $960°$

4. $\frac{8}{3}\pi$ radians

5. $-\frac{11}{4}$ radians

6. 7π radians

In Exercises 7–22, determine the value of the expression.

7. $\sin(-19\pi/4)$

8. $\sin 13\pi/2$

9. $\cos 29\pi/6$

10. $\cos(-13\pi/3)$

11. $\tan 15\pi/4$

12. $\tan(-17\pi/6)$

13. $\cot 7\pi/3$

14. $\cot(-11\pi/4)$

15. $\sec 13\pi/6$

16. $\sec(-23\pi/4)$

17. $\csc(44\pi/3)$

18. $\csc(-23\pi/2)$

19. $\sin 930°$

20. $\cos(-585°)$

21. $\tan 1110°$

22. $\sec(-750°)$

In Exercises 23–28, find the values of the six trigonometric functions at θ.

23. $\sin \theta = -\frac{2}{5}, \pi < \theta < 3\pi/2$

24. $\cos \theta = 0.9, -2\pi < \theta < -3\pi/2$

25. $\tan \theta = -8, \pi/2 < \theta < \pi$

26. $\cot \theta = -\frac{2}{3}, -\pi/2 < \theta < 0$

27. $\sec \theta = 4, 4\pi < \theta < 9\pi/2$

28. $\csc \theta = -\frac{4}{3}, -\pi < \theta < -\pi/2$

In Exercises 29–40, assume that $\sin \theta = -\frac{1}{4}$ and $\frac{3}{2}\pi < \theta < 2\pi$. Determine the values of the given expressions.

29. $\sin(\theta + \pi)$

30. $\sin(\theta - \pi/2)$

31. $\cos(\pi/2 - \theta)$

32. $\cos(\theta + 3\pi)$

33. $\tan(\theta - 3\pi/2)$

34. $\tan(4\pi - \theta)$

35. $\cot(\theta - \pi)$

36. $\cot(-\pi/2 - \theta)$

37. $\sec(\theta - \pi/2)$

38. $\sec(-\theta)$

39. $\csc(2\pi - \theta)$

40. $\csc(\theta + 3\pi/2)$

[C] In Exercises 41–46, use a calculator or Table E with interpolation to approximate the values of the given expressions.

41. $\sin 14°20'$

42. $\sin 62°50'$

43. $\cos 47°36'$

44. $\cos 185°15'$

45. $\tan 8°8'$

46. $\tan(-22°48')$

In Exercises 47–50, solve the right triangle with the given data pertaining to Figure 6.85.

47. $\beta = 10°, a = 3$

48. $\alpha = 29.1°, c = 2.1$

49. $a = 5, b = 6.7$

50. $b = 4.9, c = 17$

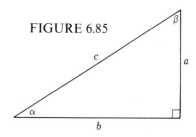

FIGURE 6.85

51. Use the information in Figure 6.86 to determine the values of all six trigonometric functions at θ.

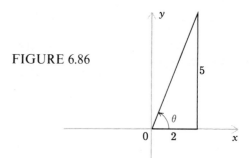

FIGURE 6.86

In Exercises 52–54, sketch the graph.

52. $1 + \sin x$

53. $\cos(x + \pi/6)$

54. $|\tan x|$

In Exercises 55–68, find the value of the given expression.

55. $\arccos \frac{1}{2}$

56. $\arcsin(-\frac{1}{2})$

57. $\arctan \sqrt{3}/3$

58. $\arctan(-1)$

59. $\arcsin(\sin 8\pi/3)$

60. $\arccos[\cos(-\pi/2)]$

61. $\arctan(\tan 7\pi/4)$

62. $\arctan[\tan(-1.7)]$

63. $\arcsin(\tan \pi/4)$

64. $\arccos(\sec \pi)$

65. $\arctan(2 \sin \pi/6)$

66. $\sin(\text{arccsc } 3)$

67. $\cos[\text{arccot}(-2)]$

68. $\tan[\text{arcsec}(-\pi/2)]$

69. For each of the numbers $x = 0.1, 0.01$, and 0.001 (in radians), calculate the value of the given expression. Then guess the limiting value of the expression as x approaches 0.

a. $\dfrac{\sin x}{x}$

b. $\dfrac{\cos x - 1}{x}$

c. $\dfrac{\tan x}{x}$

d. $x \tan^2 x$

70. For $x = 0.1, 0.01$, and 0.001 (in radians), compare the values of $\sin x$ and $x - \frac{1}{6}x^3$. (Often $x - \frac{1}{6}x^3$ is used as an approximate value of $\sin x$ for small values of x.)

71. For $x = 0.1, 0.01$, and 0.001 (in radians), compare the values of $\cos x$ and $1 - \frac{1}{2}x^2$. (Often $1 - \frac{1}{2}x^2$ is used as an approximate value of $\cos x$ for small values of x.)

72. Show that $\pi \sin \pi/6 + \arcsin 1 = \pi$.

73. Prove that $\sec(x + \pi) = -\sec x$.

74. Explain why the equation $\sin x \cos x = 1$ has no solution.

75. If the hour hand on a kitchen clock is 4 inches long and the minute hand is 6 inches long, how much farther does the tip of the minute hand travel during a 12-hour span than the tip of the hour hand?

76. When fully inflated, a tire on an automobile is 1 foot in radius. Through how many revolutions does the tire travel in a mile?

77. Chris sees a bolt of lightning strike a tower on the ground some 7000 feet away. The angle of elevation of the cloud is $40°$. How high is the cloud?

78. At the time when the moon is a half-moon, the moon angle in Figure 6.87 is $90°$. Suppose the earth angle is calculated to be $89.85°$. Determine how many times as far away from the earth the sun is as the moon.

FIGURE 6.87

79. A ski lift is divided into two parts. The first part is 1600 feet long and makes an angle of elevation of $25°$ with the horizontal. The second part is 1100 feet long and makes an angle of elevation of $40°$ with the horizontal. Determine the straight-line distance covered by the two parts of the ski lift together.

80. A pilot is instructed by the control tower to rise from ground level to an altitude of 30,000 feet by the time it has flown 10 miles. If the pilot is to achieve the goal by flying along a straight line,
a. at what angle would it rise?
b. what ground distance would the plane travel during the interim?

81. About 200 B.C., Eratosthenes of Alexandria, Egypt, devised a way of approximating the circumference of the earth. From astronomical data available to him as head librarian in Alexandria he knew that at noon on a particular June day the sun would be directly overhead in Syene, Egypt. By direct measurement he found that at the same time the sun was approximately $7°\,12'$ from the vertical in Alexandria (Figure 6.88). Using 490 miles as the distance between Syene and Alexandria, along with (3) of Section 6.1, determine an approximate value for the earth's circumference. Round your answer to the nearest mile.

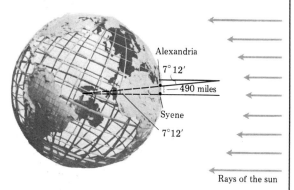

The measurement of the circumference of the earth by Eratosthenes

FIGURE 6.88

82. Suppose the angle of elevation of the sun is $30°$. Use Snell's Law (see Section 6.2) to find the angle θ at which a swimmer under water must look up in order to see the sun (Figure 6.89). Take the index of refraction μ to be 1.333.

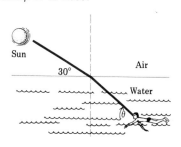

FIGURE 6.89

7

Trigonometric Identities and Equations

$\mathbf{I}$n Chapter 2 we studied equations involving polynomials, and in Chapter 5 we considered equations involving logarithms and exponentials. In this chapter we study equations involving trigonometric functions. Those equations are of two types. The first type consists of equations such as

$$\sin^2 x + \cos^2 x = 1 \quad \text{and} \quad \tan(x + y) = \frac{\tan x + \tan y}{1 - \tan x \tan y}$$

that are valid for all values of the variables for which every expression in the equation is defined. Such an equation is a **_trigonometric identity_**, and the values of the variables for which a trigonometric identity is defined are its **_admissible values_**. The initial four sections of the chapter will be devoted to trigonometric identities, which we will usually refer to just as identities. As an application of the trigonometric identities, we will show that identity (5) in Section 7.4 can be used to help us understand the nature of musical notes produced by stringed instruments such as pianos, violins, or guitars, and in particular the nature of overtones.

The other type of equation we will study in this chapter consists of **_conditional trigonometric equations_**, which are valid only for isolated values of the variable. They are exemplified by

$$\cos 2x = \cos x \quad \text{and} \quad \tan^2 x - \sec x - 1 = 0$$

In the final section of the chapter we will discuss solutions of conditional trigonometric equations.

7.1 EXAMPLES OF TRIGONOMETRIC IDENTITIES

We have already encountered a number of trigonometric identities in Chapter 6. Those that will play a prominent role in this chapter are the following:

Defining Identities

$$\tan x = \frac{\sin x}{\cos x} \qquad \sec x = \frac{1}{\cos x} \qquad \cot x = \frac{1}{\tan x}$$

$$\cot x = \frac{\cos x}{\sin x} \qquad \csc x = \frac{1}{\sin x}$$

Symmetry Identities

$$\sin(-x) = -\sin x \qquad \cot(-x) = -\cot x$$

$$\cos(-x) = \cos x \qquad \sec(-x) = \sec x$$

$$\tan(-x) = -\tan x \qquad \csc(-x) = -\csc x$$

Pythagorean Identities

$$\sin^2 x + \cos^2 x = 1$$

$$\tan^2 x + 1 = \sec^2 x$$

$$\cot^2 x + 1 = \csc^2 x$$

Each of the Defining Identities except $\cot x = 1/\tan x$ is actually the definition of the function on the left side of the equation, and this one is an immediate consequence of the definitions of $\tan x$ and $\cot x$. The Symmetry Identities follow readily from the definitions of the functions involved and are closely related to the graphs of the respective functions. Finally, the Pythagorean Identities first appeared in (5)–(7) of Section 6.2.

In each of the examples that follow we will be asked to verify a proposed identity, by which we mean to demonstrate that the given equation is valid for all admissible values of the variable. The verification usually will involve the use of identities listed above, together with algebraic manipulation.

A basic method we will employ in verifying identities involves altering only *one* side of the proposed identity and showing, perhaps in several steps, that this side is equal to the other side of the proposed identity. Normally we will pick the more complicated side for alteration. Examples 1–4 illustrate this method.

EXAMPLE 1.   Verify $\cos^4 x - \sin^4 x = \cos^2 x - \sin^2 x$.

Solution.   We alter the left side by factoring it and then using the first Pythagorean Identity, $\sin^2 x + \cos^2 x = 1$:

$$\cos^4 x - \sin^4 x = (\cos^2 x - \sin^2 x)\overbrace{(\cos^2 x + \sin^2 x)}^{=\,1}$$

$$= \cos^2 x - \sin^2 x$$

We have thus obtained the right side of the proposed identity, so the solution is complete.   □

EXAMPLE 2.   Verify $\dfrac{1 - \tan t}{1 + \tan t} = \dfrac{\cot t - 1}{\cot t + 1}$.

Solution.   We alter the right side by using the identity $\cot t = 1/\tan t$:

$$\frac{\cot t - 1}{\cot t + 1} = \frac{\dfrac{1}{\tan t} - 1}{\dfrac{1}{\tan t} + 1}$$

$$= \frac{\dfrac{1 - \tan t}{\tan t}}{\dfrac{1 + \tan t}{\tan t}} = \frac{1 - \tan t}{1 + \tan t}$$

Since the final fraction on the right is the left side of the proposed identity, the solution is complete.   □

EXAMPLE 3.   Verify $\dfrac{\sec^4 x - 1}{\tan^2 x} = \tan^2 x + 2$.

Solution.   Again we alter the left side, but this time we use the second Pythagorean Identity, $\tan^2 x + 1 = \sec^2 x$:

$$\frac{\sec^4 x - 1}{\tan^2 x} = \frac{(\sec^2 x)^2 - 1}{\tan^2 x}$$

$$= \frac{(\tan^2 x + 1)^2 - 1}{\tan^2 x}$$

$$= \frac{(\tan^4 x + 2\tan^2 x + 1) - 1}{\tan^2 x}$$

$$= \frac{\tan^4 x + 2\tan^2 x}{\tan^2 x}$$

$$= \tan^2 x + 2$$

Since we have obtained the right side of the proposed identity, the solution is finished.   □

Another way of altering one side of an identity is to write all expressions in terms of sines and cosines and then rearrange as needed.

EXAMPLE 4.   Verify $\dfrac{\cot x - \tan x}{\cos x + \sin x} = \dfrac{\cos x - \sin x}{\sin x \cos x}$.

Solution.   We start by writing the expressions on the left side in terms of sine and cosine, and only then do we simplify:

$$\frac{\cot x - \tan x}{\cos x + \sin x} = \frac{\dfrac{\cos x}{\sin x} - \dfrac{\sin x}{\cos x}}{\cos x + \sin x}$$

$$= \frac{\dfrac{\cos^2 x - \sin^2 x}{\sin x \cos x}}{\cos x + \sin x}$$

$$= \frac{\cos^2 x - \sin^2 x}{\sin x \cos x (\cos x + \sin x)}$$

$$= \frac{(\cos x - \sin x)(\cos x + \sin x)}{\sin x \cos x (\cos x + \sin x)}$$

$$= \frac{\cos x - \sin x}{\sin x \cos x}$$

Again we have obtained the right side of the proposed identity, so the identity is verified.   □

When a proposed identity contains fractions, the verification may be facilitated by clearing fractions.

EXAMPLE 5.   Verify $\dfrac{1 - \sin x}{\cos x} = \dfrac{1}{\sec x + \tan x}$.

Solution.   In order to clear fractions, we multiply both sides of the given identity by $(\cos x)(\sec x + \tan x)$.

$$\frac{1 - \sin x}{\cos x}(\cos x)(\sec x + \tan x) = \frac{1}{\sec x + \tan x}(\cos x)(\sec x + \tan x)$$

which reduces to

$$(1 - \sin x)(\sec x + \tan x) = \cos x \tag{1}$$

Now we alter the left side of (1) with the help of the Defining Identities and the first Pythagorean Identity written in the form $\cos^2 x = 1 - \sin^2 x$:

$$(1 - \sin x)(\sec x + \tan x) = (1 - \sin x)\left(\frac{1}{\cos x} + \frac{\sin x}{\cos x}\right)$$

$$= (1 - \sin x)\left(\frac{1 + \sin x}{\cos x}\right)$$

$$= \frac{1 - \sin^2 x}{\cos x}$$

$$= \frac{\cos^2 x}{\cos x}$$

$$= \cos x$$

Therefore (1), and hence the given identity, is verified. ☐

There are numerous ways of verifying proposed identities. Although one method alone will frequently not suffice, some of the more common and effective methods include:

1. Altering one side of the proposed identity by using one or more of the identities appearing at the outset of this section.
2. Writing all expressions of the proposed identity in terms of sines and cosines.
3. Clearing any fractions that appear in the proposed identity.

EXERCISES 7.1

In Exercises 1–6, verify the identity.

1. $\sin x \csc x = 1$

2. $\tan x \cot x = 1$

3. $\cot x = \csc x \cos x$

4. $\sin x = \dfrac{\tan x}{\sec x}$

5. $\tan x = \dfrac{\sec x}{\csc x}$

6. $\csc x = \cot x \sec x$

In Exercises 7–22, verify the identity.

7. $(\sec t - 1)(\sec t + 1) = \tan^2 t$

8. $\dfrac{\cos t - \sin t}{\cos t} = 1 - \tan t$

9. $\dfrac{\sin z}{\csc z} + \dfrac{\cos z}{\sec z} = 1$

10. $\tan^2 z + \sec^2 z = 2 \sec^2 z - 1$

11. $\cot^2 z + \csc^2 z + 1 = 2 \csc^2 z$.

12. $\sec^4 z - \tan^4 z = \sec^2 z + \tan^2 z$

13. $\sin x (\csc x - \sin x) = \cos^2 x$

14. $(\cos x + \sin x)^2 + (\cos x - \sin x)^2 = 2$

15. $(\cos^2 x \sin x)^2 + (\sin^2 x \cos x)^2 = \sin^2 x \cos^2 x$

16. $(1 - \cos t)^2 + \sin^2 t = 2(1 - \cos t)$

17. $(\cos t - t \sin t)^2 + (\sin t + t \cos t)^2 = 1 + t^2$

18. $\cot^2 t + \cos^2 t = -\sin^2 t + \csc^2 t$

19. $(1 - \sin^2 t) \sec t = \cos t$

20. $4 \sin^2 w + 9 \cos^2 w = 4 + 5 \cos^2 w$

21. $6 \cos^2 w + 8 \sin^2 w = 6 + 2 \sin^2 w$

22. $(1 + \tan w)^2 = \sec^2 w + 2 \tan w$

In Exercises 23–42, verify the identity.

23. $\dfrac{\sec^2 x}{\sec^2 x - 1} = \csc^2 x$ **24.** $\dfrac{\csc x}{\sec x} + \dfrac{\cos x}{\sin x} = 2 \cot x$

25. $\dfrac{\sin x}{1 - \cos x} = \dfrac{1 + \cos x}{\sin x}$

26. $\dfrac{\tan x}{\sec x - 1} = \dfrac{\sec x + 1}{\tan x}$

27. $\dfrac{1}{\csc x + \cot x} = \csc x - \cot x$

28. $\dfrac{1 - \tan^2 x}{1 - \cot^2 x} = 1 - \sec^2 x$

29. $\dfrac{1}{\tan^2 x} - \dfrac{1}{\sec^2 x} = \dfrac{\cos^4 x}{\sin^2 x}$

30. $\dfrac{\cos x - \csc x}{\sec x - \sin x} + \cot x = 0$

31. $\dfrac{\sec x - \cos x}{\sec x + \cos x} = \dfrac{\sin^2 x}{1 + \cos^2 x}$

32. $\dfrac{\cos x}{\cos x - \sin x} = \dfrac{1}{1 - \tan x}$

33. $\dfrac{1}{\tan x + \sin x} = \dfrac{\cos x \csc x}{1 + \cos x}$

34. $\dfrac{\cos x}{1 + \sin x} + \dfrac{1 + \sin x}{\cos x} = 2 \sec x$

35. $\dfrac{1}{1 + \sin x} + \dfrac{1}{1 - \sin x} = 2 \sec^2 x$

36. $\dfrac{1 + \tan z}{\sec z} = \dfrac{\cot z + 1}{\csc z}$

37. $\dfrac{\sin^2 z}{\sec z - 1} = \cos^2 z + \cos z$

38. $\dfrac{1 + \sec z}{\sin z + \tan z} = \csc z$

39. $\sec^2 t \csc^2 t = \sec^2 t + \csc^2 t$

40. $\sec t - \cos t = \sin t \tan t$

41. $\tan t + \cot t = \sec t \csc t$

42. $\sec t + \csc t = (\tan t + \cot t)(\sin t + \cos t)$

In Exercises 43–56, verify the identity.

43. $\tan^2 x - \sin^2 x = \tan^2 x \sin^2 x$

44. $\dfrac{1}{\sec x - \cos x} = \csc x \cot x$

45. $\dfrac{1 + \sec x}{\sec x} = \dfrac{\sin^2 x}{1 - \cos x}$

46. $\sec x - \tan x = \dfrac{\cos x}{1 + \sin x}$

47. $\dfrac{1 - \cos x}{\csc x} = \dfrac{\sin^3 x}{1 + \cos x}$

48. $\dfrac{1}{1 + \cos x} = \csc^2 x - \csc x \cot x$

49. $(\tan x + \cot x)(\sec x - \cos x) = \sec x \tan x$

50. $\dfrac{\sin^3 x - \cos^3 x}{\sin x - \cos x} = 1 + \sin x \cos x$

51. $\dfrac{1 - \sin x}{1 + \sin x} = (\sec x - \tan x)^2$

52. $\dfrac{\cot x - \tan x}{\cot x + \tan x} = \cos^2 x - \sin^2 x$

53. $\log(\sec x + \tan x) + \log(\sec x - \tan x) = 0$

54. $\log(\csc x + \cot x) + \log(\csc x - \cot x) = 0$

55. $\dfrac{\sin x + \cos y}{\cos x + \sin y} = \dfrac{\cos x - \sin y}{\cos y - \sin x}$

56. $\dfrac{\tan x + \tan y}{\cot x + \cot y} = \dfrac{\tan x \tan y - 1}{1 - \cot x \cot y}$

In Exercises 57–62, show that the equation is *not* a trigonometric identity by finding an admissible value of x for which the equation is not valid.

57. $\cos x \overset{?}{=} 1 - \sin x$

58. $1 + \sec^2 x \overset{?}{=} \tan^2 x$

59. $\dfrac{\cos x}{1 + \sin x} \overset{?}{=} \dfrac{1 + \sin x}{\cos x}$

60. $(\sin x + \cos x)^2 \overset{?}{=} \sin^2 x + \cos^2 x$

61. $\tan x \overset{?}{=} \sqrt{\tan^2 x}$ **62.** $\sin x \overset{?}{=} \sqrt{1 - \cos^2 x}$

In Exercises 63–64, sketch the graph of the function.

63. $\dfrac{\cos x}{1 + \sin x} + \dfrac{1 + \sin x}{\cos x}$ (*Hint:* Use Exercise 34.)

64. $(1 + \tan x)^2 - \sec^2 x$

**7.2
THE ADDITION
FORMULAS**

Sometimes we encounter trigonometric expressions such as $\sin(x + y)$, $\cos(x - y)$, and $\tan(x - y)$ that involve the sum or difference of two numbers x and y. The following identities relate such expressions to combinations of the more basic expressions $\sin x$, $\sin y$, $\cos x$, $\cos y$, $\tan x$, and $\tan y$:

Addition Formulas	
$\cos(x + y) = \cos x \cos y - \sin x \sin y$	(1)
$\cos(x - y) = \cos x \cos y + \sin x \sin y$	(2)
$\sin(x + y) = \sin x \cos y + \cos x \sin y$	(3)
$\sin(x - y) = \sin x \cos y - \cos x \sin y$	(4)
$\tan(x + y) = \dfrac{\tan x + \tan y}{1 - \tan x \tan y}$	(5)
$\tan(x - y) = \dfrac{\tan x - \tan y}{1 + \tan x \tan y}$	(6)

The identities in (1)–(6) are called the Addition Formulas for trigonometric functions. (Even though three of the identities feature the difference $x - y$, one can tnink of $x - y$ as the sum $x + (-y)$.) We will prove (2) first, then (1), and finally (3)–(6).

**The Addition
Formulas
for the Cosine**

To prove (2), we will use the fact that two angles of the same measure subtend chords of the same length on the unit circle. Thus the length of the chord joining P and Q in Figure 7.1a is equal to the length of the chord joining $(1, 0)$ and R in Figure 7.1b, since both are chords of angles of measure $x - y$. Using the coordinates of P, Q, and R appearing in Figure 7.1a, b, we find that

$$\overbrace{\left(\cos(x - y) - 1\right)^2 + \left(\sin(x - y) - 0\right)^2}^{\text{square of distance between }(1,\,0)\text{ and }R} = \overbrace{(\cos x - \cos y)^2 + (\sin x - \sin y)^2}^{\text{square of distance between }P\text{ and }Q}$$

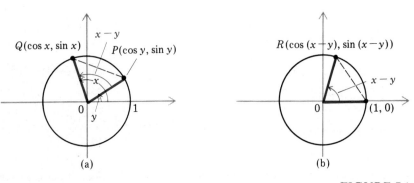

(a)   (b)

FIGURE 7.1

Squaring out the binomials and using the first Pythagorean Identity of Section 7.1, we obtain

$$\cos^2(x - y) - 2\cos(x - y) + 1 + \sin^2(x - y)$$
$$= \cos^2 x - 2\cos x \cos y + \cos^2 y + \sin^2 x - 2\sin x \sin y + \sin^2 y$$
$$1 - 2\cos(x - y) + 1 = 1 - 2\cos x \cos y + 1 - 2\sin x \sin y$$
$$-2\cos(x - y) = -2\cos x \cos y - 2\sin x \sin y$$
$$\cos(x - y) = \cos x \cos y + \sin x \sin y$$

The last equation is (2).

Addition Formula (1) follows from (2) by substituting $-y$ for y in (2) and recalling the Symmetry Identities $\cos(-y) = \cos y$ and $\sin(-y) = -\sin y$:

$$\cos(x + y) = \cos(x - (-y))$$
$$\overset{(2)}{=} \cos x \cos(-y) + \sin x \sin(-y)$$
$$= \cos x \cos y - \sin x \sin y$$

For a typical application of Addition Formulas (1) and (2), we will use the fact that $7\pi/12 = 3\pi/12 + 4\pi/12 = \pi/4 + \pi/3$ to write $\cos 7\pi/12$, whose value is not immediately recognizable, as $\cos(\pi/4 + \pi/3)$, and then apply (1).

EXAMPLE 1.   Evaluate $\cos \dfrac{7\pi}{12}$.

Solution.   Since

$$\frac{7\pi}{12} = \frac{3\pi}{12} + \frac{4\pi}{12} = \frac{\pi}{4} + \frac{\pi}{3}$$

we use (1) with $x = \pi/4$ and $y = \pi/3$:

$$\cos \frac{7\pi}{12} = \cos\left(\frac{\pi}{4} + \frac{\pi}{3}\right)$$
$$\overset{(1)}{=} \cos \frac{\pi}{4} \cos \frac{\pi}{3} - \sin \frac{\pi}{4} \sin \frac{\pi}{3}$$
$$= \left(\frac{\sqrt{2}}{2}\right)\left(\frac{1}{2}\right) - \left(\frac{\sqrt{2}}{2}\right)\left(\frac{\sqrt{3}}{2}\right)$$
$$= \frac{1}{4}(\sqrt{2} - \sqrt{6}) \quad \square$$

EXAMPLE 2.   Evaluate $\cos \dfrac{\pi}{12}$.

Solution.   Notice that

$$\frac{\pi}{12} = \frac{4\pi}{12} - \frac{3\pi}{12} = \frac{\pi}{3} - \frac{\pi}{4}$$

Therefore we can apply (2) with $x = \pi/3$ and $y = \pi/4$ to obtain

$$\cos\frac{\pi}{12} = \cos\left(\frac{\pi}{3} - \frac{\pi}{4}\right)$$

$$\stackrel{(2)}{=} \cos\frac{\pi}{3}\cos\frac{\pi}{4} + \sin\frac{\pi}{3}\sin\frac{\pi}{4}$$

$$= \left(\frac{1}{2}\right)\left(\frac{\sqrt{2}}{2}\right) + \left(\frac{\sqrt{3}}{2}\right)\left(\frac{\sqrt{2}}{2}\right)$$

$$= \frac{1}{4}(\sqrt{2} + \sqrt{6}) \quad \square$$

The Addition Formulas are valid when x and y represent degrees.

EXAMPLE 3.   Evaluate $\cos 75°$.

Solution.   Since $75° = 30° + 45°$, it follows from (1) that

$$\cos 75° = \cos(30° + 45°)$$

$$\stackrel{(1)}{=} \cos 30° \cos 45° - \sin 30° \sin 45°$$

$$= \left(\frac{\sqrt{3}}{2}\right)\left(\frac{\sqrt{2}}{2}\right) - \left(\frac{1}{2}\right)\left(\frac{\sqrt{2}}{2}\right)$$

$$= \frac{1}{4}(\sqrt{6} - \sqrt{2}) \quad \square$$

We could also have evaluated $\cos 75°$ by first noticing that $75° = 120° - 45°$ and then using (2) instead of (1).

Next we will use Addition Formula (2) in order to prove two formulas that first appeared in (7) of Section 6.6.

EXAMPLE 4.   Prove the following formulas.

   a. $\sin x = \cos\left(\frac{\pi}{2} - x\right)$ b. $\cos x = \sin\left(\frac{\pi}{2} - x\right)$

Solution.
a. Replacing x by $\pi/2$ and y by x in (2), we obtain

$$\cos\left(\frac{\pi}{2} - x\right) = \cos\frac{\pi}{2}\cos x + \sin\frac{\pi}{2}\sin x$$

$$= 0\,(\cos x) + 1\,(\sin x)$$

$$= \sin x$$

b. Replacing x by $\pi/2 - x$ in part (a), we find that

$$\cos\left[\frac{\pi}{2} - \left(\frac{\pi}{2} - x\right)\right] = \sin\left(\frac{\pi}{2} - x\right)$$

or equivalently,

$$\cos x = \sin\left(\frac{\pi}{2} - x\right) \quad \square$$

The Addition Formulas for the Sine

Now we turn to the Addition Formulas for $\sin(x + y)$ and $\sin(x - y)$. In the proof of (3) we will use (2), along with the identities derived in Example 4:

$$\sin(x + y) = \cos\left[\frac{\pi}{2} - (x + y)\right]$$

$$= \cos\left[\left(\frac{\pi}{2} - x\right) - y\right]$$

$$\overset{(2)}{=} \cos\left(\frac{\pi}{2} - x\right)\cos y + \sin\left(\frac{\pi}{2} - x\right)\sin y$$

$$= \sin x \cos y + \cos x \sin y$$

Formula (3) is thus verified. Formula (4) is verified by replacing y by $-y$ in (3) and using the Symmetry Identities (see Exercise 65).

EXAMPLE 5. Evaluate $\sin\dfrac{11\pi}{12}$.

Solution. We observe that

$$\frac{11\pi}{12} = \frac{8\pi}{12} + \frac{3\pi}{12} = \frac{2\pi}{3} + \frac{\pi}{4}$$

and then use (3):

$$\sin\frac{11\pi}{12} = \sin\left(\frac{2\pi}{3} + \frac{\pi}{4}\right)$$

$$\overset{(3)}{=} \sin\frac{2\pi}{3}\cos\frac{\pi}{4} + \cos\frac{2\pi}{3}\sin\frac{\pi}{4}$$

$$= \left(\frac{\sqrt{3}}{2}\right)\left(\frac{\sqrt{2}}{2}\right) + \left(-\frac{1}{2}\right)\left(\frac{\sqrt{2}}{2}\right)$$

$$= \frac{1}{4}(\sqrt{6} - \sqrt{2}) \quad \square$$

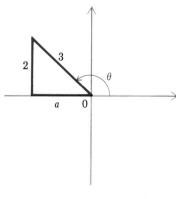

FIGURE 7.2

EXAMPLE 6.   Assume that $\sin \theta° = \frac{2}{3}$ and $90° < \theta° < 180°$. Evaluate $\sin(60° - \theta°)$.

Solution.   To find $\sin(60° - \theta°)$ we will use (4). But in order to do so we must find $\cos \theta°$. Since $\sin \theta° = \frac{2}{3}$ and $90° < \theta° < 180°$, we deduce that $\cos \theta° < 0$ and that the length a in Figure 7.2 is given by

$$a = \sqrt{3^2 - 2^2} = \sqrt{5}$$

Therefore

$$\cos \theta° = - \frac{\sqrt{5}}{3}$$

Now we use (4) with $x = 60°$ and $y = \theta°$ to conclude that

$$\sin(60° - \theta°) = \sin 60° \cos \theta° - \cos 60° \sin \theta°$$

$$= \left(\frac{\sqrt{3}}{2}\right)\left(-\frac{\sqrt{5}}{3}\right) - \left(\frac{1}{2}\right)\left(\frac{2}{3}\right)$$

$$= -\frac{\sqrt{15}}{6} - \frac{1}{3} \quad \square$$

The Addition Formula

$$\sin(x + y) = \sin x \cos y + \cos x \sin y \qquad (7)$$

can help us sketch the graph of a function of the form $p \sin x + q \cos x$ when $p \neq 0$ and $q \neq 0$. First we will find appropriate values of a and c so that

$$p \sin x + q \cos x = a \sin(x + c) \quad \text{for all } x$$

Since we know by Section 6.8 how to sketch the graph of any function of the form $a \sin(x + c)$, once we know a and c we can graph $p \sin x + q \cos x$. Let us see how all this is accomplished in a concrete example.

EXAMPLE 7.   Sketch the graph of $\sin x - \cos x$.

Solution.   We must determine numbers a and c so that

$$\sin x - \cos x = a \sin(x + c) \qquad (8)$$

By (7) we can write (8) as

$$\sin x - \cos x = a \sin x \cos c + a \cos x \sin c$$
$$= (a \cos c) \sin x + (a \sin c) \cos x$$

Evidently these equations will be valid for all values of x if

$$a \cos c = 1 \quad \text{and} \quad a \sin c = -1 \qquad (9)$$

Now we use (9) to calculate the values of a and c. To determine a we use the Pythagorean Theorem and (9) to obtain

$$a^2 = a^2(\cos^2 c + \sin^2 c) = (a \cos c)^2 + (a \sin c)^2$$
$$= 1^2 + (-1)^2 = 2$$

It follows that $a = \sqrt{2}$ or $a = -\sqrt{2}$. For simplicity we take $a = \sqrt{2}$. To find c we observe from (9) that

$$\frac{a \sin c}{a \cos c} = \frac{-1}{1}$$

so that

$$\tan c = -1$$

Thus we could take $c = 3\pi/4$ or $c = -\pi/4$. But from the two formulas in (9) and the fact that $a > 0$, we know that $\cos c > 0$ and $\sin c > 0$. This implies that $c = -\pi/4$. Substituting our values of a and c into (9), we conclude that

$$\sin x - \cos x = \sqrt{2} \sin \left(x - \frac{\pi}{4} \right)$$

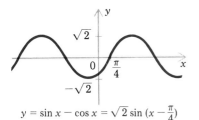

$y = \sin x - \cos x = \sqrt{2} \sin (x - \frac{\pi}{4})$

FIGURE 7.3

Consequently the graph of $\sin x - \cos x$ is the same as the graph of $\sqrt{2} \sin(x - \pi/4)$, which appears in Figure 7.3. $\square$

If we wish to determine the graph of $p \sin x + q \cos x$ for other values of p and q, we can follow the same procedure as we used in Example 7 in order to find values of a and c so that

$$p \sin x + q \cos x = a \sin (x + c) \qquad (10)$$

It turns out that (10) is valid for all values of x if

$$p = a \cos c \text{ and } q = a \sin c \qquad (11)$$

If we want a to be positive, then

$$a = \sqrt{p^2 + q^2} \quad \text{and} \quad \tan c = \frac{p}{q}$$

(provided $q \neq 0$). The given values of p and q, along with signs of $\sin c$ and $\cos c$ determined by (11), then yield the values of a and c for (10).

A physical interpretation of (10) is that any motion given by $p \sin x + q \cos x$, where x represents time, is simple harmonic motion, as described in Section 6.8.

The Addition Formulas for the Tangent

Addition Formula (5) for $\tan(x + y)$ is derived from (1) and (3) by using the fact that the tangent is the quotient of the sine and cosine and by dividing both numerator and denominator by $\cos x \cos y$:

$$\tan(x + y) = \frac{\sin(x + y)}{\cos(x + y)}$$

$$= \frac{\sin x \cos y + \cos x \sin y}{\cos x \cos y - \sin x \sin y}$$

$$= \frac{\dfrac{\sin x \cos y}{\cos x \cos y} + \dfrac{\cos x \sin y}{\cos x \cos y}}{\dfrac{\cos x \cos y}{\cos x \cos y} - \dfrac{\sin x \sin y}{\cos x \cos y}}$$

$$= \frac{\dfrac{\sin x}{\cos x} + \dfrac{\sin y}{\cos y}}{1 - \dfrac{\sin x}{\cos x}\dfrac{\sin y}{\cos y}} = \frac{\tan x + \tan y}{1 - \tan x \tan y}$$

Thus (5) is proved. Formula (6) can be proved from (5) by replacing y by $-y$ (see Exercise 66).

EXAMPLE 8.   Evaluate $\tan \dfrac{11\pi}{12}$.

Solution. Since

$$\frac{11\pi}{12} = \frac{8\pi}{12} + \frac{3\pi}{12} = \frac{2\pi}{3} + \frac{\pi}{4}$$

we find that

$$\tan \frac{11\pi}{12} = \tan\left(\frac{2\pi}{3} + \frac{\pi}{4}\right) \overset{(5)}{=} \frac{\tan 2\pi/3 + \tan \pi/4}{1 - \tan 2\pi/3 \tan \pi/4}$$

$$= \frac{-\sqrt{3} + 1}{1 - (-\sqrt{3})(1)} = \frac{1 - \sqrt{3}}{1 + \sqrt{3}} \quad \square$$

EXAMPLE 9.   Assume that $\sin x = \frac{3}{5}$ with $\pi/2 < x < \pi$, and $\cos y = -\frac{1}{4}$ with $\pi < y < 3\pi/2$. Evaluate $\tan(x - y)$.

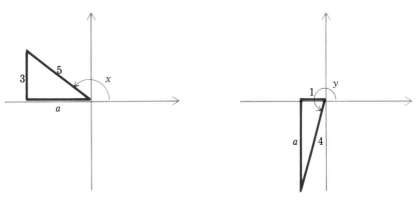

FIGURE 7.4   FIGURE 7.5

Solution.   We will use (6). But to do so we need to determine the values of $\tan x$ and $\tan y$. From the hypotheses that $\sin x = \frac{3}{5}$ and $\pi/2 < x < \pi$ we deduce that $\tan x < 0$ and that the length a in Figure 7.4 is given by

$$a = \sqrt{5^2 - 3^2} = 4$$

Therefore

$$\tan x = -\frac{3}{4}$$

In a similar way, we deduce that $\tan y > 0$ and that the length a in Figure 7.5 is given by

$$a = \sqrt{4^2 - 1^2} = \sqrt{15}$$

Therefore

$$\tan y = \sqrt{15}$$

Consequently by (6),

$$\tan(x - y) = \frac{\tan x - \tan y}{1 + \tan x \tan y} = \frac{-\frac{3}{4} - \sqrt{15}}{1 + (-\frac{3}{4})(\sqrt{15})}$$

$$= \frac{-3 - 4\sqrt{15}}{4 - 3\sqrt{15}}   \qquad \square$$

EXERCISES 7.2

In Exercises 1–12, evaluate the given expression.

1. $\sin \dfrac{\pi}{12}$

2. $\sin \dfrac{7\pi}{12}$

3. $\sin(-165°)$

4. $\sin 465°$

5. $\cos\left(-\dfrac{5\pi}{12}\right)$

6. $\cos \dfrac{13\pi}{12}$

7. $\cos 255°$

8. $\cos 105°$

9. $\tan \dfrac{\pi}{12}$

10. $\tan \dfrac{17\pi}{12}$

11. $\tan(-285°)$

12. $\tan 465°$

13. Suppose that $\sin x = -\frac{1}{4}$ with $\pi < x < 3\pi/2$. Find $\sin(\pi/6 + x)$.

14. Suppose that $\cos x = -\frac{4}{7}$ with $\pi/2 < x < \pi$. Find $\cos(\pi/4 - x)$.

15. Suppose that $\sin x = \frac{1}{5}$ with $0 < x < \pi/2$, and $\cos y = \frac{1}{4}$ with $3\pi/2 < y < 2\pi$. Find $\sin(x + y)$.

16. Suppose that $\tan x = \frac{1}{2}$ and $\sec y = 3$ with $0 < y < \pi/2$. Find $\tan(x + y)$.

17. Suppose that $\sin(x + y) = \frac{1}{3}$ with $0 < x + y < \pi/2$, and $\cos y = \frac{1}{4}$ with $0 < y < \pi/2$. Find $\sin x$.

18. Suppose that $\sin x = \frac{2}{3}$ with $0 < x < \pi/2$, and $\cos(x - y) = \frac{4}{5}$ with $0 < x - y < \pi/2$. Find $\cos y$.

In Exercises 19–30, use the Addition Formulas to verify the identity.

19. $\sin\left(\dfrac{\pi}{2} + x\right) = \cos x$

20. $\cos\left(\dfrac{\pi}{2} + x\right) = -\sin x$

21. $\sin\left(\dfrac{3\pi}{2} + x\right) = -\cos x$

22. $\cos\left(\dfrac{3\pi}{2} + x\right) = \sin x$

23. $\sin\left(\dfrac{\pi}{3} + x\right) = \dfrac{1}{2}(\sqrt{3}\cos x + \sin x)$

24. $\cos\left(\dfrac{\pi}{3} + x\right) = \dfrac{1}{2}(\cos x - \sqrt{3}\sin x)$

25. $\sin\left(\dfrac{\pi}{4} - x\right) = \dfrac{\sqrt{2}}{2}(\cos x - \sin x)$

26. $\cos\left(\dfrac{\pi}{4} + x\right) = \dfrac{\sqrt{2}}{2}(\cos x - \sin x)$

27. $\tan\left(x + \dfrac{3\pi}{4}\right) = \dfrac{\tan x - 1}{1 + \tan x}$

28. $\cot\left(x - \dfrac{3\pi}{4}\right) = \dfrac{1 - \tan x}{1 + \tan x}$

29. $\dfrac{\sin(x + \pi)}{\cos(x - \pi)} = \tan x$ 30. $\dfrac{\sin(x + \pi/2)}{\cos(x - \pi)} = -1$

In Exercises 31–52, verify the identity.

31. $\sin(x + y) + \sin(x - y) = 2 \sin x \cos y$

32. $\cos(x + y) + \cos(x - y) = 2 \cos x \cos y$

33. $\sin(x + y)\sin(x - y) = \sin^2 x \cos^2 y - \cos^2 x \sin^2 y$

34. $\sin(x + y)\sin(x - y) = \sin^2 x - \sin^2 y$

35. $\cos(x + y)\cos(x - y) = \cos^2 x - \sin^2 y$

36. $(\cos x - \sin y)^2 + (\sin x + \cos y)^2 = 2 + 2\sin(x - y)$

37. $(\cos x - \cos y)^2 + (\sin x + \sin y)^2 = 2 - 2\cos(x + y)$

38. $\dfrac{\sin x}{\sin y} + \dfrac{\cos x}{\cos y} = \csc y \sec y \sin(x + y)$

39. $\dfrac{\sin(x + y)}{\cos x \cos y} = \tan x + \tan y$

40. $\dfrac{\cos(x - y)}{\cos x \sin y} = \cot y + \tan x$

41. $\dfrac{\cos(x + y)}{\sin(x - y)} = \dfrac{\cot x - \tan y}{1 - \cot x \tan y}$

42. $\dfrac{\sin(x + y)}{\cos(x - y)} = \dfrac{\cot x + \cot y}{\cot x \cot y + 1}$

43. $\dfrac{\sin(x + y)}{\sin(x - y)} = \dfrac{\tan x + \tan y}{\tan x - \tan y}$

44. $\dfrac{\cos(x + y)}{\cos(x - y)} = \dfrac{1 - \tan x \tan y}{1 + \tan x \tan y}$

45. $\sin^2(\pi/4 + x) - \sin^2(\pi/4 - x) = \sin 2x$

46. $\dfrac{\tan(\pi/4 - x)}{\tan(\pi/4 + x)} = \dfrac{(1 - \tan x)^2}{(1 + \tan x)^2}$

47. $\tan(x + y) = \dfrac{\cot x + \cot y}{\cot x \cot y - 1}$

48. $\tan(x - y) - \tan(y - x) = \dfrac{2(\tan x - \tan y)}{1 + \tan x \tan y}$

49. $\cot(x + y) = \dfrac{\cot x \cot y - 1}{\cot x + \cot y}$

50. $\cot(x - y) = \dfrac{\cot x \cot y + 1}{\cot y - \cot x}$

51. $\sec(x + y) = \dfrac{\csc x \csc y}{\cot x \cot y - 1}$

52. $\sec(x - y) = \dfrac{\sec x \sec y}{1 + \tan x \tan y}$

53. Verify that

$$\frac{\sin(x + h) - \sin x}{h} = \cos x \, \frac{\sin h}{h} - \sin x \, \frac{1 - \cos h}{h}$$

(This formula arises in calculus.)

54. Verify that

$$\frac{\cos(x + h) - \cos x}{h} = -\sin x \, \frac{\sin h}{h} - \cos x \, \frac{1 - \cos h}{h}$$

(This formula arises in calculus.)

In Exercises 55–58, show that the equation is *not* a trigonometric identity by finding admissible values of x and y for which the equation is not valid.

55. $\sin(x + y) \overset{?}{=} \sin x + \sin y$

56. $\cos(x + y) \overset{?}{=} \cos x + \cos y$

57. $\tan(x + y) \overset{?}{=} \tan x + \tan y$

58. $\sec(x + y) \overset{?}{=} \sec x + \sec y$

In Exercises 59–64, sketch the graph of the function.

59. $\sin x + \cos x$

60. $\sin x - \sqrt{3} \cos x$

61. $2\sqrt{3} \sin x + 6 \cos x$

62. $\cos x \cos 3x - \sin x \sin 3x$

63. $\cos x \cos 3x + \sin x \sin 3x$

64. $\cos \frac{3}{5}x \cos \frac{7}{5}x - \sin \frac{3}{5}x \sin \frac{7}{5}x$

65. Prove formula (4) from (3) by replacing y with $-y$.

66. Prove formula (6) from (5) by replacing y with $-y$.

7.3 MULTIPLE-ANGLE FORMULAS

From the Addition Formulas presented in the preceding section we will obtain formulas for $\sin 2x$, $\sin 3x$, $\sin \frac{1}{2}x$, and indeed the sines of many other multiples of x, in terms of $\sin x$ and $\cos x$. Similar results will appear for the cosine and tangent functions.

The Double-Angle Formulas

The following formulas are called Double-Angle Formulas:

Double-Angle Formulas	
$\sin 2x = 2 \sin x \cos x$	(1)
$\cos 2x = \cos^2 x - \sin^2 x$	(2)
$\tan 2x = \dfrac{2 \tan x}{1 - \tan^2 x}$	(3)

Notice that the angle on the left side of each formula is $2x$, rather than x.

Formula (1) is proved by letting $y = x$ in the Addition Formula (3) of Section 7.2:

$$\sin 2x = \sin(x + x) = \sin x \cos x + \cos x \sin x = 2 \sin x \cos x$$

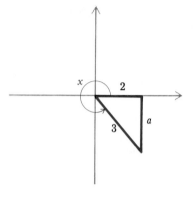

FIGURE 7.6

Formulas (2) and (3) are verified by similar methods (see Exercises 56 and 57).

EXAMPLE 1.   Suppose that $\cos x = \frac{2}{3}$ and $3\pi/2 < x < 2\pi$. Evaluate the following expressions.

    a. $\sin 2x$ b. $\cos 2x$ c. $\tan 2x$

Solution.

a. In order to use (1) we need to know the value of $\sin x$ as well as $\cos x$. Since $3\pi/2 < x < 2\pi$ and $\cos x = \frac{2}{3}$, it follows that $\sin x < 0$ and that the length a in Figure 7.6 is given by

$$a = \sqrt{3^2 - 2^2} = \sqrt{5}$$

Therefore

$$\sin x = -\frac{\sqrt{5}}{3}$$

Consequently (1) implies that

$$\sin 2x = 2 \sin x \cos x = 2\left(-\frac{\sqrt{5}}{3}\right)\left(\frac{2}{3}\right) = -\frac{4}{9}\sqrt{5}$$

b. Using (2), the given value of $\cos x$, and the value of $\sin x$ determined in part (a), we have

$$\cos 2x = \cos^2 x - \sin^2 x = \left(\frac{2}{3}\right)^2 - \left(-\frac{\sqrt{5}}{3}\right)^2 = \frac{4}{9} - \frac{5}{9} = -\frac{1}{9}$$

c. The values we have of $\sin 2x$ and $\cos 2x$ yield

$$\tan 2x = \frac{\sin 2x}{\cos 2x} = \frac{-\frac{4}{9}\sqrt{5}}{-\frac{1}{9}} = 4\sqrt{5} \quad \square$$

EXAMPLE 2.   Find a formula for $\sin 3x$ in terms of $\sin x$ and $\cos x$.

Solution.   If we think of $3x$ as $2x + x$, then Addition Formula (3) in Section 7.2 yields

$$\sin 3x = \sin(2x + x) = \sin 2x \cos x + \cos 2x \sin x$$

By virtue of Double-Angle Formulas (1) and (2),

$$\sin 2x \cos x + \cos 2x \sin x = (2 \sin x \cos x)\cos x + (\cos^2 x - \sin^2 x)\sin x$$
$$= 2 \sin x \cos^2 x + \cos^2 x \sin x - \sin^3 x$$
$$= 3 \sin x \cos^2 x - \sin^3 x$$

Therefore

$$\sin 3x = 3 \sin x \cos^2 x - \sin^3 x \quad \square$$

Using the same method, one could find a formula for $\sin 4x$, or even $\sin nx$ for any positive integer n, in terms of $\sin x$ and $\cos x$. The same holds true for $\cos nx$ and $\tan nx$.

Formula (2) can be modified to yield the auxiliary formulas

$$\sin^2 x = \frac{1 - \cos 2x}{2} \qquad (4)$$

$$\cos^2 x = \frac{1 + \cos 2x}{2} \qquad (5)$$

which are of interest in their own right. To prove (5) we start with (2) and utilize the first Pythagorean Identity written in the form $\sin^2 x = 1 - \cos^2 x$:

$$\cos 2x = \cos^2 x - \sin^2 x = \cos^2 x - (1 - \cos^2 x)$$

so that

$$\cos 2x = 2 \cos^2 x - 1 \qquad (6)$$

Therefore

$$\cos^2 x = \frac{1 + \cos 2x}{2}$$

which is (5). Formula (4) can be proved either from (2) or from (5) (see Exercise 58).

As a byproduct of the auxiliary formulas (4) and (5), the graphs of $\sin^2 x$ and $\cos^2 x$ are now very accessible, for they are simple modifications of the graph of $\cos 2x$ (Figure 7.7a, b). Were we to graph $\sin^2 x$ (or $\cos^2 x$) by plotting points and using various properties of $\sin x$ (or $\cos x$), it would be much more tedious.

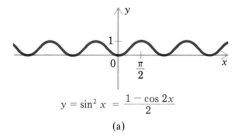

$$y = \sin^2 x = \frac{1 - \cos 2x}{2}$$

(a)

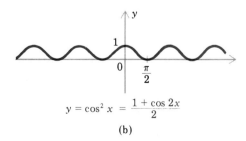

$$y = \cos^2 x = \frac{1 + \cos 2x}{2}$$

(b)

FIGURE 7.7

The Half-Angle Formulas

Formulas (4) and (5) give rise to formulas for $\sin \frac{1}{2}x$ and $\cos \frac{1}{2}x$. We first replace x by y and take square roots in (4) to obtain

$$\sin y = \sqrt{\frac{1 - \cos 2y}{2}} \quad \text{or} \quad -\sqrt{\frac{1 - \cos 2y}{2}}$$

depending on whether $\sin y \geq 0$ or $\sin y < 0$. Likewise, taking square roots in (5), we obtain

$$\cos y = \sqrt{\frac{1 + \cos 2y}{2}} \quad \text{or} \quad -\sqrt{\frac{1 + \cos 2y}{2}}$$

depending on whether $\cos y \geq 0$ or $\cos y < 0$. Then we let $y = \frac{1}{2}x$ in these formulas to obtain the so-called Half-Angle Formulas:

Half-Angle Formulas

$$\sin \frac{1}{2}x = \begin{cases} \sqrt{\dfrac{1 - \cos x}{2}} & \text{if} \quad \sin \dfrac{1}{2}x \geq 0 \\[2ex] -\sqrt{\dfrac{1 - \cos x}{2}} & \text{if} \quad \sin \dfrac{1}{2}x < 0 \end{cases} \qquad (7)$$

$$\cos \frac{1}{2}x = \begin{cases} \sqrt{\dfrac{1 + \cos x}{2}} & \text{if} \quad \cos \dfrac{1}{2}x \geq 0 \\[2ex] -\sqrt{\dfrac{1 + \cos x}{2}} & \text{if} \quad \cos \dfrac{1}{2}x < 0 \end{cases} \qquad (8)$$

Caution: Care must be taken in applying (7) or (8) in order to ensure that the correct sign is obtained. Recall that $\sin \frac{1}{2}x$ is nonnegative when the terminal side of the angle in standard position associated with $\frac{1}{2}x$ lies in the first or second quadrant and is negative otherwise. Similarly, $\cos \frac{1}{2}x$ is positive when the terminal side of the angle associated with $\frac{1}{2}x$ lies in the first or fourth quadrants and is negative otherwise.

EXAMPLE 3.   Evaluate $\sin \dfrac{\pi}{8}$.

Solution.   We have

$$\frac{\pi}{8} = \frac{1}{2} \cdot \frac{\pi}{4}$$

and $\cos \pi/4 = \sqrt{2}/2$. Since $0 < \pi/8 < \pi/2$, it follows that $\sin \pi/8 > 0$. Now we calculate $\sin \pi/8$ by using (7):

$$\sin \frac{\pi}{8} = \sin\left(\frac{1}{2} \cdot \frac{\pi}{4}\right) = \sqrt{\frac{1 - \cos \dfrac{\pi}{4}}{2}}$$

$$= \sqrt{\frac{1 - \dfrac{\sqrt{2}}{2}}{2}} = \sqrt{\frac{2 - \sqrt{2}}{4}} = \frac{1}{2}\sqrt{2 - \sqrt{2}} \quad \square$$

EXAMPLE 4.   Suppose that $\sin x = \frac{1}{4}$ and $\pi/2 < x < \pi$. Evaluate $\cos \frac{1}{2}x$.

Solution.   In order to use (8), we first need to find the value of $\cos x$. Since $\pi/2 < x < \pi$ and $\sin x = \frac{1}{4}$, it follows that $\cos x < 0$ and the length a in Figure 7.8 is given by

$$a = \sqrt{4^2 - 1^2} = \sqrt{15}$$

Therefore

$$\cos x = -\frac{\sqrt{15}}{4}$$

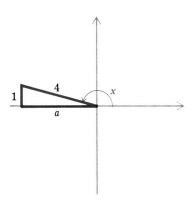

FIGURE 7.8

Because

$$\frac{\pi}{4} < \frac{1}{2}x < \frac{\pi}{2}$$

it follows that $\cos \frac{1}{2}x > 0$, so that by (8),

$$\cos \frac{1}{2}x = \sqrt{\frac{1 + \cos x}{2}} = \sqrt{\frac{1 - \dfrac{\sqrt{15}}{4}}{2}} = \sqrt{\frac{4 - \sqrt{15}}{8}} \quad \square$$

To derive a half-angle formula for the tangent function, we use (1) and (5), with x replaced by y:

$$\tan y = \frac{\sin y}{\cos y} = \frac{\sin y}{\cos y} \cdot \frac{2 \cos y}{2 \cos y}$$

$$= \frac{2 \sin y \cos y}{2 \cos^2 y} = \frac{\sin 2y}{2\left(\dfrac{1 + \cos 2y}{2}\right)} = \frac{\sin 2y}{1 + \cos 2y}$$

Letting $\frac{1}{2}x$ replace y now yields the desired Half-Angle Formula for the tangent:

$$\tan \frac{1}{2}x = \frac{\sin x}{1 + \cos x} \tag{9}$$

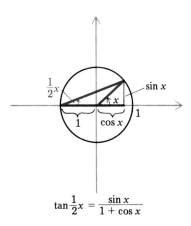

$$\tan \frac{1}{2}x = \frac{\sin x}{1 + \cos x}$$

FIGURE 7.9   In case $0 < x < \pi/2$, formula (9) can be proved geometrically (Figure 7.9).

EXAMPLE 5.   Evaluate tan 67.5°.

Solution.   Observe that $67.5° = [\frac{1}{2}(135)]°$. Since

$$\sin 135° = \frac{\sqrt{2}}{2} \quad \text{and} \quad \cos 135° = -\frac{\sqrt{2}}{2}$$

it follows from (9) that

$$\tan 67.5° = \tan\left(\frac{1}{2}(135)\right)° = \frac{\sin 135°}{1 + \cos 135°}$$

$$= \frac{\dfrac{\sqrt{2}}{2}}{1 - \dfrac{\sqrt{2}}{2}} = \frac{\sqrt{2}}{2 - \sqrt{2}} \quad \square$$

EXERCISES 7.3

In Exercises 1–6, find sin 2x, cos 2x, and tan 2x under the given conditions.

1. $\sin x = -\frac{1}{3}$ and $3\pi/2 < x < 2\pi$

2. $\cos x = \frac{3}{4}$ and $0 < x < \pi/2$

3. $\tan x = -\sqrt{2}$ and $\pi/2 < x < \pi$

4. $\cot x = \frac{2}{3}$ and $\pi < x < 3\pi/2$

5. $\sec x = -5$ and $\pi/2 < x < \pi$

6. $\csc x = 4$ and $0 < x < \pi/2$

In Exercises 7–12, find $\sin \frac{1}{2}x$, $\cos \frac{1}{2}x$, and $\tan \frac{1}{2}x$ under the given conditions.

7. $\sin x = \frac{1}{8}$ and $\pi/2 < x < \pi$

8. $\cos x = \frac{1}{4}$ and $0 < x < \pi/2$

9. $\tan x = \frac{1}{2}$ and $\pi < x < 3\pi/2$

10. $\cot x = 3$ and $2\pi < x < 5\pi/2$

11. $\sec x = -3$ and $\pi < x < 3\pi/2$

12. $\csc x = -4$ and $3\pi/2 < x < 2\pi$

In Exercises 13–24, use the Half-Angle Formulas to evaluate the expressions.

13. $\sin \dfrac{7\pi}{8}$

14. $\sin \dfrac{11\pi}{8}$

15. $\sin\left(-\dfrac{\pi}{12}\right)$

16. sin 105°

17. $\cos \dfrac{5\pi}{8}$

18. $\cos(-247.5°)$

19. $\cos \dfrac{5\pi}{12}$

20. $\cos\left(-\dfrac{19\pi}{12}\right)$

21. $\tan \dfrac{5\pi}{8}$

22. $\tan \dfrac{9\pi}{8}$

23. $\tan\left(-\dfrac{13\pi}{12}\right)$

24. tan 75°

25. Using the fact that $\cos \pi/8 = \frac{1}{2}\sqrt{2 + \sqrt{2}}$, evaluate the following.

a. $\sin \dfrac{\pi}{16}$ b. $\cos \dfrac{\pi}{16}$ c. $\tan \dfrac{\pi}{16}$

26. Using the fact that $\cos \pi/12 = \frac{1}{2}\sqrt{2 + \sqrt{3}}$, evaluate the following.

a. $\sin \dfrac{\pi}{24}$ b. $\cos \dfrac{\pi}{24}$ c. $\tan \dfrac{\pi}{24}$

In Exercises 27–38, verify the identity.

27. $\cos 2x + 2 \sin^2 x = 1$

28. $\cos 2x = \cos^4 x - \sin^4 x$

29. $\sin 2x = \dfrac{2 \sin x}{\sec x}$

30. $\cos 2x = \dfrac{\cot x - \tan x}{\cot x + \tan x}$

31. $\cot 2x = \frac{1}{2}(\cot x - \tan x)$

32. $\csc 2x = \frac{1}{2}(\cot x + \tan x)$

33. $\cot 2x = \frac{\cot^2 x - 1}{2 \cot x}$

34. $\sec 2x = \frac{1}{2 \cos^2 x - 1}$

35. $\csc 2x = \cot x - \cot 2x$

36. $\frac{\sec x}{1 + \tan x} = \frac{\cos x - \sin x}{\cos 2x}$

37. $\tan 2x - \sec 2x = \frac{\sin x - \cos x}{\sin x + \cos x}$

38. $\log(\cos x - \sin x) + \log(\cos x + \sin x) = \log \cos 2x$

In Exercises 39–42, verify the identity.

39. $\cos 3x = 4 \cos^3 x - 3 \cos x$

40. $\sin 4x = 8 \sin x \cos^3 x - 4 \sin x \cos x$

*41. $\cos 4x = 8 \cos^4 x - 8 \cos^2 x + 1$

42. $\frac{\sin 3x}{\sin x} - \frac{\cos 3x}{\cos x} = 2$ (*Hint:* Use Example 2 and Exercise 39.)

In Exercises 43–51, verify the identity.

43. $\sec^2 \frac{1}{2} x = \frac{2}{1 + \cos x}$

44. $\csc^2 \frac{1}{2} x = \frac{2}{1 - \cos x}$

45. $\sec^2 \frac{1}{2} x = \frac{2 \sec x}{\sec x + 1}$

46. $\sin^2 \frac{1}{2} x = \frac{\tan x - \sin x}{2 \tan x}$

47. $\cos^2 \frac{1}{2} x = \frac{\sec x + 1}{2 \sec x}$

48. $\tan \frac{1}{2} x = \frac{\tan x}{1 + \sec x}$

49. $\tan \frac{1}{2} x = \csc x - \cot x$

50. $\cot \frac{1}{2} x = \frac{\sin x}{1 - \cos x}$ **51.** $\cot^2 \frac{1}{2} x = \frac{\sec x + 1}{\sec x - 1}$

52. a. Verify that $\sin x = \frac{2 \tan \frac{1}{2}x}{1 + \tan^2 \frac{1}{2}x}$.

   b. Verify that $\cos x = \frac{1 - \tan^2 \frac{1}{2}x}{1 + \tan^2 \frac{1}{2}x}$.

(The identities in this exercise sometimes appear in calculus.)

53. Let $f(x) = \sin 2x \cos 2x$. Sketch the graph of f.

54. Let $f(x) = \cos^2 \frac{1}{6}x - \sin^2 \frac{1}{6}x$. Sketch the graph of f.

55. Let $f(x) = \frac{4 \tan \frac{1}{2}x}{1 - \tan^2 \frac{1}{2} x}$. Sketch the graph of f.

56. Verify (2).   **57.** Verify (3).

58. a. Prove (4) from (2).
   b. Prove (4) from (5).

7.4 THE SUM AND PRODUCT FORMULAS

In this section we will employ the Addition Formulas to prove the following identities, which are useful in calculus and in the mathematical description of certain aspects of wave propagation.

Product Formulas

$$\sin x \sin y = \frac{1}{2}[\cos(x - y) - \cos(x + y)] \tag{1}$$

$$\cos x \cos y = \frac{1}{2}[\cos(x + y) + \cos(x - y)] \tag{2}$$

$$\sin x \cos y = \frac{1}{2}[\sin(x + y) + \sin(x - y)] \tag{3}$$

Sum Formulas

$$\sin x + \sin y = 2 \sin \frac{1}{2}(x + y) \cos \frac{1}{2}(x - y) \tag{4}$$

$$\sin x - \sin y = 2 \sin \frac{1}{2}(x - y) \cos \frac{1}{2}(x + y) \tag{5}$$

$$\cos x + \cos y = 2 \cos \frac{1}{2}(x + y) \cos \frac{1}{2}(x - y) \tag{6}$$

$$\cos x - \cos y = -2 \sin \frac{1}{2}(x + y) \sin \frac{1}{2}(x - y) \tag{7}$$

Since (1)–(3) are all proved in the same way, we will prove (1) and leave (2) and (3) as exercises. By (1) and (2) of Section 7.2, we have

$$\cos(x - y) - \cos(x + y)$$
$$= (\cos x \cos y + \sin x \sin y) - (\cos x \cos y - \sin x \sin y)$$
$$= 2 \sin x \sin y$$

Dividing both sides by 2 and switching the sides, we obtain formula (1).

Formulas (4)–(7) follow from (1)–(3). We will prove (7) and leave the others as exercises. By applying (1) with x replaced by $\frac{1}{2}(x + y)$ and y replaced by $\frac{1}{2}(x - y)$, we have

$$\sin \frac{1}{2}(x + y) \sin \frac{1}{2}(x - y)$$

$$= \frac{1}{2} \left[\cos\left(\frac{1}{2}(x + y) - \frac{1}{2}(x - y)\right) - \cos\left(\frac{1}{2}(x + y) + \frac{1}{2}(x - y)\right) \right]$$

$$= \frac{1}{2}(\cos y - \cos x)$$

Multiplying both sides by -2 and reversing the order of the terms, we obtain (7).

EXAMPLE 1.   Use formula (1) to express $\sin 3x \sin 2x$ in the form $a(\cos bx - \cos cx)$.

Solution.   Applying (1) with x replaced by $3x$ and y replaced by $2x$, we have

$$\sin 3x \sin 2x = \frac{1}{2}[\cos(3x - 2x) - \cos(3x + 2x)] = \frac{1}{2}(\cos x - \cos 5x) \quad \square$$

EXAMPLE 2.   Verify the identity

$$\frac{\sin x - \sin 3x}{\cos x + \cos 3x} = -\tan x$$

Solution.   By (5) with y replaced by $3x$, we have

$$\sin x - \sin 3x = 2 \sin \frac{1}{2}(x - 3x)\cos \frac{1}{2}(x + 3x)$$

$$= 2 \sin(-x)\cos 2x$$

$$= -2 \sin x \cos 2x$$

and by (6) with y replaced by $3x$, we have

$$\cos x + \cos 3x = 2 \cos \frac{1}{2}(x + 3x)\cos \frac{1}{2}(x - 3x)$$

$$= 2 \cos 2x \cos(-x)$$

$$= 2 \cos 2x \cos x$$

Therefore

$$\frac{\sin x - \sin 3x}{\cos x + \cos 3x} = \frac{-2 \sin x \cos 2x}{2 \cos 2x \cos x} = -\frac{\sin x}{\cos x} = -\tan x$$

as desired.   □

Now we have completed our presentation of trigonometric identities and formulas. The more important ones are listed at the end of the chapter.

Musical Strings   One application of the Sum Formula occurs in the analysis of musical sounds produced by stringed instruments such as the violin or guitar. Recall from Section 6.8 that when the string is plucked or bowed, the string oscillates. The displacement at time t of the point on the string corresponding to x is given by

$$y_1 = a \sin 2\pi\left(\frac{x}{\lambda} - \frac{t}{T}\right)$$

where $|a|$ denotes the amplitude, λ the wavelength, and T the period of the motion. The motion of the string produces a wave, moving along the string. As the wave reaches the end of the string, it is reflected back along the string with the same amplitude, wavelength, and period. Considered by itself, this reflected wave would yield a displacement y_2 given by

$$y_2 = -a \sin 2\pi\left(\frac{x}{\lambda} + \frac{t}{T}\right)$$

However, the reflected wave combines with the original wave to yield a combined wave. The superposition principle from physics tells us that the displacement y of a given point x for the combined wave equals the sum $y_1 + y_2$ of the displacements y_1 and y_2 for the original and reflected waves.

In order to analyze the combined wave further, we use the Sum Formula (5):

$$y = y_1 + y_2 = a \sin 2\pi\left(\frac{x}{\lambda} - \frac{t}{T}\right) - a \sin 2\pi\left(\frac{x}{\lambda} + \frac{t}{T}\right)$$

$$\overset{(5)}{=} 2a \sin \frac{1}{2}\left[2\pi\left(\frac{x}{\lambda} - \frac{t}{T}\right) - 2\pi\left(\frac{x}{\lambda} + \frac{t}{T}\right)\right] \cos \frac{1}{2}\left[2\pi\left(\frac{x}{\lambda} - \frac{t}{T}\right) + 2\pi\left(\frac{x}{\lambda} + \frac{t}{T}\right)\right]$$

$$= -2a \cos\left(2\pi \frac{x}{\lambda}\right)\sin\left(2\pi \frac{t}{T}\right)$$

Therefore

$$y = -2a \cos\left(2\pi \frac{x}{\lambda}\right)\sin\left(2\pi \frac{t}{T}\right) \qquad (8)$$

Notice that if $\cos(2\pi x/\lambda) = 0$ in (8), then $y = 0$, which can be interpreted to mean that the point x is stationary (because $y = 0$ for any value of t). Such points are called **nodes**. In the photograph the nodes of each vibrating string are clearly visible. Because of the existence of nodes, the waves in the string appear to be stationary. This is the reason such waves are called **stationary waves**, or **standing waves**.

Because both ends of the string are fixed, and hence the ends are nodes themselves, there are exactly an integral number of "halfwaves" on the string. Thus the halfwave length $\lambda/2$ must divide the length L of the string:

$$L = n\left(\frac{\lambda}{2}\right)$$

so that

$$\lambda = \frac{2L}{n}$$

for some positive integer n. As a result, a given string with a predetermined length can vibrate only with certain wavelengths. The corresponding frequencies, which determine the sounds we hear, depend on various physical properties of the string, such as tension and mass per unit length. The frequency corresponding to a wave with two nodes (at the ends of the string) is called the **first harmonic** or **fundamental**. If n is a positive integer, then the frequency corresponding to a wave with $n + 2$ nodes is called the **nth harmonic** or **nth overtone**. The fundamental and first four overtones are shown in the photograph.

EXERCISES 7.4

In Exercises 1–8, write each expression as a sum or difference.

1. $\sin 2x \sin 4x$

2. $\sin 3x \sin 5x$

3. $\cos x \cos 3x$

4. $\cos 4x \cos 5x$

5. $\cos \frac{5}{3}x \cos \frac{1}{3}x$

6. $\sin \frac{1}{2}x \cos \frac{3}{2}x$

7. $\cos 7x \sin 3x$

8. $\sin 2x \cos 2x$

In Exercises 9–16, write each expression as a product.

9. $\sin 4x + \sin 2x$

10. $\sin x + \sin 3x$

11. $\sin 2x - \sin \frac{1}{2}x$

12. $\sin 5x - \sin 7x$

13. $\cos \frac{1}{2}x + \cos \frac{3}{2}x$

14. $\cos \frac{7}{5}x + \cos \frac{3}{5}x$

15. $\cos 7x - \cos 3x$

16. $\cos x - \cos 6x$

In Exercises 17–28, verify the identity.

17. $\dfrac{\sin x + \sin 3x}{\cos x + \cos 3x} = \tan 2x$

18. $\dfrac{\sin x + \sin 3x}{\cos x - \cos 3x} = \cot x$

19. $\dfrac{\sin x - \sin 3x}{\cos x - \cos 3x} = -\cot 2x$

20. $\dfrac{\sin x + \sin 3x}{\sin x - \sin 3x} = -\tan 2x \cot x$

21. $\dfrac{\sin 5x + \sin 3x}{\cos 3x - \cos 5x} = \cot x$

22. $\dfrac{\cos x - \cos 3x}{\cos x + \cos 3x} = \tan 2x \tan x$

23. $\dfrac{\sin 8x + \sin 3x}{\cos 8x + \cos 3x} = \tan \dfrac{11}{2} x$

24. $\dfrac{\sin 8x + \sin 3x}{\cos 8x - \cos 3x} = \cot \dfrac{5}{2} x$

25. $\dfrac{\sin 8x - \sin 3x}{\cos 8x + \cos 3x} = \tan \dfrac{5}{2} x$

26. $\dfrac{\sin 8x - \sin 3x}{\cos 8x - \cos 3x} = -\cot \dfrac{11}{2} x$

27. $\dfrac{\sin 8x + \sin 3x}{\sin 8x - \sin 3x} = \tan \dfrac{11}{2} x \cot \dfrac{5}{2} x$

28. $\dfrac{\cos 8x - \cos 3x}{\cos 8x + \cos 3x} = -\tan \dfrac{11}{2} x \tan \dfrac{5}{2} x$

In Exercises 29–34, verify the identity.

29. $\dfrac{\sin x + \sin y}{\cos x + \cos y} = \tan \dfrac{1}{2}(x + y)$

30. $\dfrac{\sin x + \sin y}{\cos x - \cos y} = -\cot \dfrac{1}{2}(x - y)$

31. $\dfrac{\sin x - \sin y}{\cos x + \cos y} = \tan \dfrac{1}{2}(x - y)$

32. $\dfrac{\sin x - \sin y}{\cos x - \cos y} = -\cot \dfrac{1}{2}(x + y)$

33. $\dfrac{\sin x + \sin y}{\sin x - \sin y} = \tan \dfrac{1}{2}(x + y) \cot \dfrac{1}{2}(x - y)$

34. $\dfrac{\cos x - \cos y}{\cos x + \cos y} = -\tan \dfrac{1}{2}(x + y) \tan \dfrac{1}{2}(x - y)$

35. Verify (2).

36. Verify (3).

37. Verify (4).

38. Verify (5).

39. Verify (6).

7.5 CONDITIONAL TRIGONOMETRIC EQUATIONS

Throughout this chapter we have studied trigonometric identities, that is, equations that are valid for all admissible values of the variables. In contrast, some trigonometric equations are valid only for special values of the variables. Such equations, which are called conditional trigonometric equations, are the topic of this section. As always, the goal will be to solve such an equation, by which we mean to find all values of the variable for which the equation is true. In this section we will use radian measure exclusively for the variables in trigonometric expressions.

We have already encountered simple conditional trigonometric equations such as $\sin x = 0$ and $\tan x = 0$. But as you will see, the equations can be much more complicated. Many of the techniques we will use to solve conditional trigonometric equations (such as factoring and using the quadratic formula) were used in solving equations in Chapter 2. Yet you should be aware of two major differences between conditional trigonometric equations and the conditional equations of Chapter 2. In the first place, trigonometric identities are available to aid us in the solutions. Second, and more important, the equations will in general have an infinite number of solutions, because the equations will involve trigonometric functions, each of which is periodic. Thus if x is a solution of an equation such as

$$\tan^2 x - \sec x - 1 = 0$$

then for any integer n, $x + 2n\pi$ is also a solution of the equation, because

$$\tan^2(x + 2n\pi) - \sec(x + 2n\pi) - 1 = \tan^2 x - \sec x - 1$$

We will generally determine the solutions that are in $[0, 2\pi)$ first, and from these we will deduce all solutions.

In each of the first three examples we will solve an equation involving a single trigonometric function.

EXAMPLE 1.   Solve the equation $\sin x = \frac{1}{2}$.

Solution.   The solutions of $\sin x = \frac{1}{2}$ belonging to $[0, 2\pi)$ are $\pi/6$ and $5\pi/6$. Therefore the total collection of solutions of $\sin x = \frac{1}{2}$ consists of $\pi/6 + 2n\pi$ and $5\pi/6 + 2n\pi$, for all integers n.   □

EXAMPLE 2.   Solve the equation $\cos 2x = -1$.

Solution.   Let $z = 2x$, so that the given equation becomes $\cos z = -1$. The only solution of $\cos z = -1$ belonging to $[0, 2\pi)$ is π, so the total collection of solutions of $\cos z = -1$ consists of $\pi + 2n\pi = (2n + 1)\pi$. Substituting $2x$ back for z, we find that the solutions of the given equation $\cos 2x = -1$ are the numbers x such that

$$2x = (2n + 1)\pi$$

that is, the numbers of the form $(2n + 1)\pi/2$, for all integers n.   □

EXAMPLE 3.   Solve the equation $\tan \frac{1}{4}x = -\sqrt{3}/3$.

Solution.   Let $z = \frac{1}{4}x$, so that the given equation becomes $\tan z = -\sqrt{3}/3$. Since the period of $\tan z$ is π, we will first find the solutions of $\tan z = -\sqrt{3}/3$ belonging to $(-\pi/2, \pi/2)$. Now the lone solution of $\tan z = -\sqrt{3}/3$ belonging to that interval is $-\pi/6$, so the total collection of solutions of $\tan z = -\sqrt{3}/3$ consists of $-\pi/6 + n\pi$ for all integers n. Since $z = \frac{1}{4}x$, the solutions of the given equation satisfy $\frac{1}{4}x = -\pi/6 + n\pi$ and thus are $4(-\pi/6 + n\pi) = -2\pi/3 + 4n\pi$, for all integers n.   □

EXAMPLE 4.   Solve the equation $2 \sin^2 x + 5 \sin x + 2 = 0$.

Solution.   Let $z = \sin x$, so that the given equation becomes $2z^2 + 5z + 2 = 0$. Now the solutions of $2z^2 + 5z + 2 = 0$ can be obtained by factoring directly or by the quadratic formula. In the latter case we obtain

$$z = \frac{-5 \pm \sqrt{5^2 - 4(2)(2)}}{2(2)} = \frac{-5 \pm \sqrt{9}}{4} = \frac{-5 \pm 3}{4} = -\tfrac{1}{2} \quad \text{or} \quad -2$$

Since $z = \sin x$ and $|\sin x| \le 1$, it follows that $\sin x = -\tfrac{1}{2}$. Thus the original equation is equivalent to the equation $\sin x = -\tfrac{1}{2}$, whose solutions in $[0, 2\pi)$ are $7\pi/6$ and $11\pi/6$. Consequently the solutions of the given equation are $7\pi/6 + 2n\pi$ and $11\pi/6 + 2n\pi$, for all integers n. $\square$

The solutions of Examples 5–9 will utilize the various identities we have encountered in the preceding sections of Chapter 7. Generally speaking, in our solutions we will want to convert a given equation into an equivalent equation (with the same solutions) in which *only one* trigonometric function and its powers appear. Then we will solve that equation instead of the original one. The conversion will normally utilize identities such as those listed at the outset of Section 7.1, as well as the Addition, Double-Angle, and Half-Angle Formulas.

EXAMPLE 5.   Solve the equation $\tan^2 x - \sec x - 1 = 0$.

Solution.   Since $\tan^2 x = \sec^2 x - 1$ by the second Pythagorean Identity, we can convert the given equation to one having only powers of $\sec x$:

$$(\sec^2 x - 1) - \sec x - 1 = 0$$

or equivalently,

$$\sec^2 x - \sec x - 2 = 0$$

But this means that

$$(\sec x + 1)(\sec x - 2) = 0$$

so that $\sec x = -1$ or $\sec x = 2$. Now the only solution of the equation $\sec x = -1$ belonging to $[0, 2\pi)$ is π, and the only such solutions of the equation $\sec x = 2$ are $\pi/3$ and $5\pi/3$. Therefore the solutions of the given equation are the numbers of the form $\pi + 2n\pi$, $\pi/3 + 2n\pi$, or $5\pi/3 + 2n\pi$, for all integers n. $\square$

EXAMPLE 6.   Solve the equation $4 \sin x \cos x = -\sqrt{3}$.

Solution.   Rather than attempting to write $\sin x$ in terms of $\cos x$ or $\cos x$ in terms of $\sin x$, we will utilize the Double-Angle Formula

$$\sin 2x = 2 \sin x \cos x$$

which converts the given equation into

$$2 \sin 2x = -\sqrt{3}$$

or equivalently,

$$\sin 2x = -\frac{\sqrt{3}}{2}$$

Letting $z = 2x$, we notice that the only solutions of $\sin z = -\sqrt{3}/2$ belonging to $[0, 2\pi)$ are $4\pi/3$ and $5\pi/3$. Therefore the collection of solutions of $\sin z = -\sqrt{3}/2$ consists of the numbers z such that

$$z = \frac{4\pi}{3} + 2n\pi \quad \text{or} \quad z = \frac{5\pi}{3} + 2n\pi$$

Substituting $2x$ back in for z, we find that the collection of solutions of $\sin 2x = -\sqrt{3}/2$ consists of the numbers x such that

$$2x = \frac{4\pi}{3} + 2n\pi \quad \text{or} \quad 2x = \frac{5\pi}{3} + 2n\pi$$

which means that the solutions of the given equation are the numbers of the form $2\pi/3 + n\pi$ or $5\pi/6 + n\pi$, for all integers n.   □

EXAMPLE 7.   Solve the equation $\cos 2x = \cos x$.

Solution.   Notice that $\cos 2x$ and $\cos x$ are two distinct functions. First we use (6) of Section 7.3:

$$\cos 2x = 2 \cos^2 x - 1$$

This allows us to rewrite the given equation as

$$2 \cos^2 x - 1 = \cos x$$

which yields

$$2 \cos^2 x - \cos x - 1 = 0$$

Factoring, we obtain

$$(2 \cos x + 1)(\cos x - 1) = 0$$

so that

$$2 \cos x + 1 = 0 \quad \text{or} \quad \cos x - 1 = 0$$

As a result,

$$\cos x = -\frac{1}{2} \quad \text{or} \quad \cos x = 1$$

Now the only solutions of the equation $\cos x = -\frac{1}{2}$ belonging to $[0, 2\pi)$ are $2\pi/3$ and $4\pi/3$, and the only such solution of the equation $\cos x = 1$ is 0. Consequently the solutions of the given equation are the numbers of the form $2n\pi$, $2\pi/3 + 2n\pi$, or $4\pi/3 + 2n\pi$ for all integers n. $\quad\square$

EXAMPLE 8. Solve the equation $\sin x = \frac{1}{2}\sqrt{3} \tan x$.

Solution. Here we again have two trigonometric functions—this time the sine and tangent functions. If we write $\tan x = (\sin x)/(\cos x)$, then the given equation becomes

$$\sin x = \frac{\sqrt{3}}{2} \frac{\sin x}{\cos x} \tag{1}$$

or equivalently,

$$\sin x - \frac{\sqrt{3}}{2} \frac{\sin x}{\cos x} = 0$$

Factoring out $\sin x$, we obtain

$$\sin x \left(1 - \frac{\sqrt{3}}{2} \frac{1}{\cos x}\right) = 0$$

This equation is satisfied if

$$\sin x = 0 \quad \text{or} \quad 1 - \frac{\sqrt{3}}{2} \frac{1}{\cos x} = 0$$

Now $\sin x = 0$ if and only if $x = n\pi$ for some integer n. Analogously, the equation

$$1 - \frac{\sqrt{3}}{2} \frac{1}{\cos x} = 0$$

is equivalent to

$$1 = \frac{\sqrt{3}}{2} \frac{1}{\cos x}, \quad \text{or} \quad \cos x = \frac{\sqrt{3}}{2}$$

We observe that the only solutions of the equation $\cos x = \sqrt{3}/2$ belonging to $[0, 2\pi)$ are $\pi/6$ and $11\pi/6$. Therefore the solutions of the given equation are the numbers of the form $n\pi$, $\pi/6 + 2n\pi$, or $11\pi/6 + 2n\pi$, for all integers n.   □

Caution: Had we canceled $\sin x$ from (1) and solved the resulting equation for x, we could have lost the solutions of the form $n\pi$. We see once again that we *must not* cancel expressions that can be 0.

EXERCISES 7.5

In Exercises 1–18, solve the equation.

1. $\sin x = 1$

2. $\sin x = -\sqrt{3}/2$

3. $\cos x = -\frac{1}{2}$

4. $\cos x = \sqrt{2}/2$

5. $\tan x = \sqrt{3}$

6. $\tan x = -1$

7. $\cot x = \sqrt{3}$

8. $\sec x = -2$

9. $\csc x = \frac{2}{3}\sqrt{3}$

10. $\csc x = -1$

11. $\sin 3x = \frac{1}{2}$

12. $\cos \frac{1}{2}x = \sqrt{3}/2$

13. $\tan \frac{1}{5}x = 1$

14. $\tan 2x = \sqrt{3}$

15. $\sec 6x = \sqrt{2}$

16. $\sin^2 x = 1$

17. $\sin^2 x/2 = \frac{1}{4}$

18. $\cos^2 x = \frac{1}{2}$

In Exercises 19–30, solve the equation.

19. $\sin x = \sqrt{3} \cos x$

20. $\cos x = \dfrac{\sqrt{2}}{2} \cot x$

21. $\tan x \csc x = -\frac{2}{3}\sqrt{3}$

22. $\cos x = -\frac{1}{4} \csc x$

23. $\sin 2x = \sin x$

24. $\tan \frac{1}{2}x = \sin x$

25. $\cot x (\tan x + 1) = 0$

26. $\sin^2 x + \sin x = 0$

27. $\cos^2 x - \sin x \cos x = 0$

28. $\sin x \cos x - \cos x + \sin x - 1 = 0$

*29. $2 \sin x \cos x - \sqrt{2} \sin x - \sqrt{2} \cos x + 1 = 0$

*30. $4 \sin x \cos x - 2\sqrt{3} \sin x - 2 \cos x + \sqrt{3} = 0$

In Exercises 31–44, solve the equation.

31. $\sin^2 x + 5 \sin x + 4 = 0$

32. $4 \cos^2 x - 4 \cos x + 1 = 0$

33. $2 \cos^2 x - \cos x - 1 = 0$

34. $\sin^2 x + \frac{1}{2} \sin x - \frac{1}{2} = 0$

35. $\cos^2 x - \frac{3}{2} \cos x - 1 = 0$

36. $\tan^2 x - 1 = 0$

37. $\tan^2 x + \frac{2}{3}\sqrt{3} \tan x - 1 = 0$

38. $\sec^2 x + 3 \sec x + 2 = 0$

39. $\csc^5 x - 4 \csc x = 0$

40. $\sin^2 x = 2 \cos x + 2$

41. $\sin x + 2 \cos^2 x = 1$

42. $\csc^2 x - \cot x - 1 = 0$

43. $\sec^2 x - \frac{3}{2} \sec x - 1 = 0$

44. $\csc^2 x - 2 \csc x = 0$

In Exercises 45–50, find the solutions of the given equation that lie in the interval $[0°, 360°)$. Use a calculator or Table E where necessary to approximate the solutions.

45. $6 \sin^2 \theta° - \sin \theta° - 1 = 0$

46. $\sin^2 \theta° + \sin \theta° - 1 = 0$

47. $15 \cos^2 \theta° + 2 \cos \theta° - 1 = 0$

48. $\tan^2 \theta° + 5 \tan \theta° + 6 = 0$

49. $\sec^2 \theta° - 3 = 0$

50. $\csc^2 \theta° + 2 \csc \theta° - 8 = 0$

KEY TERMS

trigonometric identity
admissible values

conditional trigonometric equation
Pythagorean Identities

KEY IDENTITIES AND FORMULAS

Defining Identities

$$\tan x = \frac{\sin x}{\cos x} \qquad \csc x = \frac{1}{\sin x}$$

$$\cot x = \frac{\cos x}{\sin x} \qquad \cot x = \frac{1}{\tan x}$$

$$\sec x = \frac{1}{\cos x}$$

Symmetry Identities

$$\sin(-x) = -\sin x \qquad \cot(-x) = -\cot x$$

$$\cos(-x) = \cos x \qquad \sec(-x) = \sec x$$

$$\tan(-x) = -\tan x \qquad \csc(-x) = -\csc x$$

Pythagorean Identities

$$\sin^2 x + \cos^2 x = 1 \qquad \cot^2 x + 1 = \csc^2 x$$

$$\tan^2 x + 1 = \sec^2 x$$

Addition Formulas

$$\cos(x + y) = \cos x \cos y - \sin x \sin y$$

$$\cos(x - y) = \cos x \cos y + \sin x \sin y$$

$$\sin(x + y) = \sin x \cos y + \cos x \sin y$$

$$\sin(x - y) = \sin x \cos y - \cos x \sin y$$

$$\tan(x + y) = \frac{\tan x + \tan y}{1 - \tan x \tan y}$$

$$\tan(x - y) = \frac{\tan x - \tan y}{1 + \tan x \tan y}$$

Double-Angle Formulas

$$\sin 2x = 2 \sin x \cos x$$

$$\cos 2x = \cos^2 x - \sin^2 x$$

$$\tan 2x = \frac{2 \tan x}{1 - \tan^2 x}$$

REVIEW EXERCISES

In Exercises 1–32, verify the Identity

1. $\tan x + \cot x = \dfrac{1}{\cos x \sin x}$

2. $(\sin x + \cos x)^2 = 1 + \sin 2x$

3. $(\sin x + \cos x)^2 - (\sin x - \cos x)^2 = 2 \sin 2x$

4. $\sin^2 x = \dfrac{\tan^2 x}{1 + \tan^2 x}$

5. $\sec^2 t + \tan^4 t = \sec^4 t - \tan^2 t$

6. $\sin t = (1 + \cos t)(\csc t - \cot t)$

7. $\dfrac{\sec t}{\tan t + \cot t} = \sin t$

8. $(1 + \tan^2 x)(1 + \cot^2 x) = \sec^2 x \csc^2 x$

9. $\dfrac{\sec^2 x}{\sec 2x} = \dfrac{2 \tan x}{\tan 2x}$

10. $\dfrac{1}{\sec x - \tan x} - \dfrac{1}{\sec x + \tan x} = 2 \tan x$

11. $\dfrac{1}{\csc x - \cot x} - \dfrac{1}{\csc x + \cot x} = 2 \cot x$

12. $\log |1 + \sin x| + \log |1 - \sin x| - 2 \log |\cos x| = 0$

13. $\dfrac{\tan^4 x - 1}{\sin^4 x - \cos^4 x} = \sec^4 x$

14. $\csc^2 x = \dfrac{1 + 2 \cot^2 x}{1 + \cos^2 x}$

15. $(\cos x + \sin x)(\tan x + \cot x) = \sec x + \csc x$

16. $\sin x = \sin(x + 60°) + \sin(x - 60°)$

17. $\cos x = \cos(x + 60°) + \cos(x - 60°)$

18. $\tan^2 x = \dfrac{1 - \cos 2x}{1 + \cos 2x}$

19. $\cos\left(\dfrac{\pi}{2} + x - y\right) = \sin(y - x)$

20. $\tan\left(x + \dfrac{\pi}{4}\right) = \dfrac{1 + \tan x}{1 - \tan x}$

21. $\tan x \tan y = \dfrac{\tan x + \tan y}{\cot x + \cot y}$

22. $\sin x \cos(x - y) - \cos x \sin(x - y) = \sin y$

23. $\cos x \cos(x + y) + \sin x \sin(x + y) = \cos y$

24. $\dfrac{\sin(x - y)}{\sin x \sin y} = \cot y - \cot x$

25. $\cot x = \dfrac{1 + \cos x + \cos 2x}{\sin x + \sin 2x}$

26. $\sin x + \sin 3x = 2 \sin 2x \cos x$

27. $\tan \dfrac{x}{2} = \dfrac{1 - \cos x}{\sin x}$

28. $\dfrac{\sin 2x}{\sin x} - \dfrac{\cos 2x}{\cos x} = \sec x$

29. $\dfrac{\cos 2x}{\sin x} + \dfrac{\sin 2x}{\cos x} = \csc x$

30. $\sin 4x = 4 \sin x \cos x \cos 2x$

31. $\cos 4x = 1 - 8 \cos^2 x \sin^2 x$

32. $2 \cos^2 \dfrac{x}{2} - \cos x - 1 = 0$

In Exercises 33–36, evaluate the given expression.

33. $\sin 5\pi/8$ **35.** $\sec 75°$

34. $\cos 165°$ **36.** $\tan \pi/8$

37. Suppose that $\sin x = \frac{1}{7}$ with $\pi/2 < x < \pi$. Find $\cos(x - \pi/3)$.

38. Suppose that $\tan x = \frac{3}{5}$, and $\sin y = \frac{2}{5}$ with $0 < y < \pi/2$. Find $\tan(x + y)$.

39. Suppose $\sec x = 4$ with $-\pi/2 < x < 0$. Find $\sin 2x$.

40. Suppose $\cot x = 5$ with $\pi < x < 3\pi/2$. Find $\tan \frac{1}{2}x$.

41. Suppose $x + y + z = \pi$. Show that

$$\tan x + \tan y + \tan z = \tan x \tan y \tan z$$

$\left(\textit{Hint: } z = \pi - (x + y).\right)$

42. Prove that if $\sin x = 0$, then $\sin 2x = 0$.

43. a. If $\cos x = 0$, does it follow that $\cos 2x = 0$? Explain your answer.
   b. If $\cos x = 0$, does it follow that $\cos 3x = 0$? Explain your answer.

44. Show that $\tan^2 x + \sec^2 x \stackrel{?}{=} 1$ is not an identity by finding an admissible value of x for which the equation does not hold.

45. a. Use a Half-Angle Formula to compute $\cos \pi/12$.
   b. By comparing your answer to part (a) with the result of Example 2 in Section 7.2, conclude that

$$\sqrt{2 + \sqrt{3}} = \dfrac{\sqrt{2} + \sqrt{6}}{2}$$

46. a. Express $2 \cos 5x \sin 3x$ as a sum or a difference of functions of the form $\sin ax$.
   b. Express $\cos 2x - \cos 4x$ as a product of functions of the form $\sin ax$.

47. Let $f(x) = \sqrt{3} \sin x + \cos x$. Sketch the graph of f.

48. Let $f(x) = \sin 3x \cos x - \cos 3x \sin x$. Sketch the graph of f.

In Exercises 49–58, solve the given equation.

49. $\cos x = -\sqrt{3}/2$

50. $\tan^2 x = \frac{1}{3}$

51. $\cos^2 x = 1 + \sin^2 x$

52. $\cos^2 x + 2 \cos x + \sin^2 x = 0$

53. $2 \sin^2 x + \sin x = 1$

54. $\sec^2 x - \sec x - 2 = 0$

55. $\cos 2x = \sin x$

56. $2 \sin x - \sin 2x \cos x = 0$

57. $\tan^2 x - 2 \sec^2 x = -5$

*58. $\sec x + \tan x = 1$ (*Hint:* Multiply both sides by $\sec x - \tan x$.)

8

Trigonometry and Its Applications

Since prehistoric times people have been preoccupied with calculating distances, whether it be finding the height of a tree, the distance across a lake, or the dimensions of one's own property. Each of the distances just mentioned can be ascertained by studying the angles and sides of a convenient triangle. The case in which the triangle is a right triangle was analyzed in Section 6.4. In the present chapter we turn to triangles that are not necessarily right triangles and learn how to determine the lengths of the sides and the measures of the angles of a given triangle when certain of them are already known. As before, this process is called *solving the triangle*.

Sections 8.1 and 8.2 are devoted to solving triangles by means of two famous laws, the Law of Sines and the Law of Cosines. In Section 8.3 we derive two quite different formulas for the area of an arbitrary triangle. The chapter ends with introductions to polar coordinates and vectors, each of which makes use of trigonometric functions and triangles.

As an illustration of the applications of the mathematics in this chapter, consider the following:

> Suppose that forest rangers at two park stations eight miles apart spot smoke from a forest fire in the distance. By measuring certain angles and using the methods of this chapter, we can determine which ranger is closer to the fire (see Example 4 in Section 8.1).

It is important not only in the study of trigonometry, but also in applications, that triangles can be solved whether or not they are right triangles. A triangle that is not a right triangle is said to be *oblique*. An oblique triangle has either three acute angles or two acute and one obtuse angle (Figure 8.1).

In this and the next section we discuss the two main tools—the Law of Sines and the Law of Cosines—that are especially helpful in solving oblique triangles. We will use α, β, and γ for the angles of a triangle and a, b, c for the lengths of the sides opposite α, β, and γ, respectively.

Now we give the main result of this section.

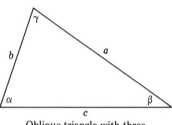

Oblique triangle with three
acute angles
(a)

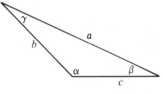

Oblique triangle with two acute
angles and one obtuse angle
(b)

FIGURE 8.1

The Law of Sines

$$\frac{\sin \alpha}{a} = \frac{\sin \beta}{b} = \frac{\sin \gamma}{c} \qquad (1)$$

To prove the Law of Sines, we first consider a triangle ABC with three acute angles and drop a perpendicular from C as in Figure 8.2. The perpendicular divides the given triangle into the smaller right triangles ADC and BCD. From triangle ADC,

$$\sin \alpha = \frac{h}{b}, \quad \text{so that} \quad h = b \sin \alpha \qquad (2)$$

and from triangle BCD,

$$\sin \beta = \frac{h}{a}, \quad \text{so that} \quad h = a \sin \beta \qquad (3)$$

Equating the two values of h in (2) and (3), we have

$$b \sin \alpha = a \sin \beta$$

so that

$$\frac{\sin \alpha}{a} = \frac{\sin \beta}{b} \qquad (4)$$

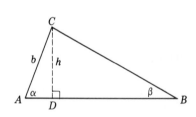

FIGURE 8.2

Similarly, it can be shown by dropping a perpendicular from A as in Figure 8.3 that

$$\frac{\sin \beta}{b} = \frac{\sin \gamma}{c} \qquad (5)$$

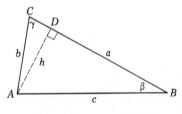

FIGURE 8.3

(see Exercise 29). Together (4) and (5) yield (1) for triangles with three acute angles.

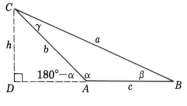

FIGURE 8.4

If a triangle ABC has an obtuse angle at α, we drop a perpendicular from C as in Figure 8.4. From triangle ACD,

$$\sin(180° - \alpha) = \frac{h}{b}$$

so that

$$h = b\sin(180° - \alpha) = b\sin\alpha \tag{6}$$

and from triangle BCD,

$$\sin\beta = \frac{h}{a}, \quad \text{so that} \quad h = a\sin\beta \tag{7}$$

Equating the values of h in (6) and (7) yields

$$b\sin\alpha = a\sin\beta$$

so that

$$\frac{\sin\alpha}{a} = \frac{\sin\beta}{b}$$

As before, one can show that

$$\frac{\sin\beta}{b} = \frac{\sin\gamma}{c}$$

so that (1) still holds for triangles with an obtuse angle. This completes the proof of the Law of Sines.

Solutions of Triangles Using the Law of Sines

The Law of Sines is applicable to solving triangles, provided that either

(a) two angles and the length of a side are known, or
(b) the lengths of two sides and the angle opposite one of these sides are known.

In the solution of a triangle by means of the Law of Sines we will use the fact that in any triangle,

$$\alpha + \beta + \gamma = 180° \tag{8}$$

Triangles that fall under case (a) can be easier to solve than those falling under case (b). Example 1 illustrates case (a).

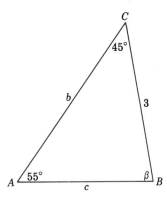

FIGURE 8.5

EXAMPLE 1.   Solve the triangle shown in Figure 8.5.

Solution.   From the figure we know that $a = 3$, $\alpha = 55°$, and $\gamma = 45°$. We need to determine b, c, and β. By (8),

$$\beta = 180° - 55° - 45° = 80°$$

so we have all the angles. With appropriate substitutions the Law of Sines becomes

$$\frac{\sin 55°}{3} = \frac{\sin 80°}{b} = \frac{\sin 45°}{c}$$

Thus

$$\frac{\sin 55°}{3} = \frac{\sin 80°}{b}, \quad \text{so that} \quad b = \frac{3 \sin 80°}{\sin 55°}$$

and

$$\frac{\sin 55°}{3} = \frac{\sin 45°}{c}, \quad \text{so that} \quad c = \frac{3 \sin 45°}{\sin 55°}$$

By calculator we find that

$$b \approx 3.607 \quad \text{and} \quad c \approx 2.590 \quad \square$$

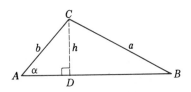

FIGURE 8.6

Case (b) involves any triangle in which the lengths a and b of two sides and the angle α opposite one of these sides are known (Figure 8.6). As we will presently show, there can be either one or two triangles with the prescribed values of a, b, and α, depending on the relationship between a, b, and α.

To determine whether the given values of a, b, and α yield one or two triangles, let h be the length of the perpendicular drawn from C to the opposite side (Figure 8.6). Since a is the hypotenuse of the right triangle BCD and h is a leg of the same triangle, it follows that $a \geq h$. Next, $h = b \sin \alpha$ because triangle ADC is a right triangle. Therefore

$$a \geq h = b \sin \alpha \tag{9}$$

so that either $a = b \sin \alpha$ or $a > b \sin \alpha$. These two possibilities, along with the possibilities $a \geq b$ or $b > a$, yield three cases, each of which determines uniquely the number of triangles that arise:

    i. $a \geq b$, in which case exactly one triangle exists (Figure 8.7a).

    ii. $b > a = b \sin \alpha$, in which case exactly one triangle exists, a right triangle with $a < b$ (Figure 8.7b).

    iii. $b > a > b \sin \alpha$, in which case two triangles exist, one with three acute angles and the other with an obtuse angle (triangles $AB_1 C$ and $AB_2 C$, respectively, in Figure 8.7c).

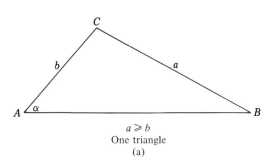
$a \geqslant b$
One triangle
(a)

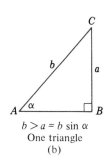
$b > a = b \sin \alpha$
One triangle
(b)

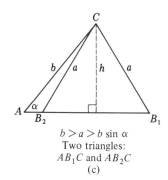
$b > a > b \sin \alpha$
Two triangles:
AB_1C and AB_2C
(c)

FIGURE 8.7

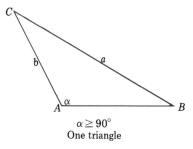
$\alpha \geq 90°$
One triangle

FIGURE 8.8

Notice that if $\alpha \geq 90°$, that is, if α is a right or obtuse angle, then $a \geq b$, so there is only one triangle (Figure 8.8). Notice also that under case (**iii**) there are two triangles; this case if often called the ***ambiguous case*** of the Law of Sines.

EXAMPLE 2.   Solve the triangle in Figure 8.9.

   Solution.   From the figure we know that $\alpha = 110°$, $a = 6$ and $b = 2$. Since $a \geq b$, the triangle satisfies (**i**), so there is only one triangle to solve. By the Law of Sines,

$$\frac{\sin 110°}{6} = \frac{\sin \beta}{2} = \frac{\sin \gamma}{c}$$

so that

$$\frac{\sin 110°}{6} = \frac{\sin \beta}{2} \quad \text{and} \quad \frac{\sin 110°}{6} = \frac{\sin \gamma}{c} \tag{10}$$

By calculator we find that

$$\sin \beta = \frac{1}{3} \sin 110° \approx 0.313$$

so that

$$\beta \approx 18.25° \ (\approx 18°15')$$

For γ we use (8):

$$\gamma = 180° - \alpha - \beta$$
$$\approx 180° - 110° - 18.25° = 51.75° \quad (\approx 51°45')$$

To find c we use the fact that $\gamma \approx 51.75°$, and deduce from the second

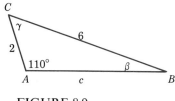

FIGURE 8.9

equation in (10) that

$$c \approx 6 \frac{\sin 51.75°}{\sin 110°}$$

By calculator we find that

$$c \approx 5.014 \quad \square$$

The next example illustrates the ambiguous case of the Law of Sines.

EXAMPLE 3.   Suppose that $a = 3.5, b = 4$, and $\alpha = 60°$. Show that two triangles result, and solve them.

Solution.   We have

$$b \sin \alpha = 4 \sin 60° = 4 \left(\frac{\sqrt{3}}{2} \right) = 2\sqrt{3} \approx 3.464$$

Therefore $b > a > b \sin \alpha$, so that **(iii)** is satisfied, and consequently there are two triangles (AB_1C and AB_2C in Figure 8.10) that meet the given specifications.

We will first solve triangle AB_1C, all of whose angles are acute (Figure 8.11). This will entail calculating β_1, γ_1, and then c_1. For β_1, we use the Law of Sines to obtain

$$\frac{\sin 60°}{3.5} = \frac{\sin \beta_1}{4}$$

so that

$$\sin \beta_1 = \frac{4 \sin 60°}{3.5} = \frac{2\sqrt{3}}{3.5}$$

By calculator we find that

$$\beta_1 \approx 81.79° \quad (\approx 81°47')$$

For γ_1 we use (8):

$$\gamma_1 = 180° - \alpha - \beta_1$$
$$\approx 180° - 60° - 81.79° = 38.21° \quad (\approx 38°13')$$

To find c_1 we apply the Law of Sines a second time to triangle AB_1C and obtain

$$\frac{\sin 60°}{3.5} = \frac{\sin \gamma_1}{c_1}$$

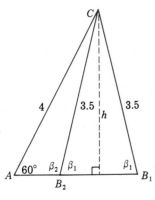

FIGURE 8.10

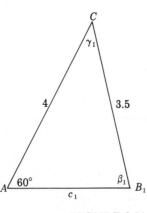

FIGURE 8.11

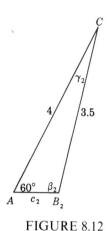

FIGURE 8.12

so that

$$c_1 = \frac{3.5 \sin \gamma_1}{\sin 60°} \approx \frac{3.5 \sin 38.21°}{\sin 60°}$$

By calculator we find that

$$c_1 \approx 2.500$$

Thus we have solved triangle AB_1C of Figure 8.11.

For triangle AB_2C, shown in Figure 8.12, we must find β_2, γ_2, and c_2. The process is simpler than the process in the first part of the solution because triangle B_2B_1C in Figure 8.10 is isosceles and consequently we can find β_2 from β_1:

$$\beta_2 = 180° - \beta_1 \approx 180° - 81.79° = 98.21° \quad (\approx 98°13')$$

To find γ_2 we apply (8) again:

$$\begin{aligned} \gamma_2 &= 180° - \alpha - \beta_2 \\ &\approx 180° - 60° - 98.21° = 21.79° \quad (\approx 21°47') \end{aligned}$$

To determine c_2, we use the Law of Sines a final time:

$$\frac{\sin 60°}{3.5} = \frac{\sin \gamma_2}{c_2}$$

so that

$$c_2 = \frac{3.5 \sin \gamma_2}{\sin 60°} \approx \frac{3.5 \sin 21.79°}{\sin 60°}$$

By calculator we find that

$$c_2 \approx 1.500$$

Therefore triangle AB_2C is solved, and the solution of the example is complete.   □

We note in passing that if we are given arbitrary positive values for a, b, and α, there may not exist any triangle with such prescribed values, since for any triangle we must have $a \geq b \sin \alpha$ by (9). (See Exercise 28.)

The Law of Sines is especially valuable in determining the distances between two given points and a third point.

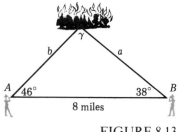

FIGURE 8.13

EXAMPLE 4.   Forest rangers at stations A and B, which are 8 miles apart, observe a fire in the distance. By using the angles given in Figure 8.13, determine how far the fire is from each of the rangers.

Solution.   We let $c = 8$, $\alpha = 46°$, and $\beta = 38°$. We need to determine the values of a and b and compare them. Before we can use the Law of Sines, we need the value of γ, which by (8) is given by

$$\gamma = 180° - \alpha - \beta = 180° - 46° - 38° = 96°$$

The Law of Sines then becomes

$$\frac{\sin 46°}{a} = \frac{\sin 38°}{b} = \frac{\sin 96°}{8}$$

In particular, this means that

$$a = \frac{8 \sin 46°}{\sin 96°} \quad \text{and} \quad b = \frac{8 \sin 38°}{\sin 96°}$$

By calculator we find that

$$a \approx 5.786 \quad \text{and} \quad b \approx 4.952$$

Therefore station B is approximately 5.8 miles from the fire and station A is approximately 5.0 miles from the fire, so station A is approximately 0.8 mile closer than station B is to the fire.   □

It can be shown that if $\alpha < \beta$, then $a < b$. Consequently in any triangle the largest (smallest) side lies opposite the largest (smallest) angle. From this observation and the angles given in Example 4, we could tell immediately that station A is closer to the fire than station B.

EXERCISES 8.1

In Exercises 1–22, determine how many triangles there are satisfying the given data, and solve any triangles that exist.

1. $\alpha = 60°$, $\beta = 55°$, $b = 8$

2. $\alpha = 72°$, $\beta = 48°$, $c = 20$

3. $\beta = 29°$, $\gamma = 122°$, $c = 6$

4. $\beta = 34°$, $\gamma = 135°$, $c = 7$

5. $\beta = 34°$, $\gamma = 23°$, $a = 5$

6. $\alpha = 51.2°$, $\gamma = 65.7°$, $a = 8.3$

7. $\alpha = 23.8°$, $\gamma = 13.3°$, $b = 10.5$

8. $\alpha = 76.2°$, $\gamma = 27.6°$, $b = 13.9$

9. $\alpha = 27°24'$, $\gamma = 61°9'$, $c = 2.6$

10. $\beta = 53°43'$, $\gamma = 87°29'$, $b = 4.2$

11. $\alpha = 37°$, $a = 1.4$, $b = 2$

12. $\alpha = 50°$, $a = 3$, $b = 1$

13. $\alpha = 20°$, $a = 2.3$, $b = 7$

14. $\alpha = 20°$, $a = 2.4$, $b = 7$

15. $\alpha = 30°$, $a = 3$, $b = 6$

16. $\alpha = 33.5°$, $a = 4.9$, $c = 7.8$

17. $\beta = 74.9°$, $b = 5$, $c = 4.5$

18. $\beta = 14.4°$, $b = 1.2$, $c = 8.7$

19. $\gamma = 26.8°$, $b = 3.1$, $c = 2.5$

20. $\gamma = 27°17'$, $a = 5.2$, $c = 3.6$

21. $\alpha = 110°45'$, $a = 4.2$, $b = 2.6$

22. $\beta = 47.6°$, $b = \sqrt{2}$, $a = 1.5$

23. Two lifeguards, who are 500 yards apart at stations A and B, spot a swimmer in distress in the ocean. If the angles are as shown in Figure 8.14, which guard is in a better position to help the swimmer, and how much closer is that guard to the swimmer than the other guard?

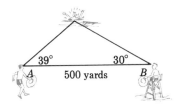

FIGURE 8.14

24. Two houses are located at points A and B, on opposite sides of a river (Figure 8.15). What is the distance between A and B?

25. The angle measurements shown in Figure 8.16 were taken on opposite sides of a tunnel through a mountain. Compute the height of the mountain.

26. The building in Figure 8.17 casts a shadow 200 feet long on the hill. How tall is side b of the building?

27. A ship leaves port in an easterly direction and later turns to the southeast. After traveling 10 miles in a southeasterly direction, the ship is 15 miles from port. How far did the ship travel in the easterly direction?

28. Show that there is no triangle with $a = 3$, $b = 5$, and $\alpha = 80°$.

29. By using Figure 8.3, prove that $(\sin \beta)/b = (\sin \gamma)/c$ if all three angles of the triangle are acute.

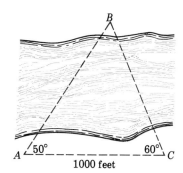

FIGURE 8.15

FIGURE 8.16

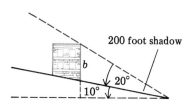

FIGURE 8.17

8.2 THE LAW OF COSINES

The companion of the Law of Sines is the Law of Cosines. It too can help us solve an arbitrary triangle when appropriate information about the sides and angles is given.

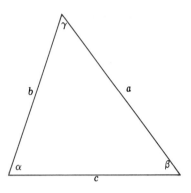

FIGURE 8.18

For an arbitrary triangle, the following three formulas, which differ only in the side of the triangle whose length is featured, comprise the Law of Cosines (see Figure 8.18):

Law of Cosines

$$a^2 = b^2 + c^2 - 2bc \cos \alpha \qquad (1)$$

$$b^2 = c^2 + a^2 - 2ca \cos \beta \qquad (2)$$

$$c^2 = a^2 + b^2 - 2ab \cos \gamma \qquad (3)$$

Notice that if the angle in any of the formulas (1)–(3) is 90°, then the triangle is a right triangle, the cosine of the angle is 0, and the corresponding formula is just the Pythagorean Theorem. As a result, the Pythagorean Theorem is a special case of the Law of Cosines.

Since (1)–(3) can be proved by the same method, we will prove only (1). To that end, let a triangle be given. Since the shape and size of the triangle are independent of the location in a coordinate system, we let the triangle be placed as in Figure 8.19. The distance *a* between the points *B* and *C* is given by the distance formula:

$$a = \sqrt{(b \cos \alpha - c)^2 + (b \sin \alpha - 0)^2}$$

Squaring both sides, rearranging, and using the fact that

$$\sin^2 \alpha + \cos^2 \alpha = 1$$

we find that

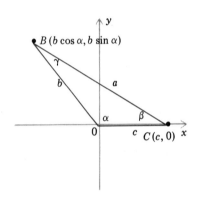

FIGURE 8.19

$$a^2 = (b \cos \alpha - c)^2 + (b \sin \alpha - 0)^2$$
$$= b^2 \cos^2 \alpha - 2bc \cos \alpha + c^2 + b^2 \sin^2 \alpha$$
$$= b^2(\cos^2 \alpha + \sin^2 \alpha) + c^2 - 2bc \cos \alpha$$
$$= b^2 + c^2 - 2bc \cos \alpha$$

This proves (1).

If we solve (1)–(3) for their respective cosine terms, we obtain the following formulas:

$$\cos \alpha = \frac{b^2 + c^2 - a^2}{2bc} \qquad (4)$$

$$\cos \beta = \frac{c^2 + a^2 - b^2}{2ca} \qquad (5)$$

$$\cos \gamma = \frac{a^2 + b^2 - c^2}{2ab} \qquad (6)$$

These formulas will be helpful in the solutions of triangles later.

Solutions of Triangles Using the Law of Cosines

The Law of Cosines may be used to solve triangles, provided that either

(a) the lengths of the three sides are known (Figure 8.20a), or
(b) the lengths of two sides and the included angle are known (Figure 8.20b).

Our first two examples illustrate (a) and (b), in turn.

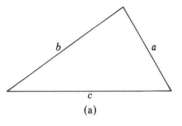

(a)   (b)

FIGURE 8.20

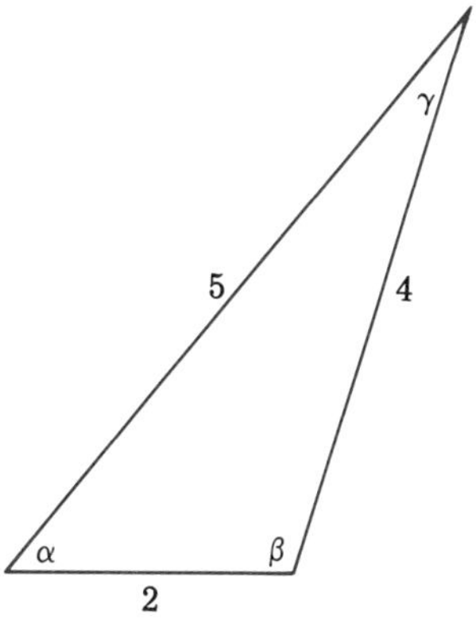

EXAMPLE 1.   Solve the triangle shown in Figure 8.21.

Solution.   We have $a = 4$, $b = 5$, and $c = 2$, and we must determine α, β, and γ. Using (4)–(6), we obtain

$$\cos \alpha = \frac{b^2 + c^2 - a^2}{2bc} = \frac{5^2 + 2^2 - 4^2}{2(5)(2)} = \frac{13}{20} = 0.65$$

$$\cos \beta = \frac{c^2 + a^2 - b^2}{2ca} = \frac{2^2 + 4^2 - 5^2}{2(2)(4)} = \frac{-5}{16} = -0.3125$$

$$\cos \gamma = \frac{a^2 + b^2 - c^2}{2ab} = \frac{4^2 + 5^2 - 2^2}{2(4)(5)} = \frac{37}{40} = 0.925$$

FIGURE 8.21

By calculator we find that

$$\alpha \approx 49.46° \ (\approx 49°28') \quad \text{and} \quad \gamma \approx 22.33 \ (\approx 22°20')$$

Consequently

$$\beta = 180° - \alpha - \gamma \approx 180° - 49.46° - 22.33°$$
$$= 108.21° \ (\approx 108°13') \quad \square$$

Caution: Because cos β is *negative*, and Table E lists only *nonnegative* values of the cosine, we could not use Table E directly to evaluate β. However, since

$$\cos(180° - \beta) = -\cos \beta = -(-0.3125) = 0.3125$$

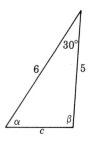

FIGURE 8.22

we could first determine an approximate value of $180° - \beta$ from Table E and from that find an approximate value of β. Our calculations would yield

$$180° - \beta \approx 71.79°$$

so that

$$\beta \approx 180° - 71.79° = 108.21° \ (\approx 108°13')$$

EXAMPLE 2.   Solve the triangle shown in Figure 8.22.

   Solution.   Because $a = 5, b = 6$, and $\gamma = 30°$, we need only find c, α, and β. Applying (3), we deduce that

$$
\begin{aligned}
c^2 &= a^2 + b^2 - 2ab \cos \gamma \\
&= 5^2 + 6^2 - 2(5)(6) \cos 30° \\
&= 61 - 60\left(\frac{\sqrt{3}}{2}\right) \\
&= 61 - 30\sqrt{3}
\end{aligned}
$$

By calculator we find that $c \approx 3.006$. Next we calculate α and β. For α we use (4):

$$\cos \alpha = \frac{b^2 + c^2 - a^2}{2bc} \approx \frac{6^2 + (61 - 30\sqrt{3}) - 5^2}{2(6)(3.006)} \approx 0.5555$$

By calculator we find that

$$\alpha \approx 56.25° \ (\approx 56°15')$$

To determine β, we can either use (5) or the values of α and γ. We choose the latter course because it entails less computation:

$$\beta = 180° - \alpha - \gamma \approx 180° - 56.25° - 30° \approx 93.75° \ (\approx 93°.45') \quad \square$$

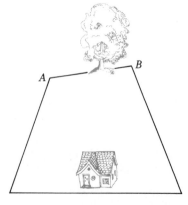

FIGURE 8.23

   The Law of Cosines is especially pertinent in surveying. Suppose we wish to determine the distance between two corners A and B of our property (Figure 8.23). Unless there is an obstacle on the line joining A and B, a tape measure suffices. But let us assume that there is a mature oak tree stretched out over the line joining A and B, as in Figure 8.23. A tape measure no longer gives an accurate estimate of the length. What a surveyor would do then is to *triangulate*, which means to locate a point C from which both A and B can be seen without obstruction, and use the triangle ABC to evaluate the length of AB (Figure 8.24). Measuring AC and BC is simple, and finding the measure of the angle at C is simple if one is armed with a transit, so finding the length of AB becomes straightforward, falling under the heading of case (b).

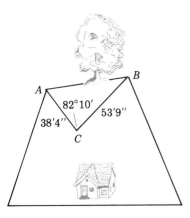

FIGURE 8.24

EXAMPLE 3.   Suppose that a surveyor places a transit at point C in Figure 8.24 and gathers the data provided in the figure. Determine the distance c between A and B.

Solution.   In order to use (3), we must have lengths in feet alone (rather than feet and inches). Thus we convert 53 feet 9 inches into 53.75 feet, and 38 feet 4 inches into approximately 38.33 feet. Now (3) applies, with $a = 53.75$, $b \approx 38.33$, $c =$ the length of the side to be determined, and $\gamma = 82°10'$. We find that

$$c^2 = a^2 + b^2 - 2ab \cos \gamma$$
$$= (53.75)^2 + (38.33)^2 - 2(53.75)(38.33)(\cos 82°10')$$

By calculator we deduce that

$$c \approx 61.62$$

This means that the distance between the corners A and B of the property is approximately 61.62 feet, that is, 61 feet and 7 inches.   □

Recapitulation   A triangle can be solved if we know one of the following.

> **i.** the lengths of all three sides (Law of Cosines)
> **ii.** the lengths of two sides and the included angle (Law of Cosines)
> **iii.** the lengths of two sides and the angle opposite one of these sides, although the solution may not be unique (ambiguous case of the Law of Sines)
> **iv.** two angles and the length of one side (Law of Sines)

The other general case, in which only the three angles are known, does not identify the triangle, as Figure 8.25 demonstrates.

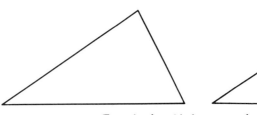

Two triangles with the same angles

FIGURE 8.25

EXERCISES 8.2

In Exercises 1–14, solve the triangle satisfying the given data.

1. $a = 6, b = 7, c = 9$

2. $a = 3, b = 12, c = 10$

3. $a = 3, b = 5, c = 7$

4. $a = 3, b = 5, c = 4.3$

5. $a = 2.3, b = 3.1, c = 3.7$

6. $a = 1, b = \frac{24}{7}, c = \frac{25}{7}$

7. $\alpha = 20°, b = 3, c = 4$

8. $\alpha = 64°, b = 7, c = 2.5$

9. $\beta = 11.7°, a = 3.2, c = 3.1$

10. $\beta = 130°$, $a = 5$, $c = 2$

11. $\gamma = 171.7°$, $a = 6.3$, $b = 4.7$

12. $\gamma = 42.8°$, $a = 14.3$, $b = 14.3$

13. $\alpha = 52°41'$, $b = 1.3$, $c = 5.4$

14. $\beta = 115°9'$, $a = 0.79$, $c = 0.61$

In Exercises 15–24 use the Law of Sines or the Law of Cosines to solve the triangle satisfying the given data.

15. $\beta = 58°30'$, $b = 5.7$, $c = 6.1$

16. $a = 0.54$, $b = 0.31$, $c = 0.33$

17. $\gamma = 99.9°$, $a = 5$, $c = 8.4$

18. $\alpha = 36°5'$, $a = 8.1$, $b = 11.4$

19. $\beta = 37°24'$, $a = 2.1$, $c = 4.8$

20. $\alpha = 96.4°$, $\gamma = 77.7°$, $a = 0.58$

21. $a = 17.1$, $b = 19.5$, $c = 10$

22. $\gamma = 14°3'$, $a = 6.4$, $b = 5.9$

23. $\beta = 151°7'$, $b = 0.3$, $c = 0.1$

24. $\beta = 6.3°$, $\gamma = 11.5°$, $c = 1.4$

25. Six dormitories are located at the vertices of a regular hexagon, as shown in Figure 8.26. How far is it from dormitory *A* to dormitory *B*?

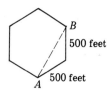

FIGURE 8.26

26. Points *A* and *C* on a gas line are separated by a fire station. Using the information in Figure 8.27, find the distance between *A* and *C*.

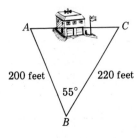

FIGURE 8.27

27. Two ships leave port at the same time, one traveling south at 20 miles per hour and the other traveling northwest at 30 miles per hour. How far apart are the ships after 30 minutes?

28. Two airplanes leave an airport at the same time. One flies north at 200 miles per hour; the other flies in a northeasterly direction at 300 miles per hour. After two hours the planes are 700 miles apart. How many degrees east from due north is the second plane flying?

29. Logs of length 10, 12, and 14 feet are placed in a triangular pattern around a campfire. What angles do the logs form?

30. In Yankee Stadium the distance between home plate and dead center field is 417 feet. Determine the distance from dead center field to third base. (*Hint:* The distance between home plate and third base is 90 feet.)

31. A carpenter plans to make a triangular frame, two sides having length 6 feet and 8 feet and forming an angle of 35° with each other. What must the length of the other side be?

32. One diagonal of a given parallelogram is 4 inches long, and the other is 7 inches long. They form an angle of 38° with one another. Determine the lengths of the sides of the parallelogram. (*Hint:* The diagonals of a parallelogram bisect each other.)

33. Find the angles of the triangle whose vertices are the points (0, 0), (3, 0), and (2, 1).

34. Find the angles of the triangle whose vertices are the points (−1, 3), (2, 2), and (0, 1).

35. A vertical flagpole is situated on a hillside that makes an angle of 15° with the horizontal. Two supporting wires are to be attached, as shown in Figure 8.28. How long should the wires be?

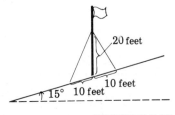

FIGURE 8.28

*36. Ship *A* is moving away from port at a speed of 5 miles per hour in the direction 30° west of north, as indicated in Figure 8.29. Ship B is 8 miles due south of port and is moving at a speed of 7 miles per hour. How many degrees west of north should B travel in

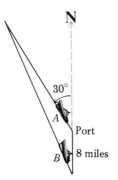

FIGURE 8.29

order to intercept A? (*Hint:* If t is the number of hours required for B to overtake A, then the triangle in Figure 8.29 has sides of length 8, 5t, and 7t miles.)

*37. **Fermat's Principle** states that in going from one point to another, light travels along the path that requires the least time. In this exercise we will demonstrate that for light traveling from one point in air to another in water, the path along which light travels may be bent rather than straight. Let us consider first the bent path from A to B, as in Figure 8.30, one meter in air and the other in water, making an initial angle of 30° with the water line at C.

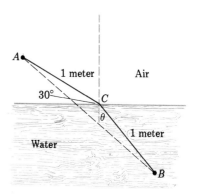

FIGURE 8.30

a. Using the speed c of light in air and the approximate speed $c/1.333$ of light in water, determine the time required for light to travel along the bent path ACB.

*b. Use Snell's Law (see Section 6.2) to determine the angle θ in Figure 8.30, next find angles A and B, and then use the Law of Cosines to determine the lengths of the segments of the straight line AB. From this information deduce the time required for light to travel in a straight line from A to B.

38. Using (1), (2), and (3), show that

$$\frac{a^2 + b^2 + c^2}{2abc} = \frac{\cos \alpha}{a} + \frac{\cos \beta}{b} + \frac{\cos \gamma}{c}$$

8.3
THE AREA
OF A TRIANGLE

Let triangle ABC have base b, height h, and area $\mathscr{A}$ (Figure 8.31). The formula

$$\mathscr{A} = \frac{1}{2}bh \tag{1}$$

is the most familiar of several formulas for the area of the triangle. The formula can be verified by noticing that triangle I in Figure 8.32 is half of

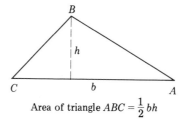

Area of triangle $ABC = \frac{1}{2}bh$

FIGURE 8.31

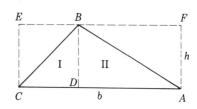

FIGURE 8.32

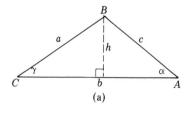
(a)

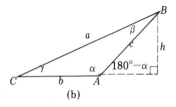
(b)

FIGURE 8.33

rectangle $CDBE$, and triangle II is half of rectangle $AFBD$, so that the area of triangle ABC is $\frac{1}{2}$ the area bh of rectangle $AFEC$.

If b and h are known, then computing the area of triangle ABC by (1) is straightforward. But if instead we know the lengths of two sides and the angle between them, or the lengths of all three sides of the triangle, and do not know h, then other formulas for the area are easier to apply. It is to these other formulas that we turn now.

First of all, suppose that for triangle ABC, b, c, and α are known (Figure 8.33a, b). By dropping a perpendicular from B, as in the figures, we see that in each case,

$$b = \text{base} \quad \text{and} \quad h = \text{height} = c \sin \alpha$$

(Note in Figure 8.33b that $h = c \sin(180° - \alpha) = c \sin \alpha$.) Therefore by (1) the area $\mathscr{A}$ of triangle ABC is given by

$$\mathscr{A} = \frac{1}{2} bh = \frac{1}{2} b(c \sin \alpha) = \frac{1}{2} bc \sin \alpha$$

By dropping perpendiculars from the other two vertices in turn, we obtain corresponding formulas involving $\sin \beta$ and $\sin \gamma$. All told, we have:

$$\mathscr{A} = \frac{1}{2} bc \sin \alpha \tag{2}$$

$$\mathscr{A} = \frac{1}{2} ac \sin \beta \tag{3}$$

$$\mathscr{A} = \frac{1}{2} ab \sin \gamma \tag{4}$$

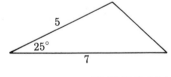

FIGURE 8.34

EXAMPLE 1.   Find the area of the triangle shown in Figure 8.34.

Solution.   Using (4) with $a = 5$, $b = 7$, and $\gamma = 25°$, we have

$$\mathscr{A} = \frac{1}{2}(5)(7) \sin 25°$$

By calculator we find that $\mathscr{A} \approx 7.396$.   □

If we know the lengths a, b, and c of all three sides of triangle ABC, then we can compute the area by means of a formula called Heron's Formula. To derive this formula we let

$$s = \frac{1}{2}(a + b + c) \tag{5}$$

and observe that

$$a + b - c = a + b + c - 2c = 2s - 2c$$

so that

$$a + b - c = 2(s - c) \tag{6}$$

Similarly, by interchanging the sides of the triangle, we have

$$b + c - a = 2(s - a) \tag{7}$$

and

$$c + a - b = 2(s - b) \tag{8}$$

From (4) above and the Law of Cosines, we find that

$$16\mathscr{A}^2 = 16\left(\frac{1}{2}ab \sin \gamma\right)^2 = 4a^2b^2 \sin^2 \gamma$$

$$= 4a^2b^2(1 - \cos^2 \gamma) \overset{\substack{\text{Law of} \\ \text{Cosines}}}{=} 4a^2b^2\left[1 - \left(\frac{a^2 + b^2 - c^2}{2ab}\right)^2\right]$$

$$= 4a^2b^2\left[1 - \frac{(a^2 + b^2 - c^2)^2}{4a^2b^2}\right]$$

$$= 4a^2b^2 - (a^2 + b^2 - c^2)^2$$

$$= [2ab + (a^2 + b^2 - c^2)][2ab - (a^2 + b^2 - c^2)]$$

$$= [(a^2 + 2ab + b^2) - c^2][c^2 - (a^2 - 2ab + b^2)]$$

$$= [(a + b)^2 - c^2][c^2 - (a - b)^2]$$

$$= [(a + b + c)(a + b - c)][(c + a - b)(c - a + b)]$$

Using (5)–(8), we can write the above in the form

$$16\mathscr{A}^2 = (2s)[2(s - c)][2(s - b)][2(s - a)]$$
$$= 16s(s - a)(s - b)(s - c)$$

It follows that

$$\mathscr{A} = \sqrt{s(s - a)(s - b)(s - c)}$$

which is called *Heron's Formula* (alternatively known as *Hero's Formula*), after the Greek mathematician Heron, who lived in the first century A.D. (Actually, the formula was known to Archimedes, who lived four centuries earlier.)

Since s is determined by the lengths of the sides of the triangle, Heron's Formula yields the area of a triangle once the lengths of all three sides are known.

EXAMPLE 2. Find the area of a triangle having sides of length 3, 5, and 6.

Solution.   If we let $a = 3$, $b = 5$, and $c = 6$, then

$$s = \frac{1}{2}(a + b + c) = \frac{1}{2}(3 + 5 + 6) = 7$$

so that by Heron's Formula,

$$\mathscr{A} = \sqrt{7(7 - 3)(7 - 5)(7 - 6)} = \sqrt{56} = 2\sqrt{14} \quad \square$$

EXERCISES 8.3

In Exercises 1–16, find the area of the triangle satisfying the given data.

1. $\alpha = 45°$, $b = 4$, $c = 3$

2. $\alpha = 27°$, $b = 8$, $c = 10$

3. $\alpha = 16°$, $b = 4.2$, $c = 4.2$

4. $\beta = 120°$, $a = 5$, $c = 10$

5. $\beta = 90°$, $a = 8$, $c = 2.5$

6. $\gamma = 153.2°$, $a = 6.3$, $b = 2.1$

7. $\gamma = 172°18'$, $a = 5.7$, $b = 8.1$

8. $\gamma = 2°$, $a = 9.3$, $b = 5.8$

9. $a = 3$, $b = 6$, $c = 7$

10. $a = 4$, $b = 3$, $c = 2$

11. $a = 1.5$, $b = 1$, $c = 1.5$

12. $a = 4$, $b = 10$, $c = 7$

13. $a = 2$, $b = 2$, $c = 2$

14. $a = 3.2$, $b = 4.5$, $c = 6.1$

15. $\alpha = 40°$, $\beta = 60°$, $c = 3.8$

16. $\beta = 108°$, $\gamma = 41°$, $a = 1.6$

17. Let the sides of a triangle have positive lengths a, b, and c. Prove that $s > a$. (Likewise, $s > b$ and $s > c$, so that $s(s - a)(s - b)(s - c) > 0$ for Heron's Formula.)

18. Suppose the coordinates of the vertices of a triangle are (x_1, y_1), (x_2, y_2), and (x_3, y_3), listed in the order in which they occur as we traverse the triangle counterclockwise. Then it can be shown that the area $\mathscr{A}$ of the triangle is

$$\mathscr{A} = \frac{1}{2}(x_1 y_2 - y_1 x_2) + \frac{1}{2}(x_2 y_3 - y_2 x_3)$$

$$+ \frac{1}{2}(x_3 y_1 - y_3 x_1)$$

Find the area $\mathscr{A}$ of the triangle having vertices
a. $(0, 0)$, $(2, 4)$, and $(-2, 2)$
b. $(-1, -2)$, $(2, 5)$, and $(5, 2)$

**8.4
POLAR
COORDINATES**

$(-1, 1)$

$\sqrt{2}$

$\frac{3\pi}{4}$

FIGURE 8.35

Until this point we have always described points in the plane by their rectangular (or Cartesian) coordinates (x, y). However, there is another commonly used system for describing points: the polar coordinate system.

To describe the polar coordinate system, consider the point P with rectangular coordinates $(-1, 1)$. The distance between P and the origin is

$$\sqrt{(-1 - 0)^2 + (1 - 0)^2}$$

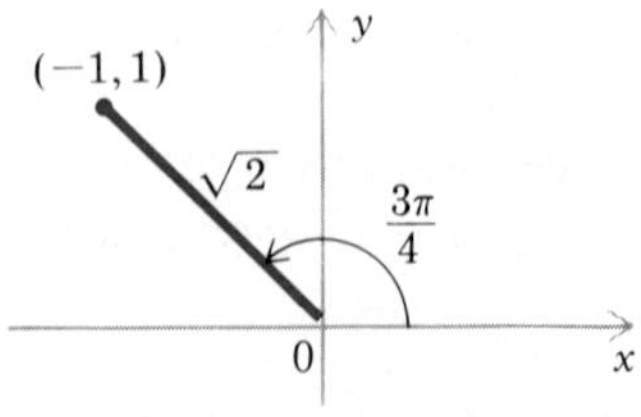

which is equal to $\sqrt{2}$ (Figure 8.35). Moreover, the line joining P and the origin makes an angle of $3\pi/4$ radians with the positive x axis. The numbers $\sqrt{2}$ and $3\pi/4$, representing the distance and the angle, suffice to locate the point P: Just start at the origin and proceed $\sqrt{2}$ units along the line that makes an angle of $3\pi/4$ radians with the positive x axis. The point P is at the end of the journey.

More generally, if we start with a given rectangular coordinate system, then to any point P in the plane except the origin we can associate an ordered pair of polar coordinates (r, θ), where r is the distance between P and the origin, and where the line segment joining P and the origin makes an angle of θ radians with the positive x axis (Figure 8.36). Since there are infinitely many choices for θ (any two differing by a multiple of 2π), P has infinitely many sets of polar coordinates. For convenience we also consider the pair $(-r, \theta + \pi)$ to be a set of polar coordinates for P. (You may think of $(-r, \theta + \pi)$ as corresponding to the point reached by proceeding a distance r in the direction opposite that of the angle $\theta + \pi$ (Figure 8.37).) Therefore if (r, θ) is one set of polar coordinates of P, then any other set will have the form

$$(r, \theta + 2n\pi), \quad \text{where } n \text{ is an integer}$$

(1)

$$(-r, \theta + \pi + 2n\pi), \quad \text{where } n \text{ is an integer}$$

To the origin we assign polar coordinates $(0, \theta)$, where θ may be any number. We have thus defined the ***polar coordinate system***. The origin is called the ***pole*** of the polar coordinate system.

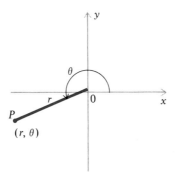

FIGURE 8.36

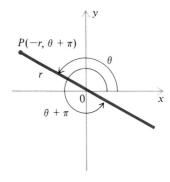

FIGURE 8.37

EXAMPLE 1. Plot the points having polar coordinates $(0, \pi/3)$, $(4, 0)$, $(-2, \pi/3)$, and $(2, -\pi/6)$.

Solution. The points are plotted in Figure 8.38. □

Graph paper for rectangular coordinates contains horizontal lines (on which y is constant) and vertical lines (on which x is constant) (Figure 8.39a). In contrast, on graph paper specially designed for polar coordinates, the polar coordinate r is constant on circles centered at the origin and the polar coordinate θ is constant on lines emanating from the origin. Figure 8.39b depicts polar coordinate graph paper.

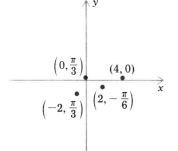

FIGURE 8.38

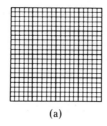
(a)

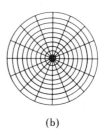
(b)

FIGURE 8.39

The dance a bee performs to communicate the location of a source of food seems to be related to polar coordinates. The orientation of the bee's body locates the direction of the food, and the intensity of the dance indicates the distance from the source.

Conversion between Cartesian and Polar Coordinates

Every point in the plane has both Cartesian and polar coordinates. We will now derive the formulas used for converting from one type of coordinates to the other.

Suppose a point P in the plane has polar coordinates (r, θ) and Cartesian coordinates (x, y). Then

$$\cos \theta = \frac{x}{r} \quad \text{and} \quad \sin \theta = \frac{y}{r}$$

so that

$$x = r \cos \theta \quad \text{and} \quad y = r \sin \theta \tag{2}$$

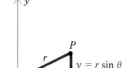

(Figure 8.40). It follows that

$$r^2 = x^2 + y^2 \quad \text{and} \quad \tan \theta = \frac{y}{x} \quad \text{for} \quad x \neq 0 \tag{3}$$

From (2) we see that x and y are uniquely determined by r and θ.

FIGURE 8.40

EXAMPLE 2.   Find the Cartesian coordinates of the point P having polar coordinates $(2, 35\pi/6)$.

Solution.   From (2) we obtain

$$x = 2 \cos \frac{35\pi}{6} = 2 \cos\left(-\frac{\pi}{6}\right) = 2 \cos \frac{\pi}{6} = 2\left(\frac{\sqrt{3}}{2}\right) = \sqrt{3}$$

and

$$y = 2 \sin \frac{35\pi}{6} = 2 \sin\left(-\frac{\pi}{6}\right) = -2 \sin \frac{\pi}{6} = -2\left(\frac{1}{2}\right) = -1$$

Therefore the Cartesian coordinates of P are $(\sqrt{3}, -1)$. $\square$

Although we cannot determine unique values for r and θ from the pair (x, y) of Cartesian coordinates by simply applying (3), it is possible to determine all sets of polar coordinates for the point with Cartesian coordinates (x, y).

EXAMPLE 3.   Find all sets of polar coordinates for the point P having Cartesian coordinates $(-7, 7\sqrt{3})$.

Solution.   First we will find the polar coordinates (r, θ) such that $r > 0$ and $0 \le \theta < 2\pi$. For that purpose we use (3), which tells us that

$$r^2 = (-7)^2 + (7\sqrt{3})^2 = 49 + 147 = 196$$

Because we desire that $r > 0$, we conclude that

$$r = \sqrt{196} = 14$$

We also know from (3) that

$$\tan \theta = \frac{7\sqrt{3}}{-7} = -\sqrt{3}$$

Therefore θ could be either $2\pi/3$ or $5\pi/3$. But since the given point $(-7, 7\sqrt{3})$ lies in the second quadrant, it follows that

$$\theta = \frac{2\pi}{3}$$

Thus one set of polar coordinates for the point with Cartesian coordinates $(-7, 7\sqrt{3})$ is $(14, 2\pi/3)$. Consequently by (1) any set of polar coordinates of P must be of the form

$$\left(14, \frac{2\pi}{3} + 2n\pi\right) \quad \text{for any integer } n$$

or

$$\left(-14, \frac{5\pi}{3} + 2n\pi\right) \quad \text{for any integer } n \quad \square$$

Caution: The solution of Example 3 shows that when we convert from Cartesian to polar coordinates, it is not enough merely to choose values of r and θ that satisfy (3). We must be sure as well that the point with polar coordinates (r, θ) lies in the correct quadrant.

Polar Graphs and Equations   The *polar graph* of an equation involving polar coordinates (r, θ) is the set of all points in the plane having a set of polar coordinates satisfying the given equation.

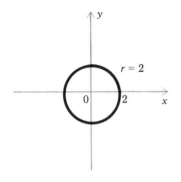

FIGURE 8.41

EXAMPLE 4.   Sketch the polar graph of the equation $r = 2$.

Solution.   Since every point that satisfies the equation $r = 2$ has distance 2 from the origin, evidently the graph is the circle with radius 2 and center at the origin (Figure 8.41).   □

More generally, if r_0 is an arbitrary positive constant, then the polar graph of the equation

$$r = r_0$$

is the circle with radius r_0 and center at the origin (Figure 8.42).

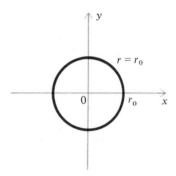

FIGURE 8.42

EXAMPLE 5.   Sketch the polar graph of the equation $\theta = \pi/3$.

Solution.   For any number r, the point with polar coordinates $(r, \pi/3)$ is on the graph of the given equation. If $r \geq 0$, then this point lies on the half-line shown in Figure 8.43a. But if $r < 0$, then the point corresponding to

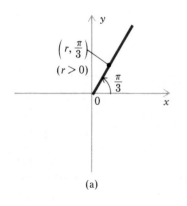

(a)

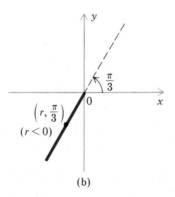

(b)

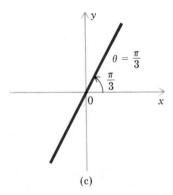

(c)

FIGURE 8.43

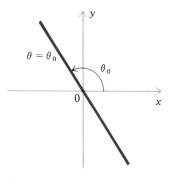

$\theta = \theta_0$

FIGURE 8.44

$(r, \pi/3)$ also has polar coordinates $(-r, \pi/3 + \pi)$, that is, $(-r, 4\pi/3)$, and since $-r > 0$, it lies on the solid half-line in Figure 8.43b. Therefore the polar graph of the equation $\theta = \pi/3$ is the line in Figure 8.43c. ☐

One can show in a similar way that for any number θ_0, the polar graph of the equation

$$\theta = \theta_0$$

is the line through the origin that makes an angle of θ_0 radians with the positive x axis (Figure 8.44).

Two other equations whose polar graphs are easy to determine are

$$r = a \cos \theta \qquad (4)$$

and

$$r = a \sin \theta \qquad (5)$$

To determine the graph of (4), we multiply both sides of (4) by r to obtain

$$r^2 = ar \cos \theta$$

Substituting for r^2 from (3), and for $r \cos \theta$ from (2), yields

$$x^2 + y^2 = ax$$

which means that

$$x^2 - ax + y^2 = 0$$

The graph of this equation becomes apparent once we complete the square:

$$\left(x^2 - ax + \frac{1}{4}a^2 - \frac{1}{4}a^2 \right) + y^2 = 0$$

$$\left(x^2 - ax + \frac{1}{4}a^2 \right) - \frac{1}{4}a^2 + y^2 = 0$$

$$\left(x - \frac{1}{2}a \right)^2 + y^2 = \frac{1}{4}a^2 = \left(\frac{1}{2}a \right)^2 \qquad (6)$$

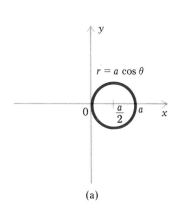

(a)

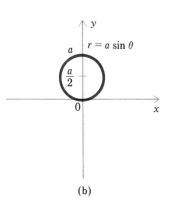

(b)

FIGURE 8.45

Notice that (6) is an equation in rectangular coordinates of the circle with radius $\frac{1}{2}|a|$ and center at the point $(\frac{1}{2}a, 0)$ (Figure 8.45a), and therefore the circle is also the graph of (4). Similarly, it is possible to prove that the graph of (5) is the circle with radius $\frac{1}{2}|a|$ and center at the point $(0, \frac{1}{2}a)$ (Figure 8.45b).

EXAMPLE 6. Sketch the polar graph of the equation $r = 3 \cos \theta$.

Solution.   Taking $a = 3$ in (4) and using Figure 8.45a as a guide, we obtain the graph sketched in Figure 8.46.   ☐

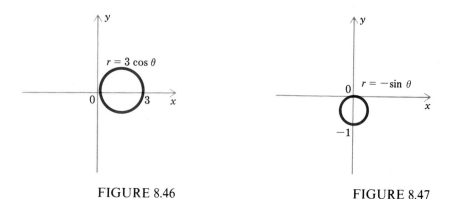

FIGURE 8.46   FIGURE 8.47

EXAMPLE 7.   Sketch the polar graph of the equation $r = -\sin\theta$.

Solution.   Taking $a = -1$ in (5) and using Figure 8.45b as a guide, we obtain the graph shown in Figure 8.47.   ☐

EXERCISES 8.4

In Exercises 1–18, find the rectangular coordinates of the point having the given polar coordinates.

1. $(0,0)$

2. $(0, \pi/6)$

3. $(-5, 0)$

4. $(4, 2\pi)$

5. $(3, \pi/2)$

6. $(\frac{1}{2}, -\pi/2)$

7. $(2.3, \pi)$

8. $(\sqrt{2}, \pi/4)$

9. $(3, \pi/3)$

10. $(-2, -\pi/6)$

11. $(1, -\pi/3)$

12. $(17, 3\pi/2)$

13. $(2, 7\pi)$

14. $(2\sqrt{2}, 5\pi/4)$

15. $(-4, -2\pi/3)$

16. $(-2, 23\pi/6)$

17. $(0.4, -19.5\pi)$

18. $(\frac{7}{3}, -8\pi)$

In Exercises 19–30, find all sets of polar coordinates of the point having the given rectangular coordinates.

19. $(5, 0)$

20. $(0, 5)$

21. $(-5, 0)$

22. $(0, -5)$

23. $(\sqrt{2}, \sqrt{2})$

24. $(3\sqrt{2}, -3\sqrt{2})$

25. $(-\frac{1}{2}, \frac{1}{2})$

26. $(-\pi, -\pi)$

27. $(\sqrt{3}, 1)$

28. $(\frac{1}{3}\sqrt{3}, -1)$

29. $(-\sqrt{2}, \sqrt{6})$

30. $(-12, -4\sqrt{3})$

In Exercises 31–45, sketch the polar graph of the given equation.

31. $r = 1$

32. $r = 4$

33. $r = -2$

34. $r = -\dfrac{1}{2}$

35. $\theta = 0$

36. $\theta = \dfrac{3\pi}{4}$

37. $\theta = -\dfrac{\pi}{6}$

38. $\theta = \dfrac{13\pi}{3}$

39. $r = \cos\theta$

40. $r = 5\cos\theta$

41. $r = -3\cos\theta$

42. $r = 4\sin\theta$

43. $r = 2\sin\theta$

44. $r = -\dfrac{1}{2}\sin\theta$

*45. $r = \cos(\theta + \pi/2)$

*46. $r = -\dfrac{1}{2}\sin(\theta + \pi/2)$

*47. $r = \theta$

8.5 VECTORS

Many physical and mathematical quantities have only magnitude and thus can be described by numbers. Examples are cost, profit, speed, height, length, area, and volume. Other quantities have both magnitude and direction. Two notable examples are velocity and force. The velocity of a moving object involves not only the speed of the object but also the direction of motion. Similarly, a force acting on an object is partly determined by its magnitude and partly by the direction in which the force acts.

Quantities that involve both magnitude and direction are described mathematically by vectors. Geometrically, a vector is a directed line segment, often described by a line segment with an arrow at one end (Figure 8.48a). For example, the velocity vector of a moving object is the directed line segment that points in the direction of motion and has length equal to the speed of the object. Similarly, the vector that describes a force is the line segment that points in the direction in which the force acts and has length equal to the magnitude of the force.

One simple way of describing a vector algebraically is to give the coordinates of the initial and terminal points of the directed line segment. But at least with velocity and force, one is normally concerned more with the direction and length of the line segment than with its initial and terminal points. Now observe that the vector appearing in Figure 8.48b represents a

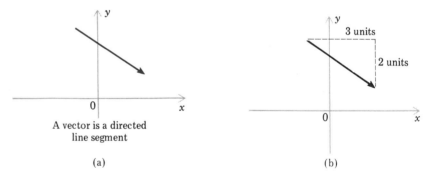

A vector is a directed line segment

(a)

(b)

FIGURE 8.48

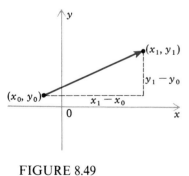

FIGURE 8.49

change of 3 units in the positive x direction and a change of 2 units in the negative y direction (which we will think of as a change of -2 units in the positive y direction). The numbers 3 and -2 identify the direction and length of the line segment, and hence the given vector can be described by the ordered pair $(3, -2)$. More generally, if a directed line segment has initial point (x_0, y_0) and terminal point (x_1, y_1) (Figure 8.49), then the corresponding vector represents a change of $x_1 - x_0$ units in the positive x direction and a change of $y_1 - y_0$ units in the positive y direction. We group these two numbers into the ordered pair $(x_1 - x_0, y_1 - y_0)$ and identify the vector with this ordered pair. In this vein, we make the following definition of a vector.

DEFINITION 8.1   A *vector* is an ordered pair (a_1, a_2) of numbers. The numbers a_1 and a_2 are called the *components* of the vector. The vector associated with the directed line segment with initial point $P = (x_0, y_0)$ and terminal point $Q = (x_1, y_1)$ is the vector $(x_1 - x_0, y_1 - y_0)$ and is denoted $\overrightarrow{PQ}$. Two vectors (a_1, a_2) and (b_1, b_2) are **equal** if $a_1 = b_1$ and $a_2 = b_2$.

EXAMPLE 1.   Let $P = (2, -1)$, $Q = (5, 3)$, $R = (-1, 6)$, and $S = (2, 10)$. Show that $\overrightarrow{PQ}$ and $\overrightarrow{RS}$ are equal.

Solution.   Applying Definition 8.1, we have

$$\overrightarrow{PQ} = (5 - 2, 3 - (-1)) = (3, 4)$$
$$\overrightarrow{RS} = (2 - (-1), 10 - 6) = (3, 4)$$

Thus $\overrightarrow{PQ} = \overrightarrow{RS}$. ☐

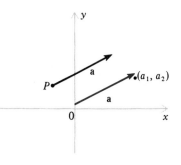

FIGURE 8.50

To distinguish vectors from numbers, which are sometimes called *scalars*, we will denote vectors by lower-case boldface letters, such as **a**, **b**, and **c**, near the beginning of the alphabet. An exception is the *zero vector* $(0, 0)$, which is denoted **0**. Since it is difficult to write a boldface letter by hand, vectors are usually written by hand by placing an arrow over a symbol or expression. Thus we would write $\vec{a}$ instead of **a**.

Each vector $\mathbf{a} = (a_1, a_2)$ can be associated with a directed line segment having an arbitrary initial point P (Figure 8.50), and different initial points give us different geometrical representations of the same vector. If P is the origin, then **a** is associated with the directed line segment from the origin to the point (a_1, a_2), or more simply, with the point (a_1, a_2). Thus we have another way of thinking of a vector—as a directed line segment starting at the origin.

A natural way to assign a length to a vector $\mathbf{a} = (a_1, a_2)$ is to assign it the length of the directed line segment from the origin to the point (a_1, a_2). In view of this, we define the length of a vector as follows.

Length (or Norm) of a Vector
$\|\mathbf{a}\| = \sqrt{a_1^2 + a_2^2}$

For example, if $\mathbf{a} = (-1, 2)$, then

$$\|\mathbf{a}\| = \sqrt{(-1)^2 + 2^2} = \sqrt{5}$$

If $P = (x_0, y_0)$ and $Q = (x_1, y_1)$ are any two given points, then $\overrightarrow{PQ} = (x_1 - x_0, y_1 - y_0)$, so that

$$\|\overrightarrow{PQ}\| = \sqrt{(x_1 - x_0)^2 + (y_1 - y_0)^2}$$

Referring to the distance formula, we see that $\|\overrightarrow{PQ}\|$ is just the distance

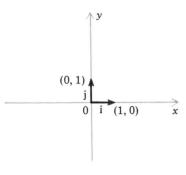

FIGURE 8.51

between P and Q. For example, if P and Q are as in Example 1, then

$$\|\overrightarrow{PQ}\| = \sqrt{(5-2)^2 + (3-(-1))^2} = \sqrt{25} = 5$$

A ***unit vector*** is a vector having length 1. There are two unit vectors that will simplify the remainder of our work with vectors:

$$\mathbf{i} = (1, 0) \quad \text{and} \quad \mathbf{j} = (0, 1)$$

(Figure 8.51). Since

$$\|\mathbf{i}\| = \sqrt{1^2 + 0^2} = 1 \quad \text{and} \quad \|\mathbf{j}\| = \sqrt{0^2 + 1^2} = 1$$

$\mathbf{i}$ and $\mathbf{j}$ are indeed unit vectors.

Combinations of Vectors

Let $\mathbf{a} = (a_1, a_2)$ and $\mathbf{b} = (b_1, b_2)$ be vectors, and let c be a number. Then the ***sum*** $\mathbf{a} + \mathbf{b}$, the ***difference*** $\mathbf{a} - \mathbf{b}$, and the ***scalar multiple*** $c\mathbf{a}$ are defined by

$$\mathbf{a} + \mathbf{b} = (a_1 + b_1, a_2 + b_2)$$
$$\mathbf{a} - \mathbf{b} = (a_1 - b_1, a_2 - b_2)$$
$$c\mathbf{a} = (ca_1, ca_2)$$

There are many laws governing these combinations. For example,

$$\mathbf{0} + \mathbf{a} = \mathbf{a} + \mathbf{0} = \mathbf{a} \qquad \mathbf{a} + \mathbf{b} = \mathbf{b} + \mathbf{a}$$
$$0\mathbf{a} = \mathbf{0} \qquad 1\mathbf{a} = \mathbf{a}$$
$$\mathbf{a} + (\mathbf{b} + \mathbf{c}) = (\mathbf{a} + \mathbf{b}) + \mathbf{c} \qquad c(\mathbf{a} + \mathbf{b}) = c\mathbf{a} + c\mathbf{b}$$
$$\mathbf{a} - \mathbf{b} = \mathbf{a} + (-1)\mathbf{b}$$

EXAMPLE 2. Let $\mathbf{a} = (2, 3)$ and $\mathbf{b} = (-1, 4)$. Find $\mathbf{a} + \mathbf{b}$, $\mathbf{a} - \mathbf{b}$, and $\frac{1}{2}\mathbf{a}$.

Solution. From the definitions we have

$$\mathbf{a} + \mathbf{b} = (2 + (-1), 3 + 4) = (1, 7)$$
$$\mathbf{a} - \mathbf{b} = (2 - (-1), 3 - 4) = (3, -1)$$
$$\frac{1}{2}\mathbf{a} = \left(\frac{1}{2}(2), \frac{1}{2}(3)\right) = \left(1, \frac{3}{2}\right) \quad \square$$

Using addition and scalar multiplication of vectors, we can express any vector $\mathbf{a} = (a_1, a_2)$ as a combination of the unit vectors $\mathbf{i}$ and $\mathbf{j}$. Indeed,

$$\mathbf{a} = (a_1, a_2) = (a_1, 0) + (0, a_2) = a_1(1, 0) + a_2(0, 1) = a_1\mathbf{i} + a_2\mathbf{j}$$

Thus

$$\mathbf{a} = a_1 \mathbf{i} + a_2 \mathbf{j} \qquad (1)$$

For example,

$$(3, -4) = 3\mathbf{i} - 4\mathbf{j} \quad \text{and} \quad \left(-\frac{1}{3}, \sqrt{2}\right) = -\frac{1}{3}\mathbf{i} + \sqrt{2}\mathbf{j}$$

We will frequently write vectors in the form of (1). For that reason we now restate our notions related to vectors in the form of (1):

$$\|a_1 \mathbf{i} + a_2 \mathbf{j}\| = \sqrt{a_1^2 + a_2^2} \qquad (2)$$

$$(a_1 \mathbf{i} + a_2 \mathbf{j}) + (b_1 \mathbf{i} + b_2 \mathbf{j}) = (a_1 + b_1)\mathbf{i} + (a_2 + b_2)\mathbf{j}$$

$$(a_1 \mathbf{i} + a_2 \mathbf{j}) - (b_1 \mathbf{i} + b_2 \mathbf{j}) = (a_1 - b_1)\mathbf{i} + (a_2 - b_2)\mathbf{j}$$

$$c(a_1 \mathbf{i} + a_2 \mathbf{j}) = ca_1 \mathbf{i} + ca_2 \mathbf{j}$$

EXAMPLE 3.   Let $\mathbf{a} = 3\mathbf{i} - \frac{5}{2}\mathbf{j}$ and $\mathbf{b} = -2\mathbf{i} - \frac{1}{2}\mathbf{j}$. Find $\mathbf{a} + \mathbf{b}$, $5\mathbf{b}$ and $\|\mathbf{a}\|$.

Solution.   By the formulas given above,

$$\mathbf{a} + \mathbf{b} = (3 + (-2))\mathbf{i} + \left(-\frac{5}{2} + \left(-\frac{1}{2}\right)\right)\mathbf{j} = \mathbf{i} - 3\mathbf{j}$$

$$5\mathbf{b} = 5(-2)\mathbf{i} + 5\left(-\frac{1}{2}\right)\mathbf{j} = -10\mathbf{i} - \frac{5}{2}\mathbf{j}$$

$$\|\mathbf{a}\| = \sqrt{(3)^2 + \left(-\frac{5}{2}\right)^2} = \sqrt{9 + \frac{25}{4}} = \sqrt{\frac{61}{4}} = \frac{1}{2}\sqrt{61} \quad \square$$

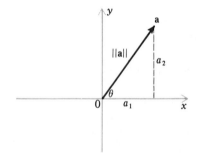

FIGURE 8.52

There is another way of representing vectors. To find it, we observe that the components a_1 and a_2 of the vector $\mathbf{a} = a_1 \mathbf{i} + a_2 \mathbf{j}$ are determined by the length of $\mathbf{a}$ and the angle θ that $\mathbf{a}$ makes with the positive x axis (Figure 8.52). Indeed, if $\mathbf{a} \neq \mathbf{0}$, then

$$\cos \theta = \frac{a_1}{\|\mathbf{a}\|} \quad \text{and} \quad \sin \theta = \frac{a_2}{\|\mathbf{a}\|}$$

so that

$$a_1 = \|\mathbf{a}\| \cos \theta \quad \text{and} \quad a_2 = \|\mathbf{a}\| \sin \theta$$

Therefore an alternative form of (1) is the following

> **Polar Form of a Vector**
>
> $$\mathbf{a} = \|\mathbf{a}\| (\cos \theta \, \mathbf{i} + \sin \theta \, \mathbf{j}) \qquad (3)$$

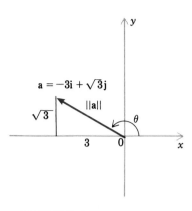

FIGURE 8.53

EXAMPLE 4.   Let $\mathbf{a} = -3\mathbf{i} + \sqrt{3}\mathbf{j}$. Express $\mathbf{a}$ in polar form.

Solution.   To write $\mathbf{a}$ in polar form we need the length $\|\mathbf{a}\|$ and the angle θ. For $\|\mathbf{a}\|$ we use (2) to obtain

$$\|\mathbf{a}\| = \sqrt{(-3)^2 + (\sqrt{3})^2} = \sqrt{9 + 3} = 2\sqrt{3}$$

To find the angle θ, we draw the triangle shown in Figure 8.53 and observe that $\tan \theta < 0$ and therefore that

$$\tan \theta = -\frac{\sqrt{3}}{3}$$

Since the point corresponding to $\mathbf{a}$ lies in the second quadrant and $\tan \theta = -\sqrt{3}/3$, we take

$$\theta = \frac{5\pi}{6}$$

Thus the polar form of $\mathbf{a}$ is given by

$$\mathbf{a} = 2\sqrt{3}\left(\cos\frac{5\pi}{6}\mathbf{i} + \sin\frac{5\pi}{6}\mathbf{j}\right) \quad \square$$

Geometric Interpretations of Vector Operations

First we will interpret the sum $\mathbf{a} + \mathbf{b}$ of two vectors $\mathbf{a}$ and $\mathbf{b}$ geometrically. We begin by letting $\mathbf{a} = a_1\mathbf{i} + a_2\mathbf{j}$ and $\mathbf{b} = b_1\mathbf{i} + b_2\mathbf{j}$. We can think of $\mathbf{a}$, $\mathbf{b}$, and $\mathbf{a} + \mathbf{b}$ as the directed line segments from the origin to the points P, Q, and R having coordinates (a_1, a_2), (b_1, b_2), and $(a_1 + b_1, a_2 + b_2)$, respectively (Figure 8.54). Now observe from the figure that if the directed line segment representing $\mathbf{b}$ is placed so that its initial point is P, then its terminal point will be R. Thus the vector $\mathbf{a} + \mathbf{b}$ can be obtained geometrically by placing the initial point of $\mathbf{b}$ on the terminal point of $\mathbf{a}$, and drawing the vector from the initial point of $\mathbf{a}$ to the terminal point of $\mathbf{b}$. The figure also tells us that

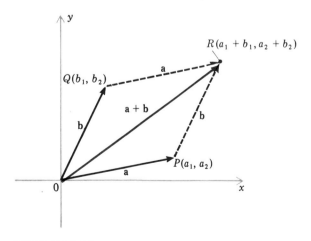

FIGURE 8.54

the two representations of **a** and **b** determine a parallelogram one of whose diagonals is **a** + **b**.

To analyze the vector *c***a** geometrically, we let **a** = $a_1\mathbf{i} + a_2\mathbf{j}$ be a vector and *c* any number. Then by the definition of *c***a**, we find that

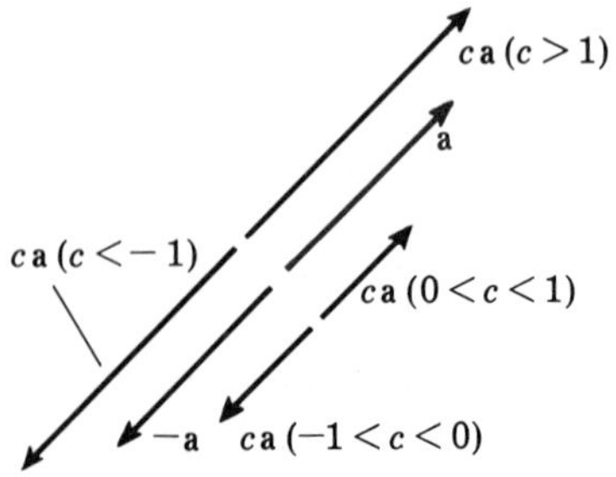

$$\|c\mathbf{a}\| = \sqrt{(ca_1)^2 + (ca_2)^2} = \sqrt{c^2(a_1^2 + a_2^2)}$$
$$= \sqrt{c^2}\sqrt{a_1^2 + a_2^2} = |c|\sqrt{a_1^2 + a_2^2} = |c|\,\|\mathbf{a}\|$$

Thus

$$\boxed{\|c\mathbf{a}\| = |c|\,\|\mathbf{a}\|}$$

It follows that when **a** is multiplied by a scalar *c*, the length of **a** is multiplied by $|c|$. If $c > 0$, the direction of **a** does not change (Figure 8.55). However, if $c < 0$, the direction of **a** is reversed (Figure 8.55). In particular, $-\mathbf{a}$ has the same length as **a** but points in the opposite direction.

For a geometric interpretation of the difference **a** − **b** of two vectors **a** and **b**, we consider **a** and **b** as directed line segments with initial points at the origin. Next we observe that

$$\mathbf{a} = \mathbf{b} + (\mathbf{a} - \mathbf{b})$$

so that **a** is the sum of the vectors **b** and **a** − **b**. Thus it follows from our previous discussion of addition of vectors that if we place the initial point of **a** − **b** at the terminal point of **b**, then the terminal point of **a** − **b** will coincide with the terminal point of **a** (Figure 8.56). In other words, the vector **a** − **b** is represented by the directed line segment that joins the terminal points of **a** and **b** and points toward **a**. The vector **a** − **b** is one diagonal of the parallelogram determined by **a** and **b** (Figure 8.57), the other diagonal being **a** + **b**.

FIGURE 8.55

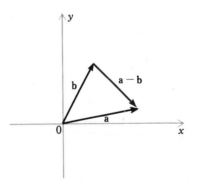

FIGURE 8.56

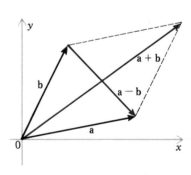

FIGURE 8.57

<u>**EXAMPLE 5.**</u>   Let **a** = 2**i** + **j** and **b** = −**i** + **j**. Construct **a** + **b** and **a** − **b** geometrically, as in Figure 8.57.

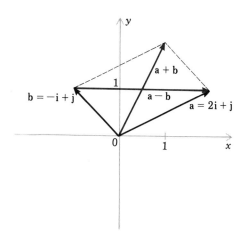

FIGURE 8.58

Solution. The constructions are shown in Figure 8.58. □

Application of Vector Addition

Many physical quantities combine according to vector addition; indeed, this is the reason we defined vector addition the way we did. For example, if several forces act simultaneously at the same point P on an object, then the object reacts as though a single force (called the **resultant force**) equal to the vector sum of the several forces is acting on the object at P.

EXAMPLE 6. Suppose a tugboat exerts a force of 1000 pounds toward the east on a ship and a wind exerts a force of 400 pounds toward the northwest on the ship. Find the resultant force **F** and its magnitude.

Solution. We set up a coordinate system as in Figure 8.59. Since $\|\mathbf{F}_1\| = 1000$ and $\mathbf{F}_1$ points along the positive x axis, it follows from (3) that

$$\mathbf{F}_1 = 1000\mathbf{i}$$

Since $\mathbf{F}_2$ points toward the northwest, it follows that $\mathbf{F}_2$ makes an angle of $3\pi/4$ with respect to the positive x axis. By hypothesis, $\|\mathbf{F}_2\| = 400$. Thus by (3),

$$\mathbf{F}_2 = \|\mathbf{F}_2\| \left(\cos \frac{3\pi}{4} \mathbf{i} + \sin \frac{3\pi}{4} \mathbf{j} \right) = 400 \left(-\frac{1}{2}\sqrt{2}\mathbf{i} + \frac{1}{2}\sqrt{2}\mathbf{j} \right)$$

$$= 200\sqrt{2}(-\mathbf{i} + \mathbf{j})$$

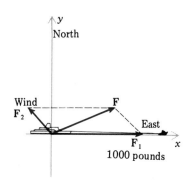

FIGURE 8.59 The resultant force **F** is $\mathbf{F}_1 + \mathbf{F}_2$, so that

$$
\begin{aligned}
\mathbf{F} = \mathbf{F}_1 + \mathbf{F}_2 &= 1000\mathbf{i} + 200\sqrt{2}(-\mathbf{i} + \mathbf{j}) \\
&= (1000 - 200\sqrt{2})\mathbf{i} + 200\sqrt{2}\mathbf{j} \\
&= 200(5 - \sqrt{2})\mathbf{i} + 200\sqrt{2}\mathbf{j}
\end{aligned}
$$

By (2), the magnitude $\|\mathbf{F}\|$ of the resultant force is given by

$$\|\mathbf{F}\| = \sqrt{[200(5 - \sqrt{2})]^2 + (200\sqrt{2})^2}$$

By calculator we find that

$$\|\mathbf{F}\| \approx 770.9$$

Thus the magnitude of the resultant force is approximately 770.9 pounds.   □

EXERCISES 8.5

In Exercises 1–6, find $\mathbf{a} + \mathbf{b}$, $\mathbf{a} - \mathbf{b}$, $2\mathbf{a} - 3\mathbf{b}$, and $-\frac{5}{2}\mathbf{a}$.

1. $\mathbf{a} = (3, 1)$, $\mathbf{b} = (-2, 4)$

2. $\mathbf{a} = (1, -4)$, $\mathbf{b} = (-2, -2)$

3. $\mathbf{a} = (-6, -3)$, $\mathbf{b} = (3, 6)$

4. $\mathbf{a} = (-5, -4)$, $\mathbf{b} = (-6, 1)$

5. $\mathbf{a} = (-2, 0)$, $\mathbf{b} = (0, 3)$

6. $\mathbf{a} = (0, 0)$, $\mathbf{b} = (-\sqrt{2}, 1)$

In Exercises 7–12, find $4\mathbf{a} - \mathbf{b}$ and $3\mathbf{a} + 2\mathbf{b}$.

7. $\mathbf{a} = 6\mathbf{i} + 2\mathbf{j}$, $\mathbf{b} = -\mathbf{i} - 2\mathbf{j}$

8. $\mathbf{a} = -3\mathbf{i} + \mathbf{j}$, $\mathbf{b} = \sqrt{3}\mathbf{i} - \sqrt{3}\mathbf{j}$

9. $\mathbf{a} = 3\mathbf{j}$, $\mathbf{b} = -5\mathbf{i} + 6\mathbf{j}$

10. $\mathbf{a} = -5\mathbf{i} - 2\mathbf{j}$, $\mathbf{b} = \mathbf{0}$

11. $\mathbf{a} = \mathbf{i} - \mathbf{j}$, $\mathbf{b} = \mathbf{i} + \mathbf{j}$

12. $\mathbf{a} = -\frac{7}{4}\mathbf{i} - \frac{4}{5}\mathbf{j}$, $\mathbf{b} = -\frac{2}{3}\mathbf{i} - \frac{6}{5}\mathbf{j}$

In Exercises 13–18, sketch vectors corresponding to $\mathbf{a}$, $\mathbf{b}$, $\mathbf{a} + \mathbf{b}$ and $\mathbf{a} - \mathbf{b}$.

13. $\mathbf{a} = (2, -1)$, $\mathbf{b} = (1, -1)$

14. $\mathbf{a} = (0, 3)$, $\mathbf{b} = (4, 0)$

15. $\mathbf{a} = (1, -3)$, $\mathbf{b} = (2, 2)$

16. $\mathbf{a} = -\mathbf{i} + \mathbf{j}$, $\mathbf{b} = 2\mathbf{i} - 3\mathbf{j}$

17. $\mathbf{a} = -\mathbf{j}$, $\mathbf{b} = 2\mathbf{i}$

18. $\mathbf{a} = 3\mathbf{i} + \mathbf{j}$, $\mathbf{b} = \mathbf{j} + 3\mathbf{i}$

In Exercises 19–24, determine the length of the vector.

19. $(3, -1)$

20. $\left(-\frac{1}{4}, -\frac{3}{4}\right)$

21. $2\mathbf{i} - \mathbf{j}$

22. $-12\mathbf{i} + 5\mathbf{j}$

23. $\sqrt{2}\mathbf{i} + \sqrt{2}\mathbf{j}$

24. $-5\mathbf{j}$

In Exercises 25–30, determine whether the given vector is a unit vector.

25. $\mathbf{i} + \mathbf{j}$

26. $\frac{3}{5}\mathbf{i} + \frac{4}{5}\mathbf{j}$

27. $3\mathbf{i} - 2\mathbf{j}$

28. $-\frac{1}{2}\mathbf{i} - \frac{1}{2}\mathbf{j}$

29. $\frac{1}{2}\sqrt{2}\mathbf{i} - \frac{1}{2}\sqrt{2}\mathbf{j}$

30. $-\frac{1}{2}\mathbf{i} + \frac{3}{2}\mathbf{j}$

In Exercises 31–36, write the vector $\mathbf{a}$ in polar form.

31. $\mathbf{a} = (-1, -1)$

32. $\mathbf{a} = \sqrt{2}\mathbf{i}$

33. $\mathbf{a} = \mathbf{i} + \sqrt{3}\mathbf{j}$

34. $\mathbf{a} = 6\mathbf{i} - 2\sqrt{3}\mathbf{j}$

35. $\mathbf{a} = -\pi\mathbf{i} + \pi\mathbf{j}$

36. $\mathbf{a} = \sqrt{15}\mathbf{i} - \sqrt{5}\mathbf{j}$

37. Show that $\mathbf{a} - \mathbf{b} = \mathbf{a} + (-\mathbf{b})$.

38. Show that $-\mathbf{a} = (-1)\mathbf{a}$.

39. Suppose $\mathbf{a}$ and $\mathbf{b}$ are vectors such that $3\mathbf{i} + 4\mathbf{j} = 2\mathbf{a} - 3\mathbf{b} = -6\mathbf{a} + \mathbf{b}$. Find $\mathbf{a}$ and $\mathbf{b}$.

The *dot product* $\mathbf{a} \cdot \mathbf{b}$ of two vectors $\mathbf{a} = a_1\mathbf{i} + a_2\mathbf{j}$ and $\mathbf{b} = b_1\mathbf{i} + b_2\mathbf{j}$ is defined by

$$\mathbf{a} \cdot \mathbf{b} = a_1 b_1 + a_2 b_2$$

In Exercises 40–43, determine the dot product of $\mathbf{a}$ and $\mathbf{b}$.

40. $\mathbf{a} = 2\mathbf{i} - 3\mathbf{j}$, $\mathbf{b} = 4\mathbf{i} + \mathbf{j}$

41. $\mathbf{a} = \frac{5}{2}\mathbf{i} - \frac{1}{3}\mathbf{j}$, $\mathbf{b} = -6\mathbf{i} - 12\mathbf{j}$

42. $\mathbf{a} = \frac{1}{2}\sqrt{3}\mathbf{i} - \mathbf{j}$, $\mathbf{b} = \sqrt{3}\mathbf{i} + \frac{3}{2}\mathbf{j}$

43. $\mathbf{a} = \frac{1}{2}\mathbf{i}$, $\mathbf{b} = 5\mathbf{j}$

44. It can be proved that the dot product $\mathbf{a} \cdot \mathbf{b}$ of $\mathbf{a}$ and $\mathbf{b}$ is also given by

$$\mathbf{a} \cdot \mathbf{b} = \|\mathbf{a}\| \, \|\mathbf{b}\| \cos \theta \qquad (4)$$

where θ is the angle with initial side pointing in the direction of $\mathbf{a}$ and terminal side pointing in the direction of $\mathbf{b}$.

a. Use (4) to show that $\mathbf{a} \cdot \mathbf{b} = 0$ if and only if $\mathbf{a} = \mathbf{0}$, $\mathbf{b} = \mathbf{0}$, or $\mathbf{a}$ and $\mathbf{b}$ are perpendicular.

b. Show that $2\mathbf{i} - 3\mathbf{j}$ and $\frac{3}{2}\mathbf{i} + \mathbf{j}$ are perpendicular.

45. If a constant force $\mathbf{F}$ acts on an object that moves along a straight line from a point P_0 to a point P_1, then the work W done on the object by the force $\mathbf{F}$ is given by

$$W = \mathbf{F} \cdot \overrightarrow{P_0 P_1}$$

Suppose a person pushes a box up a 30-foot ramp inclined at an angle of $30°$ by exerting a horizontal force $\mathbf{F}$ of magnitude 20 pounds (Figure 8.60). Determine the amount of work done on the box by $\mathbf{F}$.

46. Suppose the tugboat in Example 6 exerts its force in the northeast direction. Find the resultant force $\mathbf{F}$ and its magnitude.

47. A plane is flying 300 mph (air speed), with its nose pointed $30°$ north from the east direction. If a wind blows at 30 mph from the west, find the velocity vector, and approximate the ground speed and the actual course of the plane.

48. If two children pull on a cart with forces given in Figure 8.61, determine the resultant force on the cart and the magnitude of the force.

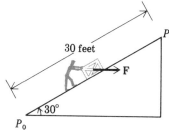

FIGURE 8.60

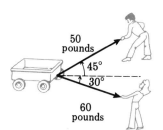

FIGURE 8.61

KEY TERMS

solving a triangle	vector
polar coordinate system	component
polar graph	length (or norm)
	unit vector

KEY FORMULAS

Law of Sines

$$\frac{\sin \alpha}{a} = \frac{\sin \beta}{b} = \frac{\sin \gamma}{c}$$

Law of Cosines

$$a^2 = b^2 + c^2 - 2bc \cos \alpha$$
$$b^2 = c^2 + a^2 - 2ca \cos \beta$$
$$c^2 = a^2 + b^2 - 2ab \cos \gamma$$

$$x = r \cos \theta \qquad y = r \sin \theta$$

$$r = x^2 + y^2 \qquad \tan \theta = \frac{y}{x} \quad \text{for} \quad x \neq 0$$

$$(a_1 \mathbf{i} + a_2 \mathbf{j}) + (b_1 \mathbf{i} + b_2 \mathbf{j}) = (a_1 + b_1)\mathbf{i} + (a_2 + b_2)\mathbf{j}$$
$$c(a_1 \mathbf{i} + a_2 \mathbf{j}) = ca_1 \mathbf{i} + ca_2 \mathbf{j}$$
$$\|\mathbf{a}\| = \sqrt{a_1^2 + a_2^2}$$
$$\mathbf{a} = \|\mathbf{a}\| (\cos \theta \, \mathbf{i} + \sin \theta \, \mathbf{j})$$

REVIEW EXERCISES

In Exercises 1–20, solve the triangle with the given data.

1. $a = 6.4, b = 6.4, c = 3.2$

2. $\gamma = 90°, a = 19.3, b = 16.8$

3. $\alpha = 32.1°, \beta = 50.5°, a = 32.1$

4. $\beta = 64°32', a = 1.5, b = 1.4$

5. $\alpha = 13°27', a = 16.2, b = 13.1$

6. $\alpha = 71°37', b = 8.4, c = 9$

7. $\beta = 124°19', a = 0.18, b = 0.23$

8. $a = 97, b = 65, c = 72$

9. $\beta = 80.3°, a = 10.3, b = 10.2$

10. $\alpha = 83°23', \gamma = 64°56', a = 0.7$

11. $a = 1.8, b = 7.2, c = 5.9$

12. $\beta = 80.8°, a = 0.35, c = 0.41$

13. $\beta = 132°46', \gamma = 26°12', b = 3.7$

14. $\beta = 134.6°, a = \sqrt{2}, c = 1.3$

15. $a = 0.8, b = 1.3, c = 1.5$

16. $\alpha = 7°11', a = 3.4, c = 6$

17. $\alpha = 60.7°, a = 10, c = 9.4$

18. $\alpha = 16.3°, \gamma = 100°, c = 16.1$

19. $\gamma = 90°, a = 90, b = 56$

20. $a = 4.8, b = 3.9, c = 3.6$

In Exercises 21–24, find the area $\mathscr{A}$ of the given triangle.

21. $\alpha = 83.2°, b = 7.1, c = 9.2$

22. $\gamma = 17°23', a = 0.6, b = 0.2$

23. $a = 1.4, b = 2.7, c = 3.5$

24. $a = 18.1, b = 9.9, c = 12.6$

In Exercises 25–28, find the rectangular coordinates of the point having the given polar coordinates.

25. $\left(\frac{3}{4}, 7\pi/4\right)$

27. $(e, -2\pi/3)$

26. $(\sqrt{3}, 17\pi/6)$

28. $(13, -9\pi/2)$

In Exercises 29–32, find all sets of polar coordinates of the point having the given rectangular coordinates.

29. $(\sqrt{5}, -\sqrt{15})$

31. $(0, \pi)$

30. $(9, 3\sqrt{3})$

32. $(-\sqrt{7}, -\sqrt{7})$

In Exercises 33–38, sketch the polar graph of the given equation.

33. $r = 3$

36. $\theta = -23\pi/4$

34. $r = -\frac{2}{3}$

37. $r = \frac{3}{2}\sin\theta$

35. $\theta = 7\pi/6$

38. $r = -2\cos\theta$

In Exercises 39–42, find $-2\mathbf{a} + 4\mathbf{b}$.

39. $\mathbf{a} = (-1, -5), \mathbf{b} = \left(\frac{2}{3}, -3\right)$

40. $\mathbf{a} = (\sqrt{3}, 0), \mathbf{b} = (0, 1)$

41. $\mathbf{a} = \mathbf{i} + 2\mathbf{j}, \mathbf{b} = 2\mathbf{i} - \mathbf{j}$

42. $\mathbf{a} = -\sqrt{2}\mathbf{j}, \mathbf{b} = \frac{1}{2}\sqrt{2}\mathbf{i} + \sqrt{2}\mathbf{j}$

In Exercises 43–48, determine the length of $\mathbf{a}$.

43. $\mathbf{a} = (6, -2\sqrt{3})$

44. $\mathbf{a} = \left(-\frac{2}{3}, \frac{2}{9}\sqrt{3}\right)$

45. $\mathbf{a} = (0, \sqrt{3}\pi - 7\sqrt{2})$

46. $\mathbf{a} = 6\mathbf{i} - 5\mathbf{j}$

47. $\mathbf{a} = -\sqrt{2}\mathbf{i} - \mathbf{j}$

48. $\mathbf{a} = 7\mathbf{i} + 24\mathbf{j}$

In Exercises 49–52, write the vector $\mathbf{a}$ in polar form.

49. $\mathbf{a} = -\sqrt{21}\mathbf{i} + \sqrt{7}\mathbf{j}$ **51.** $\mathbf{a} = \frac{1}{10}\mathbf{i} + \frac{1}{10}\sqrt{3}\mathbf{j}$

50. $\mathbf{a} = -e\mathbf{i} - e\mathbf{j}$ **52.** $\mathbf{a} = 0.43\mathbf{i} - 0.43\mathbf{j}$

53. The lengths of the sides of a triangle are 4.2, 6.1, and 5.3. Determine the largest angle in the triangle.

54. A rectangle has sides of length 4.2 and 6.5 inches. Determine the angles the diagonals of the rectangle make with each other.

55. When the angle of elevation of the sun is 63°, the shadow of a sugar maple tree is 26 feet long. Determine the height of the maple tree.

56. The Washington Monument is 555 feet tall. What is the angle of elevation of the top when viewed from a distance of 1000 feet?

57. The Washington Monument is an obelisk whose base is 55 feet long on a side. The four walls of the monument, which are trapezoidal and almost vertical, gradually slant inward to a height of 500 feet, where they are but 34.5 feet on a side. (On top of the walls is a small pyramid.) Determine the angle of elevation of the walls.

58. A side of a pyramid with square cross sections makes an angle of 68.3° with the ground. A line joining the top of the pyramid to a point on the ground 2000 feet away makes an angle of 37.1° with respect to the ground. How tall is the pyramid?

59. The pitcher's mound in baseball is 60.5 feet from home plate. When the pitcher checks a runner on first base, his head moves through the angle α depicted in Figure 8.62. Calculate α.

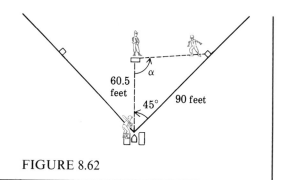

FIGURE 8.62

CUMULATIVE REVIEW II: CHAPTERS 5–8

In Exercises 1–4, evaluate the given expression.

1. $(\sqrt{3^{-4}})^{-1}$

2. $\dfrac{5^{1-\pi}5^{\pi+2}}{5^5}$

3. $\ln \dfrac{1}{e^{5/6}}$

4. $10^{-3\log_{10}5}$

In Exercises 5–11, solve the given equation.

5. $(7^x)^{2x-1} = (\sqrt{7})^{10x-8}$

6. $\log_2 x = -3$

7. $\log_x 5 = \dfrac{1}{2}$

8. $4\ln\sqrt{2x+4} = \ln 3$

9. $1.5\,e^{0.02k} = 4$

10. $e^{\frac{t\ln 2}{2}} = 14$

11. $2^{2x+1} = 3^{x-1}$

12. Approximate log 40 by using the following approximate values: $\log 2 \approx 0.3010$ and $\log 5 \approx 0.6990$.

13. Rewrite as a single logarithm and simplify:

$$\log_5(x^2 - y^2) - \log_5(x + y) - \log_5 x$$

14. Use Table C and linear interpolation to approximate the common logarithm of 62,840,000.

15. Use common logarithms to approximate

$$\frac{(1.01)^3 \times (2.03)^{1.1}}{(3.16)^2 \times \sqrt{2}}$$

16. Find the base a if the graph of $y = a^x$ contains the point $(-2, 4)$.

17. Find the base a if the graph of $y = \log_a x$ contains the point $(4, 4)$.

In Exercises 18–19, sketch the graph of the given function.

18. $f(x) = \sqrt{3}(3^{x-\frac{3}{2}})$

19. $f(x) = \log_2(x + 1)$

20. Find the length of arc on a circle of radius 2 subtended by a central angle of $(5\pi/6)$ radians.

In Exercises 21–26, find the value of the given expression.

21. $\sin\dfrac{41\pi}{6}$

22. $\tan 330°$

23. $\sec\left(-\dfrac{7\pi}{3}\right)$

24. $\csc\dfrac{5\pi}{4}$

25. $\cos 15°$

26. $\cot\dfrac{3\pi}{8}$

27. Find the values of the remaining 5 trigonometric functions at θ if $\tan\theta = \frac{1}{2}$ and $\pi < \theta < 3\pi/2$.

28. Find $\cos 2x$ if $\sin x = \frac{2}{3}$ and $\pi/2 < x < \pi$.

29. Find $\sin(x - \pi/4)$ if $\sin x = -\frac{3}{5}$ and $3\pi/2 < x < 2\pi$.

In Exercises 30–31, compute the value of the given expression.

30. $\cos[\arcsin(-\frac{1}{2})]$

31. $\arctan\left(\tan\dfrac{3\pi}{4}\right)$

32. Show that $\cot(\arcsin x) = \dfrac{\sqrt{1 - x^2}}{x}$.

In Exercises 33–34, sketch the graph of the given function.

33. $f(x) = 1 + \sin\left(x - \dfrac{\pi}{4}\right)$

34. $f(x) = 2 - \cos 2\left(x + \dfrac{\pi}{2}\right)$

In Exercises 35–39, verify the given identity.

35. $(1 + \cos^2 x)(1 + \sin^2 x) = \dfrac{8 + \sin^2 2x}{4}$

36. $(\tan x + 1)^2 + (\tan x - 1)^2 = 2\sec^2 x$

37. $\dfrac{1}{4} \sin^2 2x = \dfrac{\sin^4 x - \cos^4 x}{\tan^2 x - \cot^2 x}$

38. $\dfrac{\sin 3x}{\sin x} - \dfrac{\cos 3x}{\cos x} = 2$ **39.** $\dfrac{\cos 5x + \cos x}{\cos x - \cos 5x} = \dfrac{\cot 2x}{\tan 3x}$

In Exercises 40–42, solve the given equation.

40. $\cos t = -\dfrac{\sqrt{3}}{2}$ **42.** $\sin^2 x - \dfrac{1}{2}\sin x - \dfrac{1}{2} = 0$

41. $\cos x = \cot x$

In Exercises 43–46, solve the triangle with the given data.

43. $a = 4, b = 5, \alpha = 40°$ **45.** $a = 2, b = 5, c = 6.2$

44. $b = 5, c = 3, \alpha = 40°$ **46.** $a = 4.2, c = 6.1, \gamma = 90°$

47. Find the area $\mathscr{A}$ of the triangle whose sides have lengths 4, 8, and 10.

48. Let $\mathbf{a} = 2\mathbf{i} - 4\mathbf{j}$ and $\mathbf{b} = -5\mathbf{i} - \mathbf{j}$.

a. Find $\dfrac{1}{2}\mathbf{a} - 2\mathbf{b}$.

b. Determine which is the longer vector, $\mathbf{a}$ or $\mathbf{b}$.

49. Find all sets of polar coordinates of the point with rectangular coordinates $(6, -6\sqrt{3})$.

50. Sketch the polar graph of $r = \frac{3}{2}\cos\theta$.

51. Suppose the intensity of one sound wave is 10% greater than that of a second sound wave. Are the two sounds distinguishable to the human ear? (See Exercise 19 in Section 5.5.)

52. Earthquakes measuring 3.4 on the Richter scale preceded the major eruption of Mount St. Helens in the spring of 1980. Find the ratio of the maximal intensity of a seismic wave of such an earthquake to that of the earthquake on the California–Mexico border which occurred the same spring and which registered 6 on the Richter scale.

53. If the population of a species grows exponentially and doubles in three weeks, how long does it take for the population to multiply 10-fold?

54. At time $t = 0$ an electrical condenser with an initial charge of q_0 amperes begins to discharge. After t seconds its charge $q(t)$ is given by $q(t) = q_0 e^{-kt}$. Determine how long it takes for 95% of the charge to leave. Express your answer in terms of k.

55. The atmospheric pressure decays exponentially as a function of altitude. The pressure at sea level is known to be 14.7 pounds per square inch and is half as much at an altitude of 18,000 feet. Determine the atmospheric pressure at the top of Mt. Washington, in New Hampshire, whose altitude is 6288 feet.

56. The half-life of carbon 11 is approximately 20 minutes. If a gram of carbon 11 is present at 11 A.M., how much will be present one hour later?

57. A long-playing record makes $33\frac{1}{3}$ revolutions per minute. Through how many degrees does a point on the record travel in

a. 1 second?   b. 30 seconds?

58. Approximately how far does a point 4 inches from the center of the record mentioned in Exercise 57 travel in one minute?

59. The floor of a rectangular swimming pool is inclined lengthwise at a 9° angle from the horizontal. If the swimming pool is 100 meters long at the surface, determine the length of the floor of the pool.

60. According to one model, the relationship between the speed v of a ball just before it is batted, and the speed v' just after it is batted, is given by the formula $v' \sin\phi = v \sin\theta$, where θ and ϕ are, respectively, the angles of incidence and reflection with respect to the normal to the bat (see Figure II.1). When a baseball is batted along a foul line, we have $\theta + \phi = \pi/4$. Show that in this case,

$$v' = \frac{v\sqrt{2}}{\cot\theta - 1}$$

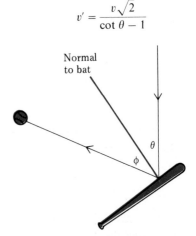

FIGURE II.1

61. A force of 6 pounds in the northeast direction, along with a force of $4\sqrt{2}$ pounds in the south direction, are applied to an object. Find the magnitude and approximate direction of the resulting force F.

Tables

TABLE A   Powers and Roots (from 1 to 100)

n	n^2	$\sqrt{n}$	n^3	$\sqrt[3]{n}$	n	n^2	$\sqrt{n}$	n^3	$\sqrt[3]{n}$
1	1	1.000	1	1.000	51	2,601	7.141	132,651	3.708
2	4	1.414	8	1.260	52	2,704	7.211	140,608	3.733
3	9	1.732	27	1.442	53	2,809	7.280	148,877	3.756
4	16	2.000	64	1.587	54	2,916	7.348	157,464	3.780
5	25	2.236	125	1.710	55	3,025	7.416	166,375	3.803
6	36	2.449	216	1.817	56	3,136	7.483	175,616	3.826
7	49	2.646	343	1.913	57	3,249	7.550	185,193	3.849
8	64	2.828	512	2.000	58	3,364	7.616	195,112	3.871
9	81	3.000	729	2.080	59	3,481	7.681	205,379	3.893
10	100	3.162	1,000	2.154	60	3,600	7.746	216,000	3.915
11	121	3.317	1,331	2.224	61	3,721	7.810	226,981	3.936
12	144	3.464	1,728	2.289	62	3,844	7.874	238,328	3.958
13	169	3.606	2,197	2.351	63	3,969	7.937	250,047	3.979
14	196	3.742	2,744	2.410	64	4,096	8.000	262,144	4.000
15	225	3.873	3,375	2.466	65	4,225	8.062	274,625	4.021
16	256	4.000	4,096	2.520	66	4,356	8.124	287,496	4.041
17	289	4.123	4,913	2.571	67	4,489	8.185	300,763	4.062
18	324	4.243	5,832	2.621	68	4,624	8.246	314,432	4.082
19	361	4.359	6,859	2.668	69	4,761	8.307	328,509	4.102
20	400	4.472	8,000	2.714	70	4,900	8.367	343,000	4.121
21	441	4.583	9,261	2.759	71	5,041	8.426	357,911	4.141
22	484	4.690	10,648	2.802	72	5,184	8.485	373,248	4.160
23	529	4.796	12,167	2.844	73	5,329	8.544	389,017	4.179
24	576	4.899	13,824	2.884	74	5,476	8.602	405,224	4.198
25	625	5.000	15,625	2.924	75	5,625	8.660	421,875	4.217
26	676	5.099	17,576	2.962	76	5,776	8.718	438,976	4.236
27	729	5.196	19,683	3.000	77	5,929	8.775	456,533	4.254
28	784	5.292	21,952	3.037	78	6,084	8.832	474,552	4.273
29	841	5.385	24,389	3.072	79	6,241	8.888	493,039	4.291
30	900	5.477	27,000	3.107	80	6,400	8.944	512,000	4.309
31	961	5.568	29,791	3.141	81	6,561	9.000	531,441	4.327
32	1,024	5.657	32,768	3.175	82	6,724	9.055	551,368	4.344
33	1,089	5.745	35,937	3.208	83	6,889	9.110	571,787	4.362
34	1,156	5.831	39,304	3.240	84	7,056	9.165	592,704	4.380
35	1,225	5.916	42,875	3.271	85	7,225	9.220	614,125	4.397
36	1,296	6.000	46,656	3.302	86	7,396	9.274	636,056	4.414
37	1,369	6.083	50,653	3.332	87	7,569	9.327	658,503	4.431
38	1,444	6.164	54,872	3.362	88	7,744	9.381	681,472	4.448
39	1,521	6.245	59,319	3.391	89	7,921	9.434	704,969	4.465
40	1,600	6.325	64,000	3.420	90	8,100	9.487	729,000	4.481
41	1,681	6.403	68,921	3.448	91	8,281	9.539	753,571	4.498
42	1,764	6.481	74,088	3.476	92	8,464	9.592	778,688	4.514
43	1,849	6.557	79,507	3.503	93	8,649	9.644	804,357	4.531
44	1,936	6.633	85,184	3.530	94	8,836	9.695	830,584	4.547
45	2,025	6.708	91,125	3.557	95	9,025	9.747	857,375	4.563
46	2,116	6.782	97,336	3.583	96	9,216	9.798	884,736	4.579
47	2,209	6.856	103,823	3.609	97	9,409	9.849	912,673	4.595
48	2,304	6.928	110,592	3.634	98	9,604	9.899	941,192	4.610
49	2,401	7.000	117,649	3.659	99	9,801	9.950	970,299	4.626
50	2,500	7.071	125,000	3.684	100	10,000	10.000	1,000,000	4.642

TABLE B   Exponential Functions e^x and e^{-x}

x	e^x	e^{-x}	x	e^x	e^{-x}
0.00	1.0000	1.0000	1.5	4.4817	0.2231
0.01	1.0101	0.9901	1.6	4.9530	0.2019
0.02	1.0202	0.9802	1.7	5.4739	0.1827
0.03	1.0305	0.9702	1.8	6.0496	0.1653
0.04	1.0408	0.9608	1.9	6.6859	0.1496
0.05	1.0513	0.9512	2.0	7.3891	0.1353
0.06	1.0618	0.9418	2.1	8.1662	0.1225
0.07	1.0725	0.9324	2.2	9.0250	0.1108
0.08	1.0833	0.9231	2.3	9.9742	0.1003
0.09	1.0942	0.9139	2.4	11.023	0.0907
0.10	1.1052	0.9048	2.5	12.182	0.0821
0.11	1.1163	0.8958	2.6	13.464	0.0743
0.12	1.1275	0.8869	2.7	14.880	0.0672
0.13	1.1388	0.8781	2.8	16.445	0.0608
0.14	1.1503	0.8694	2.9	18.174	0.0550
0.15	1.1618	0.8607	3.0	20.086	0.0498
0.16	1.1735	0.8521	3.1	22.198	0.0450
0.17	1.1853	0.8437	3.2	24.533	0.0408
0.18	1.1972	0.8353	3.3	27.113	0.0369
0.19	1.2092	0.8270	3.4	29.964	0.0334
0.20	1.2214	0.8187	3.5	33.115	0.0302
0.21	1.2337	0.8106	3.6	36.598	0.0273
0.22	1.2461	0.8025	3.7	40.447	0.0247
0.23	1.2586	0.7945	3.8	44.701	0.0224
0.24	1.2712	0.7866	3.9	49.402	0.0202
0.25	1.2840	0.7788	4.0	54.598	0.0183
0.30	1.3499	0.7408	4.1	60.340	0.0166
0.35	1.4191	0.7047	4.2	66.686	0.0150
0.40	1.4918	0.6703	4.3	73.700	0.0136
0.45	1.5683	0.6376	4.4	81.451	0.0123
0.50	1.6487	0.6065	4.5	90.017	0.0111
0.55	1.7333	0.5769	4.6	99.484	0.0101
0.60	1.8221	0.5488	4.7	109.95	0.0091
0.65	1.9155	0.5220	4.8	121.51	0.0082
0.70	2.0138	0.4966	4.9	134.29	0.0074
0.75	2.1170	0.4724	5.0	148.41	0.0067
0.80	2.2255	0.4493	5.5	244.69	0.0041
0.85	2.3396	0.4274	6.0	403.43	0.0025
0.90	2.4596	0.4066	6.5	665.14	0.0015
0.95	2.5857	0.3867	7.0	1096.6	0.0009
1.0	2.7183	0.3679	7.5	1808.0	0.0006
1.1	3.0042	0.3329	8.0	2981.0	0.0003
1.2	3.3201	0.3012	8.5	4914.8	0.0002
1.3	3.6693	0.2725	9.0	8103.1	0.0001
1.4	4.0552	0.2466	10.0	22026	0.00005

TABLE C   Common Logarithms (base 10)   A-3

x	0	1	2	3	4	5	6	7	8	9
1.0	.0000	.0043	.0086	.0128	.0170	.0212	.0253	.0294	.0334	.0374
1.1	.0414	.0453	.0492	.0531	.0569	.0607	.0645	.0682	.0719	.0755
1.2	.0792	.0828	.0864	.0899	.0934	.0969	.1004	.1038	.1072	.1106
1.3	.1139	.1173	.1206	.1239	.1271	.1303	.1335	.1367	.1399	.1430
1.4	.1461	.1492	.1523	.1553	.1584	.1614	.1644	.1673	.1703	.1732
1.5	.1761	.1790	.1818	.1847	.1875	.1903	.1931	.1959	.1987	.2014
1.6	.2041	.2068	.2095	.2122	.2148	.2175	.2201	.2227	.2253	.2279
1.7	.2304	.2330	.2355	.2380	.2405	.2430	.2455	.2480	.2504	.2529
1.8	.2553	.2577	.2601	.2625	.2648	.2672	.2695	.2718	.2742	.2765
1.9	.2788	.2810	.2833	.2856	.2878	.2900	.2923	.2945	.2967	.2989
2.0	.3010	.3032	.3054	.3075	.3096	.3118	.3139	.3160	.3181	.3201
2.1	.3222	.3243	.3263	.3284	.3304	.3324	.3345	.3365	.3385	.3404
2.2	.3424	.3444	.3464	.3483	.3502	.3522	.3541	.3560	.3579	.3598
2.3	.3617	.3636	.3655	.3674	.3692	.3711	.3729	.3747	.3766	.3784
2.4	.3802	.3820	.3838	.3856	.3874	.3892	.3909	.3927	.3945	.3962
2.5	.3979	.3997	.4014	.4031	.4048	.4065	.4082	.4099	.4116	.4133
2.6	.4150	.4166	.4183	.4200	.4216	.4232	.4249	.4265	.4281	.4298
2.7	.4314	.4330	.4346	.4362	.4378	.4393	.4409	.4425	.4440	.4456
2.8	.4472	.4487	.4502	.4518	.4533	.4548	.4564	.4579	.4594	.4609
2.9	.4624	.4639	.4654	.4669	.4683	.4698	.4713	.4728	.4742	.4757
3.0	.4771	.4786	.4800	.4814	.4829	.4843	.4857	.4871	.4886	.4900
3.1	.4914	.4928	.4942	.4955	.4969	.4983	.4997	.5011	.5024	.5038
3.2	.5051	.5065	.5079	.5092	.5105	.5119	.5132	.5145	.5159	.5172
3.3	.5185	.5198	.5211	.5224	.5237	.5250	.5263	.5276	.5289	.5302
3.4	.5315	.5328	.5340	.5353	.5366	.5378	.5391	.5403	.5416	.5428
3.5	.5441	.5453	.5465	.5478	.5490	.5502	.5514	.5527	.5539	.5551
3.6	.5563	.5575	.5587	.5599	.5611	.5623	.5635	.5647	.5658	.5670
3.7	.5682	.5694	.5705	.5717	.5729	.5740	.5752	.5763	.5775	.5786
3.8	.5798	.5809	.5821	.5832	.5843	.5855	.5866	.5877	.5888	.5899
3.9	.5911	.5922	.5933	.5944	.5955	.5966	.5977	.5988	.5999	.6010
4.0	.6021	.6031	.6042	.6053	.6064	.6075	.6085	.6096	.6107	.6117
4.1	.6128	.6138	.6149	.6160	.6170	.6180	.6191	.6201	.6212	.6222
4.2	.6232	.6243	.6253	.6263	.6274	.6284	.6294	.6304	.6314	.6325
4.3	.6335	.6345	.6355	.6365	.6375	.6385	.6395	.6405	.6415	.6425
4.4	.6435	.6444	.6454	.6464	.6474	.6484	.6493	.6503	.6513	.6522
4.5	.6532	.6542	.6551	.6561	.6571	.6580	.6590	.6599	.6609	.6618
4.6	.6628	.6637	.6646	.6656	.6665	.6675	.6684	.6693	.6702	.6712
4.7	.6721	.6730	.6739	.6749	.6758	.6767	.6776	.6785	.6794	.6803
4.8	.6812	.6821	.6830	.6839	.6848	.6857	.6866	.6875	.6884	.6893
4.9	.6902	.6911	.6920	.6928	.6937	.6946	.6955	.6964	.6972	.6981
5.0	.6990	.6998	.7007	.7016	.7024	.7033	.7042	.7050	.7059	.7067
5.1	.7076	.7084	.7093	.7101	.7110	.7118	.7126	.7135	.7143	.7152
5.2	.7160	.7168	.7177	.7185	.7193	.7202	.7210	.7218	.7226	.7235
5.3	.7243	.7251	.7259	.7267	.7275	.7284	.7292	.7300	.7308	.7316
5.4	.7324	.7332	.7340	.7348	.7356	.7364	.7372	.7380	.7388	.7396
x	0	1	2	3	4	5	6	7	8	9

TABLE C   Common Logarithms (cont.)

x	0	1	2	3	4	5	6	7	8	9
5.5	.7404	.7412	.7419	.7427	.7435	.7443	.7451	.7459	.7466	.7474
5.6	.7482	.7490	.7497	.7505	.7513	.7520	.7528	.7536	.7543	.7551
5.7	.7559	.7566	.7574	.7582	.7589	.7597	.7604	.7612	.7619	.7627
5.8	.7634	.7642	.7649	.7657	.7664	.7672	.7679	.7686	.7694	.7701
5.9	.7709	.7716	.7723	.7731	.7738	.7745	.7752	.7760	.7767	.7774
6.0	.7782	.7789	.7796	.7803	.7810	.7818	.7825	.7832	.7839	.7846
6.1	.7853	.7860	.7868	.7875	.7882	.7889	.7896	.7903	.7910	.7917
6.2	.7924	.7931	.7938	.7945	.7952	.7959	.7966	.7973	.7980	.7987
6.3	.7993	.8000	.8007	.8014	.8021	.8028	.8035	.8041	.8048	.8055
6.4	.8062	.8069	.8075	.8082	.8089	.8096	.8102	.8109	.8116	.8122
6.5	.8129	.8136	.8142	.8149	.8156	.8162	.8169	.8176	.8182	.8189
6.6	.8195	.8202	.8209	.8215	.8222	.8228	.8235	.8241	.8248	.8254
6.7	.8261	.8267	.8274	.8280	.8287	.8293	.8299	.8306	.8312	.8319
6.8	.8325	.8331	.8338	.8344	.8351	.8357	.8363	.8370	.8376	.8382
6.9	.8388	.8395	.8401	.8407	.8414	.8420	.8426	.8432	.8439	.8445
7.0	.8451	.8457	.8463	.8470	.8476	.8482	.8488	.8494	.8500	.8506
7.1	.8513	.8519	.8525	.8531	.8537	.8543	.8549	.8555	.8561	.8567
7.2	.8573	.8579	.8585	.8591	.8597	.8603	.8609	.8615	.8621	.8627
7.3	.8633	.8639	.8645	.8651	.8657	.8663	.8669	.8675	.8681	.8686
7.4	.8692	.8698	.8704	.8710	.8716	.8722	.8727	.8733	.8739	.8745
7.5	.8751	.8756	.8762	.8768	.8774	.8779	.8785	.8791	.8797	.8802
7.6	.8808	.8814	.8820	.8825	.8831	.8837	.8842	.8848	.8854	.8859
7.7	.8865	.8871	.8876	.8882	.8887	.8893	.8899	.8904	.8910	.8915
7.8	.8921	.8927	.8932	.8938	.8943	.8949	.8954	.8960	.8965	.8971
7.9	.8976	.8982	.8987	.8993	.8998	.9004	.9009	.9015	.9020	.9025
8.0	.9031	.9036	.9042	.9047	.9053	.9058	.9063	.9069	.9074	.9079
8.1	.9085	.9090	.9096	.9101	.9106	.9112	.9117	.9122	.9128	.9133
8.2	.9138	.9143	.9149	.9154	.9159	.9165	.9170	.9175	.9180	.9186
8.3	.9191	.9196	.9201	.9206	.9212	.9217	.9222	.9227	.9232	.9238
8.4	.9243	.9248	.9253	.9258	.9263	.9269	.9274	.9279	.9284	.9289
8.5	.9294	.9299	.9304	.9309	.9315	.9320	.9325	.9330	.9335	.9340
8.6	.9345	.9350	.9355	.9360	.9365	.9370	.9375	.9380	.9385	.9390
8.7	.9395	.9400	.9405	.9410	.9415	.9420	.9425	.9430	.9435	.9440
8.8	.9445	.9450	.9455	.9460	.9465	.9469	.9474	.9479	.9484	.9489
8.9	.9494	.9499	.9504	.9509	.9513	.9518	.9523	.9528	.9533	.9538
9.0	.9542	.9547	.9552	.9557	.9562	.9566	.9571	.9576	.9581	.9586
9.1	.9590	.9595	.9600	.9605	.9609	.9614	.9619	.9624	.9628	.9633
9.2	.9638	.9643	.9647	.9652	.9657	.9661	.9666	.9671	.9675	.9680
9.3	.9685	.9689	.9694	.9699	.9703	.9708	.9713	.9717	.9722	.9727
9.4	.9731	.9736	.9741	.9745	.9750	.9754	.9759	.9763	.9768	.9773
9.5	.9777	.9782	.9786	.9791	.9795	.9800	.9805	.9809	.9814	.9818
9.6	.9823	.9827	.9832	.9836	.9841	.9845	.9850	.9854	.9859	.9863
9.7	.9868	.9872	.9877	.9881	.9886	.9890	.9894	.9899	.9903	.9908
9.8	.9912	.9917	.9921	.9926	.9930	.9934	.9939	.9943	.9948	.9952
9.9	.9956	.9961	.9965	.9969	.9974	.9978	.9983	.9987	.9991	.9996
x	0	1	2	3	4	5	6	7	8	9

n	$\log_e n$	n	$\log_e n$	n	$\log_e n$
		4.5	1.5041	9.0	2.1972
0.1	-2.3026	4.6	1.5261	9.1	2.2083
0.2	-1.6094	4.7	1.5476	9.2	2.2192
0.3	-1.2040	4.8	1.5686	9.3	2.2300
0.4	-0.9163	4.9	1.5892	9.4	2.2407
0.5	-0.6931	5.0	1.6094	9.5	2.2513
0.6	-0.5108	5.1	1.6292	9.6	2.2618
0.7	-0.3567	5.2	1.6487	9.7	2.2721
0.8	-0.2231	5.3	1.6677	9.8	2.2824
0.9	-0.1054	5.4	1.6864	9.9	2.2925
1.0	0.0000	5.5	1.7047	10	2.3026
1.1	0.0953	5.6	1.7228	11	2.3979
1.2	0.1823	5.7	1.7405	12	2.4849
1.3	0.2624	5.8	1.7579	13	2.5649
1.4	0.3365	5.9	1.7750	14	2.6391
1.5	0.4055	6.0	1.7918	15	2.7081
1.6	0.4700	6.1	1.8083	16	2.7726
1.7	0.5306	6.2	1.8245	17	2.8332
1.8	0.5878	6.3	1.8405	18	2.8904
1.9	0.6419	6.4	1.8563	19	2.9444
2.0	0.6931	6.5	1.8718	20	2.9957
2.1	0.7419	6.6	1.8871	25	3.2189
2.2	0.7885	6.7	1.9021	30	3.4012
2.3	0.8329	6.8	1.9169	35	3.5553
2.4	0.8755	6.9	1.9315	40	3.6889
2.5	0.9163	7.0	1.9459	45	3.8067
2.6	0.9555	7.1	1.9601	50	3.9120
2.7	0.9933	7.2	1.9741	55	4.0073
2.8	1.0296	7.3	1.9879	60	4.0943
2.9	1.0647	7.4	2.0015	65	4.1744
3.0	1.0986	7.5	2.0149	70	4.2485
3.1	1.1314	7.6	2.0281	75	4.3175
3.2	1.1632	7.7	2.0412	80	4.3820
3.3	1.1939	7.8	2.0541	85	4.4427
3.4	1.2238	7.9	2.0669	90	4.4998
3.5	1.2528	8.0	2.0794	100	4.6052
3.6	1.2809	8.1	2.0919	110	4.7005
3.7	1.3083	8.2	2.1041	120	4.7875
3.8	1.3350	8.3	2.1163	130	4.8676
3.9	1.3610	8.4	2.1282	140	4.9416
4.0	1.3863	8.5	2.1401	150	5.0106
4.1	1.4110	8.6	2.1518	160	5.0752
4.2	1.4351	8.7	2.1633	170	5.1358
4.3	1.4586	8.8	2.1748	180	5.1930
4.4	1.4816	8.9	2.1861	190	5.2470

TABLE E   Trigonometric Functions

θ degrees	$\sin \theta$	$\cos \theta$	$\tan \theta$	$\cot \theta$	$\sec \theta$	$\csc \theta$	
0° 00′	.0000	1.0000	.0000	—	1.000	—	**90° 00′**
10	.0029	1.0000	.0029	343.8	1.000	343.8	50
20	.0058	1.0000	.0058	171.9	1.000	171.9	40
30	.0087	1.0000	.0087	114.6	1.000	114.6	30
40	.0116	.9999	.0116	85.94	1.000	85.95	20
50	.0145	.9999	.0145	68.75	1.000	68.76	10
1° 00′	.0175	.9998	.0175	57.29	1.000	57.30	**89° 00′**
10	.0204	.9998	.0204	49.10	1.000	49.11	50
20	.0233	.9997	.0233	42.96	1.000	42.98	40
30	.0262	.9997	.0262	38.19	1.000	38.20	30
40	.0291	.9996	.0291	34.37	1.000	34.38	20
50	.0320	.9995	.0320	31.24	1.001	31.26	10
2° 00′	.0349	.9994	.0349	28.64	1.001	28.65	**88° 00′**
10	.0378	.9993	.0378	26.43	1.001	26.45	50
20	.0407	.9992	.0407	24.54	1.001	24.56	40
30	.0436	.9990	.0437	22.90	1.001	22.93	30
40	.0465	.9989	.0466	21.47	1.001	21.49	20
50	.0494	.9988	.0495	20.21	1.001	20.23	10
3° 00′	.0523	.9986	.0524	19.08	1.001	19.11	**87° 00′**
10	.0552	.9985	.0553	18.07	1.002	18.10	50
20	.0581	.9983	.0582	17.17	1.002	17.20	40
30	.0610	.9981	.0612	16.35	1.002	16.38	30
40	.0640	.9980	.0641	15.60	1.002	15.64	20
50	.0669	.9978	.0670	14.92	1.002	14.96	10
4° 00′	.0698	.9976	.0699	14.30	1.002	14.34	**86° 00′**
10	.0727	.9974	.0729	13.73	1.003	13.76	50
20	.0756	.9971	.0758	13.20	1.003	13.23	40
30	.0785	.9969	.0787	12.71	1.003	12.75	30
40	.0814	.9967	.0816	12.25	1.003	12.29	20
50	.0843	.9964	.0846	11.83	1.004	11.87	10
5° 00′	.0872	.9962	.0875	11.43	1.004	11.47	**85° 00′**
10	.0901	.9959	.0904	11.06	1.004	11.10	50
20	.0929	.9957	.0934	10.71	1.004	10.76	40
30	.0958	.9954	.0963	10.39	1.005	10.43	30
40	.0987	.9951	.0992	10.08	1.005	10.13	20
50	.1016	.9948	.1022	9.788	1.005	9.839	10
6° 00′	.1045	.9945	.1051	9.514	1.006	9.567	**84° 00′**
10	.1074	.9942	.1080	9.255	1.006	9.309	50
20	.1103	.9939	.1110	9.010	1.006	9.065	40
30	.1132	.9936	.1139	8.777	1.006	8.834	30
40	.1161	.9932	.1169	8.556	1.007	8.614	20
50	.1190	.9929	.1198	8.345	1.007	8.405	10
7° 00′	.1219	.9925	.1228	8.144	1.008	8.206	**83° 00′**
	$\cos \theta$	$\sin \theta$	$\cot \theta$	$\tan \theta$	$\csc \theta$	$\sec \theta$	θ degrees

TABLE E   Trigonometric Functions (cont.)   A-7

θ degrees	sin θ	cos θ	tan θ	cot θ	sec θ	csc θ	
7° 00′	.1219	.9925	.1228	8.144	1.008	8.206	**83° 00′**
10	.1248	.9922	.1257	7.953	1.008	8.016	50
20	.1276	.9918	.1287	7.770	1.008	7.834	40
30	.1305	.9914	.1317	7.596	1.009	7.661	30
40	.1334	.9911	.1346	7.429	1.009	7.496	20
50	.1363	.9907	.1376	7.269	1.009	7.337	10
8° 00′	.1392	.9903	.1405	7.115	1.010	7.185	**82° 00′**
10	.1421	.9899	.1435	6.968	1.010	7.040	50
20	.1449	.9894	.1465	6.827	1.011	6.900	40
30	.1478	.9890	.1495	6.691	1.011	6.765	30
40	.1507	.9886	.1524	6.561	1.012	6.636	20
50	.1536	.9881	.1554	6.435	1.012	6.512	10
9° 00′	.1564	.9877	.1584	6.314	1.012	6.392	**81° 00′**
10	.1593	.9872	.1614	6.197	1.013	6.277	50
20	.1622	.9868	.1644	6.084	1.013	6.166	40
30	.1650	.9863	.1673	5.976	1.014	6.059	30
40	.1679	.9858	.1703	5.871	1.014	5.955	20
50	.1708	.9853	.1733	5.769	1.015	5.855	10
10° 00′	.1736	.9848	.1763	5.671	1.015	5.759	**80° 00′**
10	.1765	.9843	.1793	5.576	1.016	5.665	50
20	.1794	.9838	.1823	5.485	1.016	5.575	40
30	.1822	.9833	.1853	5.396	1.017	5.487	30
40	.1851	.9827	.1883	5.309	1.108	5.403	20
50	.1880	.9822	.1914	5.226	1.018	5.320	10
11° 00′	.1908	.9816	.1944	5.145	1.019	5.241	**79° 00′**
10	.1937	.9811	.1974	5.066	1.019	5.164	50
20	.1965	.9805	.2004	4.989	1.020	5.089	40
30	.1994	.9799	.2035	4.915	1.020	5.016	30
40	.2022	.9793	.2065	4.843	1.021	4.945	20
50	.2051	.9787	.2095	4.773	1.022	4.876	10
12° 00′	.2079	.9781	.2126	4.705	1.022	4.810	**78° 00′**
10	.2108	.9775	.2156	4.638	1.023	4.745	50
20	.2136	.9769	.2186	4.574	1.024	4.682	40
30	.2164	.9763	.2217	4.511	1.024	4.620	30
40	.2193	.9757	.2247	4.449	1.025	4.560	20
50	.2221	.9750	.2278	4.390	1.026	4.502	10
13° 00′	.2250	.9744	.2309	4.331	1.026	4.445	**77° 00′**
10	.2278	.9737	.2339	4.275	1.027	4.390	50
20	.2306	.9730	.2370	4.219	1.028	4.336	40
30	.2334	.9724	.2401	4.165	1.028	4.284	30
40	.2363	.9717	.2432	4.113	1.029	4.232	20
50	.2391	.9710	.2462	4.061	1.030	4.182	10
14° 00′	.2419	.9703	.2493	4.011	1.031	4.134	**76° 00′**
	cos θ	sin θ	cot θ	tan θ	csc θ	sec θ	θ degrees

TABLE E Trigonometric Functions (cont.)

θ degrees	sin θ	cos θ	tan θ	cot θ	sec θ	csc θ	
14° 00′	.2419	.9703	.2493	4.011	1.031	4.134	**76° 00′**
10	.2447	.9696	.2524	3.962	1.031	4.086	50
20	.2476	.9689	.2555	3.914	1.032	4.039	40
30	.2504	.9681	.2586	3.867	1.033	3.994	30
40	.2532	.9674	.2617	3.821	1.034	3.950	20
50	.2560	.9667	.2648	3.776	1.034	3.906	10
15° 00′	.2588	.9659	.2679	3.732	1.035	3.864	**75° 00′**
10	.2616	.9652	.2711	3.689	1.036	3.822	50
20	.2644	.9644	.2742	3.647	1.037	3.782	40
30	.2672	.9636	.2773	3.606	1.038	3.742	30
40	.2700	.9628	.2805	3.566	1.039	3.703	20
50	.2728	.9621	.2836	3.526	1.039	3.665	10
16° 00′	.2756	.9613	.2867	3.487	1.040	3.628	**74° 00′**
10	.2784	.9605	.2899	3.450	1.041	3.592	50
20	.2812	.9596	.2931	3.412	1.042	3.556	40
30	.2840	.9588	.2962	3.376	1.043	3.521	30
40	.2868	.9580	.2994	3.340	1.044	3.487	20
50	.2896	.9572	.3026	3.305	1.045	3.453	10
17° 00′	.2924	.9563	.3057	3.271	1.046	3.420	**73° 00′**
10	.2952	.9555	.3089	3.237	1.047	3.388	50
20	.2979	.9546	.3121	3.204	1.048	3.356	40
30	.3007	.9537	.3153	3.172	1.049	3.326	30
40	.3035	.9528	.3185	3.140	1.049	3.295	20
50	.3062	.9520	.3217	3.108	1.050	3.265	10
18° 00′	.3090	.9511	.3249	3.078	1.051	3.236	**72° 00′**
10	.3118	.9502	.3281	3.047	1.052	3.207	50
20	.3145	.9492	.3314	3.018	1.053	3.179	40
30	.3173	.9483	.3346	2.989	1.054	3.152	30
40	.3201	.9474	.3378	2.960	1.056	3.124	20
50	.3228	.9465	.3411	2.932	1.057	3.098	10
19° 00′	.3256	.9455	.3443	2.904	1.058	3.072	**71° 00′**
10	.3283	.9446	.3476	2.877	1.059	3.046	50
20	.3311	.9436	.3508	2.850	1.060	3.021	40
30	.3338	.9426	.3541	2.824	1.061	2.996	30
40	.3365	.9417	.3574	2.798	1.062	2.971	20
50	.3393	.9407	.3607	2.773	1.063	2.947	10
20° 00′	.3420	.9397	.3640	2.747	1.064	2.924	**70° 00′**
10	.3448	.9387	.3673	2.723	1.065	2.901	50
20	.3475	.9377	.3706	2.699	1.066	2.878	40
30	.3502	.9367	.3739	2.675	1.068	2.855	30
40	.3529	.9356	.3772	2.651	1.069	2.833	20
50	.3557	.9346	.3805	2.628	1.070	2.812	10
21° 00′	.3584	.9336	.3839	2.605	1.071	2.790	**69° 00′**
	cos θ	sin θ	cot θ	tan θ	csc θ	sec θ	θ degrees

TABLE E   Trigonometric Functions (cont.)

θ degrees	sin θ	cos θ	tan θ	cot θ	sec θ	csc θ	
21° 00′	.3584	.9336	.3839	2.605	1.071	2.790	**69° 00′**
10	.3611	.9325	.3872	2.583	1.072	2.769	50
20	.3638	.9315	.3906	2.560	1.074	2.749	40
30	.3665	.9304	.3939	2.539	1.075	2.729	30
40	.3692	.9293	.3973	2.517	1.076	2.709	20
50	.3719	.9283	.4006	2.496	1.077	2.689	10
22° 00′	.3746	.9272	.4040	2.475	1.079	2.669	**68° 00′**
10	.3773	.9261	.4074	2.455	1.080	2.650	50
20	.3800	.9250	.4108	2.434	1.081	2.632	40
30	.3827	.9239	.4142	2.414	1.082	2.613	30
40	.3854	.9228	.4176	2.394	1.084	2.595	20
50	.3881	.9216	.4210	2.375	1.085	2.577	10
23° 00′	.3907	.9205	.4245	2.356	1.086	2.559	**67° 00′**
10	.3934	.9194	.4279	2.337	1.088	2.542	50
20	.3961	.9182	.4314	2.318	1.089	2.525	40
30	.3987	.9171	.4348	2.300	1.090	2.508	30
40	.4014	.9159	.4383	2.282	1.092	2.491	20
50	.4041	.9147	.4417	2.264	1.093	2.475	10
24° 00′	.4067	.9135	.4452	2.246	1.095	2.459	**66° 00′**
10	.4094	.9124	.4487	2.229	1.096	2.443	50
20	.4120	.9112	.4522	2.211	1.097	2.427	40
30	.4147	.9100	.4557	2.194	1.099	2.411	30
40	.4173	.9088	.4592	2.177	1.100	2.396	20
50	.4200	.9075	.4628	2.161	1.102	2.381	10
25° 00′	.4226	.9063	.4663	2.145	1.103	2.366	**65° 00′**
10	.4253	.9051	.4699	2.128	1.105	2.352	50
20	.4279	.9038	.4734	2.112	1.106	2.337	40
30	.4305	.9026	.4770	2.097	1.108	2.323	30
40	.4331	.9013	.4806	2.081	1.109	2.309	20
50	.4358	.9001	.4841	2.066	1.111	2.295	10
26° 00′	.4384	.8988	.4877	2.050	1.113	2.281	**64° 00′**
10	.4410	.8975	.4913	2.035	1.114	2.268	50
20	.4436	.8962	.4950	2.020	1.116	2.254	40
30	.4462	.8949	.4986	2.006	1.117	2.241	30
40	.4488	.8936	.5022	1.991	1.119	2.228	20
50	.4514	.8923	.5059	1.977	1.121	2.215	10
27° 00′	.4540	.8910	.5095	1.963	1.122	2.203	**63° 00′**
10	.4566	.8897	.5132	1.949	1.124	2.190	50
20	.4592	.8884	.5169	1.935	1.126	2.178	40
30	.4617	.8870	.5206	1.921	1.127	2.166	30
40	.4643	.8857	.5243	1.907	1.129	2.154	20
50	.4669	.8843	.5280	1.894	1.131	2.142	10
28° 00′	.4695	.8829	.5317	1.881	1.133	2.130	**62° 00′**
	cos θ	sin θ	cot θ	tan θ	csc θ	sec θ	θ degrees

TABLE E Trigonometric Functions (cont.)

θ degrees	$\sin \theta$	$\cos \theta$	$\tan \theta$	$\cot \theta$	$\sec \theta$	$\csc \theta$	
28° 00′	.4695	.8829	.5317	1.881	1.133	2.130	**62° 00′**
10	.4720	.8816	.5354	1.868	1.134	2.118	50
20	.4746	.8802	.5392	1.855	1.136	2.107	40
30	.4772	.8788	.5430	1.842	1.138	2.096	30
40	.4797	.8774	.5467	1.829	1.140	2.085	20
50	.4823	.8760	.5505	1.816	1.142	2.074	10
29° 00′	.4848	.8746	.5543	1.804	1.143	2.063	**61° 00′**
10	.4874	.8732	.5581	1.792	1.145	2.052	50
20	.4899	.8718	.5619	1.780	1.147	2.041	40
30	.4924	.8704	.5658	1.767	1.149	2.031	30
40	.4950	.8689	.5696	1.756	1.151	2.020	20
50	.4975	.8675	.5735	1.744	1.153	2.010	10
30° 00′	.5000	.8660	.5774	1.732	1.155	2.000	**60° 00′**
10	.5025	.8646	.5812	1.720	1.157	1.990	50
20	.5050	.8631	.5851	1.709	1.159	1.980	40
30	.5075	.8616	.5890	1.698	1.161	1.970	30
40	.5100	.8601	.5930	1.686	1.163	1.961	20
50	.5125	.8587	.5969	1.675	1.165	1.951	10
31° 00′	.5150	.8572	.6009	1.664	1.167	1.942	**59° 00′**
10	.5175	.8557	.6048	1.653	1.169	1.932	50
20	.5200	.8542	.6088	1.643	1.171	1.923	40
30	.5225	.8526	.6128	1.632	1.173	1.914	30
40	.5250	.8511	.6168	1.621	1.175	1.905	20
50	.5275	.8496	.6208	1.611	1.177	1.896	10
32° 00′	.5299	.8480	.6249	1.600	1.179	1.887	**58° 00′**
10	.5324	.8465	.6289	1.590	1.181	1.878	50
20	.5348	.8450	.6330	1.580	1.184	1.870	40
30	.5373	.8434	.6371	1.570	1.186	1.861	30
40	.5398	.8418	.6412	1.560	1.188	1.853	20
50	.5422	.8403	.6453	1.550	1.190	1.844	10
33° 00′	.5446	.8387	.6494	1.540	1.192	1.836	**57° 00′**
10	.5471	.8371	.6536	1.530	1.195	1.828	50
20	.5495	.8355	.6577	1.520	1.197	1.820	40
30	.5519	.8339	.6619	1.511	1.199	1.812	30
40	.5544	.8323	.6661	1.501	1.202	1.804	20
50	.5568	.8307	.6703	1.492	1.204	1.796	10
34° 00′	.5592	.8290	.6745	1.483	1.206	1.788	**56° 00′**
10	.5616	.8274	.6787	1.473	1.209	1.781	50
20	.5640	.8258	.6830	1.464	1.211	1.773	40
30	.5664	.8241	.6873	1.455	1.213	1.766	30
40	.5688	.8225	.6916	1.446	1.216	1.758	20
50	.5712	.8208	.6959	1.437	1.218	1.751	10
35° 00′	.5736	.8192	.7002	1.428	1.221	1.743	**55° 00′**
	$\cos \theta$	$\sin \theta$	$\cot \theta$	$\tan \theta$	$\csc \theta$	$\sec \theta$	θ degrees

θ degrees	sin θ	cos θ	tan θ	cot θ	sec θ	csc θ	
35° 00′	.5736	.8192	.7002	1.428	1.221	1.743	**55° 00′**
10	.5760	.8175	.7046	1.419	1.223	1.736	50
20	.5783	.8158	.7089	1.411	1.226	1.729	40
30	.5807	.8141	.7133	1.402	1.228	1.722	30
40	.5831	.8124	.7177	1.393	1.231	1.715	20
50	.5854	.8107	.7221	1.385	1.233	1.708	10
36° 00′	.5878	.8090	.7265	1.376	1.236	1.701	**54° 00′**
10	.5901	.8073	.7310	1.368	1.239	1.695	50
20	.5925	.8056	.7355	1.360	1.241	1.688	40
30	.5948	.8039	.7400	1.351	1.244	1.681	30
40	.5972	.8021	.7445	1.343	1.247	1.675	20
50	.5995	.8004	.7490	1.335	1.249	1.668	10
37° 00′	.6018	.7986	.7536	1.327	1.252	1.662	**53° 00′**
10	.6041	.7969	.7581	1.319	1.255	1.655	50
20	.6065	.7951	.7627	1.311	1.258	1.649	40
30	.6088	.7934	.7673	1.303	1.260	1.643	30
40	.6111	.7916	.7720	1.295	1.263	1.636	20
50	.6134	.7898	.7766	1.288	1.266	1.630	10
38° 00′	.6157	.7880	.7813	1.280	1.269	1.624	**52° 00′**
10	.6180	.7862	.7860	1.272	1.272	1.618	50
20	.6202	.7844	.7907	1.265	1.275	1.612	40
30	.6225	.7826	.7954	1.257	1.278	1.606	30
40	.6248	.7808	.8002	1.250	1.281	1.601	20
50	.6271	.7790	.8050	1.242	1.284	1.595	10
39° 00′	.6293	.7771	.8098	1.235	1.287	1.589	**51° 00′**
10	.6316	.7753	.8146	1.228	1.290	1.583	50
20	.6338	.7735	.8195	1.220	1.293	1.578	40
30	.6361	.7716	.8243	1.213	1.296	1.572	30
40	.6383	.7698	.8292	1.206	1.299	1.567	20
50	.6406	.7679	.8342	1.199	1.302	1.561	10
40° 00′	.6428	.7660	.8391	1.192	1.305	1.556	**50° 00′**
10	.6450	.7642	.8441	1.185	1.309	1.550	50
20	.6472	.7623	.8491	1.178	1.312	1.545	40
30	.6494	.7604	.8541	1.171	1.315	1.540	30
40	.6517	.7585	.8591	1.164	1.318	1.535	20
50	.6539	.7566	.8642	1.157	1.322	1.529	10
41° 00′	.6561	.7547	.8693	1.150	1.325	1.524	**49° 00′**
10	.6583	.7528	.8744	1.144	1.328	1.519	50
20	.6604	.7509	.8796	1.137	1.332	1.514	40
30	.6626	.7490	.8847	1.130	1.335	1.509	30
40	.6648	.7470	.8899	1.124	1.339	1.504	20
50	.6670	.7451	.8952	1.117	1.342	1.499	10
42° 00′	.6691	.7431	.9004	1.111	1.346	1.494	**48° 00′**
	cos θ	sin θ	cot θ	tan θ	csc θ	sec θ	θ degrees

TABLE E Trigonometric Functions (cont.)

θ degrees	sin θ	cos θ	tan θ	cot θ	sec θ	csc θ	
42° 00′	.6691	.7431	.9004	1.111	1.346	1.494	**48° 00′**
10	.6713	.7412	.9057	1.104	1.349	1.490	50
20	.6734	.7392	.9110	1.098	1.353	1.485	40
30	.6756	.7373	.9163	1.091	1.356	1.480	30
40	.6777	.7353	.9217	1.085	1.360	1.476	20
50	.6799	.7333	.9271	1.079	1.364	1.471	10
43° 00′	.6820	.7314	.9325	1.072	1.367	1.466	**47° 00′**
10	.6841	.7294	.9380	1.066	1.371	1.462	50
20	.6862	.7274	.9435	1.060	1.375	1.457	40
30	.6884	.7254	.9490	1.054	1.379	1.453	30
40	.6905	.7234	.9545	1.048	1.382	1.448	20
50	.6926	.7214	.9601	1.042	1.386	1.444	10
44° 00′	.6947	.7193	.9657	1.036	1.390	1.440	**46° 00′**
10	.6967	.7173	.9713	1.030	1.394	1.435	50
20	.6988	.7153	.9770	1.024	1.398	1.431	40
30	.7009	.7133	.9827	1.018	1.402	1.427	30
40	.7030	.7112	.9884	1.012	1.406	1.423	20
50	.7050	.7092	.9942	1.006	1.410	1.418	10
45° 00′	.7071	.7071	1.0000	1.0000	1.414	1.414	**45° 00′**
	cos θ	sin θ	cot θ	tan θ	csc θ	sec θ	θ degrees

Answers to Odd-Numbered Exercises

SECTION 1.1
1. 6   **3.** 1   **5.** 22   **7.** $a^2 + 3a + 2$ **9.** $-a^2 + 1$ **11.** -2 **13.** $27 + 10\sqrt{2}$
15. $\frac{64}{9}$ **17.** $ac + bc + ad + bd - 3a - 3b + 2c + 2d - 6$ **19.** $\frac{3}{4}$ **21.** $\frac{3}{10}$
23. $-\frac{5}{7}$ **25.** $\frac{5}{8}$ **27.** $\frac{6}{5}$ **29.** $\frac{3}{4}$ **31.** l.c.d.: 24; $\frac{7}{24}$ **33.** l.c.d.: 45; $\frac{46}{45}$

35. l.c.d.: 180; $-\frac{7}{180}$ **37.** 0.022024709   **39.** 0.034176350   **41.** $-\dfrac{4a}{a^2 - 1}$

43. $\dfrac{2b - 3a + 4}{ab}$ **45.** $(a + 1)(b + 1) = ab + a + b + 1$

47. $a - (b + c) = a - b - c$ **49.** $(a + b)^3 = a^3 + 3a^2b + 3ab^2 + b^3$

51. $\dfrac{1}{a + b}$ remains $\dfrac{1}{a + b}$ **53.** 82.81

55. The decimal expansion neither terminates nor repeats.
59. a. $a = 0, b = 1, c = 1$ (other solutions possible)
    b. $a = 2, b = 2, c = 2$ (other solutions possible)

SECTION 1.2
1.

3. $x \geq \sqrt{2}$ **5.** $-1 < 6 - r < 1$ **7.** $z < 0$ **9.** $|x - 2| < 0.01$ **11.** $c \leq \frac{1}{10}$
13. $\sqrt{2} > 1$ **15.** $(-2)^2 > 3$ **17.** $0 < \pi - 3$ **19.** $\sqrt{16} = 4$ **21.** 1   **23.** 6
25. 1   **27.** 8.5   **29.** 5   **31.** 1   **33.** -4 **35.** π **37.** 2   **39.** 16   **41.** 10
43. x^2 **45.** $x^2 + 1$ **47.** $a - 4$ **49.** $a - b$ **51.** 0

SECTION 1.3
1. 1   **3.** $\frac{1}{8}$ **5.** $\frac{1}{4}$ **7.** $\frac{1}{1000}$ **9.** 4^5 **11.** -2^9 **13.** $-\frac{1}{7}$ **15.** 2^8 **17.** $\frac{1}{100}$
19. 4^7 **21.** 4^{30} **23.** $\frac{8}{9}$ **25.** $\frac{9}{4}$ **27.** $\frac{1}{147}$ **29.** a^{12} **31.** y^{-8}
33. b^5 **35.** c^8 **37.** x^3y^6 **39.** $\frac{9}{4}x^{-8}$ **41.** $r^{16}s^{14}$ **43.** t^2

45. $\dfrac{1}{ab}$ **47.** $\dfrac{a^3b^3}{(a + b)^3}$ **49.** $\dfrac{a^2b^2}{a + b}$ **51.** Positive   **53.** Negative

55. 4.832×10^2 **57.** 1.009×10^0, or 1.009 **59.** 9.999×10^{-1}
61. 2.7168876×10^7 **63.** 1.7659446×10^{-1} **65.** 3×10^{-7}
67. 1.817×10^9 **69.** 8.7445702×10^{-2} **71.** 7.8066378×10^{-3}
73. $C = 2\pi r$ **75.** $A = s^2$ **77.** $V = \frac{4}{3}\pi r^3$ **79.** $V = \frac{1}{3}\pi h r^2$
83. a. $a = 0$ and either $m \le 0$ or $n \le 0$
  b. $n \le 0$ and either $a = 0$ or $b = 0$
85. $0.00000000000000000000000000000167$
87. Approximately 1836.64508
89. a. Approximately 0.099037155
  b. Yes

SECTION 1.4

1. 4 **3.** $\frac{1}{5}$ **5.** 1.3 **7.** 0.2 **9.** $4\sqrt{3}$ **11.** $\frac{1}{12}$ **13.** $2\sqrt{10} \times 10^5$

15. $6\sqrt{2}$ **17.** 10^7 **19.** $\dfrac{\sqrt{6}}{6}$ **21.** $\frac{5}{7}$ **23.** $ab\sqrt{b}$ **25.** $4xy^2 z\sqrt{2xy}$

27. $\dfrac{p^2}{3q^4}\sqrt{\dfrac{p}{3}}$ **29.** $\dfrac{8}{r}\sqrt{\dfrac{2}{r}}$ **31.** $\dfrac{1}{c^{12}d^4\sqrt{d}}$ **33.** $x + y$ **35.** $(4a - 7b)^2$

37. $2ab^2$ **39.** $\sqrt{a} + \sqrt{b}$ **41.** 3 **43.** 0.5 **45.** 4 **47.** 3 **49.** $2\sqrt[3]{2}$

51. $2\sqrt[4]{8}$ **53.** ab^2 **55.** $-\dfrac{1}{5xy^2\sqrt[3]{z}}$ **57.** $\dfrac{x+y}{3x^3 y^2}$ **59.** t **61.** $(4x + 2y)^2$

63. $a^3 b^2$ **65.** $a^2|b|$ **67.** $x^2/|y|$ **69.** $\dfrac{\sqrt{3}}{3}$ **71.** $\sqrt{2} - 1$ **73.** $\sqrt{6} - 2$

75. $\dfrac{19 + 8\sqrt{3}}{13}$ **77.** $\dfrac{a + b + 2\sqrt{ab}}{a - b}$ **79.** $s = \sqrt[3]{V}$ **81.** $r = \sqrt[3]{\dfrac{3V}{4\pi}}$

87. $\dfrac{15}{4}\sqrt{2}$ (approximately 5.3) grams

SECTION 1.5

1. 32 **3.** $\frac{1}{25}$ **5.** $\frac{1}{256}$ **7.** $\frac{1}{32}$ **9.** $\frac{125}{27}$ **11.** 2 **13.** $250\sqrt{2}$ **15.** $\frac{8}{27}$

17. $\dfrac{1}{2\sqrt{2}}$ **19.** x **21.** $216a^{9/2}b^6$ **23.** $x^2 y^4 z^{16/3}$ **25.** $\dfrac{z^{9/4}}{8}$ **27.** $x^{2/5}y^{24/5}z^{14/5}$

29. $4a$ **31.** $\left(\dfrac{p}{q}\right)^{2/3}$ **33.** $\sqrt{p^2 + q^2}$ **35.** $\dfrac{z^2}{64(x - y)^3}$

37. $b^{13/6}$ **39.** 1.7411011 **41.** 1.4645919 **43.** 0.89921804
45. Approximately 3152.6081 **47.** It is divided by $2^{1.4}$.

SECTION 1.6

1. $3, 1, 10$ **3.** $\sqrt{5}, -4 + \sqrt{5}, -4 + \sqrt{5}$ **5.** $7x^2 - 6$
7. $-x^3 + x^2 + 2x + 4$ **9.** $8x^2 - 22x + 15$ **11.** $\frac{1}{6}x^2 + 3x + 12$
13. $7x^2 - 2x - 3$ **15.** $12x - 15$ **17.** $\sqrt{2}\,y^3 - 2y$ **19.** $2r^2 - 2r + 25$
21. $81x^4 - 162x^3 + 117x^2 - 36x + 4$ **23.** $2y^6 + 16$
25. $4x^{25} + 7x^{20} - 23x^{15} + 10x^{10}$ **27.** $4x^2 + xy + 2y^2$ **29.** $x^2 + xy - y^2$
31. $-2x^2 y^2 + 6xy - 12y^2$ **33.** $\frac{1}{4}x^2 + \frac{1}{4}xy + \frac{1}{16}y^2$
35. $64x^3 - 48x^2 y^2 + 12xy^4 - y^6$ **37.** $2xh + h^2$ **39.** $x^3 - y^3$ **41.** $r^5 s - rs^5$

43. $x^2 + 4y^2 + 9z^2 + 4xy - 12yz - 6xz$ **45.** $u + 2u^{1/2}v^{1/2} + v$

47. $\dfrac{1}{r^2} - \dfrac{2}{rs} + \dfrac{1}{s^2}$ **53.** $102x - 10x^2 - 150$

SECTION 1.7   **1.** $(x+2)(x+6)$ **3.** $(x-1)(x-6)$ **5.** $2(t-4)(t+1)$ **7.** $(y+4)(y+9)$
9. $(x-4)(x-15)$ **11.** $(7-b)(3-b)$ **13.** $(x+2)(x-2)$
15. $(a-7)^2$ **17.** $(x+\frac{1}{2})^2$ **19.** $(4+3z)(4-3z)$ **21.** $(x-\sqrt{2})^2$
23. $(2x+1)(x+3)$ **25.** $(x-3)(-x+2)$ or $(x-2)(-x+3)$ **27.** $(7x-4)(x+3)$
29. $(3t-1)(2t+6)$ **31.** $(x+y)^2$ **33.** $(x+2y)(x-2y)$
35. $(12a+b)(12a-b)$ **37.** $(x-y)^2(x+y)^2$ **39.** $(5x+y)(x-3y)$
41. $x(x+2)$ **43.** $x^6(x-1)$ **45.** $x(3x+5)(x+1)$ **47.** $(x-3)^2(x+3)^2$
49. $(x-2)(x+1)(x-1)$ **51.** $(x+1)^2(x-1)$ **53.** $(3x+2y+4)(3x+2y-3)$
55. $(x-5+2y)(x-5-2y)$ **57.** $(x+1)(x^2-x+1)$ **59.** $(x-2)(x^2+2x+4)$
61. $(2x-1)(4x^2+2x+1)$ **63.** $(2x-1)(16x^4+8x^3+4x^2+2x+1)$
65. $(2x-3y)(4x^2+6xy+9y^2)(2x+3y)(4x^2-6xy+9y^2)$
67. $(x+1)(x^4-x^3+x^2-x+1)(x-1)(x^4+x^3+x^2+x+1)$
69. $[(x-2y)^2+1](x-2y+1)(x-2y-1)$
71. Yes   **73.** No   **75.** $(31)(29)$

SECTION 1.8   **1.** All real numbers except -2 and -3; $-\frac{2}{3}$ **3.** All real numbers except 1; $-\frac{10}{7}$

5. $\dfrac{1}{x+4}$ **7.** $\dfrac{y-8}{y-3}$ **9.** $\dfrac{b-1}{4b-5}$ **11.** $\dfrac{(s+3)(s-2)(s+2)}{s-4}$ **13.** $\dfrac{x+y}{x^2+xy+y^2}$

15. $(x+1)(x-2)$ **17.** $y(y-2)$ **19.** $\dfrac{(z+3)^3}{(z+2)(z-2)(z-1)}$ **21.** $\dfrac{(x+3)^2}{(x-1)^2}$

23. $\dfrac{(b+1)(b-1)}{b-9}$ **25.** $\dfrac{(x-y)^2}{x+y}$ **27.** $\dfrac{3x-1}{x(x-1)}$ **29.** $\dfrac{3y-4}{(y+3)(y-3)}$

31. $-\dfrac{8}{(y-3)(y+8)}$ **33.** $\dfrac{20z^2+2z-13}{(6z+1)(z+4)}$ **35.** $\dfrac{5}{t-3}$

37. $\dfrac{7u(u-2)}{(3u-1)(2u-3)}$ **39.** $\dfrac{v-1}{v+2}$ **41.** $\dfrac{a^2+b^2}{ab}$ **43.** $\dfrac{b^2}{a^2+b^2}$ **45.** $\dfrac{ab^2-2b+4}{ab^3}$

47. $\dfrac{qr+pr+pq}{pqr}$ **49.** $-\dfrac{1}{x(x+h)}$ **51.** $\dfrac{x-y}{x+y}$ **53.** $\dfrac{1+x^2}{2+x}$ **55.** x

57. $\dfrac{(5z-3)(z+4)}{12(z+5)}$ **59.** $x-y$ **61.** $\dfrac{\sqrt{y}-\sqrt{x}}{\sqrt{xy}}$ **63.** $\dfrac{\sqrt{x}+\sqrt{3}}{x-3}$ **65.** $\dfrac{\sqrt{x}+9}{x-81}$

67. $\sqrt{x}-\sqrt{3}$ **69.** $\dfrac{x-2\sqrt{xy}+y}{x-y}$ **71.** $\dfrac{R_1R_2R_3}{R_2R_3+R_1R_3+R_1R_2}$

CHAPTER 1 REVIEW   **1.** $\frac{5}{12}$ **3.** $\frac{1}{27}$ **5.** 16 **7.** 0.3 **9.** 9 **11.** $-\frac{2}{3}$ **13.** $a-\sqrt{5}\geq 0$ **15.** 4.3

17. 1.59×10^5 **19.** 6.4×10^{-4} **21.** $\dfrac{1}{8a^2}$ **23.** $6\sqrt{2}$ **25.** $\frac{1}{15}$

27. $(x-5)|x-5|$ **29.** $8x^3$ **31.** $\dfrac{2a}{\sqrt[3]{b}}$ **33.** $\dfrac{2ab}{a^2+b^2}$ **35.** 1

37. $10a^2 - 22a + 4$ **39.** $34x^2 - 137x + 24$ **41.** $x - 3x^{1/3} + 3x^{-1/3} - x^{-1}$
43. $8x^6 + 60x^4y + 150x^2y^2 + 125y^3$ **45.** $\sqrt{5} + 2$ **47.** Negative
49. $(x + 3)(x - 9)$ **51.** $(t - 10)^2$ **53.** $(4x - 1)(3x - 2)$
55. $(y - 1)^3(y + 1)(y - 3)$ **57.** $(z^2 + 2)(z + 1)(z - 1)$
59. $(x + 1)(x^2 + 1)$ **61.** $(x + 7y)(x - 5y)$
63. $(4x^2 + 1)(2x + 1)(2x - 1)$ **65.** All real numbers except -4 and -5

67. -14 **69.** $\frac{28}{97}$ **71.** $\dfrac{x - 1}{x + 2}$ **73.** $\dfrac{(2x - 1)(x - 2)}{(x + 4)(3x - 1)}$ **75.** 1   **77.** $\dfrac{3x^2 - x - 6}{(x - 2)(x + 2)}$

79. $-\dfrac{2(x^2 + 5x - 3)}{x(x + 1)(x - 1)}$ **81.** $\dfrac{y}{y - x}$ **83.** $\dfrac{2y + 3x}{y - 2x}$ **85.** $\dfrac{x - 1}{x + 1}$ **87.** $S = 2s^2 + 4sh$

89. $A = \dfrac{\sqrt{3}}{4}s^2$ **91.** Jill, by \$484

SECTION 2.1   **1.** 0   **3.** $-\frac{1}{3}$ **5.** 1   **7.** 7   **9.** -3 **11.** $-\frac{2}{7}$ **13.** $\frac{1}{2}$ **15.** -5 **17.** $\frac{1}{12}$
19. $-\frac{7}{13}$ **21.** -1 **23.** $\frac{18}{7}$ **25.** $-\frac{7}{3}$ **27.** $-\frac{1}{7}$ **29.** $\frac{3}{2}$ **31.** 1, 7   **33.** $-\frac{7}{2}, \frac{3}{2}$
35. $-9, -3$ **37.** 0, 36   **39.** 1, 3   **41.** $x = y$ **43.** $x = 3y - \frac{1}{2}$
45. $x = -\frac{1}{2}y + \frac{5}{4}$ **47.** $x = 2y + 12$ **49.** $x = -\frac{7}{3}y + \frac{1}{15}$ **51.** $y = \frac{1}{2}x - \frac{7}{2}$
53. $y = \frac{2}{3}x - 4$ **57.** -2 **59.** 8   **61.** a. 30.48   b. 25   c. 39.370079
63. 212 degrees Fahrenheit   **65.** $C = \frac{5}{9}(F - 32)$ **67.** $p = \dfrac{fq}{q - f}$

SECTION 2.2   **1.** 83   **3.** 96   **5.** Karen is 8, and her father is 32.   **7.** 17 and 12
9. 4700 student tickets, 1800 nonstudent tickets
11. 100 dollar bills, 20 quarters, 50 dimes, 40 nickels   **13.** 24   **15.** \$25.50
17. 96   **19.** \$2200 in the 6% account, \$800 in the 8% account
21. 7.5% on the \$4200 investment, 6.5% on the \$2400 investment
23. Length: 30 inches; width: 20 inches   **25.** 30 meters by 50 meters   **27.** 9
29. 55,160,000 square meters   **31.** $\frac{5}{6}$ hour   **33.** 3 miles   **35.** $\frac{10}{9}$ hour
37. $\frac{36}{17}$ hour   **39.** 10 hours   **41.** 3 quarts
43. 30 pounds of 40% silver alloy, 20 pounds of 30% silver alloy
45. 1.6 gallons   **47.** \$60,000   **49.** 84

SECTION 2.3   **1.** $4, -4$ **3.** $3, -3$ **5.** -2 **7.** 25   **9.** $\frac{2}{3}$ **11.** $\sqrt{2}$ **13.** $-\frac{1}{4}$ **15.** $-4, -1$
17. $2, -5$ **19.** 7, 3   **21.** $\frac{1}{5}\sqrt{15}, -\frac{1}{5}\sqrt{15}$ **23.** $1, \frac{1}{2}$ **25.** $1 + \sqrt{7}, 1 - \sqrt{7}$
27. $5 + 3\sqrt{3}, 5 - 3\sqrt{3}$ **29.** $-1 + \frac{1}{2}\sqrt{10}, -1 - \frac{1}{2}\sqrt{10}$ **31.** $\frac{3}{2} + \frac{1}{2}\sqrt{5}, \frac{3}{2} - \frac{1}{2}\sqrt{5}$
33. None   **35.** $1 + \sqrt{5}, 1 - \sqrt{5}$ **37.** $\frac{1}{2} + \frac{1}{2}\sqrt{7}, \frac{1}{2} - \frac{1}{2}\sqrt{7}$
39. $1 + \frac{1}{3}\sqrt{6}, 1 - \frac{1}{3}\sqrt{6}$ **41.** $-\frac{1}{2}, -\frac{1}{3}$ **43.** $\frac{2}{3}(\sqrt{3} + \sqrt{6}), \frac{2}{3}(\sqrt{3} - \sqrt{6})$
45. $-\frac{1}{2}\sqrt{2} + 1, -\frac{1}{2}\sqrt{2} - 1$ **47.** None   **49.** $\frac{3}{2} + \frac{1}{2}\sqrt{29}, \frac{3}{2} - \frac{1}{2}\sqrt{29}$
51. $-2 + 2\sqrt{3}, -2 - 2\sqrt{3}$ **53.** $3 + \sqrt{14}, 3 - \sqrt{14}$ **55.** $6, -6$ **57.** 0
59. $2, -2$ **61.** 6.7949548, -1.1282881 **63.** 1.4336420, -1.0196293

65. $r = \sqrt[4]{\dfrac{A}{\pi}}$ **67.** $r = \sqrt{\dfrac{V}{\pi h}}$ **69.** $-\frac{1}{2} + \frac{1}{2}\sqrt{5}$

SECTION 2.4   **1.** $\sqrt{3}$ seconds   **3.** 3 seconds   **5.** 1 second   **7.** $\sqrt{61}$ inches
9. $-\frac{1}{2} + \frac{1}{2}\sqrt{199}$ meters, $\frac{1}{2} + \frac{1}{2}\sqrt{199}$ meters   **11.** 3000 feet
13. 40 miles per hour   **15.** 300 and 400 miles per hour   **17.** 170 miles per hour

19. 40 miles per hour in midday, 24 miles per hour in rush hour   **21.** 32, 33
23. $(0, 2\sqrt{5})$ and $(0, -2\sqrt{5})$ **25.** 9 inches by 7 inches   **27.** A 10-foot strip
29. 28 inches and 16 inches   **31.** 4 inches
33. a. $3 and $10; 70 umbrellas and 0 umbrellas, respectively
    b. $5; 50 umbrellas
35. 8 feet

SECTION 2.5
1. $-11, -4$ **3.** $7, -4$ **5.** $2, -2, \sqrt{2}, -\sqrt{2}$ **7.** $3, -3$ **9.** None
11. $\sqrt[3]{28}, -1$ **13.** 16   **15.** 36   **17.** 81   **19.** 16   **21.** 8, 1   **23.** 16
25. $\sqrt{5}, -\sqrt{5}$ **27.** $7, -2$ **29.** $\frac{1}{3}$ **31.** $\sqrt{3}, -\sqrt{3}$ **33.** 5   **35.** 1   **37.** 1
39. $\frac{3}{2}$ **41.** $\dfrac{\sqrt[3]{10}}{3}$ **43.** $2^{1/n}$ (and $-2^{1/n}$ if n is even)   **45.** -1 **47.** $\sqrt{3}, -\sqrt{3}$
49. $\sqrt[3]{23}$ **51.** $0, -1$ **53.** $0, 1, -1$ **55.** $2, -2, 4$ **57.** $3, -3$ **59.** 1
61. $0, 2, -2, 3$ **63.** a. -1 b. -1 **65.** $a = \sqrt{\left(\dfrac{Q}{E}\right)^{2/3} - 1}$ **67.** $r = \sqrt[4]{\dfrac{8\eta l V}{\pi p}}$

SECTION 2.6
1. Open, bounded   **3.** Closed, unbounded   **5.** Half-open, bounded
7. Closed, bounded   **9.** Open, bounded   **11.** Half-open, bounded
13. Open, unbounded   **15.** $(-4, 3]$ **17.** $(-1.1, -0.9)$ **19.** $(-8, \infty)$
21. $[-1, 1)$ **23.** $(-\infty, 3]$ **25.** $(-\infty, \frac{1}{4}]$ **27.** $(-\infty, -\frac{2}{7}]$ **29.** $(-\infty, \frac{5}{2})$
31. $(6, \infty)$ **33.** $(-\infty, 7]$ **35.** $(-\infty, -\frac{3}{2}]$ **37.** $(-5, -2)$ **39.** $[-3, -1)$
41. $[-23, -15]$ **43.** $(1.99, 2.01)$ **45.** $(a, a + d)$
49. Between 20 and 120, inclusive   **51.** From $0 to $5.50 inclusive
53. From $0 to $2 inclusive

SECTION 2.7
1. $(-\infty, 1], [2, \infty)$ **3.** $[-3, -1]$ **5.** $(-\infty, 3), (3, \infty)$ **7.** $(-2, 2)$
9. $[1, 3]$ **11.** $(-3, -1), (1, 3)$ **13.** $(2, 3)$ **15.** $(-\infty, -3), (5, \infty)$
17. $[-9, 0]$ **19.** $(-\frac{1}{2}, 4)$ **21.** $(-\infty, -5), (4, \infty)$ **23.** $(-2, \frac{3}{2}]$ **25.** $(\frac{1}{2}, \frac{5}{9}]$
27. $(0, \frac{3}{20})$ **29.** $(-1, 0), (1, \infty)$ **31.** $[0, \infty)$ **33.** $(2, \infty)$
35. $(-\infty, -5], [-3, 0]$ **37.** $[-3, -1], [1, 3]$ **39.** $(0, 1)$ **41.** $(-\frac{4}{3}, -\frac{2}{7}), (\frac{1}{6}, \infty)$
43. $(-\infty, 0), (1, \infty)$ **45.** $(-\frac{1}{2} - \frac{1}{2}\sqrt{5}, -\frac{1}{2} + \frac{1}{2}\sqrt{5})$ **47.** $(-1 - \frac{1}{2}\sqrt{2}, -1 + \frac{1}{2}\sqrt{2})$
49. $(-\infty, -\sqrt{2} - 4), (-\sqrt{2} + 4, \infty)$ **51.** $(-\infty, -\frac{1}{2}\sqrt{2}), (0, \frac{1}{2}\sqrt{2}), (\frac{3}{2}, \infty)$
53. $(-\infty, -2), [-1, \infty)$ **59.** $(-1, 0), (1, \infty)$ **61.** Between 1 and 2 seconds
63. Between 20 and 50   **67.** Between 0 and 2 inches

SECTION 2.8
1. $(-4, 4)$ **3.** $[-\frac{1}{5}, \frac{1}{5}]$ **5.** $(-\infty, -\frac{9}{2}], [\frac{9}{2}, \infty)$ **7.** $(-\infty, -0.01), (0.01, \infty)$
9. $(2, 8)$ **11.** $[-6, 0]$ **13.** $(-\infty, 6), (8, \infty)$ **15.** $(-\infty, \frac{5}{3}], [\frac{7}{3}, \infty)$
17. $(\frac{7}{4}, \frac{13}{4})$ **19.** $(-1, 2)$ **21.** $(-\infty, \frac{2}{3}), (\frac{2}{3}, \infty)$ **23.** $(-\infty, \infty)$ **25.** $[-\frac{1}{8}, \frac{5}{8}]$
27. $(-\infty, -\frac{10}{3}), (6, \infty)$ **29.** $(-\infty, 1], [\frac{5}{2}, \infty)$ **31.** $(-18, 6)$ **33.** $(-6, 6)$
35. $(3, 4], [6, 7)$ **37.** $[-\frac{7}{6}, -\frac{5}{6}], [-\frac{1}{2}, -\frac{1}{6}]$
39. a. $(-2\sqrt{5}, 2\sqrt{5})$ b. $(-\infty, -2\sqrt{5}), (2\sqrt{5}, \infty)$

CHAPTER 2 REVIEW
1. 26   **3.** 3   **5.** $-\frac{2}{19}$ **7.** $\frac{1}{2}, -\frac{5}{3}$ **9.** $1, \frac{7}{3}$ **11.** $\sqrt{3}, -\sqrt{3}$ **13.** None
15. $\frac{1}{4} + \frac{1}{4}\sqrt{17}, \frac{1}{4} - \frac{1}{4}\sqrt{17}$ **17.** $3, -3$ **19.** $-3 + \sqrt{6}, -3 - \sqrt{6}$
21. $\sqrt{2}, -\sqrt{2}$ **23.** $\frac{1}{2}, 1$ **25.** $-4 - 2\sqrt{5}$ **27.** $0, -3 + 3\sqrt{2}, -3 - 3\sqrt{2}$
29. $0, \frac{1}{2}, 1, -1$ **31.** $2, -2, 2\sqrt{2}, -2\sqrt{2}$ **33.** $2\sqrt{3}, -2\sqrt{3}, 2\sqrt{5}, -2\sqrt{5}$

35. $(-\infty, \frac{5}{2}]$ **37.** $(-\infty, \frac{9}{2})$ **39.** $[-\frac{1}{2}, \infty)$ **41.** $(-\infty, 4), (5, \infty)$ **43.** $(-2, -\frac{1}{3})$
45. $[-\frac{1}{2}, \frac{2}{5})$ **47.** $(-\infty, -\frac{7}{6}], (3, \infty)$ **49.** $(-\infty, -\sqrt{2}], [\sqrt{2}, \infty)$ **51.** $[-4, 9]$
53. -1 **55.** $(3, 8)$ **57.** $58°C$ and approximately $-88.3°C$ **59.** 13 miles
61. $\frac{9}{2}$ hours   **63.** Between 1 and 2 seconds   **65.** 4   **67.** $10,000   **69.** $4\sqrt{2}$ feet

71. $46\frac{2}{3}\%$ **73.** a. 7.5   b. $\frac{4}{3}$ **75.** $r = \dfrac{A_n - P}{Pn}$ **77.** $x = \dfrac{Qy}{17{,}860y - 1.798Q}$

SECTION 3.1

1.   **3.**   **5.**

7. x and y axes

9.   **11.**   **13.**

15.   **17.**   **19.**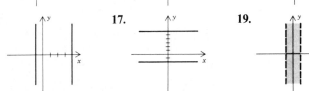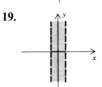

21. Second and fourth quadrants, not including axes   **23.** 2   **25.** 13   **27.** $2\sqrt{10}$
29. 13   **31.** $\sqrt{29 - 2\sqrt{2}}$ **33.** $5a$ **35.** $(6, 6)$ **37.** $(0, 7)$ **39.** $(0.7, 2.85)$
41. $(0, a)$ **43.** $(4 + \sqrt{3}, 0)$ and $(4 - \sqrt{3}, 0)$ **45.** $(\frac{1}{3}a + \frac{2}{3}c, \frac{1}{3}b + \frac{2}{3}d)$
51. $90\sqrt{2}$ feet (approximately 127.28 feet)

SECTION 3.2

1. $-14; -14; -14$ **3.** $4; 8; t^4 - 3t^2 + 4$ **5.** $2; \dfrac{4a^2 - 2a}{a^2 - 0.09}$ **7.** $2; \sqrt{7}; |x|$

9. $-504; 2z^2 - z^9$ **11.** $5; 0$ **13.** $11; 3$ **15.** $\frac{4}{5}; \frac{5}{6}$ **17.** $-2; -6$

19. $\dfrac{2}{3}; \dfrac{3\sqrt{2}}{3\sqrt{2} - 1}$ **21.** All real numbers   **23.** All real numbers except $0, \frac{5}{2}$

25. All real numbers   **27.** All real numbers except $1, -6$ **29.** All real numbers
31. All negative numbers   **33.** All numbers in $(-\infty, -2]$ and $[2, \infty)$
35. All numbers in $(-\infty, 0]$ and $(1, \infty)$ **37.** All real numbers
39. All numbers in $[1, \infty)$ **41.** All real numbers except 0   **43.** c **45.** No
51. a. 107.90807   b. -2.3467185 c. 6218.8918
53. a. 2.8792360   b. 2.1858128   c. 2.9712669
55. a. $A = s^2$ for $s \geq 0$ b. $A = \frac{1}{2}d^2$ for $d \geq 0$

SECTION 3.3

1. $p = 4s$ **3.** $S = 6s^2$ **5.** $E = hv$ **7.** $R = cv^2$ **9.** a. $A = \dfrac{\sqrt{3}}{4}s^2$ b. 6 inches

11. a. $l_f = \dfrac{1250}{381} l_m$ **b.** $l_y = cl_m$, where $c = \dfrac{1250}{1143}$

13. Approximately 2.498×10^{19} ergs

15. a. $K = \frac{1}{2}mv^2$ **b.** Velocity of lighter one is $\sqrt{3}$ times as great

17. $\lambda = \dfrac{c}{v}$ **19.** $F = \dfrac{cq_1 q_2}{r^2}$

SECTION 3.4

1. y intercept: 3;
x intercept: -3

3. y intercept: 0.7;
x intercept: 0.7

5. y intercept: 5;
x intercepts: 5, -5

7. y intercept: 2;
x intercept: 2

9. y intercept: 3;
x intercepts: $\sqrt{3}, -\sqrt{3}$

11. no y intercept;
x intercept: -1

13. y intercept: -1;
x intercept: 1

15. y intercept: 0;
x intercept: 0

17. y intercept: 0;
x intercept: 0

19. y intercept: 1; x intercept: -1 **21.** y intercept: 0; x intercepts: 0.4

23. y intercept: 9; x intercept: 3   **25.** y intercept: -1; no x intercept

27. y intercept: $\frac{1}{2}$; x intercept: -1 **29.** y intercept: $\sqrt{2}$; x intercept: 2

31. y axis   **33.** Neither   **35.** y axis   **37.** y axis   **39.** Neither

41. Odd   **43.** Neither   **45.** Neither   **47.** Even   **49.** a, c, g, h

51. The graph of f is one unit to the right of the graph of g.

53.

SECTION 3.5

1. 3   **3.** $-\frac{5}{2}$ **5.** $\frac{2}{3}$ **7.** $y - (-4) = 0(x - 3)$ (or $y + 4 = 0$)   **9.** $y - 0 = \frac{2}{3}(x - 3)$

11. $y - \frac{3}{2} = -1(x - 0)$ **13.** $y - 1 = 2x$ **15.** $y + 2 = -4(x + 1)$

17. $x = 2$ **19.** $y = 0x - 2$ (or $y = -2$)   **21.** $y = -\frac{4}{5}x - 1$ **23.** $y = -\sqrt{3}x + \frac{1}{2}$

25. $m = 0$; $b = 3$ **27.** $m = 3$; $b = -\frac{1}{2}$ **29.** $m = -\frac{2}{3}$; $b = 2$ **31.** $m = 3$; $b = 0$

33. Perpendicular   **35.** Perpendicular   **37.** Neither   **39.** Parallel
41. Neither   **43.** Perpendicular
45.

47.

49.

51.

53. b. $y - y_1 = \dfrac{y_2 - y_1}{x_2 - x_1}(x - x_1)$ **55.** a. $\frac{1}{3}$ b. -5 c. $-\frac{3}{4}$

57. a. $\dfrac{x}{-1} + \dfrac{y}{2} = 1$ b. $\dfrac{x}{3} + \dfrac{y}{\frac{1}{2}} = 1$ c. $\dfrac{x}{0.7} + \dfrac{y}{0.3} = 1$ d. $\dfrac{x}{-2} + \dfrac{y}{-3} = 1$

59. a, c, f   **61.** $-\frac{4}{3}$ **63.** $f(x) = -2x - 6$ **65.** $f(x) = 4x + 4$
67. $y - 5 = -\frac{4}{3}(x + 2)$ **69.** $(-1, 2)$ **75.** a. $K = C + 273.15$
b. $K = \frac{5}{9}(F - 32) + 273.15$ c. 373.15 d. 293.15 **77.** $V = \frac{1}{64}\pi h + \frac{1}{384}\pi$

SECTION 3.6   **1.** $f(x) = x - 1, g(x) = 2x$ **3.** $f(x) = x^4, g(x) = \dfrac{3}{x}$ **5.** $f(t) = \sqrt{t}, g(t) = \dfrac{1}{2t}$

7. Domain: all real numbers; rule: $(g \circ f)(x) = 2x^2 - 3x + 2$

9. Domain: $(0, \infty)$; rule: $(g \circ f)(x) = \sqrt{\dfrac{1}{2x}}$

11. Domain: $[1, \infty)$; rule: $(g \circ f)(x) = \sqrt{x - 1} + 1$

13. Domain: all real numbers except $0, \frac{1}{2}, -\frac{1}{2}$; rule: $(g \circ f)(x) = \dfrac{4x^2}{1 - 4x^2}$

15. Domain: all real numbers except 0; rule: $(g \circ f)(x) = x$

17. Domain: all real numbers except -3 and -1; rule: $(g \circ f)(x) = -\dfrac{4x + 2}{x + 3}$

19. Domain: $(-\infty, -\sqrt{3}]$ and $[\sqrt{3}, \infty)$; rule: $(g \circ f)(x) = \sqrt{x^2 - 3}$
25. $(9, 10]$ **27.** m **29.** $V(r(t)) = \frac{9}{2}\pi t^6$

SECTION 3.7   **1.** $f^{-1}(x) = \frac{1}{2}x$ **3.** $f^{-1}(x) = 3 - 2x$ **5.** $f^{-1}(x) = x^{1/3}$
7. $f^{-1}(x) = [\frac{1}{3}(x + 5)]^{1/3}$ **9.** $f^{-1}(x) = (1 - x)^{1/5}$
11. $f^{-1}(x) = x^2$ for $x \geq 0$

13. $f^{-1}(x) = x^2 + 3$ for $x \geq 0$ **15.** $f^{-1}(x) = \dfrac{1}{x^2}$ for $x > 0$

17. $f^{-1}(x) = 2 - \dfrac{1}{2x^3}$ **19.** $f^{-1}(x) = \dfrac{x+1}{1-x}$ **21.** $f^{-1}(x) = \left(\dfrac{3x+1}{2-x}\right)^{1/3}$

23.   **25.**

27.   **29.** $f = f^{-1}$ **31.**

35. a. $s = 2A^{1/2}/3^{1/4}$ b. $s = 8/3^{1/4}$

SECTION 3.8   **1.** $x^2 + y^2 = 9$ **3.** $(x+1)^2 + (y-4)^2 = 4$ **5.** $x^2 + y^2 = 17$
7. $(x+2)^2 + (y-3)^2 = 25$

9. y intercept: 3;
symmetry with respect to y axis

11. y intercept: 0; x intercept: 0;
symmetry with respect to x axis

13. x intercept: 1;
symmetry with respect to x axis

15. y intercept: 0; x intercept: 0;
symmetry with respect to origin

17. Circle: y intercepts and x intercepts: 2 and -2; all symmetries
19. y intercept: 0; x intercept: 0;
no symmetry

21. y intercepts: $\frac{1}{2}$ and $-\frac{1}{2}$; x intercepts: 1 and -1 **23.** x intercept: 3
25. x intercepts: 1 and -1 **27.** y intercepts: $\sqrt{5}$ and $-\sqrt{5}$; x intercept: 5
29. y intercepts: 3 and -2 **31.** y intercepts: 1 and -3; x intercepts: 1 and 3

CHAPTER 3 REVIEW

1. a. $3\sqrt{5}$ b. $\frac{1}{2}\sqrt{82}$ **3.** All real numbers except -3
5. All real numbers except $-2 + \sqrt{11}$ and $-2 - \sqrt{11}$
7. All numbers in $\left[\frac{2}{3}, \infty\right)$ **9.** All numbers in $(-\infty, -5]$ and $[5, \infty)$
13. $-\frac{1}{3}$ **15.** a. All numbers in $(-\infty, -2]$ and $[2, \infty)$

17. y intercept: -5;
x intercept: $\frac{5}{2}$; no symmetry

19. y intercept: -6;
symmetry with respect to y axis
x intercepts: $\sqrt{6}$ and $-\sqrt{6}$;

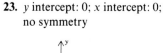

21. y intercept: 2;
x intercepts: 2 and -2;
symmetry with respect to y axis

23. y intercept: 0; x intercept: 0;
no symmetry

25. No intercepts;
symmetry with respect to origin

27. y intercept: $\sqrt{3}$; x intercept:
$-\sqrt{2}$; no symmetry

29. y intercepts: 3 and -5;
x intercepts: 3 and 5; no symmetry

31. y intercepts: $-3 + \sqrt{3}$
and $-3 - \sqrt{3}$; no symmetry

33. y intercepts: $\sqrt{3}$ and $-\sqrt{3}$; x intercept: -3;
symmetry with respect to x axis

35. $(f \circ g)(x) = 1/x^2$ for all x except 0
$(g \circ f)(x) = 125/x^2$ for all x except 0
37. $(f \circ g)(x) = (2x + 3)^{3/2}$ for $x \geq -\frac{3}{2}$
$(g \circ f)(x) = \sqrt{2x^3 + 3}$ for all $x \geq -\left(\frac{3}{2}\right)^{1/3}$

39. $(f \circ g)(x) = \sqrt{x^2 - 4}$ for all x in $(-\infty, -2]$ and $[2, \infty)$
$(g \circ f)(x) = \sqrt{x^2 - 4}$ for all x in $(-\infty, -2]$ and $[2, \infty)$

43. No   **45.** $f^{-1}(x) = \dfrac{1}{x} - 3$ **47.** $f^{-1}(x) = \dfrac{2x + 1}{4 - 3x}$

49. $2\sqrt{a + b}$ **51.** $(7 + 2\sqrt{5}, 2 + \sqrt{5})$ and $(7 - 2\sqrt{5}, 2 - \sqrt{5})$
53. $\left(x - \frac{1}{2}\right)^2 + \left(y + \frac{1}{4}\right)^2 = \frac{67}{16}$ **55.** $y + 3 = -\frac{1}{2}(x + 1)$

57. $6y - 8x = 7$ **59.** No

65. a. $L = \dfrac{cI}{D^2}$ b. Multiplied by 9

SECTION 4.1   **1.** Vertex: $(0, 0)$; y intercept: 0; x intercept: 0

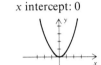

3. Vertex: $(0, 1)$; y intercept: 1

5. Vertex: $(0, 8)$; y intercept: 8; x intercepts: 2 and -2

7. Vertex: $(1, 0)$; y intercept: $-\frac{3}{2}$; x intercept: 1

9. Vertex: $\left(-\dfrac{1}{3}, -\dfrac{1}{3}\right)$; y intercept: $-\dfrac{1}{6}$; x intercepts: $-\dfrac{1}{3} + \dfrac{\sqrt{2}}{3}$ and $-\dfrac{1}{3} - \dfrac{\sqrt{2}}{3}$

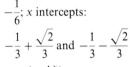

11. Vertex: $(-1, -1)$; y intercept: 0; x intercepts: 0 and -2

13. Vertex: $(1, 1)$; y intercept: 2

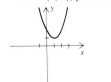

15. Vertex: $(3, 4)$; y intercept: -5; x intercepts: 5 and 1

17. Vertex: $\left(-\frac{1}{3}, \frac{4}{3}\right)$; y intercept: 1;
x intercepts: $\frac{1}{3}$ and -1

19. Vertex: $\left(-\frac{1}{4}, -\frac{9}{8}\right)$; y intercept: -1;
x intercepts: $\frac{1}{2}$ and -1

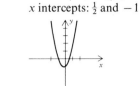

21. $y = \frac{1}{2}(x-1)^2$ **23.** $y = -(x-1)^2 + 1$ **25. a.** $g(x) = -2x^2 + 4$
b. $g(x) = -2x^2$ **c.** $g(x) = -2\left(x - \frac{1}{2}\right)^2 + 3$ **d.** $g(x) = -2(x+4)^2 + 3$
27. 3   **29.** 8   **33.** 12 feet

SECTION 4.2   **1.** Minimum value: 5   **3.** Minimum value: -20 **5.** Maximum value: 1
7. Minimum value: -28 **9.** Minimum value: $-\frac{9}{4}$ **11.** 1 mile by $\frac{1}{2}$ mile
13. In half   **15.** 3 inches each   **17.** 4 by 4   **19.** -8 and -8
21. 8 and 24   **23.** $(2, -2)$ **25.** $\left(\frac{1}{2}\sqrt{2}, \frac{1}{2}\right)$ and $\left(-\frac{1}{2}\sqrt{2}, \frac{1}{2}\right)$ **27.** \$2.50
29. 100 feet; 1.5 seconds

SECTION 4.3   **1.** y intercept: 1;
x intercept: -1

3. y intercept: $\frac{1}{4}$;
x intercept: $-\frac{1}{2}$

5. y intercept: -2;
x intercept: -1

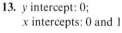

7. y intercept: 1;
x intercept: -1

9. y intercept: 2;
x intercepts: $-1, 1$, and 2

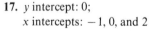

11. y intercept: 0;
x intercepts: $-2, -1$, and 0

13. y intercept: 0;
x intercepts: 0 and 1

15. y intercept: -1;
x intercepts: 1 and -1

17. y intercept: 0;
x intercepts: $-1, 0$, and 2

19. y intercept: 0;
    x intercepts: -1, 0, and 1

21. y intercept: 0;
    x intercepts: 0 and 1

23. y intercept: 4;
    x intercepts: -2,
    -1, 1, and 2

25. 2 **27.** -9

SECTION 4.4

1. No intercepts; vertical asymptote: $x = 0$; horizontal asymptote: $y = 0$

3. y intercept: 2; no x intercepts; vertical asymptote: $x = 2$; horizontal asymptote: $y = 0$

5. y intercept: -1; no x intercepts; vertical asymptote: $x = -1$; horizontal asymptote: $y = 0$

7. y intercept: -4; no x intercepts; vertical asymptote: $x = -\frac{1}{2}$; horizontal asymptote: $y = 0$

9. y intercept: 0; x intercept: 0; vertical asymptote: $x = 2$; horizontal asymptote: $y = 1$

11. y intercept: 0; x intercept: 0; vertical asymptotes: $x = 2$ and $x = -2$; horizontal asymptote: $y = 0$

13. y intercept: -1; no x intercepts; vertical asymptotes: $x = 1$ and $x = -2$; horizontal asymptote: $y = 0$

15. y intercept: 0; x intercept: 0; vertical asymptotes: $x = -3$ and $x = -1$; horizontal asymptote: $y = 0$

17. y intercept: 0;
x intercept: 0;
vertical asymptotes:
$x = 2$ and $x = -\frac{1}{2}$; horizontal
asymptote: $y = \frac{1}{2}$

19. No y intercept;
x intercept: $-\frac{1}{2}$;
vertical asymptote: $x = 0$;
horizontal asymptote:
$y = 4$

21. y intercept: $-\frac{1}{4}$;
t intercept: $-\frac{1}{3}$;
no vertical asymptotes;
horizontal asymptote: $y = 0$

23. y intercept: $\frac{3}{4}$;
x intercept: -1;
vertical asymptote: $x = -4$;
horizontal asymptote: $y = 0$

SECTION 4.5   **1.** Vertex: $(0, 0)$;
axis: $y = 0$

3. Vertex: $(-5, 1)$;
axis: $y = 1$

5. Vertex: $\left(-\frac{1}{4}, -\frac{1}{2}\right)$;
axis: $y = -\frac{1}{2}$

7. Vertex: $\left(\frac{1}{2}, \frac{1}{2}\right)$;
axis: $y = \frac{1}{2}$

9. Vertex: $(3, 2)$;
axis: $x = 3$

11. Vertex: $(-1, 3)$;
axis: $y = 3$

13. Vertex: $(-2, -3)$;
axis: $x = -2$

15. $y = \frac{5}{4}x^2$ **17.** $y - 5 = \frac{1}{2}(x + 2)^2$ **19.** $y - 4 = (x - 3)^2$
21. The lines $x - 3y + 4 = 0$ and $x + 3y - 2 = 0$ **23.** The point $\left(-\frac{1}{2}, \frac{1}{3}\right)$
25. -15 **27.** $(3, -\frac{1}{2})$ and $(3, 1)$ **29.** $y = \dfrac{316}{(1750)^2}x^2$

SECTION 4.6

1. Center: (0, 0); major axis between (−4, 0) and (4, 0); minor axis between (0, −2) and (0, 2); vertices: (−4, 0), (4, 0)

3. Circle: center (0, 0), radius $\sqrt{3}$

5. Center: (0, 0); major axis between (0, −3) and (0, 3); minor axis between (−2, 0) and (2, 0); vertices: (0, −3), (0, 3)

7. Center: (4, −3); major axis between (4, −7) and (4, 1); minor axis between (1, −3) and (7, −3); vertices: (4, −7), (4, 1)

9. Center: (−2, −1); major axis between (−6, −1) and (2, −1); minor axis between (−2, −3) and (−2, 1); vertices: (−6, −1), (2, −1)

11. Center: (1, −2); major axis between (1, −6) and (1, 2); minor axis between (−1, −2) and (3, −2); vertices: (1, −6), (1, 2)

13. Center: (−3, 3); major axis between (−3, −2) and (−3, 8); minor axis between (−7, 3) and (1, 3); vertices: (−3, −2), (−3, 8)

15. $\dfrac{x^2}{4} + \dfrac{y^2}{9} = 1$ **17.** $\dfrac{x^2}{4} + \dfrac{y^2}{16} = 1$ **19.** $\dfrac{(x-3)^2}{49} + \dfrac{(y-2)^2}{25} = 1$

21. $\dfrac{(x+2)^2}{1} + \dfrac{(y+5)^2}{4} = 1$ **23.** $4(x + \frac{3}{2})^2 + \dfrac{4\left(y - \frac{1}{2}\right)^2}{25} = 1$

25. $\dfrac{4x^2}{9} + \dfrac{(y-4)^2}{1} = 1$ and $\dfrac{x^2}{1} + \dfrac{4(y-4)^2}{9} = 1$

29. $4 + \frac{3}{2}\sqrt{3}$ and $4 - \frac{3}{2}\sqrt{3}$ **31.** b. $\left(\frac{24}{5}, -\frac{12}{5}\right)$

33. a. $\frac{1}{2}\sqrt{3}$ b. $\frac{1}{2}\sqrt{3}$ c. $\frac{1}{3}\sqrt{5}$ d. 0

SECTION 4.7

1. Center: $(0, 0)$; vertices: $(-1, 0), (1, 0)$; asymptotes: $y = x$, $y = -x$

3. Center: $(0, 0)$; vertices: $(-3, 0), (3, 0)$; asymptotes: $y = \frac{2}{3}x$, $y = -\frac{2}{3}x$

5. Center: $(0, 0)$; vertices: $(0, -2), (0, 2)$; asymptotes: $y = \frac{2}{5}x$, $y = -\frac{2}{5}x$

7. Center: $(-2, -1)$; vertices: $(-6, -1)$, $(2, -1)$; asymptotes: $y + 1 = \frac{5}{4}(x + 2)$, $y + 1 = -\frac{5}{4}(x + 2)$

9. Center: $(0, -1)$; vertices: $(0, -2), (0, 0)$; asymptotes: $y + 1 = \frac{1}{2}x$, $y + 1 = -\frac{1}{2}x$

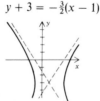

11. Center: $(3, 0)$; vertices: $(3, -3), (3, 3)$; asymptotes: $y = \frac{3}{4}(x - 3)$, $y = -\frac{3}{4}(x - 3)$

13. Center: $(1, -3)$; vertices: $(-1, -3)$, $(3, -3)$; asymptotes: $y + 3 = \frac{3}{2}(x - 1)$, $y + 3 = -\frac{3}{2}(x - 1)$

15. $\dfrac{x^2}{4} - \dfrac{y^2}{36} = 1$ **17.** $\dfrac{y^2}{16} - \dfrac{11x^2}{256} = 1$

19. $\dfrac{(x - 2)^2}{4} - \dfrac{(y + 3)^2}{4} = 1$ **21.** $\frac{1}{2}\sqrt{3}$ and $-\frac{1}{2}\sqrt{3}$

23. b. $\left(\frac{13}{6}, -\frac{5}{4}\right)$ **25.** a. $\sqrt{5}$ b. $\frac{1}{2}\sqrt{5}$ c. $\frac{1}{2}\sqrt{13}$ d. $\sqrt{2}$

27. Ellipse; center: $(0, 0)$
29. Parabola; vertex: $\left(-\frac{3}{2}, \frac{1}{2}\right)$ 31. Ellipse; center: $(-4, 3)$
33. Ellipse; center: $(-3, 0)$ 35. Hyperbola; center: $(2, -2)$
37. Hyperbola; center: $(3, 4)$

CHAPTER 4 REVIEW

1. Vertex: $(0, -4)$; axis: $x = 0$;
   y intercept: -4;
   x intercepts: 4 and -4

3. Vertex: $(-2, -3)$;
   axis: $x = -2$;
   y intercept: -7;
   no x intercepts

5. y intercept: 2;
   x intercept: $2^{1/3}$

7. y intercept:
   2; no x intercepts

9. y intercept: 0;
   x intercepts: 0 and 2

11. y intercept: 9;
    x intercepts: -3,
    -1, 1, and 3

13. y intercept: -1;
    x intercept: -1;
    vertical asymptote: $x = 1$;
    horizontal asymptote: $y = 1$

15. No y intercept;
    x intercept: -2;
    vertical asymptotes:
    $x = 0$ and $x = -4$;
    horizontal asymptote: $y = 0$

17. Circle: center $(-1, 2)$, radius $\sqrt{3}$

19. Center: $\left(\frac{1}{2}, -\frac{1}{2}\right)$; vertices:
    $\left(-\frac{3}{2}, -\frac{1}{2}\right)$ and $\left(\frac{5}{2}, -\frac{1}{2}\right)$;
    asymptotes: $y + \frac{1}{2} = x - \frac{1}{2}$
    and $y + \frac{1}{2} = -\left(x - \frac{1}{2}\right)$

21. Center: $(-3, -4)$; vertices:
    $\left(-3, -4 - 2\sqrt{3}\right)$, $\left(-3, -4 + 2\sqrt{3}\right)$;
    asymptotes: $y + 4 = \frac{2\sqrt{3}}{3}(x + 3)$

    and $y + 4 = -\frac{2\sqrt{3}}{3}(x + 3)$

23. $-\frac{1}{3}$ **25.** $-9, -1$ **27.** \$350, and 350 people   **29.** 4096 feet

31. For the square: $\dfrac{8}{\pi + 4}$ feet; for the circle: $\dfrac{2\pi}{\pi + 4}$ feet

33. Length: $1 - \frac{1}{2}\sqrt{2}$; width: $\frac{1}{4}\sqrt{2} + \frac{1}{2}$

CUMULATIVE REVIEW I: CHAPTERS 1–4

1. $-\frac{3}{4}\sqrt[3]{3}$ **3.** $x^{39/4}y$ **5.** $(x - 3)^2(2x - 3)^2(4x^2 - 27)$ **7.** $\dfrac{4 - 2x^3}{3x^3 + x}$

9. $\dfrac{12x}{(x + 2)^2(x - 2)}$ **11.** $\frac{5}{4}, \frac{1}{4}$ **13.** $-2 \pm \frac{1}{2}\sqrt{26}$ **15.** $(-\infty, 5)$

17. $\left(-\infty, -\frac{2}{5}\right], \left(\frac{1}{2}, \infty\right)$ **19.** b. $y + \frac{1}{3} = -\frac{3}{2}(x + 1)$

21. a. $[-2, 2]$ b. $\sqrt{2}$ c. $[0, 2]$ **23.** $f^{-1}(x) = \dfrac{3 - x}{x + 2}$

25. 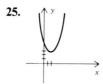   **27.**

29.   **31.**

33. $\dfrac{(x + 1)^2}{4} + \dfrac{(y + 2)^2}{9} = 1$ **35.** $a = 2b$ **37.** c. $\frac{1}{7} + \frac{1}{91}$ **39.** $C = ct^{3/2}$

41. $V = (24 - 2x)(16 - 2x)x$ **43.** $\frac{1}{75}$ cup   **45.** 6 hours
47. a. $40\sqrt{6}$ feet per second   b. $\frac{5}{2}\sqrt{6}$ seconds
49. a. $[0, 26]$ b. 5 feet

SECTION 5.1   **1.**   **3.**   **5.** The line $y = 1$ **7.**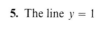

9. 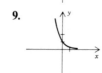   **11.**   **13.**

15. π^2 **17.** $5^{\sqrt{6}}$ **19.** $\frac{1}{4}$ **21.** 3^π **23.** 2   **25.** 7   **27.** $\frac{1}{3}$ **29.** $\frac{1}{4}$
31. $\frac{1}{64}$ **33.** 1.4049476   **35.** 0.30119421   **37.** 1.6487213   **39.** 11   **41.** 22
43. 0 and $\frac{12}{5}$ **45.** $-2\sqrt{2} - 1$ **47.** $\sqrt{5}$ and $-\sqrt{5}$

49.

(a)   (b)

51. $e^{(e^e)}$ **53.** b. $e^{1.5} \approx 4.48168907$, sum ≈ 4.481686598

57. a. Approximately 6.76×10^{-6} atmospheres

b. Approximately 3.07×10^{-3} atmospheres

c. Approximately 5.99×10^{-1} atmospheres

d. Approximately 1.12 atmospheres

59. Approximately 7.5212062×10^{-4} coulombs

SECTION 5.2   **1.** $\log_{10} 10 = 1$ **3.** $\log_{1/2} 32 = -5$ **5.** $\log_x 16 = -4$ **7.** $10^3 = 1000$
9. $4^{2.5} = 32$ **11.** $e^2 = x$ **13.** 3   **15.** 4   **17.** -4 **19.** 0   **21.** $\frac{3}{2}$
23. -1 **25.** $\frac{1}{3}$ **27.** -2 **29.** -1 **31.** -4 **33.** $\frac{5}{2}$ **35.** $\frac{4}{3}$ **37.** 6
39. 4   **41.** 5   **43.** 1   **45.** 16   **47.** 2   **49.** $\frac{1}{16}$ **51.** 1   **53.** 12   **55.** 67
57. 4   **59.** $\frac{2}{3}\sqrt{3}$ **61.** 11, -11 **63.** 8   **65.** 4   **67.** 5   **69.** 2, -2 **71.** 2
73. 9   **75.** $\frac{1}{3}$ **77.** $\frac{1}{4}$

79.   **81.**

83.   **85.**

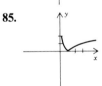

87. No, since $\log_{10} \frac{1}{6} < 0$

SECTION 5.3   **1.** 0.7781   **3.** 1.0791   **5.** -0.4771 **7.** 0.5229   **9.** -0.2498 **11.** -0.6990
13. 0.73855   **15.** 2.9957   **17.** -1.9459 **19.** 3.8918   **21.** 1.6095
23. 3.1987   **25.** 0.8797   **27.** 1.7590   **29.** 15   **31.** $\frac{7}{12}$ **33.** 1.6021
35. -0.6021 **37.** 5.7318   **39.** -6.2682 **41.** 27.318   **43.** 0.69897000
45. -2.0710923 **47.** -3.97029191 **49.** $\log_2 5$ **51.** $\log_2 \dfrac{x^6}{y}$ **53.** $\log_{10} \dfrac{x^6}{y}$
55. $\log_a x + \log_a y - \log_a z$ **57.** $3 \log_a z + \frac{1}{2}\log_a x + \frac{1}{2}\log_a y$
59. $\frac{1}{2}\log_a x + \frac{5}{2}\log_a y - \frac{3}{2}\log_a z$ **61.** 4   **63.** $\frac{1}{2}$ **65.** 3   **67.** 2.8073
69. 1.6309   **71.** 0.6309   **73.** 2.3219281   **75.** 1.3671030   **77.** -0.56347377
79. -9.3857357

SECTION 5.4   **1.** 0.0043   **3.** 2.7160   **5.** $-2 + 0.9917$ **7.** 4.7185   **9.** 0.0746
11. $-1 + 0.1593$ **13.** 0.5160   **15.** 7.1858   **17.** $-3 + 0.1575$ **19.** 4.1016
21. 7.51   **23.** 99,600   **25.** 631,000,000   **27.** 0.00515   **29.** 9.445   **31.** 155.4

33. 0.02083   **35.** 0.4043   **37.** 70.40   **39.** 3.870   **41.** 1.673×10^9
43. 0.002243   **45.** a. 89.78 kilograms   b. 10.22 kilograms   c. 0.7471 meters
d. 0.6194 meters

SECTION 5.5   **1.** $\dfrac{\log 5}{\log 2}$ **3.** $\dfrac{1}{2}\left(\dfrac{\log 51}{\log 5}\right)$ **5.** $\dfrac{\log 5}{\log 4} - 2$ **7.** $\dfrac{3\log 5 + 2\log 3}{5\log 3 - 2\log 5}$

9. $0, \dfrac{\log 3}{\log 2}$ **11.** $-2, \dfrac{\log 4}{\log 5} + 2$ **15.** 10^{-7} watts per square meter

17. 20 decibels   **19.** $10^{0.06} \approx 1.1481536$ **21.** $10^{-0.1} \approx 0.79432823$
23. $10^{8.9} \approx 794{,}328{,}230$ **25.** a. 0   b. 3   c. 4   **27.** $\log 2 \approx 0.3$
29. Highly acidic   **31.** Wheat

SECTION 5.6   **1.** $\frac{1}{3}\ln 4$ **3.** $\dfrac{1}{1.2}\ln 2$ **5.** 3   **7.** $0, \ln 2$ **9.** 0

13. a. $f(t) = 1000\, e^{(\ln 1.537)t}$ b. $\dfrac{\ln 2}{\ln 1.537} \approx 1.61$ weeks   **15.** No

17. At 32.3°C   **19.** 5.223 billion   **23.** Approximately 20.35%
25. 28,987 years old   **27.** Approximately 79%   **29.** Approximately 1.2%
31. Approximately 65.25 million years   **33.** a 29.92 inches of mercury
b. Approximately 24.50 inches of mercury
c. Approximately 17.28 inches of mercury   **35.** No

CHAPTER 5 REVIEW   **1.**   **3.**

5. 2   **7.** $6 + \ln 6$ **9.** $-\frac{1}{2} + \frac{1}{2}\sqrt{5}, -\frac{1}{2} - \frac{1}{2}\sqrt{5}$ **11.** $\frac{1}{8}$ **13.** $\frac{7}{5}$

15. $-\dfrac{\log 7}{\log 2}$ **17.** $0, \left(\dfrac{\log 2}{\log 3}\right)^{1/3}$ **19.** $\frac{1}{5}\ln 4.37$ **21.** -4 **23.** 5

25. 2.6233   **27.** $-3 + 0.8865$ **29.** 0.6160   **31.** 8.8051   **33.** 3.56
35. 4.733   **37.** 2.079   **39.** 0.4343   **41.** $x = e^{(y-a)/b}$ **45.** $a = \frac{1}{5}, c = -2$
47. 8.1   **49.** a. Approximately 23.1 years   b. Approximately 5.78 years earlier
51. Approximately 26.58 minutes

SECTION 6.1   **1.**   **3.**   **5.**   **7.**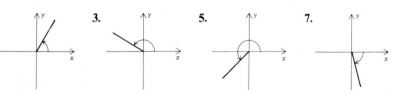

9. **11.** **13.**

15. **17.** **19.**

21. **23.** **25.**

27. **29.** **31.**

33. 0 **35.** $-\pi/4$ **37.** $13\pi/6$ **39.** $-\pi$ **41.** $49\pi/6$ **43.** $-2\pi/5$ **45.** $\pi/40$
47. $405°$ **49.** $1080°$ **51.** $1980°$ **53.** $-585°$ **55.** $70°$ **57.** $(324/\pi)°$
59. Second **61.** Third **63.** Fourth **65.** Second **67.** Fourth
69. $2\pi/3$ radians: b, c, h, i; $4\pi/3$ radians: a, d, e, f, g, j **71.** a. $\pi/8$ b. $\pi/24$
  c. $\pi/2$ **73.** $11.25°, \pi/16$ **75.** a. $\pi/12$ b. $\pi/720$ c. $\pi/43{,}200$
77. 25π feet (approximately 78.5 feet)

SECTION 6.2 **1.** $\sqrt{2} - \frac{3}{2}\sqrt{3}$ **3.** 0 **5.** $2 + \frac{2}{3}\sqrt{3}$ **7.** a, d, and h; e and g
9. $\sin \theta = \frac{1}{37}\sqrt{37}$; $\cos \theta = \frac{6}{37}\sqrt{37}$; $\tan \theta = \frac{1}{6}$; $\cot \theta = 6$; $\sec \theta = \sqrt{37}/6$;
  $\csc \theta = \sqrt{37}$ **11.** $\sin \theta = \frac{3}{5}$; $\cos \theta = \frac{4}{5}$; $\tan \theta = \frac{3}{4}$; $\cot \theta = \frac{4}{3}$; $\sec \theta = \frac{5}{4}$;
  $\csc \theta = \frac{5}{3}$ **13.** $\sin \theta = \frac{12}{13}$; $\cos \theta = \frac{5}{13}$; $\tan \theta = \frac{12}{5}$; $\cot \theta = \frac{5}{12}$;
  $\sec \theta = \frac{13}{5}$; $\csc \theta = \frac{13}{12}$ **15.** $\sin \theta = \frac{1}{2}\sqrt{2}$; $\cos \theta = \frac{1}{2}\sqrt{2}$; $\tan \theta = 1$;
  $\cot \theta = 1$; $\sec \theta = \sqrt{2}$; $\csc \theta = \sqrt{2}$ **17.** $\sin \theta = \frac{24}{25}$; $\cos \theta = \frac{7}{25}$; $\tan \theta = \frac{24}{7}$;
  $\cot \theta = \frac{7}{24}$; $\sec \theta = \frac{25}{7}$; $\csc \theta = \frac{25}{24}$ **19.** $\sin \theta = \frac{1}{6}$; $\cos \theta = \frac{1}{6}\sqrt{35}$;
  $\tan \theta = \frac{1}{35}\sqrt{35}$; $\cot \theta = \sqrt{35}$; $\sec \theta = \frac{6}{35}\sqrt{35}$; $\csc \theta = 6$
21. $\cos \theta = \sqrt{21}/5$; $\tan \theta = \frac{2}{21}\sqrt{21}$; $\cot \theta = \sqrt{21}/2$; $\sec \theta = \frac{5}{21}\sqrt{21}$;
  $\csc \theta = \frac{5}{2}$ **23.** $\sin \theta = \sqrt{7}/4$; $\tan \theta = \sqrt{7}/3$; $\cot \theta = \frac{3}{7}\sqrt{7}$; $\sec \theta = \frac{4}{3}$;
  $\csc \theta = \frac{4}{7}\sqrt{7}$ **25.** $\sin \theta = \frac{2}{5}\sqrt{5}$; $\cos \theta = \frac{1}{5}\sqrt{5}$; $\cot \theta = \frac{1}{2}$; $\sec \theta = \sqrt{5}$;
  $\csc \theta = \sqrt{5}/2$ **27.** $\sin \theta = \frac{10}{269}\sqrt{269}$; $\cos \theta = \frac{13}{269}\sqrt{269}$; $\tan \theta = \frac{10}{13}$;
  $\sec \theta = \frac{1}{13}\sqrt{269}$; $\csc \theta = \frac{1}{10}\sqrt{269}$ **29.** $\sin \theta = \sqrt{21}/11$; $\cos \theta = \frac{10}{11}$;
  $\tan \theta = \sqrt{21}/10$; $\cot \theta = \frac{10}{21}\sqrt{21}$; $\csc \theta = \frac{11}{21}\sqrt{21}$ **31.** $\sin \theta = \frac{3}{5}$;
  $\cos \theta = \frac{4}{5}$; $\tan \theta = \frac{3}{4}$; $\cot \theta = \frac{4}{3}$; $\sec \theta = \frac{5}{4}$

33. $\left(\dfrac{\pi}{4}, \dfrac{\pi}{2}\right)$ **43.** No, since then $\mu = \dfrac{\sin i}{\sin r} = \dfrac{\sqrt{2}/2}{1/2} = \sqrt{2}$, whereas $\mu \le 1.340$.

SECTION 6.3   **1.** 0.13917310   **3.** 0.89100652   **5.** 0.24932800   **7.** 1.1503684   **9.** 1.0001523
11. 1.2360680   **13.** 0.62023549   **15.** 0.96045583   **17.** 1.5566639
19. 0.89386968   **21.** 57.274049   **23.** 1.0000001   **25.** a. 0.09983342   b. 0.0001
c. 0.000001   d. 0.99500417   e. 0.9999995   f. 1   **29.** 0.6648   **31.** 0.2022
33. 11.06   **35.** 0.0262   **37.** 6.166   **39.** 2.010   **41.** 0.9246   **43.** 0.1877
45. 23.39   **47.** 3.991   **49.** 1.666   **51.** 1.018   **53.** $33°\,18'$ **55.** $79°\,22'$
57. $10°\,22'$ **59.** Approximately $234°\,36'$ **61.** a. Approximately $14°\,52'\,2''$
b. Approximately $36°\,48'\,26''$ c. Approximately $48°\,34'\,1''$
63. a. Approximately 47.023 inches   b. Approximately 50.059 inches
c. Approximately 50.263 inches   d. Approximately 50.265 inches
65. Maximum height: approximately 39.0625 feet; range: approximately 270.63 feet

SECTION 6.4   **1.** $a \approx 3.864$, $b \approx 1.035$, $\alpha = 75°$ **3.** $a \approx 2.384$, $c \approx 3.111$, $\alpha = 50°$
5. $b \approx 4.037$, $c \approx 5.152$, $\beta = 51.6°$
7. $b \approx 8$, $\alpha \approx 61.93°$ ($\approx 61°\,56'$), $\beta \approx 28.07°$ ($\approx 28°\,4'$)
9. $a = 1$, $\alpha \approx 18.43°$ ($\approx 18°\,26'$), $\beta \approx 71.57°$ ($\approx 71°\,34'$)
11. $c \approx 0.5207$, $\alpha \approx 39.81°$ ($\approx 39°\,48'$), $\beta \approx 50.19°$ ($\approx 50°\,12'$)
13. Approximately 38.39 feet   **15.** Approximately $48.59°$ ($\approx 48°\,35'$)
17. Approximately 4.668 miles   **19.** Approximately 77.65 feet
21. Approximately 362.0 feet   **23.** Approximately $1.30°$ ($\approx 1°\,18'$)
25. Approximately 482 feet

SECTION 6.5   **1.** sine, cosine, tangent, secant   **3.** sine, cosine, cotangent, cosecant
5. sine, cosine, cotangent, cosecant   **7.** sine, cosine, cotangent, cosecant
9. $30°$ **11.** $\pi/4$ **13.** $\sin 135° = \sqrt{2}/2$, $\cos 135° = -\sqrt{2}/2$, $\tan 135° = -1$,
$\cot 135° = -1$, $\sec 135° = -\sqrt{2}$, $\csc 135° = \sqrt{2}$ **15.** $\sin 5\pi/3 = -\sqrt{3}/2$,
$\cos 5\pi/3 = \frac{1}{2}$, $\tan 5\pi/3 = -\sqrt{3}$, $\cot 5\pi/3 = -\sqrt{3}/3$, $\sec 5\pi/3 = 2$,
$\csc 5\pi/3 = -\frac{2}{3}\sqrt{3}$
17. $\sin(-120°) = -\sqrt{3}/2$, $\cos(-120°) = -\frac{1}{2}$, $\tan(-120°) = \sqrt{3}$,
$\cot(-120°) = \sqrt{3}/3$, $\sec(-120°) = -2$, $\csc(-120°) = -2\sqrt{3}/3$
19. $\sin(-3\pi/4) = -\sqrt{2}/2$, $\cos(-3\pi/4) = -\sqrt{2}/2$, $\tan(-3\pi/4) = 1$,
$\cot(-3\pi/4) = 1$, $\sec(-3\pi/4) = -\sqrt{2}$, $\csc(-3\pi/4) = -\sqrt{2}$ **21.** $\sin 750° = \frac{1}{2}$,
$\cos 750° = \sqrt{3}/2$, $\tan 750° = \sqrt{3}/3$, $\cot 750° = \sqrt{3}$, $\sec 750° = \frac{2}{3}\sqrt{3}$,
$\csc 750° = 2$ **23.** $\sin(41\pi/6) = \frac{1}{2}$, $\cos(41\pi/6) = -\sqrt{3}/2$, $\tan(41\pi/6) = -\sqrt{3}/3$,
$\cot(41\pi/6) = -\sqrt{3}$, $\sec(41\pi/6) = -2\sqrt{3}/3$, $\csc(41\pi/6) = 2$ **25.** $\cos\theta = -\frac{4}{5}$,
$\tan\theta = \frac{3}{4}$, $\cot\theta = \frac{4}{3}$, $\sec\theta = -\frac{5}{4}$, $\csc\theta = -\frac{5}{3}$ **27.** $\sin\theta = -\sqrt{6}/3$,
$\cos\theta = -\sqrt{3}/3$, $\cot\theta = \sqrt{2}/2$, $\sec\theta = -\sqrt{3}$, $\csc\theta = -\sqrt{6}/2$
29. $\sin\theta = \sqrt{6}/3$, $\cos\theta = -\sqrt{3}/3$, $\tan\theta = -\sqrt{2}$, $\cot\theta = -\sqrt{2}/2$,
$\csc\theta = \sqrt{6}/2$ **31.** False   **33.** False   **35.** True   **37.** True
39. $\sqrt{2(1 - \cos s \cos t - \sin s \sin t)}$

SECTION 6.6   **1.** $\sqrt{3}/2$ **3.** $\frac{1}{2}$ **5.** $-\sqrt{2}$ **7.** $\sqrt{2}/2$ **9.** $-\sqrt{2}/2$ **11.** $2\sqrt{3}/3$ **13.** $-\sqrt{3}$
15. 2   **17.** $\sqrt{3}/2$ **19.** $\sqrt{3}/3$ **21.** $\frac{1}{3}$ **23.** $-\frac{1}{3}$ **25.** $\frac{1}{3}$ **27.** $-2\sqrt{2}/3$
29. $\frac{1}{3}$ **31.** $-\frac{2}{5}$ **33.** $-\frac{2}{5}$ **35.** $-\sqrt{21}/5$ **37.** $-\frac{2}{5}$ **39.** 2   **41.** $-\frac{1}{2}$
43. 2   **45.** $-\frac{1}{2}\sqrt{2 - \sqrt{2}}$ **47.** $-\frac{1}{2}\sqrt{2 - \sqrt{2}}$ **49.** $-\frac{1}{2}\sqrt{2 - \sqrt{2}}$
51. $1 - \sqrt{2}$ **53.** $\sqrt{2} + 1$ **55.** π **57.** $\sqrt{2}\pi$ **59.** π **61.** π

SECTION 6.7

1.   3.   5.

5.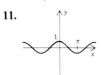

7.   9.   11.

13. 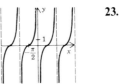   15.   17.

19.   21.   23.

25.   27.   29.

SECTION 6.8

1.   3.   5.

7.   9.   11.

13. $4\sin(3x - \pi)$ 15. $\frac{3}{2}\sin(2x - \pi/3)$

SECTION 6.9 **1.** $-\pi/2$ **3.** $\pi/2$ **5.** 0 **7.** $\pi/2$ **9.** $3\pi/4$ **11.** 0
13. $2\pi/3$ **15.** $\pi/2$ **17.** $-\pi/2$ **19.** 0.30469265
21. 1.0003592 **23.** 0.10955953 **25.** 0.81631614
27. 1.1410209 **29.** -0.21440486 **31.** $\pi/6$ **33.** $-\pi/3$
35. $2\pi/3$ **37.** $3\pi/4$ **39.** $\pi/13$ **41.** $5\pi/6$ **43.** $\pi/6$
45. 0 **47.** $\frac{1}{4}$ **49.** $-\frac{1}{2}$ **51.** $\sqrt{2}/2$ **53.** -1 **55.** $\frac{3}{5}$
57. $-\frac{4}{3}$ **59.** $3\sqrt{2}/4$ **61.** $\pi/2$ **63.** $\pi/6$
71.   **73.**

**CHAPTER 6
REVIEW**

1.   **3.**   **5.**

7. $-\sqrt{2}/2$ **9.** $-\sqrt{3}/2$ **11.** -1 **13.** $\sqrt{3}/3$ **15.** $2\sqrt{3}/3$ **17.** $2\sqrt{3}/3$
19. $-\frac{1}{2}$ **21.** $\sqrt{3}/3$
23. $\sin\theta = -\frac{2}{5}$, $\cos\theta = -\sqrt{21}/5$, $\tan\theta = 2\sqrt{21}/21$,
$\cot\theta = \sqrt{21}/2$, $\sec\theta = -5\sqrt{21}/21$, $\csc\theta = -\frac{5}{2}$
25. $\sin\theta = 8\sqrt{65}/65$, $\cos\theta = -\sqrt{65}/65$, $\tan\theta = -8$, $\cot\theta = -\frac{1}{8}$,
$\sec\theta = -\sqrt{65}$, $\csc\theta = \sqrt{65}/8$
27. $\sin\theta = \sqrt{15}/4$, $\cos\theta = \frac{1}{4}$, $\tan\theta = \sqrt{15}$, $\cot\theta = \sqrt{15}/15$,
$\sec\theta = 4$, $\csc\theta = 4\sqrt{15}/15$
29. $\frac{1}{4}$ **31.** $-\frac{1}{4}$ **33.** $\sqrt{15}$ **35.** $-\sqrt{15}$ **37.** -4 **39.** 4
41. 0.24756272 **43.** 0.67430239 **45.** 0.14291469
47. $b \approx 0.5290$, $c \approx 3.046$, $\alpha = 80°$
49. $c \approx 8.360$, $\alpha \approx 36.73°$ ($\approx 36°\,44'$), $\beta \approx 53.27°$ ($\approx 53°\,16'$)
51. $\sin\theta = 5\sqrt{29}/29$, $\cos\theta = 2\sqrt{29}/29$, $\tan\theta = \frac{5}{2}$, $\cot\theta = \frac{2}{5}$, $\sec\theta = \sqrt{29}/2$,
$\csc\theta = \sqrt{29}/5$
53.

55. $\pi/3$ **57.** $\pi/6$ **59.** $\pi/3$ **61.** $-\pi/4$ **63.** $\pi/2$ **65.** $\pi/4$ **67.** $-2\sqrt{5}/5$
69. a. 1 b. 0 c. 1 d. 0 **75.** 136π inches **77.** Approximately 5874 feet
79. Approximately 2677.70 feet **81.** 24,500 miles

SECTION 7.1 **57.** $\pi/4$ **59.** $\pi/4$ **61.** $-\pi/4$ **63.** Graph of $2\sec x$

SECTION 7.2 **1.** $\frac{1}{4}(\sqrt{6} - \sqrt{2})$ **3.** $\frac{1}{4}(\sqrt{2} - \sqrt{6})$ **5.** $\frac{1}{4}(\sqrt{6} - \sqrt{2})$ **7.** $\frac{1}{4}(\sqrt{2} - \sqrt{6})$

9. $\dfrac{\sqrt{3} - 1}{\sqrt{3} + 1}$ **11.** $\dfrac{\sqrt{3} + 1}{\sqrt{3} - 1}$ **13.** $-\frac{1}{8}(\sqrt{15} + \sqrt{3})$

15. $\frac{1}{20}(1 - 6\sqrt{10})$ **17.** $\frac{1}{12}(1 - 2\sqrt{30})$ **55.** $x = \pi/2,\ y = \pi/2$
57. $x = \pi/6,\ y = \pi/6$ **59.** Graph of $\sqrt{2}\sin(x + \pi/4)$
61. Graph of $4\sqrt{3}\sin(x + \pi/3)$ **63.** Graph of $\cos 2x$

SECTION 7.3   **1.** $\sin 2x = -4\sqrt{2}/9,\ \cos 2x = \frac{7}{9},\ \tan 2x = -4\sqrt{2}/7$
3. $\sin 2x = -2\sqrt{2}/3,\ \cos 2x = -\frac{1}{3},\ \tan 2x = 2\sqrt{2}$
5. $\sin 2x = -4\sqrt{6}/25,\ \cos 2x = -\frac{23}{25},\ \tan 2x = 4\sqrt{6}/23$

7. $\sin \frac{1}{2}x = \frac{1}{4}\sqrt{8 + 3\sqrt{7}},\ \cos \frac{1}{2}x = \frac{1}{4}\sqrt{8 - 3\sqrt{7}},\ \tan \frac{1}{2}x = \dfrac{1}{8 - 3\sqrt{7}}$

9. $\sin \frac{1}{2}x = \sqrt{\dfrac{5 + 2\sqrt{5}}{10}},\ \cos \frac{1}{2}x = -\sqrt{\dfrac{5 - 2\sqrt{5}}{10}},\ \tan \frac{1}{2}x = \dfrac{1}{2 - \sqrt{5}}$

11. $\sin \frac{1}{2}x = \sqrt{6}/3,\ \cos \frac{1}{2}x = -\sqrt{3}/3,\ \tan \frac{1}{2}x = -\sqrt{2}$ **13.** $\frac{1}{2}\sqrt{2 - \sqrt{2}}$

15. $-\frac{1}{2}\sqrt{2 - \sqrt{3}}$ **17.** $-\frac{1}{2}\sqrt{2 - \sqrt{2}}$ **19.** $\frac{1}{2}\sqrt{2 - \sqrt{3}}$ **21.** $\dfrac{\sqrt{2}}{\sqrt{2} - 2}$

23. $-\dfrac{1}{2 + \sqrt{3}}$ **25. a.** $\frac{1}{2}\sqrt{2 - \sqrt{2 + \sqrt{2}}}$ **b.** $\frac{1}{2}\sqrt{2 + \sqrt{2 + \sqrt{2}}}$

c. $\dfrac{\sqrt{2 - \sqrt{2 + \sqrt{2}}}}{\sqrt{2 + \sqrt{2 + \sqrt{2}}}}$, or $\dfrac{\sqrt{2 - \sqrt{2}}}{2 + \sqrt{2 + \sqrt{2}}}$

53. Graph of $\frac{1}{2}\sin 4x$ **55.** Graph of $2 \tan x$

SECTION 7.4   **1.** $\frac{1}{2}(\cos 2x - \cos 6x)$ **3.** $\frac{1}{2}(\cos 4x + \cos 2x)$ **5.** $\frac{1}{2}(\cos 2x + \cos 4x/3)$
7. $\frac{1}{2}(\sin 10x - \sin 4x)$ **9.** $2 \sin 3x \cos x$ **11.** $2(\sin 3x/4)(\cos 5x/4)$
13. $2 \cos x \cos x/2$ **15.** $-2 \sin 5x \sin 2x$

SECTION 7.5   **1.** $\dfrac{\pi}{2} + 2n\pi$ **3.** $\dfrac{2\pi}{3} + 2n\pi, \dfrac{4\pi}{3} + 2n\pi$ **5.** $\dfrac{\pi}{3} + n\pi$ **7.** $\dfrac{\pi}{6} + n\pi$

9. $\dfrac{\pi}{3} + 2n\pi, \dfrac{2\pi}{3} + 2n\pi$ **11.** $\dfrac{\pi}{18} + \dfrac{2n\pi}{3}, \dfrac{5\pi}{18} + \dfrac{2n\pi}{3}$ **13.** $\dfrac{5\pi}{4} + 5n\pi$

15. $\dfrac{\pi}{24} + \dfrac{n\pi}{3}, -\dfrac{\pi}{24} + \dfrac{n\pi}{3}$ **17.** $\dfrac{\pi}{3} + 2n\pi, \dfrac{5\pi}{3} + 2n\pi$ **19.** $\dfrac{\pi}{3} + n\pi$

21. $\dfrac{5\pi}{6} + 2n\pi, \dfrac{7\pi}{6} + 2n\pi$ **23.** $n\pi, \dfrac{\pi}{3} + 2n\pi, -\dfrac{\pi}{3} + 2n\pi$ **25.** $-\dfrac{\pi}{4} + n\pi$

27. $\dfrac{\pi}{2} + n\pi$ **29.** $\dfrac{\pi}{4} + 2n\pi, \dfrac{3\pi}{4} + n\pi$ **31.** $\dfrac{3\pi}{2} + 2n\pi$

33. $\dfrac{2\pi}{3} + 2n\pi, \dfrac{4\pi}{3} + 2n\pi, 2n\pi$ **35.** $\dfrac{2\pi}{3} + 2n\pi, \dfrac{4\pi}{3} + 2n\pi$

37. $\dfrac{\pi}{6} + n\pi, -\dfrac{\pi}{3} + n\pi$ **39.** $\dfrac{\pi}{4} + \dfrac{n\pi}{2}$

41. $-\dfrac{\pi}{6} + 2n\pi, \dfrac{7\pi}{6} + 2n\pi, \dfrac{\pi}{2} + 2n\pi$ **43.** $\dfrac{\pi}{3} + 2n\pi, -\dfrac{\pi}{3} + 2n\pi$

45. $30°, 150°, 199° 28', 340° 32'$ **47.** $78° 28', 281° 32', 109° 28', 250° 32'$
49. $54° 44', 305° 16', 125° 16', 234° 44'$

CHAPTER 7 REVIEW

33. $\frac{1}{2}\sqrt{2 + \sqrt{2}}$ **35.** $\dfrac{4}{\sqrt{6} - \sqrt{2}}$, or $\dfrac{2}{\sqrt{2 - \sqrt{3}}}$ **37.** $-3\sqrt{3}/14$

39. $-\sqrt{15}/8$ **43.** a. No   b. Yes   **45.** a. $\frac{1}{2}\sqrt{2 + \sqrt{3}}$

47. Graph of $2 \sin(x + \pi/6)$ **49.** $\dfrac{5\pi}{6} + 2n\pi, \dfrac{7\pi}{6} + 2n\pi$ **51.** $n\pi$

53. $\dfrac{\pi}{6} + 2n\pi, \dfrac{5\pi}{6} + 2n\pi, \dfrac{3\pi}{2} + 2n\pi$ **55.** $\dfrac{\pi}{6} + 2n\pi, \dfrac{5\pi}{6} + 2n\pi, \dfrac{3\pi}{2} + 2n\pi$

57. $\dfrac{\pi}{3} + n\pi, -\dfrac{\pi}{3} + n\pi$

SECTION 8.1

1. $\gamma = 65°, a \approx 8.458, c \approx 8.851$ **3.** $\alpha = 29°, a \approx 3.430, b \approx 3.430$
5. $\alpha = 123°, b \approx 3.334, c \approx 2.329$ **7.** $\beta = 142.9°, a \approx 7.024, c \approx 4.004$
9. $\beta = 91° 27', a \approx 1.366, b \approx 2.967$
11. Two triangles: $\beta \approx 59.29° (\approx 59° 17'), \gamma \approx 83.71° (\approx 83° 43'), c \approx 2.312$;
    $\beta \approx 120.71° (\approx 120° 43'), \gamma \approx 22.29° (\approx 22° 17'), c \approx 0.8822$
13. No triangle   **15.** $\beta = 90°, \gamma = 60°, c = 3\sqrt{3} \approx 5.196$
17. $\alpha \approx 44.77° (\approx 44° 46'), \gamma \approx 60.33° (\approx 60° 20'), a \approx 3.647$
19. Two triangles: $\alpha \approx 119.21° (\approx 119° 12'), \beta \approx 33.99° (\approx 33° 59'), a \approx 4.840$;
    $\alpha \approx 7.19° (\approx 7° 12'), \beta \approx 146.01° (\approx 146° 1'), a \approx 0.6942$
21. $\beta \approx 35.37° (\approx 35° 22'), \gamma \approx 33.88° (\approx 33° 53'), c \approx 2.504$
23. The guard at A is closer, by approximately 69.26 yards.
25. Approximately 0.9493 miles   **27.** Approximately 6.158 miles

SECTION 8.2

1. $\alpha \approx 41.75° (\approx 41° 45'), \beta \approx 50.98° (\approx 50° 59'), \gamma \approx 87.27° (\approx 87° 16')$
3. $\alpha \approx 21.79° (\approx 21° 47'), \beta \approx 38.21° (\approx 38° 13'), \gamma \approx 120.0°$
5. $\alpha \approx 38.27° (\approx 38° 16'), \beta \approx 56.60° (\approx 56° 36'), \gamma \approx 85.13° (\approx 85° 8')$
7. $a \approx 1.564, \beta \approx 40.99° (\approx 40° 59'), \gamma \approx 119.01° (\approx 119° 1')$
9. $b \approx 0.6498, \alpha \approx 92.96° (\approx 92° 57'), \gamma \approx 75.34° (\approx 75° 20')$
11. $c \approx 10.97, \alpha \approx 4.75° (\approx 4° 45'), \beta \approx 3.55° (\approx 3° 33')$
13. $a \approx 4.726, \beta \approx 12.64° (\approx 12° 38'), \gamma \approx 114.68° (\approx 114° 41')$
15. Two triangles: $a \approx 5.519, \alpha \approx 55.65° (\approx 55° 39'), \gamma \approx 65.85° (\approx 65° 51')$;
    $a \approx 0.8552, \alpha \approx 7.35° (\approx 7° 21'), \gamma \approx 114.15° (\approx 114° 9')$
17. $b \approx 5.945, \alpha \approx 35.90° (\approx 35° 54'), \beta \approx 44.20° (\approx 44° 12')$
19. $b \approx 3.382, \alpha \approx 22.16° (\approx 22° 10'), \gamma \approx 120.44° (\approx 120° 26')$
21. $\alpha \approx 61.21° (\approx 61° 13'), \beta \approx 87.96° (\approx 87° 58'), \gamma \approx 30.83° (\approx 30° 50')$
23. $a \approx 0.2084, \alpha \approx 19.61° (\approx 19° 37'), \gamma \approx 9.27° (\approx 9° 16')$
25. $500\sqrt{3}$ feet (≈ 866 feet)   **27.** Approximately 23.18 miles
29. Approximately $44.42° (\approx 44° 25'), 57.12° (\approx 57° 7')$, and $78.46° (\approx 78° 28')$
31. Approximately 4.62 feet
33. The angle at $(0, 0) \approx 26.57° (\approx 26° 34')$; the angle at $(3, 0) = 45°$; the angle at $(2, 1) \approx 108.43° (\approx 108° 26')$.
35. Approximately 19.91 feet and 24.57 feet

37. a. $\dfrac{1}{c}(2.333)$ b. Approximately $\dfrac{1}{c}(2.367)$

SECTION 8.3   **1.** $\mathscr{A} = 3\sqrt{2} \approx 4.243$ **3.** $\mathscr{A} \approx 2.431$ **5.** $\mathscr{A} \approx 10$ **7.** $\mathscr{A} \approx 3.093$
9. $\mathscr{A} = 4\sqrt{5} \approx 8.944$ **11.** $\mathscr{A} = 1/\sqrt{2} \approx 0.7071$ **13.** $\mathscr{A} = \sqrt{3} \approx 1.732$
15. $\mathscr{A} \approx 4.081$

SECTION 8.4   **1.** $(0, 0)$ **3.** $(-5, 0)$ **5.** $(0, 3)$ **7.** $(-2.3, 0)$ **9.** $(\frac{3}{2}, 3\sqrt{3}/2)$
11. $(\frac{1}{2}, -\sqrt{3}/2)$ **13.** $(-2, 0)$ **15.** $(2, 2\sqrt{3})$ **17.** $(0, 0.4)$
19. $(5, 2n\pi)$ and $(-5, \pi + 2n\pi)$ for any integer n
21. $(5, \pi + 2n\pi)$ and $(-5, 2n\pi)$ for any integer n

23. $\left(2, \dfrac{\pi}{4} + 2n\pi\right)$ and $\left(-2, \dfrac{5\pi}{4} + 2n\pi\right)$ for any integer n

25. $\left(\dfrac{\sqrt{2}}{2}, \dfrac{3\pi}{4} + 2n\pi\right)$ and $\left(\dfrac{-\sqrt{2}}{2}, \dfrac{7\pi}{4} + 2n\pi\right)$ for any integer n

27. $\left(2, \dfrac{\pi}{6} + 2n\pi\right)$ and $\left(-2, \dfrac{7\pi}{6} + 2n\pi\right)$ for any integer n

29. $\left(2\sqrt{2}, \dfrac{2\pi}{3} + 2n\pi\right)$ and $\left(-2\sqrt{2}, \dfrac{5\pi}{3} + 2n\pi\right)$ for any integer n

31.   **33.**   **35.**

37.   **39.**   **41.**

43.   **45.**   **47.**
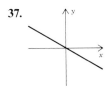 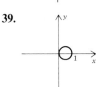 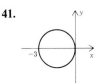
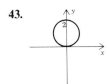 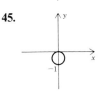 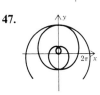

SECTION 8.5   **1.** $\mathbf{a} + \mathbf{b} = (1, 5)$, $\mathbf{a} - \mathbf{b} = (5, -3)$, $2\mathbf{a} - 3\mathbf{b} = (12, -10)$, $-\frac{5}{2}\mathbf{a} = \left(-\frac{15}{2}, -\frac{5}{2}\right)$
3. $\mathbf{a} + \mathbf{b} = (-3, 3)$, $\mathbf{a} - \mathbf{b} = (-9, -9)$, $2\mathbf{a} - 3\mathbf{b} = (-21, -24)$, $-\frac{5}{2}\mathbf{a} = \left(15, \frac{15}{2}\right)$
5. $\mathbf{a} + \mathbf{b} = (-2, 3)$, $\mathbf{a} - \mathbf{b} = (-2, -3)$, $2\mathbf{a} - 3\mathbf{b} = (-4, -9)$, $-\frac{5}{2}\mathbf{a} = (5, 0)$
7. $4\mathbf{a} - \mathbf{b} = 25\mathbf{i} + 10\mathbf{j}$, $3\mathbf{a} + 2\mathbf{b} = 16\mathbf{i} + 2\mathbf{j}$
9. $4\mathbf{a} - \mathbf{b} = 5\mathbf{i} + 6\mathbf{j}$, $3\mathbf{a} + 2\mathbf{b} = -10\mathbf{i} + 21\mathbf{j}$
11. $4\mathbf{a} - \mathbf{b} = 3\mathbf{i} - 5\mathbf{j}$, $3\mathbf{a} + 2\mathbf{b} = 5\mathbf{i} - \mathbf{j}$ **19.** $\sqrt{10}$ **21.** $\sqrt{5}$ **23.** 2 **25.** No

27. No   **29.** Yes   **31.** $\sqrt{2}\left(\cos\dfrac{5\pi}{4}\mathbf{i} + \sin\dfrac{5\pi}{4}\mathbf{j}\right)$ **33.** $2\left(\cos\dfrac{\pi}{3}\mathbf{i} + \sin\dfrac{\pi}{3}\mathbf{j}\right)$

35. $\pi\sqrt{2}\left(\cos\dfrac{3\pi}{4}\mathbf{i} + \sin\dfrac{3\pi}{4}\mathbf{j}\right)$ **39.** $\mathbf{a} = -\frac{3}{4}\mathbf{i} - \mathbf{j}, \mathbf{b} = -\frac{3}{2}\mathbf{i} - 2\mathbf{j}$ **41.** -11

43. 0   **45.** $300\sqrt{3} \approx 519.6$ foot-pounds

47. Velocity vector: $(150\sqrt{3} + 30)\mathbf{i} + 150\mathbf{j}$;

ground speed: $30\sqrt{101 + 10\sqrt{3}} \approx 326.3$ miles per hour

Actual course: approximately $27.37°$ ($\approx 27° 22'$) north of east

CHAPTER 8 REVIEW

1. $\alpha \approx 75.52°$ ($\approx 75° 31'$), $\beta \approx 75.52°$ ($\approx 75° 31'$), $\gamma \approx 28.96°$ ($\approx 28° 58'$)

3. $b \approx 46.61$, $c \approx 59.9$, $\gamma = 97.4°$ ($= 97° 24'$)

5. $c \approx 28.65$, $\beta \approx 10.84°$ ($\approx 10° 50'$), $\gamma \approx 155.71°$ ($\approx 155° 43'$)

7. $c \approx 0.07401$, $\alpha \approx 40.27°$ ($\approx 40° 16'$), $\gamma \approx 15.41°$ ($\approx 15° 25'$)

9. Two triangles: $c \approx 2.716$, $\alpha \approx 84.48°$ ($\approx 84° 29'$), $\gamma \approx 15.22°$ ($\approx 15° 13'$);

$c \approx 0.7548$, $\alpha \approx 95.52°$ ($\approx 95° 31'$), $\gamma \approx 4.18°$ ($\approx 4° 11'$)

11. $\alpha \approx 10.96°$ ($\approx 10° 58'$), $\beta \approx 130.48°$ ($\approx 130° 29'$), $\gamma \approx 38.56°$ ($\approx 38° 34'$)

13. $a \approx 1.809$, $c \approx 2.225$, $\alpha \approx 21.03°$ ($\approx 21° 2'$)

15. $\alpha \approx 32.20°$ ($\approx 32° 12'$), $\beta = 60°$, $\gamma \approx 87.80°$ ($\approx 87° 48'$)

17. $b \approx 10.33$, $\beta \approx 64.24°$ ($\approx 64° 14'$), $\gamma \approx 55.06°$ ($\approx 55° 4'$)

19. $c = 106$, $\alpha \approx 58.11°$ ($\approx 58° 7'$), $\beta \approx 31.89°$ ($\approx 31° 53'$)

21. $\mathscr{A} \approx 32.43$ **23.** $\mathscr{A} \approx 1.735$ **25.** $\left(\frac{3}{4}\sqrt{2}, -\frac{3}{4}\sqrt{2}\right)$ **27.** $\left(-\frac{1}{2}e, -\frac{1}{2}\sqrt{3}e\right)$

29. $\left(2\sqrt{5}, -\frac{\pi}{3} + 2n\pi\right)$ and $\left(-2\sqrt{5}, \frac{2\pi}{3} + 2n\pi\right)$ for any integer n

31. $\left(\pi, \frac{\pi}{2} + 2n\pi\right)$ and $\left(-\pi, \frac{3\pi}{2} + 2n\pi\right)$ for any integer n

33.

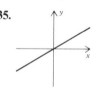

35.

37.

39. $\left(\frac{14}{3}, -2\right)$ **41.** $6\mathbf{i} - 8\mathbf{j}$ **43.** $4\sqrt{3}$ **45.** $7\sqrt{2} - \sqrt{3}\pi$ **47.** $\sqrt{3}$

49. $2\sqrt{7}\left(\cos\frac{5\pi}{6}\mathbf{i} + \sin\frac{5\pi}{6}\mathbf{j}\right)$ **51.** $\frac{1}{5}\left(\cos\frac{\pi}{3}\mathbf{i} + \sin\frac{\pi}{3}\mathbf{j}\right)$

53. Approximately $78.97°$ ($\approx 78° 58'$)   **55.** Approximately 51.03 feet

57. Approximately $88.83°$ ($\approx 88° 50'$)   **59.** Approximately $92.82°$ ($\approx 92° 49'$)

CUMULATIVE REVIEW II: CHAPTERS 5–8

1. 9   **3.** $-\frac{5}{6}$ **5.** 2, 1   **7.** 25   **9.** $50 \ln\frac{8}{3}$ **11.** $\dfrac{\ln 3 + \ln 2}{\ln 3 - 2\ln 2}$

13. $\log_5\dfrac{x - y}{x}$ **15.** 0.1589   **17.** $\sqrt{2}$

19.

21. $\frac{1}{2}$ **23.** 2   **25.** $\frac{1}{2}\sqrt{2 + \sqrt{3}}$

27. $\sin\theta = -\frac{1}{5}\sqrt{5}$, $\cos\theta = -\frac{2}{5}\sqrt{5}$, $\cot\theta = 2$, $\sec\theta = -\frac{1}{2}\sqrt{5}$, $\csc\theta = -\sqrt{5}$

29. $-\frac{7}{10}\sqrt{2}$ **31.** $-\pi/4$ **33.**

41. $\frac{\pi}{2} + n\pi$ for any integer n

43. Two triangles: $c \approx 6.212$, $\beta \approx 53.46°$ ($\approx 53° \, 28'$), $\gamma \approx 86.54°$ ($\approx 86° \, 32'$); $c \approx 1.449$, $\beta \approx 126.54°$ ($\approx 126° \, 32'$), $\gamma \approx 13.46°$ ($\approx 13° \, 28'$)

45. $\alpha \approx 16.52°$ ($\approx 16° \, 31'$), $\beta \approx 45.31°$ ($\approx 45° \, 19'$), $\gamma \approx 118.17°$ ($\approx 118° \, 10'$)

47. $\sqrt{231}$ **49.** $\left(12, -\frac{\pi}{3} + 2n\pi\right)$ and $\left(-12, \frac{2\pi}{3} + 2n\pi\right)$ for any integer n

51. No   **53.** $3\dfrac{\ln 10}{\ln 2} \approx 9.97$ weeks

55. Approximately 11.54 pounds per square inch   **57.** a. 200   b. 6000

59. Approximately 101.25 meters

61. Magnitude: $2\sqrt{5}$ pounds; direction: approximately $18.43°$, or $18° \, 26'$, south of east

Index

RIGHT TRIANGLES

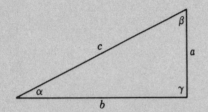

$$\sin \alpha = \frac{a}{c} \qquad \tan \alpha = \frac{a}{b} \qquad \sec \alpha = \frac{c}{b}$$

$$\cos \alpha = \frac{b}{c} \qquad \cot \alpha = \frac{b}{a} \qquad \csc \alpha = \frac{c}{a}$$

Pythagorean Theorem: $c^2 = a^2 + b^2$

OBLIQUE TRIANGLES

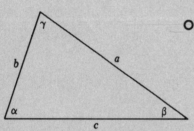

$$\alpha + \beta + \gamma = 180°$$

Law of Sines: $\dfrac{\sin \alpha}{a} = \dfrac{\sin \beta}{b} = \dfrac{\sin \gamma}{c}$

Law of Cosines: $a^2 = b^2 + c^2 - 2bc \cos \alpha$

ANGLE CONVERSION

$$\theta° = \left(\frac{\pi}{180} \cdot \theta\right) \text{ radians} \qquad t \text{ radians} = \left(\frac{180}{\pi} \cdot t\right)°$$